THE GALACTIC CENTER
4th ESO/CTIO Workshop

A SERIES OF BOOKS ON RECENT DEVELOPMENTS IN ASTRONOMY AND ASTROPHYSICS

A.S.P. CONFERENCE SERIES PUBLICATIONS COMMITTEE

Dr. Sallie L. Baliunas, Chair
Dr. John P. Huchra
Dr. Roberta M. Humphreys
Dr. Catherine A. Pilachowski

© Copyright 1996 Astronomical Society of the Pacific
390 Ashton Avenue, San Francisco, California 94112

All rights reserved

Printed by BookCrafters, Inc.

First published 1996

Library of Congress Catalog Card Number: 96-85665
ISBN 1-886733-22-8

D. Harold McNamara, Managing Editor of Conference Series
206 KMB
Brigham Young University
Provo, UT 84602-4463
801-378-2298

pasp@astro.byu.edu
Fax 801-378-2265

A SERIES OF BOOKS ON RECENT DEVELOPMENTS IN ASTRONOMY AND ASTROPHYSICS

Vol. 1-Progress and Opportunities in Southern Hemisphere Optical Astronomy: The CTIO 25th Anniversary Symposium
ed. V. M. Blanco and M. M. Phillips   ISBN 0-937707-18-X

Vol. 2-Proceedings of a Workshop on Optical Surveys for Quasars
ed. P. S. Osmer, A. C. Porter, R. F. Green, and C. B. Foltz   ISBN 0-937707-19-8

Vol. 3-Fiber Optics in Astronomy
ed. S. C. Barden   ISBN 0-937707-20-1

Vol. 4-The Extragalactic Distance Scale: Proceedings of the ASP 100th Anniversary Symposium
ed. S. van den Bergh and C. J. Pritchet   ISBN 0-937707-21-X

Vol. 5-The Minnesota Lectures on Clusters of Galaxies and Large-Scale Structure
ed. J. M. Dickey   ISBN 0-937707-22-8

Vol. 6-Synthesis Imaging in Radio Astronomy: A Collection of Lectures from the Third NRAO Synthesis Imaging Summer School
ed. R. A. Perley, F. R. Schwab, and A. H. Bridle   ISBN 0-937707-23-6

Vol. 7-Properties of Hot Luminous Stars: Boulder-Munich Workshop
ed. C. D. Garmany   ISBN 0-937707-24-4

Vol. 8-CCDs in Astronomy
ed. G. H. Jacoby   ISBN 0-937707-25-2

Vol. 9-Cool Stars, Stellar Systems, and the Sun. Sixth Cambridge Workshop
ed. G. Wallerstein   ISBN 0-937707-27-9

Vol. 10-The Evolution of the Universe of Galaxies. The Edwin Hubble Centennial Symposium
ed. R. G. Kron   ISBN 0-937707-28-7

Vol. 11-Confrontation Between Stellar Pulsation and Evolution
ed. C. Cacciari and G. Clementini   ISBN 0-937707-30-9

Vol. 12-The Evolution of the Interstellar Medium
ed. L. Blitz   ISBN 0-937707-31-7

Vol. 13-The Formation and Evolution of Star Clusters
ed. K. Janes   ISBN 0-937707-32-5

Vol. 14-Astrophysics with Infrared Arrays
ed. R. Elston   ISBN 0-937707-33-3

Vol. 15-Large-Scale Structures and Peculiar Motions in the Universe
ed. D. W. Latham and L. A. N. da Costa   ISBN 0-937707-34-1

Vol. 16-Atoms, Ions and Molecules: New Results in Spectral Line Astrophysics
ed. A. D. Haschick and P. T. P. Ho   ISBN 0-937707-35-X

Vol. 17-Light Pollution, Radio Interference, and Space Debris
ed. D. L. Crawford   ISBN 0-937707-36-8

Vol. 18-The Interpretation of Modern Synthesis Observations of Spiral Galaxies
ed. N. Duric and P. C. Crane   ISBN 0-937707-37-6

Vol. 19-Radio Interferometry: Theory, Techniques, and Application, IAU Colloquium 131
ed. T. J. Cornwell and R. A. Perley   ISBN 0-937707-38-4

Vol. 20-Frontiers of Stellar Evolution, celebrating the 50th Anniversary of McDonald Observatory
ed. D. L. Lambert   ISBN 0-937707-39-2

Vol. 21-The Space Distribution of Quasars
ed. D. Crampton   ISBN 0-937707-40-6

Vol. 22-Nonisotropic and Variable Outflows from Stars
ed. L. Drissen, C. Leitherer, and A. Nota   ISBN 0-937707-41-4

Vol. 23-Astronomical CCD Observing and Reduction Techniques
ed. S. B. Howell   ISBN 0-937707-42-4

Vol. 24-Cosmology and Large-Scale Structure in the Universe
ed. R. R. de Carvalho   ISBN 0-937707-43-0

Vol. 25-Astronomical Data Analysis Software and Systems I
ed. D. M. Worrall, C. Biemesderfer, and J. Barnes   ISBN 0-937707-44-9

Vol. 26-Cool Stars, Stellar Systems, and the Sun, Seventh Cambridge Workshop
ed. M. S. Giampapa and J. A. Bookbinder   ISBN 0-937707-45-7

Vol. 27-The Solar Cycle
ed. K. L. Harvey   ISBN 0-937707-46-5

Vol. 28-Automated Telescopes for Photometry and Imaging
ed. S. J. Adelman, R. J. Dukes, Jr., and C. J. Adelman   ISBN 0-937707-47-3

Vol. 29-Workshop on Cataclysmic Variable Stars
ed. N. Vogt   ISBN 0-937707-48-1

Vol. 30-Variable Stars and Galaxies, in honor of M. S. Feast on his retirement
ed. B. Warner   ISBN 0-937707-49-X

Vol. 31-Relationships Between Active Galactic Nuclei and Starburst Galaxies
ed. A. V. Filippenko   ISBN 0-937707-50-3

Vol. 32-Complementary Approaches to Double and Multiple Star Research, IAU Collouquium 135
ed. H. A. McAlister and W. I. Hartkopf   ISBN 0-937707-51-1

Vol. 33-Research Amateur Astronomy
ed. S. J. Edberg   ISBN 0-937707-52-X

Vol. 34-Robotic Telescopes in the 1990s
ed. A. V. Filippenko   ISBN 0-937707-53-8

Vol. 35-Massive Stars: Their Lives in the Interstellar Medium
ed. J. P. Cassinelli and E. B. Churchwell   ISBN 0-937707-54-6

Vol. 36-Planets and Pulsars
ed. J. A. Phillips, S. E. Thorsett, and S. R. Kulkarni   ISBN 0-937707-55-4

Vol. 37-Fiber Optics in Astronomy II
ed. P. M. Gray   ISBN 0-937707-56-2

Vol. 38-New Frontiers in Binary Star Research
ed. K. C. Leung and I. S. Nha   ISBN 0-937707-57-0

Vol. 39-The Minnesota Lectures on the Structure and Dynamics of the Milky Way
ed. Roberta M. Humphreys   ISBN 0-937707-58-9

Vol. 40-Inside the Stars, IAU Colloquium 137
ed. Werner W. Weiss and Annie Baglin   ISBN 0-937707-59-7

Vol. 41-Astronomical Infrared Spectroscopy: Future Observational Directions
ed. Sun Kwok   ISBN 0-937707-60-0

Vol. 42-GONG 1992: Seismic Investigation of the Sun and Stars
ed. Timothy M. Brown   ISBN 0-937707-61-9

Vol. 43-Sky Surveys: Protostars to Protogalaxies
ed. B. T. Soifer   ISBN 0-937707-62-7

Vol. 44-Peculiar Versus Normal Phenomena in A-Type and Related Stars
ed. M. M. Dworetsky, F. Castelli, and R. Faraggiana   ISBN 0-937707-63-5

Vol. 45-Luminous High-Latitude Stars
ed. D. D. Sasselov   ISBN 0-937707-64-3

Vol. 46-The Magnetic and Velocity Fields of Solar Active Regions, IAU Colloquium 141
ed. H. Zirin, G. Ai, and H. Wang   ISBN 0-937707-65-1

Vol. 47-Third Decinnial US-USSR Conference on SETI
ed. G. Seth Shostak   ISBN 0-937707-66-X

Vol. 48-The Globular Cluster-Galaxy Connection
ed. Graeme H. Smith and Jean P. Brodie   ISBN 0-937707-67-8

Vol. 49-Galaxy Evolution: The Milky Way Perspective
ed. Steven R. Majewski   ISBN 0-937707-68-6

Vol. 50-Structure and Dynamics of Globular Clusters
ed. S. G. Djorgovski and G. Meylan   ISBN 0-937707-69-4

Vol. 51-Observational Cosmology
ed. G. Chincarini, A. Iovino, T. Maccacaro, and D. Maccagni   ISBN 0-937707-70-8

Vol. 52-Astronomical Data Analysis Software and Systems II
ed. R. J. Hanisch, J. V. Brissenden, and Jeannette Barnes   ISBN 0-937707-71-6

Vol. 53-Blue Stragglers
ed. Rex A. Saffer   ISBN 0-937707-72-4

Vol. 54-The First Stromlo Symposium: The Physics of Active Galaxies
ed. Geoffrey V. Bicknell, Michael A. Dopita, and Peter J. Quinn ISBN 0-937707-73-2

Vol. 55-Optical Astronomy from the Earth and Moon
ed. Diane M. Pyper and Ronald J. Angione   ISBN 0-937707-74-0

Vol. 56-Interacting Binary Stars
ed. Allen W. Shafter   ISBN 0-937707-75-9

Vol. 57-Stellar and Circumstellar Astrophysics
ed. George Wallerstein and Alberto Noriega-Crespo   ISBN 0-937707-76-7

Vol. 58-The First Symposium on the Infrared Cirrus and Diffuse Interstellar Clouds
ed. Roc M. Cutri and William B. Latter   ISBN 0-937707-77-5

Vol. 59-Astronomy with Millimeter and Submillimeter Wave Interferometry
ed. M. Ishiguro and Wm. J. Welch   ISBN 0-937707-78-3

Vol. 60-The MK Process at 50 Years: A Powerful Tool for Astrophysical Insight
ed. C. J. Corbally, R. O. Gray, and R. F. Garrison   ISBN 0-937707-79-1

Vol. 61-Astronomical Data Analysis Software and Systems III
ed. Dennis R. Crabtree, R. J. Hanisch, and Jeannette Barnes   ISBN 0-937707-80-5

Vol. 62-The Nature and Evolutionary Status of Herbig Ae / Be Stars
ed. P. S. Thé, M. R. Pérez, and E. P. J. van den Heuvel   ISBN 0-937707-81-3

Vol. 63-Seventy-Five Years of Hirayama Asteroid Families: The role of Collisions in the Solar System History
ed. R. Binzel, Y. Kozai, and T. Hirayama   ISBN 0-937707-82-1

Vol. 64-Cool Stars, Stellar Systems, and the Sun, Eighth Cambridge Workshop
ed. Jean-Pierre Caillault   ISBN 0-937707-83-X

Vol. 65-Clouds, Cores, and Low Mass Stars
ed. Dan P. Clemens and Richard Barvainis　　　　　　　ISBN 0-937707-84-8

Vol. 66- Physics of the Gaseous and Stellar Disks of the Galaxy
ed. Ivan R. King　　　　　　ISBN 0-937707-85-6

Vol. 67-Unveiling Large-Scale Structures Behind the Milky Way
ed. C. Balkowski and R. C. Kraan-Korteweg　　　　　　ISBN 0-937707-86-4

Vol. 68-Solar Active Region Evolution: Comparing Models with Observations
ed. K. S. Balasubramaniam and George W. Simon　　　　　　ISBN 0-937707-87-2

Vol. 69-Reverberation Mapping of the Broad-Line Region in Active Galactic Nuclei
ed. P. M. Gondhalekar, K. Horne, and B. M. Peterson　　　　　　ISBN 0-937707-88-0

Vol. 70-Groups of Galaxies
ed. Otto G. Richter and Kirk Borne　　　　　　ISBN 0-937707-89-9

Vol. 71-Tridimensional Optical Spectroscopic Methods in Astrophysics
ed. G. Comte and M. Marcelin　　　　　　ISBN 0-937707-90-2

Vol. 72-Millisecond Pulsars—A Decade of Surprise, ed. A. A. Fruchter
M. Tavani, and D. C. Backer　　　　　　ISBN 0-937707-91-0

Vol. 73-Airborne Astronomy Symposium on the Galactic Ecosystem: From Gas to Stars to Dust
ed. M. R. Haas, J. A. Davidson, and E. F. Erickson　　　　　　ISBN 0-937707-92-9

Vol. 74-Progress in the Search for Extraterrestrial Life,
ed. G. Seth Shostak　　　　　　ISBN 0-937707-93-7

Vol. 75-Multi-Feed Systems for Radio Telescopes
ed. D. T. Emerson and J. M. Payne　　　　　　ISBN 0-937707-94-5

Vol. 76-GONG '94: Helio- and Astero-Seismology from the Earth and Space
ed. Roger K. Ulrich, Edward J. Rhodes, Jr., and Werner Däppen　　　　　　ISBN 0-937707-95-3

Vol. 77-Astronomical Data Analysis Software and Systems IV
ed. R. A. Shaw, H. E. Payne, and J. J. E. Hayes　　　　　　ISBN 0-937707-96-1

Vol. 78-Astrophysical Applications of Powerful New Databases
ed. S. J. Adelman and W. L. Wiese　　　　　　ISBN 0-937707-97-X

Vol. 79-Robotic Telescopes: Current Capabilities, Present Developments, and Future
Prospects for Automated Astronomy
ed. Gregory W. Henry and Joel A. Eaton　　　　　　ISBN 0-937707-98-8

Vol. 80-The Physics of the Interstellar Medium and Intergalactic Medium
ed. A. Ferrara, C. F. McKee, C. Heiles, and P. R. Shapiro　　　　　　ISBN 0-937707-99-6

Vol. 81-Laboratory and Astronomical High Resolution Spectra
ed. A. J. Sauval, R. Blomme, and N. Grevesse　　　　　　ISBN 1-886733-01-5

Vol. 82-Very Long Baseline Interferometry and the VLBA,
ed. J. A. Zensus, P. K. Diamond, and P. J. Napier　　　　　　ISBN 1-886733-02-3

Vol. 83-Astrophysical Applications of Stellar Pulsation, IAU Colloquium 155
ed. R. S. Stobie and P. A. Whitelock　　　　　　ISBN 1-886733-03-1

Vol. 84-The Future Utilisation of Schmidt Telescopes, IAU Colloquium 148
ed. Jessica Chapman, Russell Cannon, Sandra Harrison, and Bambang Hidayat　　ISBN 1-886733-05-8

Vol. 85-Cape Workshop on Magnetic Cataclysmic Variables
ed. D. A. H. Buckley and B. Warner　　　　　　ISBN 1-886733-06-6

Vol. 86-Fresh Views of Elliptical Galaxies
ed. Alberto Buzzoni, Alvio Renzini, and Alfonso Serrano   ISBN 1-886733-07-4

Vol. 87-New Observing Modes for the Next Century
ed. Todd Boroson, John Davies, and Ian Robson   ISBN 1-886733-08-2

Vol. 88-Clusters, Lensing, and the Future of the Universe
ed. Virginia Trimble and Andreas Reisenegger   ISBN 1-886733-09-0

Vol. 89-Astronomy Education: Current Developments, Future Coordination
ed. John R. Percy   ISBN 1-886733-10-4

Vol. 90-The Origins, Evolution, and Destinies of Binary Stars in Clusters
ed. E. F. Milone and J. -C. Mermilliod   ISBN 1-886733-11-2

Vol. 91-Barred Galaxies, IAU Colloquium 157
ed. R. Buta, D. A. Crocker, and B. G. Elmegreen   ISBN 1-886733-12-0

Vol. 92-Formation of the Galactic Halo--Inside and Out
ed. H. L. Morrison and A. Sarajedini   ISBN 1-886733-13-9

Vol. 93-Radio Emission from the Stars and the Sun
ed. A. R. Taylor and J. M. Paredes   ISBN 1-886733-14-7

Vol. 94-Mapping, Measuring, and Modelling the Universe
ed. Peter Coles, Vicent Martinez, and Maria-Jesus Pons-Borderia   ISBN 1-886733-15-5

Vol. 95-Solar Drivers of Interplanetary and Terrestrial Disturbances
ed. K.S. Balasubramaniam, S. L. Keil, and R. N. Smartt   ISBN 1-886733-16-3

Vol. 96-Hydrogen-Deficient Stars
ed. C. S. Jeffery and U. Heber   ISBN 1-886733-17-1

Vol. 97-Polarimetry of the Interstellar Medium
ed. W. G. Roberge and D. C. B. Whittet   ISBN 1-886733-18-X

Vol. 98- From Stars to Galaxies: The Impact of Stellar Physics on Galaxy Evolution
ed. Claus Leitherer, Uta Fritze-von Alvensleben, and John Huchra   ISBN 1-886733-19-8

Vol. 99- Cosmic Abundances
ed. Stephen S. Holt and George Sonneborn   ISBN 1-886733-20-1

Vol. 100- Energy Transport in Radio Galaxies and Quasars
ed. P. E. Hardee, A. H. Bridle, and J. A. Zensus   ISBN 1-886733-21-X

Vol. 101- Astronomical Data Analysis Software and Systems V
ed. George H. Jacoby and Jeannette Barnes   ISSN   1080-7926

Vol. 102- The Galactic Center, 4th ESO/CTIO Workshop
ed. Roland Gredel   ISBN 1-886733-22-8

Inquiries concerning these volumes should be directed to the:
 Astronomical Society of the Pacific
 CONFERENCE SERIES
 390 Ashton Avenue
 San Francisco, CA 94112-1722
 415-337-1100
 e-mail asp@stars.sfsu.edu

ASTRONOMICAL SOCIETY OF THE PACIFIC
CONFERENCE SERIES

Volume 102

THE GALACTIC CENTER
4th ESO/CTIO Workshop

La Serena, Chile, 10-15 March 1996

Edited by
Roland Gredel

Table of Contents

Preface xvii

Conference participants xviii

Conference photo xxiv

Part 1. Kinematics of the Gas and Dust

Section A. Large Scale Structure

The Galactic Center at Low Radio Frequencies 3
    H. Alvarez, J. Aparici, and J. May

Molecular Clouds near the Galactic Center 8
    John Bally

Neutral Carbon and the Dense Interstellar Medium in the Inner 300 pc of the Galaxy 16
    D. T. Jaffe, R. Plume, Neal J. Evans II, and John Bally

Large Scale Observations of the Far-Infrared [CII] Line Emission from the Galactic Center 20
    Takao Nakagawa, Yasuo Doi, Yukari Yamashita Yui, Haruyuki Okuda, Kenji Mochizuki, Hiroshi Shibai, Tetsuo Nishimura, and Frank J. Low

H_2 Emission from the Inner 400 Parsecs of the Galaxy: II. The UV-Excited H_2 28
    Soojong Pak, D. T. Jaffe, and L. D. Keller

Large Scale Surface Photometry of the Galactic Center 36
    L. Schmidtobreick, C. Tappert, W. Schlosser, P. Koczet, S. Wiemann, M. Jütte, B. Hoffmann, Th. Schmidt-Kaler, and S. Kimeswenger

The Formation of Large Scale Structures Via the Parker Instability . . . 40
    A. Santillán, M. A. Martos, J. Franco, and D. P. Cox

Section B. Molecular Clouds at the Galactic Center

SiO Emission from the Galactic Center Molecular Clouds 47
    Jesús Martín-Pintado, Pablo de Vicente, Asuncion Fuente, and Pere Planesas

The Molecular Gas in the Galactic Center Region based on $C^{18}O$ Measurements 54
    G. Dahmen, S. Hüttemeister, T.L. Wilson, R. Mauersberger, A. Linhart, L. Bronfman, A.R. Tieftrunk, K. Meyer, W. Wiedenhöver, T.M. Dame, E.S. Palmer, J. May, J.Aparici, and F. MacAuliffe

CS (J=2-1) Observations of the Sgr B Complex and FIR 21 60
    F. Yusef-Zadeh, K.I. Uchida, D.M. Mehringer, D. Roberts, L.-A. Nyman, S. Casement, and M. Lindqvist

A Hot Ring in the Sgr B2 Molecular Cloud 64
    Pablo de Vicente, Jesús Martín-Pintado, and Tom L. Wilson

Kinematical, Physical and Evolutionary Status of the Galactic Center Giant Molecular Cloud G1.6-0.025 68
    Andrej M. Sobolev

Section C. Sgr A and the Radio Arc

Kinematics of the Sgr A Cloud Complex 77
    Robert Zylka

Millimetre Spectral Line Observations of the Galactic Centre 85
    Aa. Sandqvist

Detection of OH (1720 MHz) Masers in Sgr A East 93
    A.J. Green, D.A. Frail, W.M. Goss, F. Yusef-Zadeh, and D.A. Roberts

Physical Condition of UV Heated Warm Gas around the Galactic Center 101
    Yukari Y. Yui, Takao Nakagawa, Haruyuki Okuda, Hiroshi Shibai, Norihisa Hiromoto, and Yasuo Doi

KWIC Imaging of the Galactic Center 106
    H.M. Latvakoski, G.J. Stacey, T.L. Hayward, and G.E.Gull

Far-Infrared Fine Structure Line Observations of the Galactic Center Radio Arc 114
    A. Poglitsch, R. Genzel, S.C. Madden, T. Nikola, R. Timmermann, N. Geis, and C.H. Townes

Interstellar Extinction in the Vicinity of Galactic Center Thermal Radio Emission Regions 122
    Angela S. Cotera, Janet P. Simpson, Edwin F. Erickson, Sean W.J. Colgan, and Michael G. Burton

The Turbulent Galactic Center: Extremely Strong Scattering near Sgr A* 127
    T. Joseph W. Lazio and James M. Cordes

Section D. The Central Parsec and the CND

Kinematics of the Ionized Gas in Sgr A West: The Minicavity and the Western Arc 137
    D.A. Roberts, F. Yusef-Zadeh, and W.M. Goss

The −185 km s^{-1} Cloud Interacting with the Galactic Center 146
    Jun-Hui Zhao, W.M. Goss, and P.T.P. Ho
The 1720 MHz OH Emission from G359.1−0.5 and the CND 151
    F. Yusef-Zadeh, B.T. Robinson, D.A. Roberts, W.M. Goss, D.A. Frail, and A. Green
H$_2$ Emission from the Galactic Centre Molecular Ring 163
    S.K. Ramsay Howat, C.M. Mountain, and T.R. Geballe
Molecular Line Maps of the Galactic Centre Circumnuclear Disk 171
    Glenn J. White
10μm Imaging Polarimetry 179
    David Aitken, Craig Smith, Toby Moore, and Patrick Roche

Part 2. The Stellar Population

Section A. The Central Stellar Cluster

Astrometry in the Galactic Center Region 189
    Michael R. Rosa
Observations of the Galactic Center with SHARP: First Stellar Proper Motions 196
    A. Eckart and R. Genzel
Quantitative Spectroscopy of the HeI Cluster 203
    Francisco Najarro, Rolf P. Kudritzki, Alfred Krabbe, Reinhard Genzel, Dieter Lutz, and D. John Hillier
Chemical Abundances of Stars at the Galactic Center 212
    John S. Carr, K. Sellgren, and Suchitra C. Balachandran
Origin of the Peculiar HeI Stars 220
    Peter Tamblyn, Fulvio Melia, and G.H. Rieke
2.2 μm Keck Images of the Galaxy's Central Stellar Cluster at 0$''$.05 Resolution 228
    B.L. Klein, A.M. Ghez, M. Morris, and E.E. Becklin
Fourier Imaging Spectroscopy of the Galactic Center 232
    D.A. Simons and J.P. Maillard
Consequences of Extended Dark Mass in the Galactic Center Cluster . . 236
    Peter J. McGregor, Geoffrey V. Bicknell, and Prasenjit Saha

Section B. Star Formation and Hot Stars

Masers and Star Formation near the Galactic Centre 247
    J. L. Caswell
Object #17 and Quintuplet Clusters near the Galactic Center 255
    Tetsuya Nagata

Hot Stars in the Quintuplet 263
    Donald F. Figer, Mark Morris, and Ian S. McLean
The Chemical Composition near the Galactic Centre - A Study of Four
    Blue Supergiants.. 271
    S.J. Smartt, P.L. Dufton, and D.J. Lennon

Section C. Evolved Stars and Stellar Remnants

Really Cool Stars at the Galactic Center 277
    R.D. Blum, K. Sellgren, and D.L. DePoy
Interstellar Extinction and the Luminosity Function of Galactic Center
    Stars 285
    K. Sellgren, R.D. Blum, and D.L. DePoy
The Nature of OH/IR Stars in the Galactic Centre Region 289
    Joris Blommaert, Wil van der Veen, Harm Habing, Huib-Jan van
    Langevelde, and Loránt Sjouwerman
A new Sample of OH/IR Stars in the Galactic Center 294
    Loránt Sjouwerman, Anders Winnberg, Huib Jan van Langevelde, Harm
    Habing, and Michael Lindqvist
Spectroscopy of New Planetary Nebulae close to the Galactic Center .. 299
    G.C. Van de Steene and G.H. Jacoby

Section D. Surveys

The 15 μm ISOGAL Survey of the Inner Galactic Disk 305
    Alain Omont
Variable Stars close to the Galactic Centre 312
    I.S. Glass, S. Matsumoto, T. Ono, and K. Sekiguchi
The Stellar Content of the Bulge: NTT and HST Photometry 320
    A. Vallenari, C. Chiosi, G. Bertelli, and Y.K. Ng

Section E. Dynamical Studies

Dynamics of the Core Region 327
    R.H. Miller
The Dynamics of the Galactic Center: Origin of the Mini-Spiral 335
    A.M. Fridman, O.V. Khoruzhii, V.V. Lyakhovich, L. Ozernoy, O.K.
    Sil'chenko, and L. Blitz
A Large Proper-Motion Survey in Plaut's Low-Extinction Window ... 345
    R.A. Méndez, R.M. Rich, W.F. van Altena, T.M. Girard, S. van den
    Bergh, and S.R. Majewski
Galactic Cores as Separate Stellar Subsystems 353
    Olga K. Sil'chenko

On Masses of Equilibrium Configurations 357
    L.V. Verozub
MOND and the Feeding of the Monster: The Conveyor Belt 362
    Fernando J. Selman

Part 3. The Central Engine and High Energy Phenomena

Section A. Sagittarius A*

The Discovery of the Radio Source Sagittarius A (Sgr A) 369
    W.M. Goss and R.X. McGee
The Spectrum of Sgr A* and the Central Parsec 380
    Peter G. Mezger
The Interpretation of the Sgr A* Spectrum 389
    Wolfgang J. Duschl and Thomas Beckert
Activity and Radiative Characteristics of the Central Engine in the Galactic Center 395
    F. Melia
Bondi-Hoyle Accretion onto Sgr A* 403
    R. F. Coker and F. Melia
The First Detection of a Source Coincident with Sgr A* at 8.7 μm ... 407
    Susan R. Stolovy, T.L. Hayward, and Terry Herter
ROSAT/X-ray Studies of the Galactic Center 415
    P. Predehl and H. Zinnecker

Section B. High Energy Phenomena

ASCA Observations of the Galactic Center 423
    Yoshitomo. Maeda and Katsuji. Koyama
Six Years of Hard X-Ray and Soft γ-Ray Observations of the Galactic Center with the Sigma Telescope 431
    M. Vargas, A. Goldwurm, J. Paul, J. Ballet, L. Bouchet, J.P. Roques, E. Jourdain, V. Borrel, R. Sunyaev, M. Gilfanov, E. Churazov, S. Kuznetsov, S. Trudolubov, M. Revnivtsev, N. Khavenson, and A. Dyachkov
A Search for Infrared Positronium Line Emission from the Great Annihilator near the Galactic Centre 439
    P.J. Puxley and G.K. Skinner
G357.1–00.2: A Peculiar Nonthermal Radio Source near the Galactic Centre 443
    Andrew D. Gray

The Interaction of the G359.54+0.18 Nonthermal Filaments with the Ambient Medium 447
    J. Staguhn, J. Stutzki, F. Yusef-Zadeh, and K.I. Uchida

Section C. Comparison with other Nuclei of Galaxies

Sgr A* and its Siblings in nearby Galaxies 453
    Heino Falcke
Stars, Stellar Remnants, and the Black Hole in the Galactic Center . . . 462
    G.H. Rieke and M.J. Rieke
Comparison between the Galactic Center and Activity in NGC 4258 (M106) 471
    E.M. Burbidge
The Role of the Galactic Center on our Understanding of Violent Activity in the Nuclei of Galaxies 480
    G. Burbidge and F. Hoyle

Part 4. Late Papers

Mid-Infrared Emission and Luminous Sources in the Central Parsec . . 491
    Dan Gezari
OH/IR Stars - Dynamical Studies 500
    Anders Winnberg

Author index 508

Subject index 511

Preface

This volume contains the manuscripts presented at the workshop on "The Galactic Center", held March 10–15, 1996 in La Serena, Chile. The workshop was the 4th international meeting jointly organized by the European Southern Observatory (ESO) and Cerro Tololo Inter American Observatory (CTIO). A total of 114 scientists from 18 countries attended, including many students and post-docs.

The workshop was organized into individual sessions covering the kinematics of the gas and dust at the Galactic Center, its stellar population and dynamical studies, the central engine and high energy phenomena. The observational data presented covered the full spectrum, all the wavelengths accessible from the ground to airborne observations and results obtained from the ASCA, ROSAT, and ISO satellites, and from the HST and the Sigma telescope on board of the Russian GRANAT space observatory.

The conference would not have been possible without the help of many individuals. The scientific programme was put together by Jorge Melnick, Bob Schommer and Roland Gredel, following helpful advise from characters such as Paul Ho, Hans Zinnecker, and Bo Reipurth. The local organization was efficiently done by Elaine Mac-Auliffe, Claudia Contreras, Renato Vargas, and David Gonzalez, and Edmund Janssen helped with the design of the poster. And last but not least, the hospitality of the Intendencia of La Serena is greatly appreciated.

Roland Gredel
European Southern Observatory

May 1996

Cover Illustration : A coadded image made from ten separate Brackett-gamma images, using the CFHT imaging Fourier Transform Spectrometer, together with spectra of various sources in the stellar cluster calibrated in velocity around the Brackett-gamma line. Provided by D.A. Simons and J.-P. Maillard (see their paper on page 232).

Participant List

Aguilera, Claudio A., Pontificia Universidad Catolica, Av Vicu na Mackenna 4860, Casilla 104, Santiago 22, Santiago, Chile ⟨ guilera@astro.puc.cl ⟩

Aitken, Dave, University of Hertfordshire, Division of Physical Sciences, 5 Dewsbury Cottages, York, Y01 1HB, UK ⟨ D.aitken@roe.ac.uk ⟩

Alvarado, Franklin, Instituto Isaac Newton, Fernandez Concha 472, Casilla 8-9 Correo 9, Santiago, Chile ⟨ falvarad@sc.eso.org ⟩

Alvarez, Hector, Departamento de Astronomia, U. de Chile, Casilla 36-D, Santiago, Chile ⟨ hector@das.uchile.cl ⟩

Arenas, Jose, Dpto. de Fisica, U. de Concepcion, Casilla 4009, Concepcion, Chile ⟨ jarenas@gauss.cfm.udec.cl ⟩

Athey, Alex, Pomona College Physics Dept, Claremont, CA 91711, U.S.A. ⟨ aathey@pomona.edu ⟩

Bally, John, University of Colorado, Boulder, C323 A, Duane Physics, Colorado Blvd., Boulder, CO 80309, U.S.A. ⟨ bally@nebula.colorado.edu ⟩

Barrera, Luis Humberto, Universidad Catolica del Norte, Dpto. de Astronomia, Avda. Angamos 0610, Antofagasta, Chile ⟨ lbarrera@socompa.cecun.ucn.cl ⟩

Blommaert, Joris, ESA Villafranca Satellite Tracking Station, PO Box 50727, 28080 Madrid Aptdo 50727, Spain ⟨ jblommae@iso.vilspa.esa.es ⟩

Blum, Robert, University of Colorado, JILA Campus box 440, University of Colorado Boulder, CO 80309, U.S.A. ⟨ rblum@casa.colorado.edu ⟩

Burbidge, Margaret, University of California, San Diego, Center for Astrophysics & Space Sciences 0111, La Jolla, CA 92093-0111, U.S.A. ⟨ btravell@ucsd.edu ⟩

Burbidge, Geoffrey, University of California, San Diego, Center for Astrophysics & Space Sciences, 0111, La Jolla, CA 92093-0111, U.S.A. ⟨ btravell@ucsd.edu ⟩

Carr, John, Naval Research Lab., Code 7217, Washington, D.C. 20375-5351, U.S.A. ⟨ carr@mriga.nrl.navy.mil ⟩

Caswell, J.L., ATNF, CSIRO, PO Box 76, Epping, NSW, Australia 2121 ⟨ jcaswell@atnf.csiro.au ⟩

Coker, Robert, Univeristy of Arizona Physics Dept., PAS 81, 1118 E 4th St, Tucson AZ 85721, U.S.A. ⟨ rfc@soliton.physics.arizona.edu ⟩

Cotera, Angela, NASA/Jet Propulsion Laboratory 100-22, 4800 Oak Grove, Pasadena, CA 91109, U.S.A. ⟨ cotera@halley.la.asu.edu ⟩

Covarrubias, Ricardo, Observatorio Interamericano de Cerro Tololo, Colina el Pino s/n, La Serena, Chile ⟨ riccov@ctio.noao.edu ⟩

d'Ans, Barthelemy, Instituto Peruano de Astronomia, Malecón de la Marina 1080-401, Miraflores, Lima 18, Perú

Participant List

Dahmen, Gereon, Physics Department Queen Mary & Westfield College, University of London, Mile End Road, London E1 4NS, UK ⟨ G.Dahmen@qmw.ac.uk ⟩

Duschl, Wolfgang J., Institut für Theoretische Astrophysik, Tiergartenstr. 15, 69121 Heidelberg, Germany ⟨ wjd@platon.ita.uni-heidelberg.de ⟩

Eckart, Andreas, Max-Planck-Institut für extraterrestrische Physik, Giessenbachstrasse, 85748 Garching, Germany ⟨ eckart@mpe.mpe-garching.mpg.de ⟩

Elston, Richard, Observatorio Interamericano de Cerro Tololo, Colina el Pino s/n, La Serena, Chile ⟨ elston@ctio.noao.edu ⟩

Falcke, Heino, Astronomy Department, University of Maryland, College Park, MD 20742, U.S.A. ⟨ hfalcke@astro.umd.edu ⟩

Figer, Donald F., UCLA, Division of Astronomy, LA, CA 90095, U.S.A. ⟨ figer@astro.ucla.edu ⟩

Fridman, Alexei, Institute of Astronomy, 48 Pyatnitzkaya St, Moscow, 109017, RUSSIA ⟨ afridman@inasan.rssi.ru or afridman@astro.free.net ⟩

Garay, Guido, Departamento de Astronomia, Universidad de Chile, Casilla 36-D, Santiago, Chile ⟨ guido@das.uchile.cl ⟩

Geballe, Thomas, Joint Astronomy Centre, 660 N. A'ohoku Place, University Park, Hilo, HI 96720, U.S.A. ⟨ tom@jach.hawaii.edu ⟩

Genzel, Reinhard, Max-Planck-Institut für extraterrestrische Physik, Giessenbachstrasse, 85748 Garching, Germany ⟨ genzel@mpe-garching.mpg.de ⟩

Gezari, Dan, NASA/Goddard Laboratory for Astronomy and Solar Physics, Code 685, Greenbelt, MD 20771, U.S.A. ⟨ gezari@stars.gsfc.nasa.gov ⟩

Glass, Ian S., SAAO, PO Box 9, Observatory 7935, S. Africa ⟨ isg@saao.ac.za ⟩

Gomez M., Ximena, Departamento de Astronomia, Universidad de Chile, Casilla 36-D, Santiago, Chile ⟨ xgomez@calan.das.uchile.cl ⟩

Goss, Miller, National Radio Astronomy Observatory, P.O. Box 0, Socorro, NM 87801, U.S.A. ⟨ mgoss@aoc.nrao.edu ⟩

Gray, Andrew D., DRAO, P.O. Box 248, Penticton, BC, V2A 6K3, Canada ⟨ agray@cygnus.drao.nrc.ca ⟩

Gredel, Roland, European Southern Observatory, Casilla 19001, Santiago 19, Chile ⟨ rgredel@eso.org ⟩

Green, Anne, Astrophysics Dept., School of Physics, University of Sydney, NSW 2006, Australia ⟨ agreen@physics.usyd.edu.au ⟩

Hasegawa, Tetsuo, Institute of Astronomy, The University of Tokyo, 2-21-1 Osawa, Mitaka, Tokyo 181, Japan ⟨ tetsuo@ghz.mtk.ioa.s.u-tokyo.ac.jp ⟩

Ho, Paul, Smithsonian Astrophysical Observatory, 60 Garden St., MS 78, Cambridge, MA 02138, U.S.A. ⟨ ho@cfaho1.harvard.edu ⟩

Heathcote, Stephen, Observatorio Interamericano de Cerro Tololo, Colina el Pino s/n, La Serena, Chile ⟨ sheathcote@ctio.noao.edu ⟩

Participant List

Hüttemeister, Susanne, CfA, Radio- and Geoastronomy, MS 72, 60 Garden Street, Cambridge, MA 02138, U.S.A. ⟨ huette@cfa.harvard.edu ⟩

Jaffe, Daniel, University of Texas, Dept. of Astronomy, RLM 15.308, Austin, TX 78712, U.S.A. ⟨ dtj@astro.as.utexas.edu ⟩

Klein, Beth, U.C.L.A., Physics and Astronomy, 405 Hilgard Ave., Los Angeles, CA. 90095, U.S.A. ⟨ kleinb@mira.astro.ucla.edu ⟩

Kohnenkamp, Ive, Universidad de Chile, Departamento de Astronomia, Casilla 36-D, Santiago ⟨ ikohnenk@das.uchile.cl ⟩

Krabbe, Alfred, Max-Planck Institut für Extraterrestrische Physik, Giessenbachstr., 85740 - Garching, Germany ⟨ krabbe@mpe.mpe-garching.mpg.de ⟩

Latvakoski, Harri, Cornell University, Astronomy Dept., 225 Space Sciences, Ithaca NY 14853, U.S.A. ⟨ harri@tristan.tn.cornell.edu ⟩

Lazio, T. Joseph W., Cornell University, 514 Space Sciences Bldg., Ithaca, NY 14853-6801, U.S.A. ⟨ lazio@spacenet.tn.cornell.edu ⟩

Leiton, Roger, Universidad de La Serena, La Serena, Chile ⟨ leiton@ctios1.ctio.noao.edu ⟩

Lemke, Roland, European Southern Observatory, Casilla 19001, Santiago 19, Chile ⟨ rlemke@eso.org ⟩

Lo, Kwok-Yung, University of Illinois, Dept. of Astronomy, 1002 W. Green Street, Urbana, IL 61801, U.S.A. ⟨ kyl@astro.uiuc.edu ⟩

Loup, Cécile, Institut d'Astrophysique de Paris, CNRS, 98b Blvd Arago, Paris, France ⟨ loup@iap.fr ⟩

Maeda, Yoshitomo, , Kitashirakawa-Oiwake-chyo Sakyo-ku, Kyoto 606-01, Japan ⟨ maeda@cr.scphys.kyoto-u.ac.jp ⟩

Maillard, Jean-Pierre, Institut d'Astrophysique de Paris, CNRS, 98b Blvd Arago, Paris, France ⟨ maillard@iap.fr ⟩

Marr, Jonathan, Physics Department, Union College, Schenectady, NY 12308, U.S.A. ⟨ marrj@gar.union.edu ⟩

Martin-Pintado, Jesus, Centro Astronomico de Yebes, Apartado 148, 19080, Spain ⟨ martin@cay.es ⟩

Masset, Frederic, CEA, Service d'Astrophysique, Gif sur Yvette 91191, France ⟨ tagger@ariane.saclay.cea.fr ⟩

McGregor, Peter J., Mount Stromlo and Siding Spring Obs., Private Bag, PO, Weston Creek ACT 2611, Australia ⟨ peter@mso.anu.edu.au ⟩

Melia, Fulvio, University of Arizona Astronomy Dept., Cherry Street, Tucson AZ 85721, U.S.A. ⟨ melia@as.arizona.edu ⟩

Melnick, Jorge, European Southern Observatory, Casilla 19001, Santiago 19, Chile ⟨ jmelnick@eso.org ⟩

Mendez, Rene A., European Southern Observatory, Karl-Schwarzschild Str. 2, 85748 Garching, Germany ⟨ rmendez@eso.org ⟩

Mennickent, Ronald, Dpto. de Fisica, U. de Concepcion, Casilla 4009, Concepcion, Chile ⟨ rmennick@gauss.cfm.udec.cl ⟩

Mezger, Peter, Max-Planck-Institut für Radioastronomie, Auf dem Hügel 69, 53121 Bonn, Germany ⟨ pmezger@mpifr-bonn.mpg.de ⟩

Miller, Richard H., Astronomy Center, 5640 S. Ellis Ave., Chicago 60637, U.S.A. ⟨ rhm@oddjob.uchicago.edu ⟩

Mirabel, Felix, Service d'astrophysique, CE-Saclay, 91911 Gif/Yvette, France ⟨ mirabel@ariane.saclay.cea.fr ⟩

Morris, Mark, UCLA, Physics and Astronomy, 405 Hilgard Ave, Box 951562, Los Angeles, 90095-1562, U.S.A. ⟨ morris@osprey.astro.ucla.edu ⟩

Mountain, Matt, Gemini Project, 950 N. Cherry Ave., Tucson, AZ 85719, U.S.A. ⟨ mmountain@gemini.edu ⟩

Nagata, Tetsuya, Dept. Physics, Nagoya Univ., Furo-cho, Chikusa-ku, Nagoya 464-01, Japan ⟨ nagata@zlab.phys.nagoya-u.ac.jp ⟩

Najarro, Francisco, Universitäts-Sternwarte München, Scheinerstraße 1, D-81679 München, Germany ⟨ paco@usm.uni-muenchen.de ⟩

Nakagawa, Takao, The Institute of Space and Astronautical Science, Yoshinodai 3-1-1, Sagamihara, Kanagawa 229, Japan ⟨ nakagawa@astro.isas.ac.jp ⟩

Omont, Alain, Institut d'Astrophysique de Paris, CNRS, 98b Blvd Arago, Paris, France ⟨ omont@iap.fr ⟩

Pak, Soojong, Department of Astronomy, The University of Texas at Austin, Austin, TX 78712-1083, U.S.A. ⟨ soojong@astro.as.utexas.edu ⟩

Phillips, Mark, Observatorio Interamericano de Cerro Tololo, Colina el Pino s/n, La Serena, Chile ⟨ phillips@ctio.noao.edu ⟩

Poglitsch, Albrecht, Max-Planck-Institut für extraterrestrische Physik, Giessenbachstr., 85748 Garching, Germany ⟨ alpog@fifi.mpe-garching.mpg.de ⟩

Probst, Ronald, Observatorio Interamericano de Cerro Tololo, Colina el Pino s/n, La Serena, Chile ⟨ probst@ctio.noao.edu ⟩

Puxley, Phil, Royal Observatory, Blackford Hill, Edinburgh EH9 3HJ, UK ⟨ pjp@roe.ac.uk ⟩

Ramsay Howat, Suzanne, Royal Observatory, Blackford Hill, Edinburgh EH9 3HJ, UK ⟨ skr@roe.ac.uk ⟩

Reipurth, Bo, European Southern Observatory, Casilla 19001, Santiago 19, Chile ⟨ breipurt@eso.org ⟩

Rieke, George, Steward Observatory, University of Arizona, Tucson, AZ 85721, U.S.A. ⟨ grieke@as.arizona.edu ⟩

Rieke, Marcia, Steward Observatory, University of Arizona, Tucson, AZ 85721, U.S.A. ⟨ mrieke@as.arizona.edu ⟩

Rivinius, Thomas, Landessternwarte Koenigstuhl, 69117 Heidelberg, Germany ⟨ T.Rivinius@lsw.uni-heidelberg.de ⟩

Roche, Patrick, Nuclear and Astrophyisics Building, Keble Road, Oxford, OX1 3RH, UK, p.roche@physics.ox.ac.uk ⟨ Oxford University, Dept. of Physics ⟩

Rosa, Michael, ST-ECF, Karl-Schwarzschild Str. 2, 85748 Garching, Germany ⟨ mrosa@eso.org ⟩

Rubio, Monica, Departamento de Astronomia, Universidad de Chile, Casilla 36-D, Santiago, Chile ⟨ mrubio@das.uchile.cl ⟩

Sandqvist, Aage, Stockholm Observatory, S-133 36 Saltsjobaden, Sweden ⟨ sandqvis@astro.su.se ⟩

Santillan, Alfredo, Instituto de Astronomia UNAM, Ciudad Universitaria, cp 04510 Mexico, D.F. Mexico ⟨ alfredo@astroscu.unam.mx ⟩

Schmidtobreick, Linda, Astronomisches Institut der Ruhr-Universität Bochum, D-44780 Bochum, Germany ⟨ lkl@astro.ruhr-uni-bochum.de ⟩

Schneider, Jodi, St. John's College, 808 New York Avenue, Cherry Hill, NJ 08002-3114, U.S.A. ⟨ jschneid@geom.umn.edu ⟩

Schommer, Robert, Observatorio Interamericano de Cerro Tololo, Colina el Pino s/n, La Serena, Chile ⟨ rschommer@ctio.noao.edu ⟩

Sellgren, Kris, Astronomy Dept., Ohio State University, 174 West 18th Av, Columbus, OH 43210, U.S.A. ⟨ sellgren@payne.mps.ohio-state.edu ⟩

Selman, Fernando, , Las Malvas 228, Santiago, Chile

Sil'chenko, Olga, Sternberg Astronomical Institute, University av. 13,, 119899 Moscow, Russia ⟨ olga@sai.msu.su ⟩

Simon, Guy, DASGAL Observatoire de Paris, 61, Avenue de l'Observatoire, 75014 Paris, France ⟨ simon@obspm.fr ⟩

Simons, Doug, Gemini Project, 950 N. Cherry Ave., Tucson, AZ 85719, U.S.A. ⟨ dsimons@gemini.edu ⟩

Sjouwerman, Lorant, Onsala Space Observatory, 439 92 Onsala, Sweden ⟨ sjouwerm@oso.chalmers.se ⟩

Skinner, Gerry, University of Birmingham, Space Research Group, Edgbaston, Birmingham B15 2TT, UK ⟨ gks@star.sr.bham.ac.uk ⟩

Smartt, Stephen J., Department of Pure and Applied Physics, Queens University of Belfast, Belfast, BT7 1NN, Northern Ireland ⟨ S.Smartt@qub.ac.uk ⟩

Smith, Malcolm, Observatorio Interamericano de Cerro Tololo, Colina el Pino s/n, La Serena, Chile ⟨ mgs@ctio.noao.edu ⟩

Sobolev, A. M., Astronomical Observatory, Ural State University, Lenin Ave 51, Ekaterinburg 620083, Russia ⟨ Andrej.Sobolev@usu.ru ⟩

Soto, Alejandro, Dartmouth College, 4085 Hinman, Hanover, NH 03755-4025, U.S.A. ⟨ mahikan@dartmouth.edu ⟩

Staguhn, Johannes, 1. Physikal. Institut, Universität Köln, Zülpicher Strasse 77, 50937 Köln, Germany ⟨ staguhn@ph1.uni-koeln.de ⟩

Stolovy, Susan, Cornell University, 206 Space Sciences Building, Ithaca, NY 14853, U.S.A. ⟨ stolovy@astrosun.tn.cornell.edu ⟩

Tamblyn, Peter, Steward Observatory, University of Arizona, Tucson, AZ 85721, U.S.A. ⟨ ptamblyn@as.arizona.edu ⟩

Van de Steene, Griet, European Southern Observatory, Casilla 19001, Santiago 19, Chile ⟨gsteene@eso.org⟩

Turner, Elizabeth, Georgia Institute of Technology, 325497 GT Station, Atlanta, GA 30332-1060, U.S.A. ⟨gt5497b@prism.gatech.edu⟩

van Hoof, Peter, Kapteyn Astronomical Institute, P.O. Box 800, Groningen 9700 AV, The Netherlands ⟨p.van.hoof@astro.rug.nl⟩

Vallenari, Antonella, Osservatorio Astronomico, Vicolo Osservatorio 5, 35122-Padova, Italy ⟨vallenari@astrpd.pd.astro.it⟩

Vargas, Marielle, Centre d'etude de Saclay, Batiment 709, Ormes des Merisiers, 91191 Gif sur Yvette, France ⟨mv@salyut.saclay.cea.fr⟩

Walker, Alistair, Observatorio Interamericano de Cerro Tololo, Colina el Pino s/n, La Serena, Chile ⟨awalker@ctio.noao.edu⟩

White, Glenn J., Physics Department, Queen Mary & Westfield College, University of London, Mile End Road, London E1 4NS, UK ⟨G.J.White@qmw.ac.uk⟩

Winnberg, Anders, Onsala Space Observatory, 439 92 Onsala, Sweden ⟨anders@oso.chalmers.se⟩

Yan, Lin, European Southern Observatory, Karl-Schwarzschild Str. 2, 85748 Garching, Germany ⟨lyan@eso.org⟩

Yui, Yukari Yamashita, COE Communication Research Laboratory, Nukuikita-machi 4-2-1, Koganei, Tokyo, 184, Japan ⟨yukari@crl.go.jp⟩

Yusef-Zadeh, Farhad, Dept of Physics and Astronomy, 2145 Sheridan Road, Northwestern University, Evanston, Il 60208, U.S.A. ⟨zadeh@ossenu.astro.nwu.edu⟩

Zhao, Jun-Hui, Institute of Astronomy & Astrophysics, Academia Sinica, P.O. Box 1-87, Nankang, Taipei, Taiwan 115 ⟨zhao@biaa5.biaa.sinica.edu⟩

Zinnecker, Hans, Astrophysikalisches Institut Potsdam, An der Sternwarte 16, 14482 Potsdam, Germany ⟨hzinnecker@aip.de⟩

Zylka, Robert, Institut für Theoretische Astrophysik, Tiergartenstr. 15, 69121 Heidelberg, Germany ⟨zylka@ita.uni-heidelberg.de⟩

Conference Photo

Part 1. Kinematics of the Gas and Dust
Section A. Large Scale Structure

The Galactic Center at Low Radio Frequencies

H. Alvarez, J. Aparici and J. May
Departamento de Astronomía, Universidad de Chile, Casilla 36-D, Santiago, Chile

Abstract. A region about 10° around the Galactic Centre (G.C.) has been studied at frequencies equal and lower than 408 MHz. The observations can be explained by a model consisting of three sources approximately concentric with the G.C.: an HII region uniformly mixed with a nonthermal source (the narrow source) of about the same size; the third source is a nonthermal object several degrees wide (the broad source). The broad source is readily seen at most frequencies in profiles along the galactic plane, however the presence of the narrow source is suggested only at 408 MHz. We have determined the temperature, spectrum of the nonthermal sources.

1. Introduction

The University of Chile has recently finished a 45 MHz continuum survey of the southern sky that covers from the south pole up to +19° declination (Alvarez et al. 1996). The data were taken with a filled array having a resolution of 4.6° R.A. × 2.4° Dec. Details of the instrument can be found in May et al. (1984), and Alvarez et al. (1994).

We studied an area about 10° × 10° around the galactic center and at frequencies equal or lower than 408 MHz. In this frequency range, which we have arbitrarily defined as the low frequency range, there are several surveys that cover the area under examination and reach down to 19.7 MHz. Unfortunately not all of them are suitable to our study. For example, some were made with too low a resolution, while others were obtained with the synthesis technique which removes the low spatial frequencies, etc... Out of all the surveys found we selected those at 408 MHz (Haslam et al. 1982), 85.7 MHz (Hill et al. 1958) and 45 MHz (Alvarez et al. 1996). The first two surveys have resolutions of 0.8°.

2. Analysis

At 408 MHz the map shows a maximum peak at the center. However, at 85.7 MHz and below, the area presents a characteristic structure consisting of a trough close to the center (l =0, b =0) and flanked by two intense peaks. This structure was noticed by Mills forty years ago (Mills 1956) and he advanced the hypothesis that it is produced by the absorption of a bright background radiation in front or inside an HII region. This idea has been mentioned in the literature however, as far as we know, it has not been pursued. The existence of the bright source, several degrees wide, is seen prominently in temperature profiles along the galactic equator, specially at 85.7 and 45 MHz, frequencies at which most of the emission is nonthermal. This is the source that we will identify later on as the "broad" source. Close to the position of the absorption area there is an HII region which we will assume is the one that produces the absorption. The physical properties of the region have been studied by Matthews et al (1973), Jones and Finlay (1974), and Downes and Maxwell (1968).

To study the absorption hypothesis we have obtained temperature profiles along a line connecting the centers of the flanking peaks, at the different frequencies. This line is, in general, slightly inclined to the galactic equator and very close to it. The resulting profiles are then double peaked curves. We have defined the "expected profile" by fitting a Gaussian curve to the wings of the observed profile which we assume to be the non absorbed part of it. At 85.7 an 45 MHz the data can be fitted by one Gaussian, however at 408 MHz a good fit requires two Gaussians. The Gaussian profiles are superposed on a diffuse continuum background at temperature T_B. The amplitude of the Gaussian and the depth of the absorption are defined as T_A and Δ, respectively. The minimum of the trough is approximately at $l = -0.2°$, $b = 0.1°$, however for simplicity we will assume it to be at the center.

In a first attempt we studied Mills' simple model of an HII region inside a bright source, and both concentric with the galactic center. We assumed the bright source to be nonthermal. To his model we added a diffuse background. We predicted the absorption depth at 85.7 MHz using the equation of transfer (for the antenna pointing to the center), the 408 MHz data and adopting an electronic temperature $T_e = 10^4 K$ from Matthews et al. (1973). The prediction resulted 60% larger than the actually observed depth or, in other words, the trough observed at 85.7 MHz was too shallow. The data are good enough to preclude an error of such magnitude, therefore we have postulated the existence of a second source at the center and we will call it the "narrow" source. We will assume it to be nonthermal since we need to increase the radiation strength at lower frequencies in order to explain the shallowness of the trough. The existence of this narrow source would help to explain an excess of radiation found by Little (1974) at 408 MHz, excess he could not account for. The new model consists of three spherical sources concentric with the galactic centre. Two of the sources will be assumed nonthermal: the broad source, several degrees wide and the narrow source about one degree or less. The third source is an HII region uniformly mixed with the narrow source, and of about its size. Following the conclusions of Matthews et al. (1973) we have placed the HII region at the galactic centre. The size of the HII region is adopted from Jones and Finlay (1974) while the size of the narrow source is bound by the width of the narrow Gaussian fitted to the 408 MHz profiles (3°). The model is completed with the SgrA complex which, given the size of the beam involved, is considered as a point source. Finally, the whole system is immersed in a diffuse continuum background. Our aim is to determine the spectrum of the broad and narrow sources and to find out if they turn out nonthermal, as it was assumed in the model. To do this we write down the equation of transfer for the model assuming that the antenna beam is pointing towards the center. Since the beam widths are the same for 85.7 and 408 MHz the equation is also the same for both frequencies. However the equation is slightly different for 45 MHz because at this frequency the beam is wider. We have assumed that the angular size of the HII region and the narrow source are the same and that it does not change with frequency.

For 85.7 and 408 MHz the equation is:

$$\Delta = (1 - e^{-\tau})\left[\frac{1}{2}T_A + \frac{1}{2}\left(\frac{2L - \theta R}{L + R - \theta R}\right)T_B - T_e - \frac{1}{\tau}T_C\right] - T_S e^{-\frac{\tau}{2}}$$

For 45 MHz the equation is:

$$\Delta = k(1 - e^{-\tau})\left[\frac{1}{2}T_A + \frac{1}{2}\left(\frac{2L - \theta R}{L + R - \theta R}\right)T_B - T_e - \frac{1}{\tau}T_C\right] - T_S e^{-\frac{\tau}{2}}$$

We want to calculate T_C. The quantities not defined earlier are: τ, the optical depth; T_S, the contribution from SgrA complex; θ, the angular size of the source; L, the radius of the Galaxy (20 kpc) and R, the distance to the galactic center (8.5 kpc). Also $k = \Omega_r/\Omega_B$; where Ω_r and Ω_B are the solid angles of the thermal source and of the beam, respectively. The rest of the quantities are obtain as follows: the optical depth at 408 MHz is determined from an extrapolation of microwave temperatures between 1.41 and 14.5 GHz quoted by Downes and Maxwell (1966); the angular sizes of the sources are known or measured; T_A, T_B, and Δ are measured; T_e, as before, is adopted from Matthews et al. (1973) and, finally, T_S is estimated from the spectrum given by Pedlar et al. (1989).

Figure 1 shows the temperature spectrum of the narrow source. The points at 408 and 85.7 MHz are quite reliable, however the point at 45 MHz is not so well determined because of the difficulty to resolve the trough; the temperature at this frequency could be lower. The spectrum has a spectral index −2.4 which indicates a nonthermal nature. This is confirmed by a comparison with the spectrum of W28 which is a strong supernova remnant not too far from the galactic center ($l \approx 6°, b \approx 0°$) and apparently suffers little or no absorption. Then the narrow source has resulted nonthermal, in agreement with the initial assumption.

Figure 1 also shows the temperature spectrum of the broad source. Here we have added a point at 30 MHz from the survey by Matthewson et al. (1965), made with an 11° beam. The spectral index −2.7 at the high frequency end justifies the initial assumption of the nonthermal nature of the source. The size of this broad source, given by the FWHM of the fitted Gaussian and corrected for beam smoothing, is quite large and increases from 3° at 408 MHz, up to 8° at 45 MHz. In terms of linear dimensions this corresponds to radii between 200 and 600 pc. In the model, and for simplicity, this source was assumed spherical, however the contours are fairly elliptical with major axis approximately lying on the equator. Also, the center of the source was assumed to be at $l \approx 0°, b \approx 0°$, while actually it is at $l \approx 2°, b \approx 0°$. We believe that these simplifications do not change the basic results reached in this paper. The nature of the broad source we do not know. Because of its large size it does not fit easily into the category of classical galactic nonthermal sources. We believe it does not correspond to some kind of transient source, as a supernova remnant, but rather it is a feature that comes from the formation of the Galaxy. In a comparison with the optical galactic bulge the broad source could be thought of as a radio bulge.

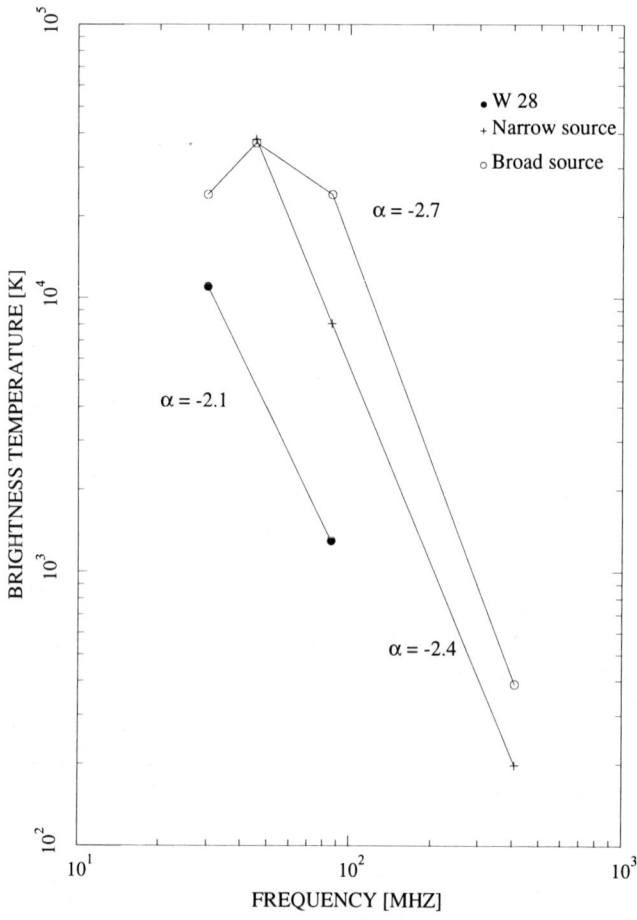

Figure 1. Temperature spectrum of the narrow source, the broad source and the supernova remnant W28.

Acknowledgments. We thank Dr. Patricia Reich for helping with the processing of he data and for making comments on the manuscript. This work was funded by FONDECYT, Chile, through grant 1930750.

References

Alvarez, H., Aparici, J., May, J. & Olmos, F. 1994, Experimental Astronomy, 5, 315

Alvarez, H., Aparici, J., May, J. & Olmos, F. 1996, in preparation

Downes, D. & Maxwell, A. 1996, ApJ, 146, 653

Haslam, C.G.T., Salter, C.J., Stoffel, H. & Wilson W.E. 1982, A&AS, 47, 1

Hill, E.R., Slee, O.B. & Mills, B.Y. 1958, Aust. J. Phys., 11, 530

Little, A.G. 1974, in I.A.U. Symp. 60, Galactic Radio Astronomy, F.J. Kerr and S.C. Simonson, III (Dordrecht: Reidel), 491

Matthews, H.E., Davies, R.D. & Pedlar, A. 1973, MNRAS, 165, 173

May, J., Reyes, F., Aparici, J., Bitran M., Alvarez, H. & Olmos, F. 1984, A&A, 140, 377

Mills, B.Y. 1956, The Observatory, 76, 65

Jones, B.B, & Finlay, E.A. 1974, Aust. J. Phys., 25, 687

Pedlar, A., Anantharamaiah, K.P., Ekers, R.D., Goss, W.N., van Gorkow, J.H., Schwars, V.J., & Zhao, H.H. 1989, ApJ, 342, 769

The Galactic Center
ASP Conference Series, Vol. 102, 1996
Roland Gredel, ed.

Molecular Clouds near the Galactic Center

John Bally [1]

Department of Astrophysical, Planetary, and Atmospheric Sciences,
University of Colorado, Boulder, CO 80309

Abstract. The properties of molecular clouds in the inner 500 parsec region of the Galaxy are reviewed. Evidence for a bar inclined at approximately $20 deg$ with respect to our line-of-sight is summarized. The first results of a J=1-0 ^{12}CO survey with the Bell Laboratories 7 meter antenna are presented, including evidence for vertical walls of molecular gas that provide evidence for superbubble "blowout" energized by massive OB associations. Finally, a possible nearby analog to the nonthermal radio filaments found in the galactic center, HH 222 located in the Orion molecular cloud complex, is discussed.

1. Properties of Molecular Gas in the Galactic Center

The Galactic center region contains the highest concentration of molecular gas in the galaxy. This gas is distinguished from clouds in the disk of the Milky Way by several parameters: [1] The gas has an order of magnitude larger line width than GMCs in the disk with $\Delta v \approx$10 to 30 km/s. [2] The gas is more than a factor of 10 brighter in high dipole moment molecules such as CS and HCN, implying mean densities greater than 10^4 cm^{-3}. [3] Molecular gas in the inner 500 parsecs of the Galaxy shows large systematic deviations from circular motion about the center. [4] The gas is distributed asymmetrically in space and velocity about the dynamic center of the Galaxy (Bally et al. 1987, 1988; Morris & Serabyn 1996).

The molecular clouds that dominate the Galactic center ISM are by necessity quite dense (greater than 10^4 cm^{-3}) in order to survive tidal shearing in the strong potential gradient surrounding the Galactic center. The tidal constraint requires that any gravitationally bound cloud have a minimum density given by

$$\overline{n(H)}_{cloud} \geq \frac{3V_{circ}^2}{2\pi\mu m_H G R^2} \approx 1.4 \times 10^4 (cm^{-3})[V_{200km/s}]^2 [R_{100pc}]^{-2} \quad (1)$$

where $V_{200km/s}$ is the orbital velocity, V_{circ}, in units of 200 km/s and R_{100pc} is the galactocentric distance, R, in units of 100 parsecs. If such clouds are gravitationally bound and approximately virialized, the line widths must be of order $\Delta v \approx \sqrt{GM/r_c} = \sqrt{2}(V_{circ} r_c/R)$ where r_c is the cloud radius. This

---

[1] Center for Astrophysics and Space Astronomy

implies that for a 10 pc radius cloud at $R = 100$ pc, $\overline{n(H)}_{cloud} \geq 10^4$ cm^{-3} with $\Delta v \geq 28$ km/s. Thus, gravitationally bound clouds near Galactic center are easy to distinguish from foreground or background disk clouds. The tidal constraint on gravitationally bound clouds near the GC implies that mean densities are sufficiently large to collisionally excite high dipole moment molecules such as CS and HCN and that they have relatively large line widths. Most of this high density material is confined to within $\pm$ 0.°2 of the plane and to within several degrees in longitude about the Galactic center; it is often referred to as the Galactic center disk population.

2. Gas Dynamics: Evidence for a Bar in the Inner Galaxy

Large amplitude non-circular motions are found in the inner 500 parsec region of the Galaxy. Molecular spectra towards $l = 0°$ exhibit emission that spans more than 300 km/s, and longitude-velocity ($l - v$) diagrams (cf. Bally et al. 1987; 1988; Blitz at el. 1993) show a parallelogram of high velocity CO emission extending from $l = 1.°5$ to $l = -0.°5$ (named the "Expanding Molecular Ring" [EMR]). As the name suggests, this feature was first interpreted as evidence for large a scale radial expansion of the molecular gas in the GC. The kinetic energy associated with such expansion would be $E \approx 10^{54} M_7 v_{100}^2$ ergs for M_7 in units of 10^7 M$_\odot$ and v_{100} in units of 100 km/s.

An alternative explanation for the non-circular motions is that the gas near the galactic center is forced to move along highly elongated orbits by a tri-axial gravitational potential produced by a bar-like distribution of stars in the inner kiloparsec of the Milky Way. It is easy to see how a bar can produce an $l - v$ diagram that looks like the EMR. It is assumed that gas can only move on non-intersecting orbits, otherwise shocks form and dissipate angular momentum, forcing the gas to move inward. As the gravitational potential of the inner Galaxy is distorted from a point symmetric (or oblate spheroidal) potential to a barred (tri-axial) potential with its major axis lying in galactic plane, stable circular orbits, the only non-intersecting orbits for gas in the plane of symmetry (the galactic disk), degenerate into two families of stable elongated orbits called the $x1$ and $x2$ orbits (Athanassoula 1988; see Figure 1 of Binney et al. 1991). The $x1$ orbits have major axes aligned with the bar (in a frame of reference co-rotating with the bar) while the $x2$ orbits have their major axes aligned along the bar's minor axis. If the bar weakens and becomes point symmetric far beyond some length scale, these orbits degenerate into the usual circular orbits.

As one moves in toward the bar, circular orbits devolve into the $x1$ family. With decreasing galactocentric distance, the $x1$ orbits develop cusps at the two orbital end points most distant from the center and become self-intersecting (see Binney et al. 1991). Any gas migrating onto the outermost self-intersecting $x1$ orbit will dissipate orbital angular momentum and within an orbit time, fall inward until it finds itself on an interior $x2$ orbit with lower net orbital angular momentum where once again the orbit is non-intersecting.

Inclination of the major axis of the elongated $x1$ orbits with respect to our line-of-sight explains both the observed forbidden velocities and the large deviations from circular motion. Binney et al. (1991) model the EMR by assuming a tri-axial potential in the inner few kiloparsec region of the Galaxy. The large velocity gradients that mark the ends of the EMR in $l - v$ diagrams strongly confine the orientation of the major axis of the bar to be inclined $18° \pm 4°$ with respect to our line-of-sight to the GC. At this inclination angle, our line of sight lies tangent to the "straight" segment of the innermost non self-intersecting $x1$ orbit, where within less than $0.5°$ of longitude, we see gas that is at its nearest point of approach to the GC (where the gas moves with the largest velocity) and at the point along this orbit where it is most distant from the GC (and moving with the lowest velocity). Thus, we observe a very large range of radial velocities within a very small range in longitude, just what is needed to explain the high and low longitude ends of the "parallelogram" in the $l - v$ diagrams that make up the EMR. This orientation of the bar is consistent with the near-infrared isophotal distortion of the inner bulge above and below the mid-plane of the Milky Way which shows that the positive longitude side of the bulge is closer to us than the negative longitude side (see Blitz et al. 1993 and references therein).

Analysis of $l - v$ diagrams show that the densest gas in the GC lies in the interior of the EMR, in the region where the $x2$ orbits are expected. When gas passes through the orbital angular momentum dissipating self-intersecting innermost $x1$ orbits, shock dissipation is expected to make the gas denser and more resistant to tidal shearing. Thus, a barred model of the inner Galaxy explains the relative location of the massive star forming complexes SgrA, Sgr B1, Sgr B2, and Sgr C in the interior of the EMR. It also provides a "minimum energy" explanation for the large radial motions of the EMR without requiring large releases of kinetic energy in the GC in the recent past. The radial motion of the EMR is maintained by the radial forcing imposed on it by a rotating bar. This model, however, does not explain the displacement of the centroid of the GC molecular gas towards positive longitudes ($l \approx 0.°7$).

3. Out of the Plane Molecular Features: Supershell Blowout?

The existence of organized out-of-plane gaseous features have been known for some time. Radio continuum observations reveal linear features on scales of order 10 to 100 parsecs such as the Sofue-Handa (Sofue & Handa 1984) Galactic center lobe defined by two spurs of emission at $l = 0.°18$ and $-0.°3$ and extend above and below the Galactic plane for nearly $1°$. It has been suggested that the Galactic center lobe is powered by winds and outflow from the inner 10 pc region of the GC (the nuclear cluster) or by the expulsion of gas from the plane by the twisting of magnetic fields.

In 1991, before leaving Bell Laboratories, I completed a fully sampled 115 GHz J=1-0 ^{12}CO survey of the GC region ($l = -2.°3$ to $4deg$; $b = -0.deg8$ to $0.°8$) with the 7 meter antenna that contains over 33,000 spectra on a $1'$ grid (Uchida, Morris, & Bally 1996; data available at http:// casa.colorado.edu/ ~bally/ GC/ gc.html). The most striking emission features seen in the ^{12}CO data cube are a dozen vertical CO filaments (elongated in the latitude direction) that extend well beyond the dense and flattened layer of the Galactic center disk gas. Some out-of-plane vertical CO filaments are identified in Table 1. Figure 1 shows an image of the GC in a range of radial velocities that contains several vertical CO features, including the "Rabbit Ears" near $l = 1.°3$. Most of these features are located at greater projected distances from the GC than the non-

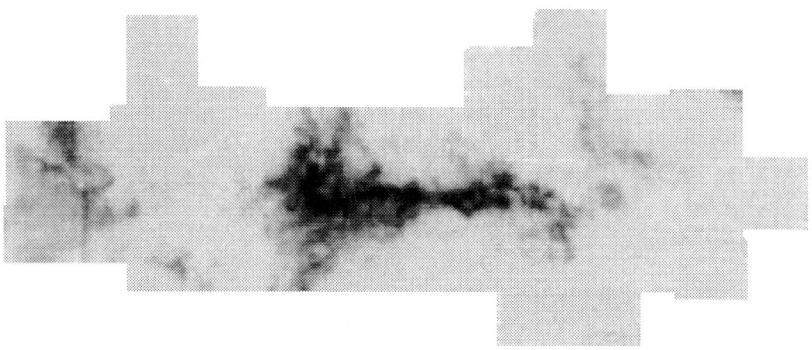

Figure 1. ^{12}CO J=1-0 emission from $l = 357.°8$ to $l = 4.°0$ between $v_{lsr} = 60$ and 90 km/s showing several vertical CO emitting filaments listed in Table 1. The image shows part of the $l = 3°$ complex (left), the "Rabbit Ears" near $l = 1.°3$ (left center), and the vertical strucure near $l = 359.°3$ (right of center).

thermal radio filaments. Except for the cloud at -359.4+0.18, where both a CO vertical filament and a non-thermal filament coincide, the two populations of sources are distinct.

Table 1.   Vertical Molecular Features Near the Galactic Center.

l_1	b_1	l_2	b_2	v_{min}	v_{max}	comments
358.82	-0.18	358.87	-0.62	-50	-90	Uchida et al. (1994) Bubble
359.28	0.23	359.45	0.93	50	80	Uchida (1992)
359.50	-0.32	359.28	-0.80	-170	-120	Uchida et al. (1994) Bubble
359.58	0.17	359.52	0.68	120	145	
1.27	0.07	0.02	0.57	30	80	Western Rabbit Ear
1.33	0.18	0.02	-0.73	90	120	Below Rabbit Ears
1.40	0.08	0.02	0.58	60	90	Eastern Rabbit Ear
1.58	-0.27	0.03	-0.70	20	60	
1.85	0.00	0.03	0.33	20	50	High l end of big bubble
2.08	-0.08	0.03	-0.65	-40	-70	Diagonal filament
2.25	-0.72	0.04	0.47	-40	-10	Diagonal filament next to $l = 3$
3.03	0.03	0.05	0.52	0	200	$l = 3$ feature
3.58	0.28	0.06	-0.03	30	90	Diagonal filament

The vertical CO filaments have masses, estimated from the velocity integrated CO surface brightness, using the X factor appropriate to the Solar vicinity (admittedly a shaky assumption), in the range 10^4 to a few times 10^6 M$_\odot$. The largest vertical filament in Table 1, the $l = 3°$ feature, consists of many individual clouds and lies at the extreme high end of this mass scale. It may be a complex of clouds that lies on an $x1$ orbit with an orbital plane that is slightly inclined with respect to the mean gas layer so that it has a pronounced vertical extent.

Most vertical CO filaments (except the $l = 3°$ feature) may be the walls of old superbubbles (see Shull & Saken 1994 for a recent review and references to earlier work) blown into the GC gas by large OB associations whose massive stars inject energy through UV ionization, stellar winds, and supernova explosions. OB associations with 10 to over 100 massive stars can inject over 10^{52} to 10^{53} ergs of kinetic energy, sufficient to cause the resulting bubble to "blow out" of the galactic gas layer (Mac Low et al. 1989). In the GC environment, shear due to the differential rotation of the Galaxy elongates the bubble in the galactic plane. Over typical lifetimes of superbubbles (active energy injection phase may last $\approx$ 30 Myr), a bubble can be sheared over a significant portion of an orbit. The strong large scale poloidal magnetic fields found near the GC can inhibit bubble expansion in the midplane and enhance blow-out in the Z-direction. A simple model of superbubble evolution in the GC environment is presented by Uchida et al. (1992.) Most important for the vertical CO filaments, blowout of the bubble from the galactic plane will produce vertical walls of swept-up gas that can trace the interface between the bubble interior and the surrounding ISM. For a sheared superbubble blowing out of the Galaxy and ramming a uniform ISM, two vertical walls of CO are expected where the line of sight from our vantage point lies tangent to the inner (low R_0) and outer (high R_0) edges of the bubble. The pair of vertical CO filaments near $l = 1.°3$, $b = 0.°25$ (the "Rabbit Ears") may be just such sheared supershell walls produced by a bubble blowing out of the gas disk near $l = 1.°3$.

Vertical CO filaments are likely to be tracers of past star formation near the GC. By the time a superbubble has reached sufficient size to have swept up 10^6 M$_\odot$ or more and to have blown out of the galactic gas layer, most of its massive stars will have died in supernova explosions. There may be very little dense ionized gas, and no significant infrared signature of the parent OB association. Thus, by the time vertical CO filaments are produced an OB association is likely to be in excess of 30 Myr old. Supershell models indicate that at this age, the walls are expanding with a velocity of order 3 to 10 km/s. In another 30 Myrs or so, random cloud motions are likely to erase the coherent and recognizable structure of these "fossil" supershells. Therefore, the expected lifetime of vertical CO walls is around 30 Myr. Not all OB associations are likely to produced recognizable vertical CO walls. Assuming a factor of two correction for incompleteness, the dozen or so recognized filaments imply that massive OB associations must have formed at a rate of order one per every few Myr over the past 30 to 60 Myr (the assumed lifetime of these structures).

Today, there are at least a half dozen OB associations forming in the GC gas as revealed by IR and radio continuum emission. These young OB associations include Sgr A, Sgr B1 and B2, Sgr C, the "Pistol", and Sgr D. Assuming a formation time scale of 10 Myr (typical of the Solar vicinity), the current OB association formation rate is estimated to be about one every 2 Myr, which would result in the production of about 15 OB associations in 30 Myr, consistent with the rate required to explain the observed number of vertical CO filaments.

Energy release by massive OB associations in the GC may produce the asymmetry in the location of the molecular gas. Ultraviolet radiation produced by massive stars may dissociate and ionize nearby gas and the kinetic energy generated by the rocket effect, stellar winds, and collective acceleration by supernova explosions may produce large non-gravitational acceleration of molecular clouds near the GC, producing the observed longitude asymmetry.

4. HH 222 in Orion: A Nearby Analog to the Non-Thermal Filaments?

The nature of the dozens of non-thermal filaments found within 0.°5 of the GC, which provide the evidence for a poloidal B-field, remains a mystery. We have identified a potential nearby analog to these streamers. A filamentary optical emission nebula, which has a non-thermal radio counterpart, lies at the northern tip of the L1641 cloud, 2° south of the Great Nebula in Orion (M 42). Located at a distance of 500 pc, this object is known as Herbig-Haro object HH 222, the "V 571 Orionis Nebula", or the "Orion Streamers" (Reipurth 1996).

The shock excited spectrum of this filamentary nebula was first recognized by Reipurth & Sandell (1985), and the non-thermal radio source was discovered by Yusef-Zadeh et al. (1990). Recent work by my student David Devine and I has shown that the compact radio core coincides (to within 1″) with a [S II] bright knot near the brightest portion of the streamers. The radio core is polarized and has a non-thermal spectral index. Extended radio continuum emission traces the filaments of Hα and [S II] emission. Unlike most Herbig-Haro objects, which are collisionally excited nebulae powered by outflows from young stars, HH 222 has very low radial velocities, and no detected proper motions. Furthermore, the filaments appear to trace the surface of a nearby molecular cloud core. Unlike most HH objects, the optical filaments are very smooth over their length of several arc minutes and are less than several arc seconds wide. These characteristics make HH 22 unlike any other Herbig-Haro object known. Figure 2 shows a recent ESO NTT narrow band (Hα + [S II]) image obtained by Bo Reipurth and myself.

We propose that this unusual nebula is powered by the impact of 10 to 10^3 eV charged particles ("non-thermal particles") that are gyrating about and guided by the ambient interstellar B-field which collisionally excites gas at the surface of the molecular cloud. These particles are likely to be produced in or near the radio core, but can drift for several parsecs to produce the filamentary optical emission. Thus, HH 222 may be considered to be an *"interstellar aurora"*. The production of these non-thermal particles may be associated with the production of a relativistic population of electrons which produce the synchrotron radio source. The essence of this model is that *a localized source of accelerated particles can produce a filamentary radio source and a collisionally excited optical nebula that can extend for parsecs from the site of particle acceleration.*

How are the non-thermal and relativistic particles produced? The compact radio core and associated [S II] bright knot are likely to be the site of particle acceleration. There are no stars at this location in either visual (< 21 mag at R) on near-IR images (< 18 mag at K). These limits rule out a pre-main sequence star as the source (V 571 Ori is over 30″ away). Either a compact object (white dwarf or neutron star) or a process which can operate in the diffuse ISM must be responsible. Although we can not rule out the former, we can construct reasonable models that explain the observations for the latter by invoking particle acceleration triggered by the passage of a magnetized shock through a two component fluid consisting of ordinary gas plus a relativistic (cosmic ray) fluid.

We propose that a shock is produced near (and localized to) the compact core by either the impact of a stellar outflow (possibly from nearby V 571 Ori) or by a "wind" impacting the northern tip of the L1641 cloud. Such a wind may be produced by the outward motion of hot gas from the interior of the Orion OB association located about $5°$ to the north and may be powered by the collective effects of photo-ablation, massive star winds, and expansion of the supernova heated interior of the Orion superbubble.

In our model of HH 222, the optical emission arises not from the shock but from stationary gas collisionally excited by the impact of particles accelerated in the shock. The shock itself may be confined to the region of the compact radio continuum source and the vicinity of the [S II] knot. Radio continuum emission is produced by boosted relativistic particles. Two types of particle acceleration can work in a strong magnetized shock. The shock may accelerate the thermal particles and boost the pre-existing relativistic fluid by *direct compression of the magnetized fluid*. Conservation of the "first adiabatic invariant" – essentially the conservation of angular momentum of gyrating particles – implies that the orbital velocity about the B-field is boosted as the gyro-radius is decreased by compression of the cooling fluid behind the shock. The relativistic particle energies are boosted, producing a large enhancement in synchrotron emissivity. Gyrating thermal particles can undergo Fermi acceleration by passing back and forth through the shock ("mirror effect") or be scattered off density irregularities or Alfven waves, resulting in the production of a high energy non-Gaussian (non-thermal) tail of 0.01 to a few keV particles.

Alternatively, the shock may trigger a magnetic reconnection event near the surface of the cloud. The wind or outflow may advect an interstellar B-field with a different orientation than the B-field near the cloud. The ram pressure of this wind or outflow may bring these different field orientations into close proximity, producing a "sharp structure" where opposing field lines are forced near each other. Annihilation of the field may lower the magnetic pressure and energy, and this re-configuration at a magnetic X-point may lead to a tear in the magnetic geometry – a reconnection event similar in nature to what may occur on the Sun. Reconnection will accelerate thermal particles to produce the 1 keV tail needed for impact excitation of the optical emission, and boost the extant cosmic rays to enhance the synchrotron emissivity. The observed total (radio, IR, optical) luminosity of HH 222 can be powered for about 10^4 years by reconnection assuming a 10 μgauss B-field and that reconnection releases $B^2/8\pi$ of energy per unit volume.

Acknowledgments. We thank David Devine and Youssef Billawala for assistance with the images and catalog of filaments. This research was supported in part by NASA grant NAGW-4590 (Origins) and NASA grant NAGW-3192 (LTSA).

References

Athanassoula, E. 1988 in *Proceedings of the Joint Varenna-Abastumani International School & Workshop on Plasma Physics* ESA Sp-285, Vol I, p. 341. ed. T. D. Guyenne & J. J. Hunt (ESA, Paris)

Bally, J., Stark, A. A., Wilson, R. W. and Henkel, C. 1987, ApJS, 65, 13

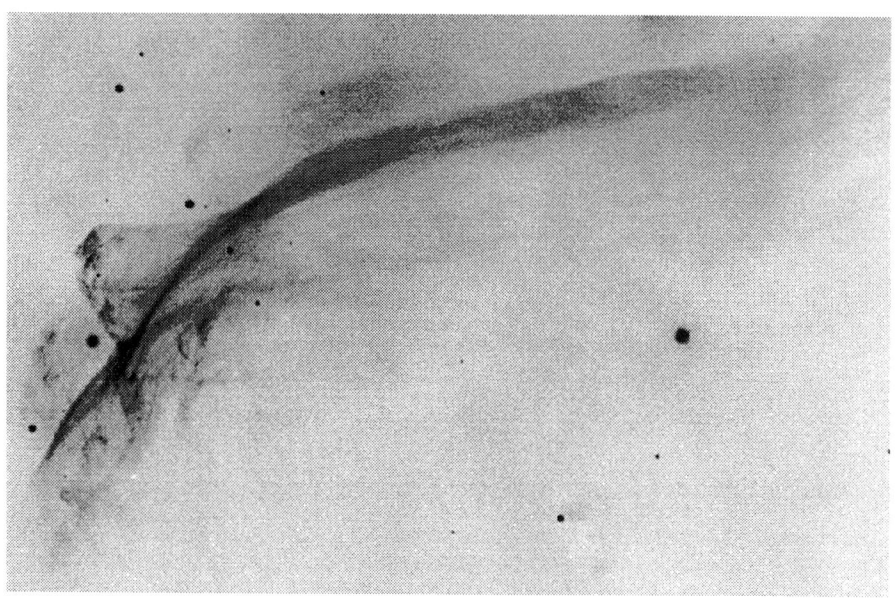

Figure 2. An ESO NTT narrow band Hα + [S II] image showing the filamentary nebula in Orion, HH 222 [$\alpha(1950) = 05^h 33^m 15.^s 6$, $\delta(1950) = -06°24'54''$]. North is to the left and west is at the top.

Bally, J., Stark, A. A., Wilson, R. W., and Henkel, C. 1988, ApJ, 324, 223
Blitz, L., Binney, J., Lo, K. Y., Bally, J., & Ho, P. T. P. 1993 Nature, 361, 417
Binney, J., Gerhard, O. E., Stark, A. A., Bally, J., & Uchida, K. 1991, MNRAS, 252, 210
Mac Low, Mc Cray, R., & M. L. Norman 1989, ApJ, ApJ, 337, 141
Morris, M. & Serabyn, E. 1996, Ann.Rev.Astron.&Astrophys., (in press)
Reipurth, B. 1996 *A general Catalog of Herbig-Haro Objects* published electronically, reipurth@eso.org
Reipurth, B., & Sandell, G. 1986, A&A, 150, 307
Shull, J. M., & Saken, J. 1995, ApJ, 444, 663
Sofue, Y., & Handa, T. 1984, Nature, 310, 568
Uchida, K., Morris, M., Serabyn, E., & Bally 1994, ApJ, 421, 505
Uchida, K., Morris, M., Bally, J., Pound, M., & Yusef-Zadeh, F. 1992, ApJ, 398, 128
Uchida, K., Morris, M., & Bally, J. 1996 ApJ, (in preparation)
Yusef-Zadeh, F., Cornwell, T. J., Reipurth, B., & Roth, M. 1990, ApJ, 348, L61

The Galactic Center
ASP Conference Series, Vol. 102, 1996
Roland Gredel, ed.

Neutral Carbon and the Dense Interstellar Medium in the Inner 300 pc of the Galaxy

D.T. Jaffe, R. Plume[1], Neal J. Evans II

Department of Astronomy, University of Texas, Austin, TX 78712

John Bally

Department of Astrophysics, Planetary & Atmospheric Science, University of Colorado, Boulder, CO 80309

1. Introduction

We present observations of the 492 GHz transition of CI over the inner 400 pc of the Galactic center. A layer of neutral carbon forms at the surfaces of molecular clouds wherever far-UV radiation strikes the clouds. Emission in the ground state fine-structure lines of CI traces this material. We can therefore use CI and molecular line observations to learn about the structure of neutral clouds in the Galactic center and the influence of UV radiation upon them. We compare the CI emission to ^{13}CO and CS emission and discuss the implications of the results for the nature of interstellar clouds in the inner Galaxy.

2. Results and Discussion

Figure 1 shows longitude-velocity diagrams for CI $^3P_1 \rightarrow ^3P_0$ and CS J=2→1 (hereafter CI and CS). The diagrams, as well as a comparison of individual spectra, show that CI has a very similar distribution in both longitude and velocity to emission in CS (as well as ^{13}CO J=1→0 (hereafter ^{13}CO), not shown). The mean properties of the Galactic center gas do not change much with position. Both the integrated CI/CS and CI/^{13}CO intensity ratios and the average values for 10 km s^{-1} bins at each longitude remain almost constant as a function of position $((\int T_A^* \, dV)_{CI}/(\int T_A^* \, dV)_{CS} \simeq 1.8, (\int T_A^* \, dV)_{CI}/(\int T_A^* \, dV)_{^{13}CO\,J=1\rightarrow 0} \simeq 0.7)$.

To produce figure 2, we have divided the spectrum at each longitude into 10 km s^{-1} bins and plotted the ratio of CI/CS line strength versus the CS line strength integrated over each bin. The mean value of the CI/CS is again $\simeq 1.8$ for regions where $(\int T_A^* \, dV)_{CS}$ is ≥ 5 K km s^{-1} per 10 km s^{-1} interval. The ratio slowly rises at lower CS line strengths. For comparison, we present CI/CS data for the L1630/NGC 2024 molecular cloud (Plume et al. 1996). The mean CI/CS ratio in L1630 is higher than in the Galactic center and shows a similar rise at low CS line strength.

---

[1]Currently at Harvard-Smithsonian center for Astrophysics, 60 Garden St., Cambridge, MA 02138

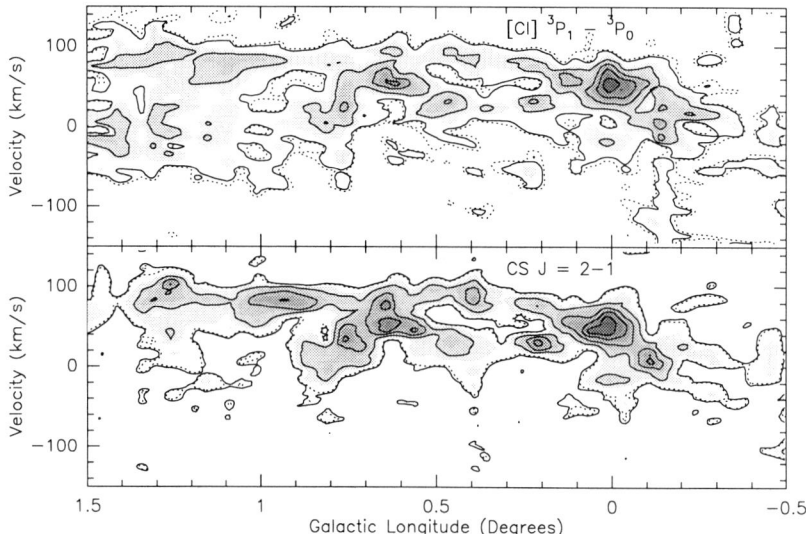

Figure 1. Longitude-velocity diagrams for the Galactic center at $b = -0.05°$. We took the CI $^3P_1 \rightarrow {}^3P_0$ spectra at 3' intervals in 1995 March and June with the Univ. of Texas re-imager (beamsize 2') on the CSO (Plume & Jaffe 1995). The CS data are from Bally et al. (1987), smoothed to a 2' beamsize. Contour levels are 10%, 30%, 50%, 70% and 90% of the peak intensity. The dotted contour represents the 3σ level for the spectra, when smoothed to a resolution of 10 km s^{-1}.

The dispersion in the CI/CS ratio seen in figure 2 is very large but drops with increasing CS line strength. For regions in the Galactic center with weak CS lines, the dispersion about the mean is almost a factor of 2. A plot of CI/^{13}CO versus ^{13}CO line strength appears similar. The large dispersion of CI/CS line strength in the Galactic center implies that clouds have a wide range of physical conditions and geometries. One can understand the lower dispersion in the ratio at higher intensity as the result of each position–velocity bin containing a mixture of gas with different properties. A mélange of different types of clouds is more likely to look like a typical cloud when it contains a larger number of individual objects.

The CI/^{13}CO and CI/CS ratios in the Galactic center fall within the *range* of ratios observed toward individual points in clouds in the disk. The systematic differences between the Galactic center and L1630 must reflect changes in the mixture of substructures within each beam. The lower mean CI/CS ratio in the Galactic center may result largely from a greater admixture of high density substructures in the clouds in the inner Galaxy. Higher density results in lower CI column densities in the photodissociated gas (Hollenbach, Takahashi, & Tielens (1991), Van Dishoeck & Black (1988)) and in more effective excitation of the CS J=2 level.

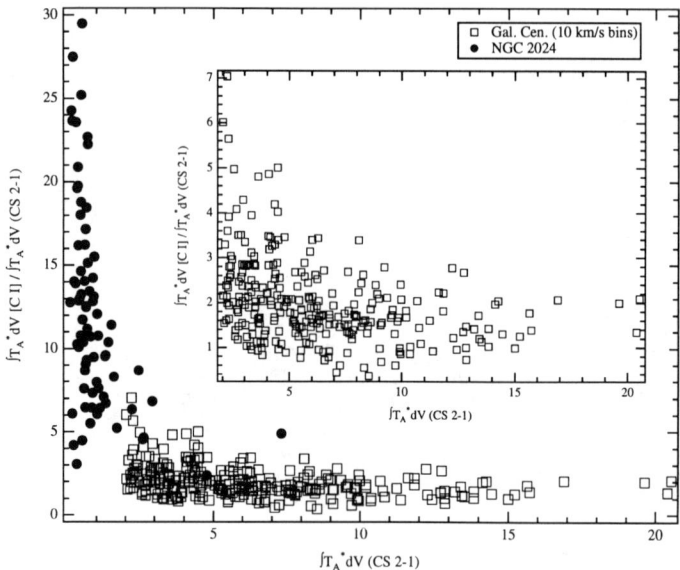

Figure 2. Ratio of CI $^3P_1 \rightarrow {}^3P_0$ to CS J=2→1 line strengths plotted against the strength of the CS J=2→1 line. The open squares show values from the Galactic center for individual 10 km s^{-1} intervals. The inset displays the Galactic center data in an expanded scale. The solid dots represent the CI/CS ratio integrated over the entire line for points in the inner 2.5 pc of the L1630/NGC 2024 molecular cloud.

Table 1 compares the ratios of various molecular line strengths to CI and ^{13}CO in the Galactic center, L1630, and in several nearby galactic nuclei. There is an apparent contradiction between the lower ^{13}CO/CI ratios and the higher CS/^{13}CO ratios in the galactic nuclei, compared to L1630 and the Galactic center. In a medium composed of identical clouds with photodissociated surfaces, lower ^{13}CO/CI ratios should imply lower column densities in individual structures and/or lower densities. These changes should lead to lower CS/^{13}CO ratios. The enormous point to point variations in the line ratios in the inner Galaxy hint that heterogeneity might explain the apparent contradiction. Variations in CS abundance may also be a factor.

The bulk properties of the Galactic center and of IC 342, M82, and NGC 253 on comparable scales differ significantly. It is not possible to take a mixture of average disk clouds and average Galactic center clouds and synthesize an object with the CI/CS and ^{13}CO/CS ratios that the other galactic nuclei have. One possible way to explain the results is to synthesize the galaxies out of parts of the galactic sources. Both the Galactic center and clouds like L1630 are heterogeneous enough to make this possible. The key element is a healthy dose of low column density gas. We have invoked a component of this sort already to explain the high ^{12}CO/^{13}CO ratios in IC 342 where the low-J ^{13}CO lines appear optically thick (Wall & Jaffe 1990).

Acknowledgments. This work was supported in part by NSF Grants AST-9117373 and 9317567 and by the David and Lucile Packard Foundation.

Table 1. CI and Molecular Line Intensities

		Relative to CI			Relative to ^{13}CO			
		^{12}CO	^{13}CO	CS	CI	^{12}CO	CS	Ref.
L1630	(0.35×0.35 pc)	3.1	1.1	0.3	0.9	2.8	0.2	1,2
L1630	(2.5×2.5 pc)	10.5	1.5	0.1	0.7	7.0	0.07	1,2
G.C.	(6 pc)	9.3	1.3	0.6	0.8	7.2	0.5	3
G.C.	(300 pc)	14.2	1.9	0.6	0.5	7.5	0.3	3
IC 342		4.2	0.5	0.2	2.0	8.6	0.4	4-6
M 82		1.9	0.3	0.3	3.2	6.1	1.1	6-9
NGC 253		3.8		0.2				6, 10

[1]Plume et al. 1996, [2]Lada et al. 1991, [3]This Paper, [4]Buettgenbach et al. 1992
[5]Eckart et al. 1990, [6]Mauersberger & Henkel 1989, [7]Schilke et al. 1993, White et al. 1994
[8]Nakai et al. 1987, [9]Loiseau et al. 1990, [10]Israel, White, & Baas 1995

References

Bally, J., Stark, A.A., Wilson, R.W., & Henkel, C. 1987, ApJS, 65, 13
Büttgenbach, T.H., Keene, J., Phillips, T.G., & Walker, C.K. 1992, ApJ, 397, L15
Eckart, A., Downes, D., Genzel, R., Harris, A.I., Jaffe, D.T., and Wild, W. 1990, ApJ, 348, 434
Hollenbach, D.J., Takahashi, T., & Tielens, A.G.G.M. 1991, ApJ, 377, 192.
Israel, F.P., White, G.J., and Baas, F. 1995, A&A, 302, 343
Lada, E.A., Bally, J., & Stark, A. A. 1991, ApJ, 368, 432
Loiseau, N., Nakai, N., Sofue, Y., Wielebinski, R., Reuter, H.-P., and Klein, U. 1990, A&A, 228, 331
Mauersberger, R., & Henkel, C. 1989, A&A, 223, 79
Nakai, N., Hayashi, M., Handa, T., Sofue, Y., Hasegawa, T., & Sasaki, M. 1987, PASJ, 39, 685
Plume, R., & Jaffe, D.T. 1995, PASP, 107, 488
Plume, R., Jaffe, D.T., & Keene, J. 1994, ApJ, 425, L49
Plume, R., Jaffe, D.T., Tatematsu, K., & Evans N.J. II 1996, in prep.
Schilke, P., Carlstrom, J.E., Keene, J., & Phillips, T.G. 1993, ApJ, 417, L67
Van Dishoeck, E.F., & Black, J.H. 1988, ApJ, 334, 771
Wall, W.F., & Jaffe, D.T. 1990, ApJ, 361, L45
White, G.J., Ellison, B., Claude, S., Dent, W.R.F., and Matheson, D.N. 1994, A&A, 284, L23

Large Scale Far-Infrared [C II] Line Emission from the Galactic Center

Takao Nakagawa, Yasuo Doi[1], Yukari Yamashita Yui[1], Haruyuki Okuda, Kenji Mochizuki[2], and Hiroshi Shibai

The Institute of Space and Astronautical Science, 3-1-1 Yoshinodai, Sagamihara, Kanagawa 229, Japan

Tetsuo Nishimura[3], and Frank J. Low

Steward Observatory, University of Arizona, Tucson, AZ 85721, USA

Abstract. We made large scale observations of the 158 μm [C II] line emission from the Galactic plane ($260° \lesssim l \lesssim 60°$, $|b| \lesssim 3°$) using the Balloon-borne Infrared Carbon Explorer (BICE) with the spatial resolution of $15'$.

In the Galactic disk, the spatial distribution of the [C II] line correlates very well with that of far-infrared continuum. However, in the Galactic center, the spatial distribution of the [C II] line is quite different from that of far-infrared continuum. Moreover, the ratio of the [C II] line emission to far-infrared continuum ($I_{\mathrm{[CII]}}/I_{\mathrm{FIR}}$) is systematically low within the central several hundred parsecs of the Galaxy.

The observational results indicate that the abundance of the C^+ ions themselves is low in the Galactic center. We attribute this low abundance mainly to soft UV radiation with fewer C-ionizing photons. This soft radiation field, together with the pervasively high molecular gas density, makes the molecular self-shielding more effective in the Galactic center. The self-shielding further reduces the abundance of C^+ ions, and raises the temperature of molecular gas at the $C^+/C/CO$ transition zone.

1. Introduction

The characteristics of interstellar gas in the center is quite different from those of interstellar gas in the Galactic disk (Güsten 1989): (1) the interstellar gas at the center is dominantly molecular, and (2) the clouds there are denser and warmer ($n \sim 10^4$ cm^{-3} and $T \sim 70$ K) than typical clouds in the disk ($n \sim 10^{2.5}$ cm^{-3} and $T \sim 15$ K).

---

[1] Present Address: Communications Research Laboratory, Nukui-kitamachi 4-2-1, Koganei, Tokyo 184, Japan

[2] Department of Astronomy, The University of Tokyo, Hongo 7-1-1, Bunkyo-ku, Tokyo 113, Japan

[3] Present Address: National Astronomical Observatory, Osawa 2-21-1, Mitaka, Tokyo 181, Japan

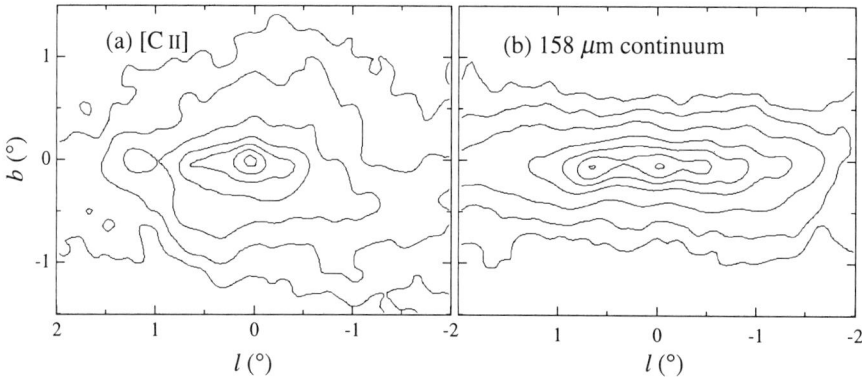

Figure 1. Far-infrared images of the Galactic center: (a)[C II] line image. Contour levels are 0.5, 1, 2, 4, 6, 8, and 10 $\times 10^{-4}$ ergs s^{-1} cm^{-2} st^{-1}. (b) far-infrared (158 μm) continuum image. Contour levels are 1, 2, 4, 8, 12, 16, and 20 $\times 10^9$ Jy st^{-1}.

Some of these characteristics can be naturally explained. The clouds must be sufficiently dense to withstand the tidal stress caused by the steep potential well in the region (Stark 1989), and the pervasively high density makes molecular forms more preferable than atomic forms.

However, the origin of the high temperature and the source of the luminosity, especially on a large scale, is still controversial (e.g., Morris 1989). The relative paucity of the usual indicators of star-formation activity (e.g., H_2O masers, and SNRs) suggests that the star-formation rate per unit mass in the Galactic center on a large scale may be lower than that of the Galactic disc (Morris 1989), and that young OB stars are not the dominant source of infrared luminosity (Cox & Laureijs 1989). In order to reveal the sources of luminosity and heating mechanisms, it is essential to study the energy budget of the clouds on a large scale.

The far-infrared [C II] fine structure line ($^2P_{3/2} \rightarrow\, ^2P_{1/2}$, 157.7409 μm, Cooksy et al. 1986) is a particularly useful probe to study the energy budget of neutral clouds, since the line is the dominant coolant of neutral interstellar gas (Hollenbach et al. 1991), and one of the brightest emission lines in the Galaxy (Shibai et al. 1991; Bennet et al. 1994). Previous observations of the [C II] line toward the Galactic center, however, were limited to small regions (e.g., Poglitsch et al. 1991; Mizutani et al. 1994), or had very low angular resolution (Bennet et al. 1994).

We made a large-scale survey of the [C II] 158 μm line emission from the Galactic plane using a balloon-borne telescope (Nakagawa 1993). Our observation covers a wide area ($260° \lesssim l \lesssim 60°$, $|b| \lesssim 3°$), including the Galactic center, with sufficiently good spatial resolution (15'). In this paper, we concentrate on large-scale (several hundred parsecs) [C II] line emission from the Galactic center. The ratio of the [C II] line emission to the far-infrared continuum emis-

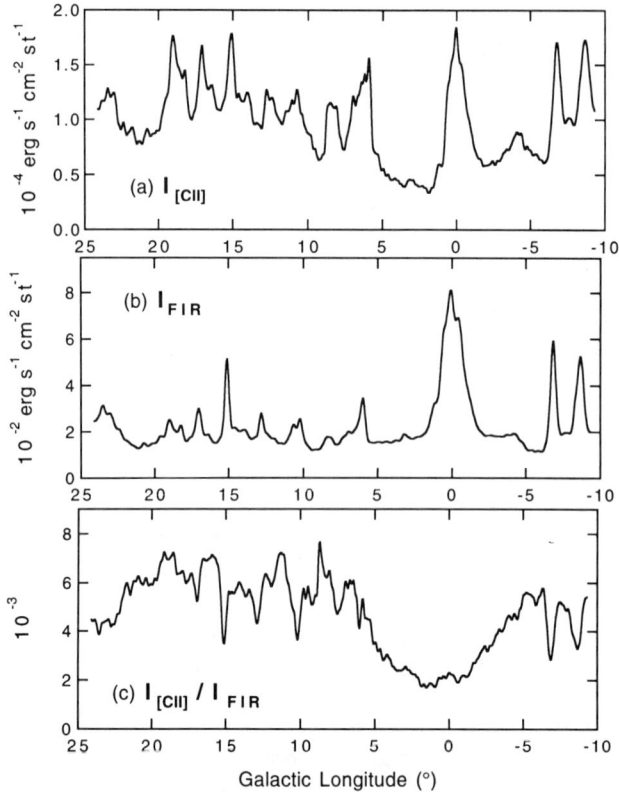

Figure 2. The longitudinal distributions of (a) $I_{\rm [CII]}$ (this work), (b) $I_{\rm FIR}$ (IRAS), and (c) $I_{\rm [CII]}/I_{\rm FIR}$. These are obtained by averaging signals at $|b| \leq 2°$. $I_{\rm FIR}$ is the flux between 40 - 120 μm derived from the IRAS 60 μm and 100 μm data.

sion is systematically low at the Galactic center than in the Galactic plane (see also Nakagawa et al. 1995). Origins of this low ratio and their implications, especially the relation with warm molecular clouds, are briefly discussed.

2. Observations

We observed the far-infrared [C II] line emission from the Galactic plane with the Balloon-borne Infrared Carbon Explorer (BICE, Nakagawa 1993), which is a system customized to make large-scale spectroscopic observations in the far-infrared. The velocity resolution (Δv) is 175 km s^{-1}, and the beam size is 12$.\!\!^\prime$4 (FWHM) with an effective solid angle of 1.5×10^{-5} ster.

To make a large-scale observations of the Galactic plane, we launched balloons at various places: at the National Scientific Balloon Facility in Palestine, Texas, USA, in 1991, at the Alice Springs Balloon Base, Australia, in 1992, and at the Sanriku Balloon Center, Japan, in 1994. The Galactic center region was observed in the 1991 flight.

In order to observe extended [C II] line emission efficiently, we used a fast spectral scanning method (Nakagawa 1993) instead of the conventional spatial chopping method. The observed area covers roughly $260° \lesssim l \lesssim 60°$ and $|b| \lesssim 4°$, but the the astronomical emission is assumed to be negligible at $|b| > 3°$ to remove instrumental and atmospheric emission. The final map was spatially smoothed to $15'$ (FWHM).

We observed M17 as a [C II] flux calibrator during each flight. The [C II] intensity of M17 was estimated from the map by Matsuhara et al. (1989). The calibration uncertainty is 30 %, and the detection limit of the final smoothed map is 1.5×10^{-5} ergs s^{-1} cm^{-2} ster^{-1} (3σ).

3. Results

Among our surveyed areas, we concentrate on the Galactic center region in this paper. Fig.1 shows the [C II] 158 μm line emission map of the Galactic center region together with the far-infrared continuum emission at 158 μm, both of which were obtained by the BICE observations.

In most of the Galactic plane, the spatial distribution of the [C II] line correlates very well with that of far-infrared continuum (e.g., Okuda et al. 1994). However, in the Galactic center, the spatial distribution of the [C II] line is quite different from that of far-infrared continuum (Fig.1).

Moreover, the [C II] line emission is relatively weak toward the Galactic center. Fig.2 shows the longitudinal distributions, which were obtained by averaging the signals over $|b| \leq 2°$ at each longitude. Fig.2a shows our observed longitudinal distribution of the far-infrared [C II] line emission ($I_{\rm [CII]}$). Also shown are the longitudinal distribution of far-infrared continuum emission ($I_{\rm FIR}$, Fig.2b).

The [C II] distribution is quite different from others. Although the Galactic center is the dominant peak both in the far-infrared continuum (Fig.2b), it is not in the [C II] line (Fig.2a). This situation is better illustrated in Fig.2c, which shows the $I_{\rm [CII]}/I_{\rm FIR}$ ratio; the ratio is almost constant (0.6 %) along the Galactic disk but is systematically low (down to 0.2 %) around the Galactic center. The longitudinal range of the dip is very wide, covering roughly $-3° \lesssim l \lesssim 5°$ (-450 pc to +750 pc).

4. [C II] Line Deficit toward the Galactic Center

4.1. Weak [C II] Line Emission toward the Galactic Center

Since the optical depth of the [C II] line toward the Galactic center is not large enough to explain the low $I_{\rm [CII]}/I_{\rm FIR}$ ratio (Nakagawa et al. 1995), the intrinsic intensity of the [C II] line must be weak in the Galactic center.

The $I_{\rm [CII]}/I_{\rm FIR}$ ratio is also small for the regions such as those around active star-forming regions. In this subsection, we discuss if the Galactic center is simply an ensemble of these regions.

Photodissociation regions (PDRs) have been thought to be a strong source of [C II] emission (e.g., Hollenbach et al. 1991; Shibai et al. 1991). Wolfire et al.

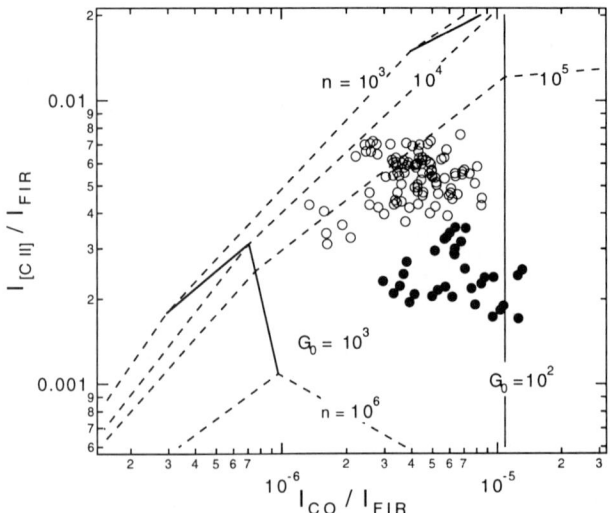

Figure 3. The ratio $I_{\rm [CII]}/I_{\rm FIR}$ vs the ratio $I_{\rm CO}/I_{\rm FIR}$. The data are derived from the longitudinal distribution as in Fig.2. Filled circles are for the Galactic center ($-3° \leq l \leq 5°$), and open circles are for the Galactic plane in $348° \leq l \leq 25°$. The [CII] data are from this work, and the CO data are from Dame et al. (1987). The lines are theoretical curves from Wolfire et al. (1989).

(1989) modeled PDRs as a one-dimensional molecular cloud with the constant density (n) illuminated by FUV (6 - 13 eV) flux of G_0, which is normalized by the solar neighborhood value. They plotted $I_{\rm [CII]}/I_{\rm FIR}$ vs $I_{\rm CO}/I_{\rm FIR}$ assuming that all of CO, [C II], and FIR continuum emission originates mainly from PDRs.

We plotted our results in the same figure (Fig.3). Most of our observed points in the Galactic disk corresponds to the regions with moderate FUV flux density ($G_0 \sim 10^2$).

Active star-forming region / H II region complexes lie toward the lower left in Fig.3, which indicates larger G_0. This is naturally understood since these regions are expected to be in a high flux radiation field from embedded young stars. These regions also have lower $I_{\rm [CII]}/I_{\rm FIR}$ ratios. Model calculations (Hollenbach et al. 1991) showed that, in high-density regions with large G_0, the [C II] line intensity is rather saturated due to its lower excitation energy and lower critical density, and other lines such as [O I] 63 μm line become the dominant coolants. Moreover, dust grains become positively charged in the regions with large G_0, and the total gas heating efficiency, which is mainly due to the dust photoelectric heating, also decreases (Hollenbach et al. 1991). Hence the $I_{\rm [CII]}/I_{\rm FIR}$ ratio decreases around each H II regions with high G_0.

However, the Galactic center data occupy different regions from active star-forming regions in Fig.3; the $I_{\rm CO}/I_{\rm FIR}$ ratio is much larger for the Galactic center. Moreover, although the low $I_{\rm [CII]}/I_{\rm FIR}$ ratio is confined to small regions for Galactic star-forming regions, the low ratio is ubiquitous in the Galactic center on a large scale. In the Galactic center, the $I_{\rm [CII]}/I_{\rm FIR}$ ratio is low even at the regions without any indication of star-forming activity.

Hence, the low $I_{\rm [CII]}/I_{\rm FIR}$ ratio in the Galactic center cannot be attributed to a simple ensemble of active star-forming regions, but some characteristics unique to the Galactic center must decrease the ratio.

4.2. Soft Interstellar Radiation Field

As discussed in the introduction, there are some lines of evidence which indicate that the current star-forming activity in the Galactic center may be relatively lower than that of the Galactic disk, and that young OB stars may not be the dominant energy source in the center. Cox & Laureijs (1989) estimated the Infrared Excess (IRE, the ratio of infrared luminosity to the Lyman α luminosity) of the Galactic center region ($2° \times 3°$) on the basis of IRAS observations. Most compact sources associated with the center showed IRE $\sim$ 10, which is typical also for disk H II regions. On the other hand, the diffuse component in the center, which dominates the infrared luminosity, shows a much higher IRE $\sim$ 30. Hence they concluded that the dominant heating source for the dust on a large scale is not young OB stars but rather the population of cool stars - K and M giants - which comprises the Galactic nucleus. Previous balloon-borne observations of diffuse far-infrared thermal emission from the Galactic center also suggest a deficiency of O-stars compared to the solar vicinity (Boissé et al. 1981; Odenwald & Fazio 1984).

On the basis of these IRE differences between the center and the disk H II regions, we assume that (1) most of the infrared luminosity of the *disk* is attributed to young OB stars, and that (2) only one third of the infrared luminosity of the diffuse component in the *center* is due to OB stars while two thirds is attributed to cool stars. Radiation from cool stars can heat the dust, but it contains few carbon-ionizing photons. Hence this relative softness of the interstellar radiation field in the Galactic center reduces the C^+ abundance and can roughly explain the small $I_{\rm [CII]}/I_{\rm FIR}$ ratio, which is also one third of the Galactic disk value.

4.3. Molecular Self-Shielding

The soft interstellar radiation field has an additional effect on the abundance of the C^+ ions. FUV photons more energetic than the ionization potential of hydrogen (13.6 eV) are absorbed within H II regions and less energetic FUV photons go into PDRs. Since the ionization potential of carbon (11.3 eV) is lower than 13.6 eV, carbon is easily ionized at the surface of PDRs, where incident FUV radiation is strong. Within the PDRs, the FUV photons are gradually attenuated, and the C^+ becomes C or CO at some depth from the surface.

The depth of the $C^+/C/CO$ transition zone is determined mainly by two mechanisms: dust extinction and molecular self-shielding (Hollenbach et al. 1991). If there are many FUV photons available relative to gas particles ($G_0/n > 10^{-2}$, Hollenbach et al. 1991), the dust extinction determines the depth ($A_V = 2 \sim 4$ mag) of the transition zone. In this case, C^+ regions and far-infrared continuum emitting regions have roughly the same thickness. On the other hand, if there are fewer photons available ($G_0/n < 10^{-2}$, Hollenbach et al. 1991), the molecular self-shielding becomes important. The molecular gas density in the

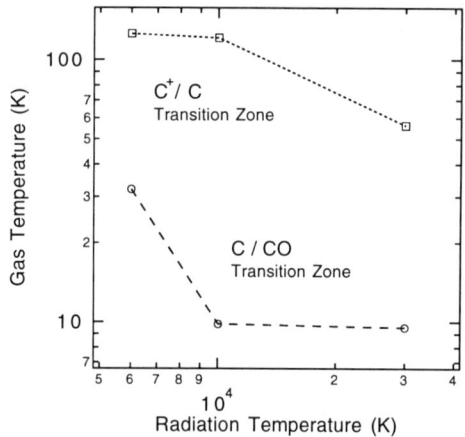

Figure 4. The gas temperature as a function of the effective temperature of the radiation. This is adopted from the model by Spaans et al. (1994) with $n = 10^3$ cm^{-3} and $G_0 = 10^3$.

Galactic center is high on a large scale (e.g., Bally et al. 1987), typically 10^4 cm^{-3}, which is one to two orders of magnitudes larger than those of the disk clouds (Güsten 1989). This high density makes molecular self-shielding more effective in the center than in the disk. The softness of the radiation field in the Galactic center region also makes the molecular self-shielding more effective, since fewer photons are available to photodissociate CO and to ionize C (Spaans et al. 1994). The molecular self-shielding moves the C$^+$/C/CO transition layer closer to the surface of the cloud. Then, the C$^+$ regions become thinner, and the $I_{\rm [CII]}/I_{\rm FIR}$ ratio also decreases.

Hence we conclude that, in the Galactic center region, the combination of pervasively high gas density and soft radiation field makes molecular self-shielding more important in determining the position of the C$^+$/C/CO transition zone than in disk clouds and the molecular self-shielding also decreases the $I_{\rm [CII]}/I_{\rm FIR}$ ratio.

4.4. Warm Molecular Gas in the Galactic Center

Hence the combination of the soft radiation field and the effective molecular self-shielding increase the amount of warm molecular gas.

Although the Galactic center is not the prominent peak in the [C II] line emission, it shows the prominent peak in the CO (1-0) line (Dame et al. 1987), high-J CO line (J = 2-1, 3-2, 4-3, and 5-4; Bennet et al. 1994), and [C I] line (609 μm and 370 μm; Bennet et al. 1994) emission. These results indicate that there is a large amount of warm neutral gas in the Galactic center. Excitation studies of symmetric top molecules (e.g., NH$_3$, CH$_3$CCH, and CH$_3$CN) also suggest uniformly high gas temperatures (50 - 100 K) throughout the central 500 pc irrespective of the local environment (Güsten 1985).

Spaans et al. (1994) showed that, due to molecular self-shielding, the gas temperature (T_{gas}) at the $C^+/C/CO$ transition zone becomes higher as the effective temperature of the radiation field (T_{eff}) decreases (Fig.4).

The exact nature of the gas-heating mechanism at the Galactic center has been controversial (Güsten 1989; Morris et al. 1989). We suggest here that the combination of the soft-radiation field and the molecular self-shielding in the Galactic center can effectively heat molecular gas on a large scale. This interpretation is consistent with the observed low $I_{\mathrm{[CII]}}/I_{\mathrm{FIR}}$ ratio in the Galactic center as we report in this paper.

References

Bally, J., Stark, A. A., Wilson, R. W., & Henkel, C. 1987, ApJS, 65, 13

Bennett, C. L. et al. 1994, ApJ, 434, 587

Boissé, P., Gispert, T., Coron, N., Wijinbergen, J. J., Serra, G., Ryter, C., & Puget, J. L. 1981, A&A, 94, 265

Cooksy, A. L., Blake, G.A., & Saykally, R. J. 1986, ApJ, 305, L89

Cox, P., & Laureijs, R. 1989, in IAU Symp. 136, The Center of the Galaxy, ed. M. Morris, (Dordrecht: Kluwer), 121

Dame, T. M., et al. 1987, ApJ, 322, 706

Güsten, R., Walmsley, C. M., Ungerechts, H., & Churchwell, E. 1985, A&A, 142, 381

Güsten, R. 1989, in IAU Symp. 136, The Center of the Galaxy, ed. M. Morris, (Dordrecht: Kluwer), 89

Hollenbach, D., Tielens, A. G. G. M., & Takahashi, T. 1991, ApJ, 377, 192

Matsuhara, H., et al. 1989, ApJ, 339, L67

Mizutani, K., et al. 1994, ApJS, 91, 613

Morris, M. 1989, in IAU Symp. 136, The Center of the Galaxy, ed. M. Morris, (Dordrecht: Kluwer), 171

Nakagawa, T. 1993, in Astronomical Infrared Spectroscopy: Future Observational Directions, ed. S. Kwok (San Francisco: ASP), 373

Nakagawa, T., et al. 1995, ApJ, 455, L35

Odenwald, S. F., & Fazio, G. G. 1984, ApJ, 283, 601

Okuda, H., et al. 1994, IR Phys, 35, 391

Poglitsch, A., et al. 1991, ApJ, 374, L33

Shibai, H. et al. 1991, ApJ, 374, 522

Spaans, M., Tielens, A. G. G. M., van Dishoeck, E. F., & Bakes, E. L. O., 1994, ApJ, 437, 270

Stark, A. A., Bally, J., Wilson, R.W., & Pound, M.W. 1989,in IAU Symp. 136, The Center of the Galaxy, ed. M. Morris, (Dordrecht: Kluwer), 129

Wolfire, M. G., Hollenbach, D., & Tielens, A.G.G.M. 1989, ApJ, 344, 770

H_2 Emission from the Inner 400 Parsecs of the Galaxy II. The UV–Excited H_2

Soojong Pak[1], D. T. Jaffe, and L. D. Keller[1]

Astronomy Department, University of Texas, Austin, TX 78712

Abstract. We have observed near–IR H_2 line emission on large scales in the Galactic center. Paper I discussed our 400 pc long strip map and 50 pc map of the H_2 $v = 1 \rightarrow 0$ $S(1)$ line. In this paper, we present observations of the higher vibrational lines (H_2 $v = 2 \rightarrow 1$ $S(1)$ and $v = 3 \rightarrow 2$ $S(3)$) at selected positions and conclude that strong far–UV radiations excites the H_2. We compare the H_2 $v = 1 \rightarrow 0$ $S(1)$ emission to far–IR continuum emission and show that the ratio of these two quantities in the Galactic center equals the ratio seen in the starburst galaxies, M82 and NGC 253, and in ultraluminous infrared bright galaxies.

1. Introduction

The central kpc regions in starburst galaxies and ultraluminous IR bright galaxies are powerful emitters of near–IR H_2 emission (Puxley, Hawarden, & Mountain 1990; Goldader et al. 1995). Ro–vibrational lines of H_2 can trace both photon–dominated regions (PDRs), where far–UV photons excite the H_2, and shocked regions, where the H_2 is thermally excited. Vigorous star formation in these galaxies produces large numbers of UV photons which fluorescently excite H_2, while subsequent supernovae shock–excite the H_2.

We have used the University of Texas near–IR Fabry–Perot Spectrometer, to survey giant molecular clouds (GMCs) on $1 - 10$ pc scales (Luhman et at. 1994; Luhman & Jaffe 1996; Luhman et al. 1996). In Orion A, for example, the H_2 $v = 1 \rightarrow 0$ $S(1)$ line emission extends up to 8 pc (1°) from the central UV source, θ^1 Ori C. The detection of higher vibrational state H_2 lines, e.g., $v = 6 \rightarrow 4$ $Q(1)$ and $v = 2 \rightarrow 1$ $S(1)$, showed that far–UV photons excite the H_2. Although the shock–excited H_2 emission is intense in the Orion $BN - KL$ region, the emission region is relatively compact ($\sim 1'$). The total H_2 luminosity in the $BN - KL$ region is only $\sim 1\%$ of the Orion PDR H_2 luminosity. Similarly, UV–excited H_2 dominates the large–scale H_2 emission from other GMCs.

We have observed the H_2 emission in the inner ~ 400 pc ($\sim 3°$) of our Galaxy in order to investigate H_2 emission on a more global scale and to compare the Galactic center with central ~ 1 kpc regions in external galaxies. The physical conditions in the interstellar medium of the Galactic center are signifi-

---

[1] Visiting Astronomer, Cerro Tololo Inter-American Observatory, National Optical Astronomy Observatory, which are operated by the Association of Universities for Research in Astronomy, under contract with the National Science Foundation

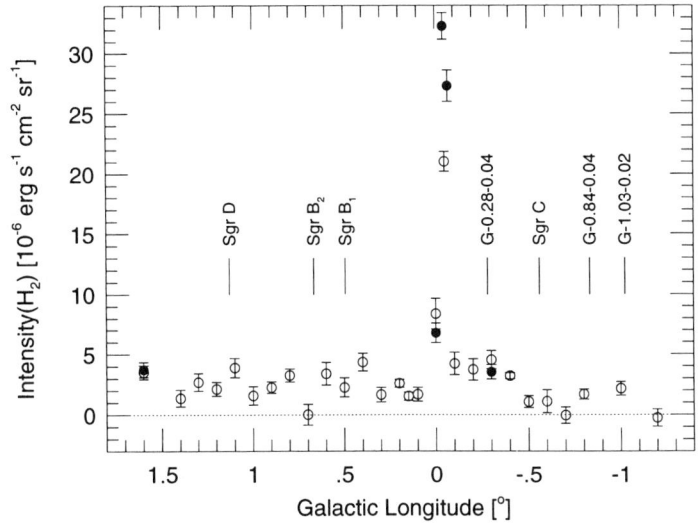

Figure 1. Observed intensity distribution of H_2 $v = 1 \to 0$ $S(1)$ ($\lambda = 2.121$ μm) along the Galactic plane at $b = -0°.05$. The open circles were taken at the McDonald 0.9 m telescope with a $3'.3$ beam (Paper I) and the filled circles at the CTIO 1.5 m telescope with a $1'.35$ beam. The intensities have not been corrected for interstellar extinction. The error bars represent 1σ measurement uncertainties.

cantly different from those in the solar neighborhood. The thin disk (diameter of 450 pc, height of 40 pc) of dense interstellar material in the Galactic center contains $M(H_2) > 2 \times 10^7 M_\odot$ (Güsten 1989; Hasegawa et al. 1996). The molecular clouds in the Galactic center have higher density, higher metallicity, and higher internal velocity dispersion than the clouds in the solar neighborhood (Blitz et al. 1993). There is strong radio continuum radiation from giant H II regions (Sgr A, Sgr B, Sgr C, and Sgr D) and extended low-density (ELD) ionized gas. The spectral index in the areas away from the discrete H II regions shows that thermal bremsstrahlung from ionized gas can account for about half of the emission from the extended gas (Sofue 1985). Another indicator of the intense UV radiation in the central 400 pc is strong far-IR continuum emission (Odenwald & Fazio 1984). About 90% of the far-UV energy is absorbed by dust and reradiated in the far-IR. From the far-IR intensity, we estimate that the far-UV radiation field is $\sim 10^3$ times the value in the solar neighborhood ($I_o = 4 \times 10^{-4}$ ergs s^{-1} cm^{-2} sr^{-1}, Draine 1978). The energetic conditions in the Galactic center mean that the center can provide a unique view of the interaction between stellar UV radiation and molecular clouds, and serve as a nearby model for the nuclei of galaxies.

In paper I (Pak, Jaffe, & Keller 1996) we showed the distribution of H_2 $v = 1 \to 0$ $S(1)$ emission along a 400 pc-long strip and in the inner 50 pc of the Galactic center. We detected H_2 emission throughout the surveyed region. The typical dereddened ($A_K = 2.5$ mag) H_2 $v = 1 \to 0$ $S(1)$ intensity, $\sim 3 \times 10^{-5}$

ergs s^{-1} sr^{-1}, is similar to the surface brightness in Galactic PDRs (Luhman & Jaffe 1996). In this Paper, we present observations of several H$_2$ lines, discuss the excitation mechanism, and compare the Galactic center observations to observations of other galaxies.

2. Observations and Results

We observed three H$_2$ emission lines: $v = 1 \to 0$ $S(1)$ ($\lambda = 2.121$ μm), $v = 2 \to 1$ $S(1)$ ($\lambda = 2.247$ μm), and $v = 3 \to 2$ $S(3)$ ($\lambda = 2.201$ μm), at the Cerro Tololo Inter–American Observatory 1.5 m telescope in 1995 July and October. We used the University of Texas Near–Infrared Fabry–Perot Spectrometer. The instrument was specially designed to observe very extended, low surface brightness objects, and has a single channel InSb detector with surface area of 1 mm to maximize the beam size (Luhman et al. 1995). The telescope ($f/30$), a collimator (effective focal length 686 mm), and a field lens (effective focal length 20mm) produce a beam diameter of 1.′35 (equivalent disk).

The Fabry–Perot interferometer operates in 94th order ($\lambda_o = 2.121\mu$m) with an effective finesse of 26, yielding a spectral resolution of 125 km s^{-1} (FWHM). Scans covered in 15 sequential steps, ±300 km s^{-1} centered at $V_{LSR} \simeq 0$ km s^{-1}. In order to subtract background and telluric OH line emission, we chopped the secondary mirror to $\Delta b = +16'$ or $-16'$ at 0.5 Hz.

We observed five positions: $(l, b) = (-0°.0433, -0°.0462), (-0°.0683, -0°.0462)$, $(0°.00, -0°.05), (-0°.30, -0°.05)$, and $(+1°.60, -0°.05)$. In Figure 1, we plot the new H$_2$ $v = 1 \to 0$ $S(1)$ data overlaid on the data from Paper I and compare the two data sets. The 3.′3 beam of the McDonald 0.9 m telescope centered at Sgr A* ($l = -0°.0558, b = -0°.0462$) covers the whole circumnuclear gas ring (Gatley et al. 1986), while, with the 1.′35 beam of the CTIO 1.5 m telescope, we observed the $+\Delta l$ H$_2$ peak ($-0°.0433, -0°.0462$) and the $-\Delta l$ H$_2$ peak ($-0°.0683, -0°.0462$). The difference between the 3.′3 beam data and the 1.′35 beam data toward Sgr A is an effect of different beam sizes because the H$_2$ emission sources are relatively compact. In the large–scale emission beyond Sgr A, the two data sets agree to within the errors, indicating that the H$_2$ emission varies slowly on $1' - 3'$ scales.

3. Extinction Correction

The extinction in K-band toward the Galactic center is significant. Figure 2a shows the classification of the extinction into *foreground extinction* by material in spiral arms at $R = 4 - 8$ kpc, and *Galactic center extinction* by material in the Galactic center clouds. Catchpole, Whitelock, & Glass (1990) measured the foreground extinction as $A_K \simeq 2.5$ mag.

A discussion of the Galactic center extinction requires a different approach because individual clouds in the Galactic center are almost opaque in the near-IR ($A_K = 10-30$ mag for typical clouds of $D \simeq 10$ pc and $n(H_2) \simeq 10^4$ cm^{-3}). If the UV–excited H$_2$ emission arises on the cloud surfaces, we need only consider the effects of shadowing by other Galactic center clouds (see Figure 2b). From millimeter observations of ^{12}CO $J = 1 \to 0$ emission, we can estimate the velocity–integrated area filling factor of clouds, f. If the millimeter telescope

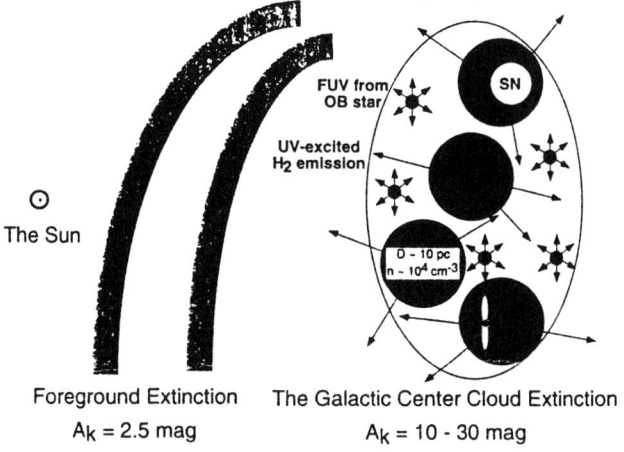

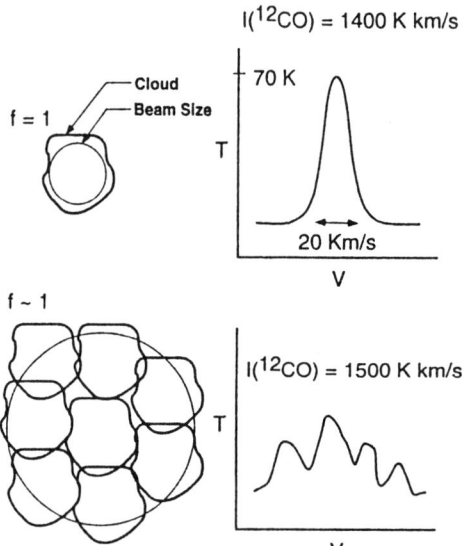

Figure 2. (a) Top–view schematic of the distribution of interstellar material in two foreground spiral arms (foreground extinction) and in the GMCs in the inner ~ 400 pc of the Galaxy (Galactic center extinction). (b) Schematic diagram of small and large beam observations in the Galactic center. ^{12}CO $J = 1 \rightarrow 0$ spectrum of a typical cloud is beam diluted. The velocity–integrated intensity including the clouds in the beam at other velocities is ~ 1500 K km s^{-1}, which indicates that the area filling factor, f, is ~ 1. The cloud components do not usually overlap along the line–of–sight.

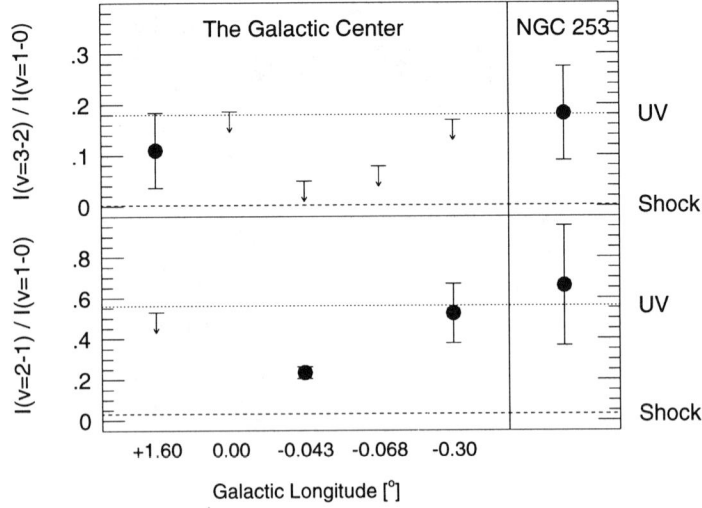

Figure 3. Observed H$_2$ line ratios at positions along the Galactic Plane ($b = -0°.05$) and in the central 1 kpc of NGC 253. The dotted lines are modeled ratios of UV–excited H$_2$ lines, (Black & van Dishoeck 1987), and shock–excited H$_2$ lines ($V_{shock} = 30$ km s^{-1}; Draine, Roberge, & Dalgarno 1983). The arrows show the 3σ limits where we did not detect the higher vibrational level lines.

beam size is smaller than the individual clouds and covers only one cloud along the line–of–sight, the area filling factor, f, is 1. The upper diagram in Figure 2b shows an expected ^{12}CO $J = 1 \rightarrow 0$ spectrum of typical clouds in the Galactic center which have kinetic temperature of ~ 70 K and line widths of ~ 20 km s^{-1} (Güsten 1989). In general, the clouds have different sizes and may overlap along the line–of–sight. The lower diagram in Figure 2b shows an observed typical ^{12}CO $J = 1 \rightarrow 0$ spectrum where the velocity–integrated intensity is ~ 1500 K km s^{-1} (Bally et al. 1987; Bally et al. 1988). The value f is the ratio of the observed velocity–integrated intensity of ^{12}CO $J = 1 \rightarrow 0$ to the single typical cloud intensity (70 K $\times$ 20 km s^{-1}). The f toward the Galactic center clouds is ~ 1, implying that there is little or no overlap along a typical line–of–sight. If $f \leq 1$, we only miss the near-IR H$_2$ flux from the back sides of the clouds. If $f > 1$, H$_2$ radiation is blocked by the foreground clouds, and the ratio of the observed H$_2$ flux to the emitted flux is inversely proportional to f. Since $f \simeq 1$, we use the foreground values, $A_K = 2.5$, for the extinction correction.

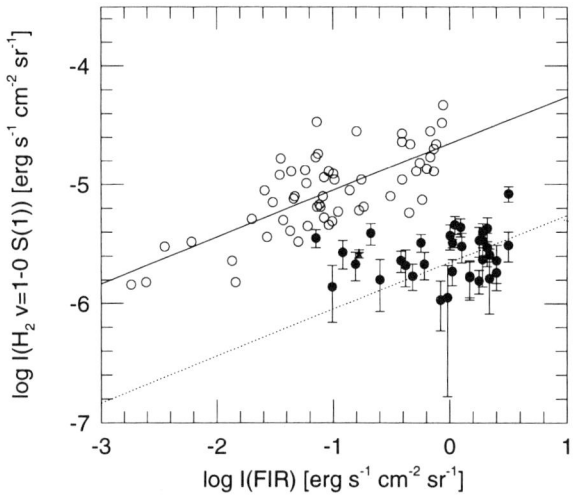

Figure 4. I_{FIR} versus $I_{H2 v=1\to 0\ S(1)}$ for the Galactic PDRs and the Galactic center. The open circles are from Orion A and B, ρ Ophiuchi, and G236+39 (Luhman & Jaffe 1996), and the filled circles are from the Galactic center (Paper I). The Galactic center data are not corrected for extinction. The solid line ($\log I_{H2\ v=1\to 0\ S(1)} = -4.65 + 0.39 \log I_{FIR}$) is derived from the Galactic PDR data using a least squares method, and the dotted line shows the vertically shifted solid line by $\Delta \log I_{H2} = -1$.

4. H_2 Excitation Mechanism

4.1. H_2 Line Ratios

In UV-excited H_2, the branching ratios in the downward cascade determine the relative strengths of the near-IR lines. On the other hand, the energy level populations of shock-excited H_2 are thermalized. We use the line intensity ratios of higher vibrational level lines to the $v = 1 \to 0\ S(1)$ line in order to identify the H_2 excitation mechanism.

In Figure 3, the observed ratios in the large-scale Galactic center and the central 1 kpc region of NGC 253 imply that the H_2 emission may result from UV-excitation. In the circumnuclear gas ring ($l = -0°.0433$ and $-0°.0683$), the UV-excited H_2 energy levels are partially thermalized because of the relatively high density (Sternberg & Dalgarno 1989; see also Ramsay-Howat, Mountain, & Geballe 1996 for the H_2 observations in the circumnuclear gas ring). The determination of line ratios consistent with UV excitation in the large-scale Galactic center and NGC 253 means the gas is not dense enough for collisions to significantly alter the radiative cascade, $n(H_2) < 10^5$ cm^{-3} (Luhman et al. 1996).

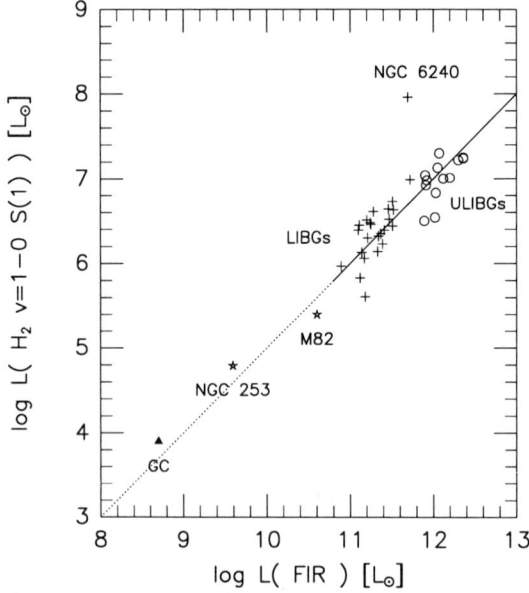

Figure 5. L_{FIR} versus $L_{H2\ v=1\to 0\ S(1)}$ of various kinds of galaxies. The solid line ($\log L_{H2\ v=1\to 0\ S(1)} = -5 + \log L_{FIR}$) is derived from data of ultraluminous IR bright galaxies (open circles) and luminous IR bright galaxies (plus signs, Goldader et al. 1995) The dotted line shows extrapolation from the solid line. The H_2 data of M82 were taken at the McDonald 2.7 m telescope and the H_2 data of NGC 253 at the CTIO 1.5 m telescope, both with the UT FPS.

4.2. I_{FIR} versus I_{H2}

If large-scale H_2 emission arises in the surface layers of the clouds where far-UV photons can excite the molecules, the dust, which absorbs the bulk of the incident flux, ought to radiate in the far-IR continuum as well. If we de-redden the Galactic center H_2 observations by $A_K = 2.5$ mag, the Galactic center results are consistent with the empirical far-IR vs. H_2 relationship derived for the UV-excited surfaces of clouds in the galactic disk (see Figure 4).

5. Comparison with other Galaxies

We extrapolate from our 400 pc long strip to the total $H_2\ v = 1 \to 0\ S(1)$ luminosity of the Galactic Center by assuming that the scale height of the H_2 emission equals that of the far-IR radiation ($h \simeq 0°.2$, Odenwald & Fazio 1984) and that $A_K = 2.5$ mag and $f \simeq 1$. The $H_2\ v = 1 \to 0\ S(1)$ luminosity in the inner 400 pc diameter of the Galaxy is $8.0 \times 10^3\ L_\odot$.

For ultraluminous and luminous infrared bright galaxies ($L_{IR} \gtrsim 10^{11}\ L_\odot$), Goldader et al. (1995) showed the correlation between L_{FIR} and $L_{H2\ v=1\to 0\ S(1)}$.

We can extend the relationship to nearby starburst galaxies like M82 and NGC 253, and to the Galactic center (see Figure 5). The strong correlation between the far-IR and H_2 luminosity for various classes of galaxies indicates that the far–UV radiation may excite large scale H_2 emission in all of these sources.

Acknowledgments. This work was supported by NSF grant AST 9117373 and by David and Lucile Packard Foundation. We thank M. Luhman and T. Benedict for contributions to the Fabry–Perot Spectrometer Project, and J. Elias, B. Gregory, and the staff of the CTIO for their assistance in setting up our instrument.

References

Bally, J., Stark, A. A., Wilson, R. W., & Henkel, C. 1987, ApJS, 65, 13

Bally, J., Stark, A. A., Wilson, R. W., & Henkel, C. 1988, ApJ, 324, 223

Black, J. H., & van Dishoeck, E. F. 1987, ApJ, 322, 412

Blitz, L., Binney, J., Lo, K. Y., Bally, J., Ho, P. T. P. 1993, Nature, 361, 417

Catchpole, R. M., Whitelock, P. A., & Glass, I. S. 1990, MNRAS, 247, 479

Draine, B.T. 1978, ApJS, 36, 595

Draine, B. T., Roberge, W. G., & Dalgarno, A. 1983, ApJ, 264, 485

Gatley, I., Jones, T. J., Hyland, A. R., Wade, R., Geballe, T. R., & Krisciunas, K. 1986, MNRAS, 222, 299

Goldader, J. D., Joseph, R. D., Doyon, R., & Sanders, D. B. 1995, ApJ, 444, 97

Güsten, R. 1989, in IAU Symp. 136, The Center of the Galaxy, ed. M. Morris (Dordrecht: Kluwer), 89

Hasegawa, T., Oka, T., Handa, T., Hayashi, M., & Sakamoto, S. 1996, this volume

Luhman, M.L., & Jaffe, D.T. 1996, ApJ, 483(May 20 issue)

Luhman, M. L., Jaffe, D. T., Keller, L. D., & Pak, S. 1994, ApJ, 436, L185

Luhman, M. L., Jaffe, D. T., Keller, L. D., & Pak, S. 1995, PASP, 107, 184

Luhman, M. L., Jaffe, D. T., Sternberg, A., Herrmann, F., & Poglitsch, A. 1996, ApJ, in preparation

Odenwald, S. F., & Fazio, G. G. 1984, ApJ, 283, 601

Pak, S., Jaffe, D. T., & Keller, L. D. 1996, ApJ, 457, L43 (Paper I)

Puxley, P. J., Hawarden, T. G., & Mountain, C. M. 1990 ApJ, 364, 77

Ramsay-Howat, S., Mountain, C. M., & Geballe, T. R. 1996, this volume

Sofue, Y. 1985, PASJ, 37, 697

Sternberg, A., & Dalgarno, A. 1989, ApJ, 338, 197

Large Scale Surface Photometry of the Galactic Center

L. Schmidtobreick, C. Tappert

Grupo de Astrofísica, P. Universidad Católica, Casilla 104, Santiago 22, Chile & Astronomisches Institut der Ruhr-Universität Bochum, 44780 Bochum, Germany

W. Schlosser, P. Koczet, S. Wiemann, M. Jütte, B. Hoffmann, Th. Schmidt-Kaler

Astronomisches Institut der Ruhr-Universität Bochum, 44780 Bochum, Germany

S. Kimeswenger

Institut für Astronomie der Leopold Franzens Universität Innsbruck, Technikerstraße 25, A-6020 Innsbruck, Austria

Abstract. Photographic images of the Milky Way were obtained in various passbands between 121nm and 360nm. They have been transformed to maps of the Milky Way. Similar surface photometries have been created in U, B, V, and R earlier. The region around the Galactic Center has been cut out of all these surface photometries. The comparing colour maps give information about the morphology of the Galactic Center, especially about the contribution of dust and gas.

1. Introduction

Photographic images of the Milky Way were obtained with the GAUSS-Camera (GAlactical Ultrawideangle Schmidt System) as part of the D2-Mission, flown on the space-shuttle Columbia in 1993. Similar photometries of the southern Milky Way have been created in Johnson U, B, V, R (excluding $H\alpha$)(Hoffmann et al 1990, Hoffmann et al 1993, Kimeswenger et al 1993, Tappert et al 1993). They are based on photographic plates which were obtained with the Bochum Super Wide Angle Camera at La Silla in 1971 (Schlosser & Schmidt-Kaler, 1977). For this representation, the region $-20° \leq l \leq 20°$ und $-20° \leq b \leq 20°$ has been cut out of all these photometries and the resulting colour maps (see figure 1).

2. The GAUSS-Camera

The GAUSS-Camera has been developed to obtain large scale structures. Its field of view is about $140°$ due to a special construction: The primary mirror is a hyperboloid that may diverge the incoming light but is responsible for the large field of view. The imaging mirror is spheric with a classical Schmidt-Corrector

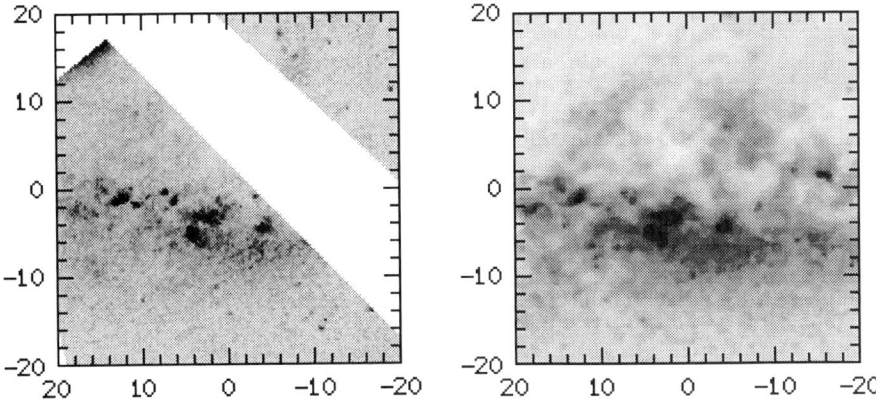

Figure 1. Examples for maps of the central galactic region in U (left, $white = 0...300 S_{10} \leq black$) and in B (right, $white = 0...400 S_{10} \leq black$).

of a lithiumfluorid-crystal. By a fitting combination of the two curvatures of these mirrors a flat image plane was received.
A ring diaphragm was employed for broad band photometries and for all colour images. Interference filters centered on 122nm, 170nm, 208nm, 280nm, and 356nm separated different pass bands in the ultraviolet spectral range.

3. Reduction

The GAUSS-images have been digitized with a resolution of $10 \mu m$/pixel by using a PDS-machine of the Astronomisches Institut der Universität Münster. The reduction has been done with procedures based on MIDAS.
To derivate the geometry of the images, the deviation between the plate coordinates (x,y) and the galactic coordinates (l,b) was split into two components - along the radius r and along a somehow to define azimuth ϕ - and was described by using r and ϕ as polar coordinates. It was found that the deviation along ϕ is given by a setoff, whereas along r a polynom of third degree is sufficient. A set of 120 stars has been chosen to optimize the parameters of this polynom, the setoff and the coordinates of the tangential point (x_0, y_0, l_0, b_0). Finally, a standard deviation of about $4 \mu m$, that is 1' or about a third pixel could be received.
The calibration of the images has been done with given photometries of about 800 stars per image. Their pixel of maximal density has been related to the catalogued brightness to get a relative density curve. A Moffat formula has been fitted to these data. By integrating each star, its relative brightness was received and could be related to the catalogued flux to get an absolute calibration curve. The S_{10} value was computed by using the definition of $1 S_{10}$ which is the surface brightness of 1 star of mag 10 in a square degree. The size of each star is determined by the pointspread function and is represented by its equivalent width which was computed by integrating the relative brightness of the star and dividing the result by the maximum. With the knowledge of how often a star

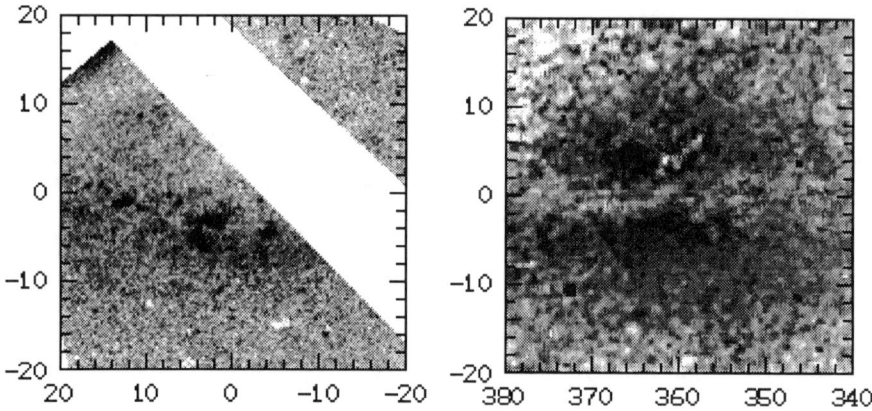

Figure 2. Examples for colour maps of the central galactic region are shown here. Left: $(280nm - U) + 2.5 \cdot \lg(1\frac{kgm^2}{s^3 sr S_{10}})$ ($white \leq 20...22\frac{kgm^2}{s^3 sr S_{10}} \leq black$); Right: (B-V)($white \leq 0.4...1.8 \leq black$).

fits into a square degree, the brightness U was transformed into absolute S_{10} values.
The images have been corrected for zodiacal light and shuttle glow (Jütte et al, 1995), disturbing bright stars have been removed.
The reduction of the images obtained by the Bochum Super Wide Angle Camera has been described in several papers (Hoffmann et al 1990, Hoffmann et al 1993, Kimeswenger et al 1993, Tappert et al 1993, Tappert et al 1995) before.

4. Two-colour diagrams

Plotting two-colour indices of surface brightnesses yields quite different results compared to normal two-colour diagrams of stars due to the fact that every pixel contains integrated light from different sources. Therefore by creating a two-colour diagram of a somehow defined field, one usually gains a rather homogeneous distribution whose shape and compactness depends on one hand on the size of the examined area and on the other hand on its galactic position. The deviation from the 'true' position in the diagram (which means the average true colour of the field) can be explained by the following effects:
- The interstellar extinction along the line of sight (normal colour excess)
- Effects of population, metallicity, ...
- The integrated straylight of other sources shifting the colour in a way that depends on the albedo of the interstellar dust but isn't well known up to now

For this representation, 60 fields of different size have been selected by considering structures shown in the V-map. The average ((280nm-U)/(U-B)) and ((U-B)/(B-V)) coordinates have been examined, they show a large distribution.

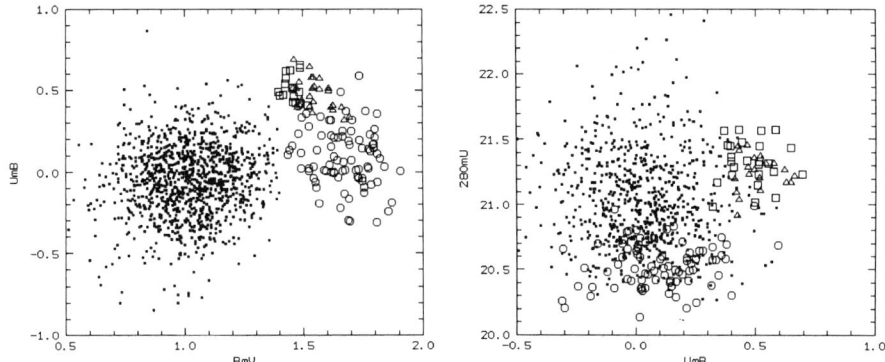

Figure 3. In these diagrams the distributions in four examined fields are presented. ◯: Galactic Bulge ($367°.25 \geq l \geq 364°$; $4° \leq b \leq 5°.25$), △: Baade's Window ($364°.5 \geq l \geq 363°.5$; $-5°.5 \leq b \leq -4°.75$), □: Baade's Window ($362°.5 \geq l \geq 361°.5$; $-3°.75 \leq b \leq -2°.75$) and · : Dust Lane.

As an example, the distribution in four of these fields is plotted in figure 3. For the interpretation of the colour diagrams, further investigations have to be done. It is planned to observe stellar populations in some regions to get information about their age and the distribution along the line of sight to select the effects mentioned above and give a more complete view of the Milky Way.

References

Hoffmann, B., Schlosser, W., Schmidt-Kaler, Th., Tappert, C. 1993, AG Abstract Ser., 9, 171

Jütte, M., Schmidtobreick, L., Wiemann, S., Schlosser, W., Koczet, P. 1995, AG Abstract Ser., 11, 202

Kimeswenger, S. 1989, Master Thesis, Ruhruniversität Bochum

Kimeswenger, S., Hoffmann, B., Schlosser, W., Schmidt-Kaler, Th. 1993, A&AS, 97, 517

Schlosser, W., Schmidt-Kaler, Th. 1977, Vist. Astron., 21, 447

Schmidtobreick, L., Wiemann, S., Jütte, M., Schlosser, W., Koczet, P. 1995, AG Abstract Ser., 11, 129

Tappert, C., Hoffmann, B., Schlosser, W., Schmidt-Kaler, Th. 1995, RevMexAA (Serie de Conferencias), 2, 124

The Formation of Large Scale Structures Via the Parker Instability

A. Santillán[1], M.A. Martos, and J. Franco

Instituto de Astronomía, UNAM, Apdo. Postal 70-264, 04510 México, D.F.

D. P. Cox

Department of Physics, University of Wisconsin-Madison, 1150 University Av. Madison, WI 53706, USA.

Abstract. We present two-dimensional magnetohydrodynamic numerical simulations of the Parker instability in a thick galactic disk. Calculations at various perturbation wavelengths show that the minimum unstable wavelength is between 2 kpc and 3 kpc and the time scale for growth is $\sim 3 \times 10^8$ yr. For the minimum unstable wavelengths (3–5 kpc), we find a stable solution consisting of a periodic array of dense sheet of gas extending perpendicular to the galactic plane. Longer wavelengths (≥ 5 kpc) give rise to an instability, in which the sheets coalesce and lead to the formation of large structures in the galactic midplane.

1. Model

In our calculations for the study of the Parker instability in a thick, gaseous disk, we make the following assumptions: (a) the medium behaves as an ideal gas, (b) the gas motions occur isothermally, (c) the magnetic field is frozen into the gas, i.e., the magnetic flux remains constant, (d) we do not take into account differential rotation or self–gravity, and (e) the numerical simulations are made in two-dimensional Cartesian coordinates (x, z), where the x coordinate represents the azimuthal direction of the Galaxy and the z coordinate is in the direction perpendicular to the galactic plane.

Our model of the thick disk has been presented in Martos and Cox (1994). It is a plane–parallel disk, in which magnetic and thermal pressures support the initial hydrostatic equilibrium against the non–uniform gravitational field provided by the stars. When due account is taken of the observed neutral and ionized gas components at high latitudes, as well as the slow decay of magnetic and cosmic ray pressures with the height z above the plane, the resultant scale height of the disk is of the order of kiloparsecs. The initial structure is determined by the hydrostatic equilibrium equation:

---

[1] Dirección General de Servicios de Cómputo Académico, UNAM, Apdo. Postal 20-059, 04510 México, D.F.

$$\frac{d}{dz}\left(P_{\text{th}}(z) + P_{\text{b}}(z)\right) = -m_{\text{eff}} n(z) g(z) \tag{1}$$

where $n(z)$ is the numerical density of the interstellar material, $g(z)$ the gravitational field produced by the stars, $P_{\text{th}}(z)$ the thermal pressure, and $P_{\text{b}}(z)$ the magnetic pressure. These quantities are given respectively by the following equations

$$n(z) = 0.6 e^{-\frac{z^2}{2(70\text{pc})^2}} + 0.3 e^{-\frac{z^2}{2(135\text{pc})^2}} + 0.07 e^{-\frac{z^2}{2(135\text{pc})^2}} + 0.1 e^{-\frac{|z|}{400\text{pc}}} + 0.3 e^{-\frac{|z|}{900\text{pc}}} \text{ cm}^{-3} \tag{2}$$

$$g(z) = 8 \times 10^{-9} \left(1 - .52 e^{-\frac{|z|}{325\text{pc}}} - .48 e^{-\frac{|z|}{900\text{pc}}}\right) \text{ cm s}^{-2}, \tag{3}$$

$$P_{\text{th}}(z) = n(z) kT \tag{4}$$

$$P_{\text{b}}(z) = \frac{B_x^2(z)}{8\pi} \tag{5}$$

The different terms of (2) correspond to the contributions of H_2, cold HI, warm HI in clouds, warm intercloud HI, and warm diffuse HII respectively, with scale heights appropriate for the solar neighborhood (Boulares and Cox 1990). Equation (3) gives a good fit to the gravitational field proposed by Bienaymé, Robin and Crézé (1987). Equation (4) is the ideal gas assumption. The magnetic field of (5) is initially parallel to the x axis with an intensity B_x of 5 μG in the plane of the disk ($z = 0$); it decreases only slightly with z up to 1 kpc as determined by (1) and (5).

We calculate the disk temperature using known values at the midplane,

$$T = \frac{P(0) - P_{\text{b}}(0)}{n(0) k} \tag{6}$$

giving a temperature of 10 900 K.

1.1. Perturbations

We perturb the initial hydrostatic configuration with a small amplitude perturbation in the z component of the velocity given by

$$V_z(x) = V_0 \sin\left(\frac{2\pi x}{\lambda_x}\right) \tag{7}$$

where V_0 is the perturbation amplitude of 1 km s^{-1}, and λ_x is the wavelength of the perturbation.

1.2. Boundary Conditions and Numerical Procedure

The numerical calculations were performed using the Cray supercomputer of the UNAM, with the 3-D MHD code ZEUS (Stone and Norman 1992a, 1992b). The code ZEUS is a three-dimensional ideal (non-resistive, non viscous, adiabatic) non-relativistic magnetohydrodynamical fluid solver which solves the following coupled partial differential equations as a function of time and space:

$$\frac{\partial \rho}{\partial t} + \nabla \cdot (\rho \mathbf{v}) = 0 \tag{8}$$

$$\frac{\partial \mathbf{S}}{\partial t} + \nabla \cdot (\mathbf{S}\mathbf{v}) = -\nabla p - \rho \nabla \Phi + \mathbf{J} \times \mathbf{B} \tag{9}$$

$$\frac{\partial e}{\partial t} + \nabla \cdot (e\mathbf{v}) = -p \nabla \cdot \mathbf{v} \tag{10}$$

$$\frac{\partial \mathbf{B}}{\partial t} = \nabla \times (\mathbf{v} \times \mathbf{B}) \tag{11}$$

where ρ is the matter density, $\mathbf{v}$ the velocity flow field, $\mathbf{S} = (\rho\, \mathbf{v})$ the momentum vector field, p the thermal pressure, Φ the gravitational potential, $\mathbf{J}$ the current density, $\mathbf{B}$ the magnetic induction and e the internal energy density.

Typically, we had a resolution in our grid of 128×64 zones and the boundary conditions used for a variety of experiments were periodic in the x direction and out conditions in the z direction.

2. Results

The Parker instability as usually considered, leads to fairly dense sheets of gas perpendicular to the galactic plane (Lerche 1967). They are vertically supported by magnetic tension, on a deeply induced field structure. Along the field lines leading away from these sheets, there is eventually a hydrostatic thermally-supported gradient in density. There is also a depression in the galactic disk below the sheet, where the magnetic tension supports part of the overlying weight. An example of this structure is shown in Figure 1f for an initial galactic "structure" described by (1) to (5).

Calculations at various perturbation wavelengths show that for this initial configuration, the minimum unstable wavelength is between 2 kpc and 3 kpc. The time scale for growth to the conditions shown is 3×10^8 yr.

A rather different structure appears in an experiment with $\lambda_{\text{pert}} \geq 5$ kpc. The superalfvenic downflows, on either side of the eventually high density zone, shock well before they are close to one another (Figure 1c). Rather than the usual single high density sheet fed by shocks on both sides, two separate high density sheets occur approximately normal to the flow (Figure 1a). These two horns are driven together by the ram pressure, coalescing dramatically.

The single sheet thus formed has a significant planeward velocity and descends well into the galactic disk. At this time, the distortion of the magnetic field in the upper part of the sheet is such that it is discontinuous in direction within the resolution of the grid. We observe the onset of magnetic reconection,

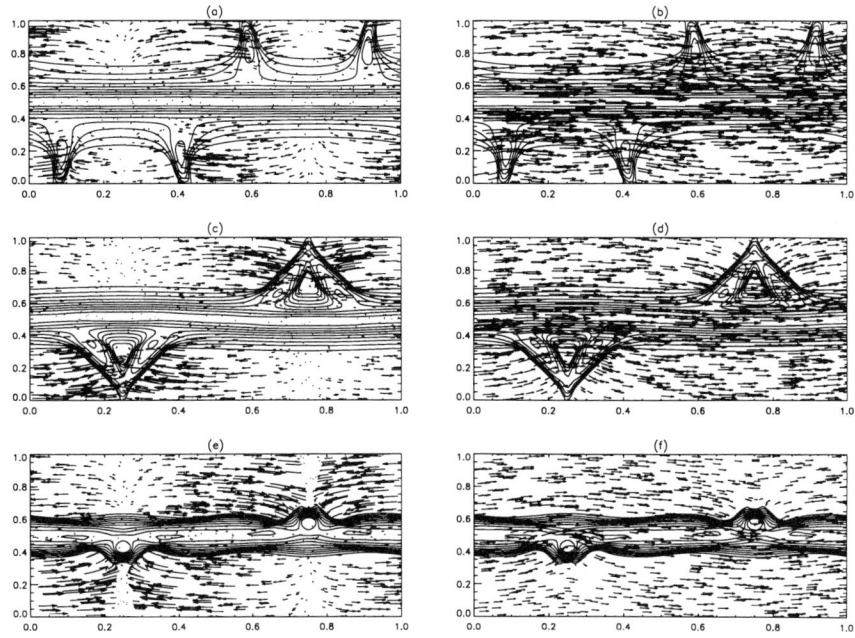

Figure 1. The sequence shows isodensity contours and the velocity fields, indicated by arrows (a,c,e) and isodensity contours and the magnetic fields, indicated by arrows (b,d,f), at three selected times: 2.5×10^8, 3.2×10^8 and 5.4×10^8 The maximum velocity values are 37, 54, and 50 km s^{-1}, respectively. The horizontal axis has a total length of 6 kpc and the vertical axis, with midplane at the center, ranges from -1.5 to 1.5 kpc.

but not a physically meaningful rate. ZEUS does not contain the resistivity required.

However, the resulting structures are extremely interesting and could well have astrophysical applications. In particular, the escape of magnetic field lines from the dense region, analogous to ambipolar diffusion, allows the gas to achieve a very dense equilibrium structure, lying on the plane rather than being redistributed to a vertical sheet.

There is factor of 10 density differential at given height z, between the pre and post-perturbations states. There are thermal pressure concentrations reaching $p_{\text{th}}/k \sim 5 \times 10^4$ cm^{-3} K which ar e confined neither by surrounding gas of similar p_{th}/k, nor by external magnetic pressure. They are stable and have circumferencial magnetic fields which lead to confinement by magnetic tension.

Acknowledgments. We would like to express our gratitude to E. Parker and S. Kurtz for useful discussions. This work has been partially supported by UNAM-CRAY through the grant SC-004296 and DGAPA–UNAM through the grand IN105894.

References

Bienaymé, O., Robin, A. C. & Crézé, M. 1987, A&A, 180
Boulares, A., & Cox, D. P. 1990, ApJ, 365, 544
Lerche, I. 1967, ApJ, 149, 395
Martos, M. A., & Cox, D. P. 1994, *Numerical Simulations in Astrophysics*, ed. J. Franco, S. Lizano, L. Aguilar, E. Daltabuit, (Cambridge Univ. Press: Cambridge), 229
Stone, J.M., & Norman, M.L. 1992a, ApJS, 80, 753
Stone, J.M., & Norman, M.L. 1992b, ApJS, 80, 791

Part 1. Kinematics of the Gas and Dust
Section B. Molecular Clouds at the Galactic Center

SiO Emission from the Galactic Center Molecular Clouds

Jesús Martín-Pintado, Pablo de Vicente, Asuncion Fuente, Pere Planesas

Observatorio Astronómico Nacional (IGN), Campus Universitario, Apartado 1143, E-28800 Alcalá de Henares, Spain

Abstract. The first results of a survey of the $J = 1 \to 0$ SiO emission in the Galactic Center region are presented. The spatial distribution of the SiO emission is very fragmented . This is in contrast to the spatial distribution of other high dipole moment molecules like CS whose emission is nearly uniform . The SiO clouds have typical sizes between 4 and 20 pc. The intensities of the J=2→1 , J=3→2 and J=5→4 transitions of SiO in the Sgr B2 and Sgr A clouds imply H_2 densities for these clouds of a few 10^4 cm^{-3} and SiO fractional abundance of $\sim 10^{-9}$. For the molecular clouds without SiO emission, the SiO fractional abundance is $\lesssim 10^{-10}$. The characteristics (size and H_2 densities) of the SiO emission in the Galactic Center are very different from those observed in the galactic disk. The implications of the SiO emission in the molecular clouds of the Galactic Center are briefly discuss . The particular chemistry in these clouds is probably related to fast shocks in the Galactic Center region.

1. Introduction

It is well known that the molecular clouds in the Galactic Center (GC) show different physical conditions than the molecular clouds in the galactic disk (see e.g. Güsten 1988). The GC clouds exhibit high kinetic temperatures (Tk $\gtrsim 80$ K) (Güsten, Walmsley & Pauls 1981; Morris et al. 1983; Hüttermeister et al. 1993), well above the dust color temperatures, ~ 30 K, derived from the FIR emission (Odenwald & Fazio 1984). In order to explain the high gas kinetic temperatures and the low dust temperatures found in the GC, several heating mechanisms which act only on the gas have been proposed (Wilson et al. 1982; Güsten et al. 1985). In particular, the large width ($\gtrsim 15$ km s^{-1}) of the molecular lines in these clouds and its possible correlation with the kinetic temperature suggest that dissipation of turbulence driven by differential galactic rotation can be an attractive heating mechanism for the GC molecular clouds (Wilson et al. 1982; Güsten et al. 1985).

If supersonic turbulence is an efficient heating mechanism for the GC clouds, one would expect that the associated shocks will also influence the chemical composition of these molecular clouds, and molecular species which are formed by shock chemistry should show enhanced abundances. From observations of the molecular clouds in the disk, It is well established that SiO is an unambiguous tracer of high temperature and/or shock chemistry (Downes et al. 1982; Ziurys,

Fribeg & Irvine 1989; Martín-Pintado, Bachiller& Fuente 1992). Observations of this molecule can help to understand the nature of the GC molecular clouds and elucidate the heating mechanisms. We present the first results of a lager scale survey of the J = 1 → 0 line of SiO toward GC region combined with high angular resolution observations of the J=2→1 , J=3→2 and J=5→4 lines of SiO toward the Sgr A and Sgr B2 molecular clouds.

2. Observations

We have used the 14-m telescope of the Centro Astronómico de Yebes (Spain) to carry out the large scale mapping of the Galactic Center in the J = 1 → 0 line of SiO. The half power beamwidth (HPBW) of telescope at the rest frequency of the J = 1 → 0 line of SiO was 2′. The receiver had a double side band noise temperature of 75 K. As spectrometer we used a 512 channel acusto-optic device which provided a velocity resolution of 0.74 km s^{-1}. The calibration was made by using the standard chopper wheel method and the line intensities were converted to main beam brightness temperature.

The observations of the J=2→1 , J=3→2 and J=5→4 lines of SiO towards the Sgr A and Sgr B2 molecular clouds were made with the IRAM 30-m telescope at Pico Veleta (Spain) and the observing procedure have been described by Martín-Pintado et al (1992). The HPBW of the telescope at these wavelengths were 26, 17 and 14″ for the three SiO lines respectively. The three transitions were observed simultaneously using SIS receivers with temperatures ranging from 110 to 150 K. As spectrometers, two 1MHzx512 channel filter banks were used, providing a velocity resolution of 3.5, 2.3 and 1.4 km s^{-1} for the J=2→1 , J=3→2 and J=5→4 lines respectively. The calibration of the data was made by observing the sky, a cold and a hot load. The line intensities are in units of T_A^*.

3. Results

3.1. Large scale distribution

The SiO profiles are broad with typical line widths of 30 to 60 km s^{-1}. Most of the SiO emission is concentrated for radial velocities between −10 and 90 km s^{-1}. Fig. 1 shows the integrated intensity map of the J = 1 → 0 line of SiO. The SiO emission shows a very fragmented distribution. The SiO clouds have typical sizes between 4 and 20 pc. This is in contrast with the emission from other high density tracers like CS which is fairly uniform over the region (Bally et al. 1987). At large scale, the SiO emission, however, shows similar spatial distribution that the hot gas observed in the (3,3) line of NH$_3$ (Morris et al. 1983). There are three major complexes of SiO clouds. The first one, located south of Sgr A, surrounds the southern edge of the radio continuum emission from Sgr A East. The second one, toward the star forming regions Sgr B1 and Sgr B2. The third SiO complex is found towards the Galactic Center nonthermal radio arc at l∼0.2°. It is remarkable that the SiO emission is not only restricted to the molecular cloud M0.20-0.03 (Serabyn & Güusten 1991; Lindqvist et al. 1995), but it shows the same morphology than the non-thermal filaments over

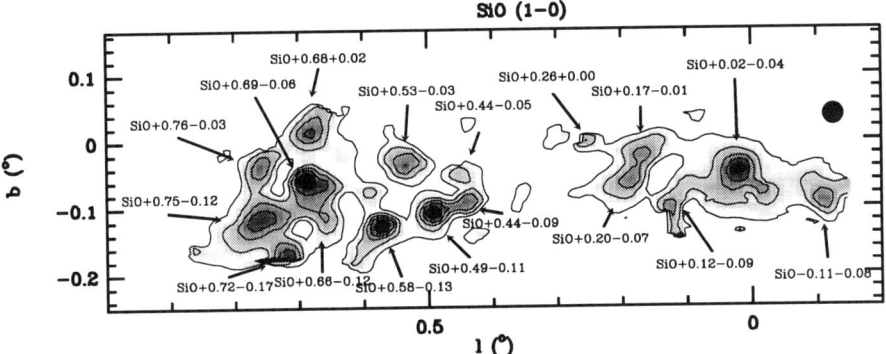

Figure 1. Integrated line intensity map of the $J = 1 \to 0$ SiO line toward the galactic center. The beam size is shown as a filled circle in the upper right corner. The contour levels are 9 to 94.5 km s^{-1} by steps of 17.1 K km s^{-1}

scales of several parsecs. Surprisingly, SiO emission is not detected at negative radial velocities toward the thermal arched filaments to our limit of 0.15 K.

3.2. High angular resolution SiO maps of Sgr A and Sgr B2

The high angular resolution (26″) maps of the $J=2\to1$ line of SiO towards Sgr A and Sgr B2 presented in Fig 2 and 3, respectively, show that the SiO emission is very extended and exhibits a rather smooth distribution. Towards Sgr A (see Fig. 2), the SiO emission is dominated by a ridge of 8 pc×16 pc extending along the galactic plane with two main clouds, M-0.13-0.08 and M+0.20-0.07, and a condensation south of Sgr A*. The SiO emission does not show the prominent cold and dense dust condensations observed at 1.3 mm (Mezger et al. 1989). Like at large scale, the morphology of the SiO emission resembles that of the hot gas observed in the (3,3) line of NH$_3$(Güsten, Walmsley & Pauls 1981; Ho et al. 1991). The SiO emission toward Sgr B2 extends over a region of at least ∼24×24 pc and presents a fragmented distribution. The typical sizes of the SiO condensations are $\gtrsim 1'$ ($\gtrsim 2$ pc). From the integrated emission one can recognize several shell-like structures. The largest one, centered on Sgr B2M surrounding the hot ring observed in CH$_3$CN (de Vicente et al. 1996). The SiO profiles in the envelope of Sgr B2 are very broad with with linewidths to zero intensity of up 100 km s^{-1}.

4. The physical conditions in the SiO clouds

Typically, the intensity of the $J=5\to4$ line of SiO in the GC clouds is 6-10 times weaker than that of the $J=2\to1$ line (Martín-Pintado et al. 1996). This is contrast with the line intensity ratios of ∼1 found in star forming re-

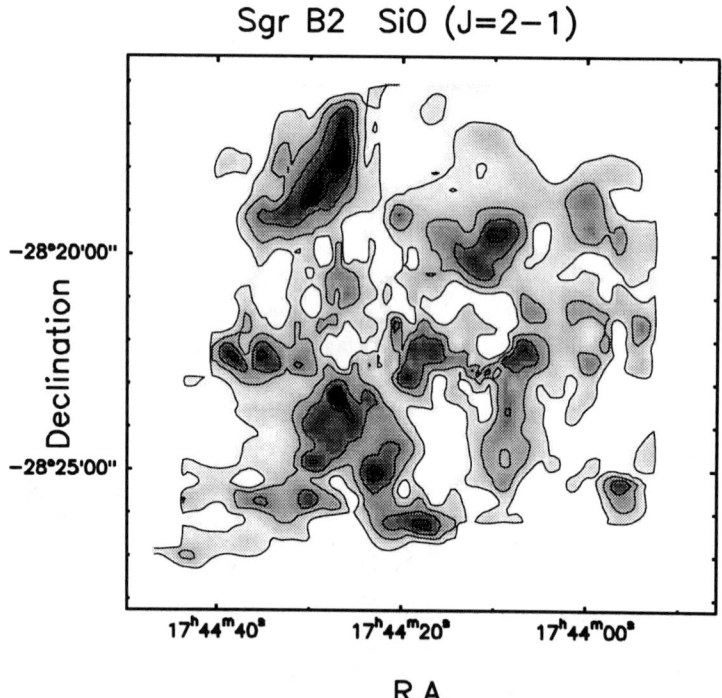

Figure 2. Integrated line intensity map of the J=2→1 SiO line toward the Sgr B2 molecular cloud. The contour levels are: 9.23 18.46 27.69 36.92 and 41.53 K km s^{-1}.

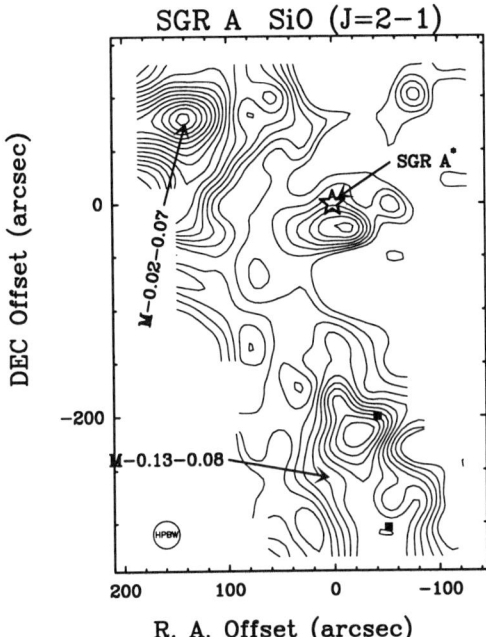

Figure 3. Integrated intensity map of the J=2→1 SiO line of the Sgr A molecular cloud. The beam size is shown as an open circle in the lower left corner. The open star shows the location of Sgr A* and the filled squares the FIR sources observed in the continuum emission at 1.3mm (Mezger et al. 1989). The offsets are relative to Sgr A*. The lowest contour level and the interval are 2 K km s^{-1}.

gions in the galactic disk (Martín-Pintado, Bachiller &Fuente 1992). The low J=5→4 /J=2→1 line intensity ratios in the GC imply that the SiO emission in these clouds arise from regions with lower H_2 densities than those in the galactic disk. For the typical kinetic temperatures, $\gtrsim$50 K, of the GC molecular clouds (Güsten, Walmsley & Pauls 1981; Morris et al. 1983; Hüttermeister et al. 1993), model calculation of the excitation of SiO indicate that H_2 densities of a few 10^4 cm^{-3} can explain the observed line intensity ratios. The fractional abundance of SiO for the SiO clouds which is similar to those of HCO^+ and CS, $\sim 10^{-9}$. The fractional abundance of SiO is, at least, one order of magnitude smaller, $\lesssim 10^{-10}$, for the clouds where SiO emission has not been detected. Similar low SiO abundances are also found for the molecular material associated to the thermal arched filaments in the GC.

5. Discussion

The peculiar chemistry of SiO in the molecular clouds of the galactic disk has led to the conclusion that Si is strongly depleted in the molecular clouds and

SiO only appears in small regions where shock disruption of grains releases Si or SiO to gas phase (Martín-Pintado et al. 1992). The fragmented distribution of the SiO emission and the variations of the SiO abundance in the GC clouds suggest that a peculiar chemistry for SiO is also taking place in this region of the galaxy. Chemistry based only on high temperatures (Ziurys, Fribeg & Irvine 1989; Langer & Glassgold 1990, Turner 1992; MacKay 1995) cannot account for the low SiO abundance, $\lesssim 10^{-10}$, of the high temperature gas ($\gtrsim 80$ K, Serabyn & Güsten 1987) associated to the thermal arched filaments. The low SiO abundance in the thermal arched filaments is very likely related to heating of these filaments which is thought to be mainly due to UV radiation from massive stars (Poglitsch et al. 1991).

It has been proposed that the high temperatures in the GC clouds are due to the dissipation of turbulence produced by cloud-cloud collisions which are expected to be more frequent in the GC region than in the disk (Wilson et al. 1982; Güsten et al. 1995). Low velocity shocks (~ 10 km s^{-1}) associated to cloud-cloud collision have also been claimed to explain the large SiO abundance derived from the narrow absorption lines toward Sgr B2M (Hüttemeister et al. 1995; Peng et al. 1996). However, the SiO emission in the envelope of Sgr B2 shows much broader profiles with radial velocities of up to $\pm$ 50 km s^{-1} (Martín-Pintado et al. 1996) indicating the presence of shocks faster than 30 km s^{-1}. Large scale cloud collision with velocities of ~ 30 km s^{-1} in Sgr B2 has been also suggested to explain the morphology and kinematics of the ^{13}CO emission (Hasegawa et al. 1994). For shocks velocities larger than 30 km s^{-1}, grain destruction becomes important (Seab &Shull 1983; Tielens et al. 1994) and Si and/or SiO can be released to gas phase. It is therefore likely that the large abundance of SiO in the GC molecular clouds are due, like in the galactic disk, to relatively fast shocks.

The origin of the fast shocks in the GC molecular clouds are very likely to be different for the three SiO complexes. For the Sgr A molecular clouds, there have been evidences that these molecular clouds are interacting with nearby supernova remnants (Ho et al. 1991; Mezger et al. 1989). The clouds SiO+0.17-0.01 and SiO+0.20-0.07 follow the GC nonthermal arc filaments suggesting a direct association of the SiO clouds with the nonthermal structures. It has been proposed that these molecular clouds should play an important role as the source of relativistic in the Galactic Center radio arc (Serabyn & Morris 1994). In this scenario strong shocks are expected in the region where particle acceleration takes place. The origin of the shocks for the SiO clouds in the Sgr B1 and Sgr B2 clouds can be due to large scale cloud collision (Hasegawa et al. 1994), expanding bubbles driving by HII regions (Soufe 1990, de Vicente et al. 1996) and sources with powerful stellar winds like Wolf-Rayet (WR) stars (Martín-Pintado et al. 1996).

Acknowledgments. This work has been partially supported by the Spanish CICYT under grant number PB93-048.

References

Bally, J., Stark, A.A., Wilson, R.W. & Henkel, C. 1987, A & A136, 243

de Vicente, P., Martín-Pintado, J., Wilson, T.L., 1996, A & A, in press

Downes, D., Genzel, R., Hjalmarson, A., Nyman, L.A. & Ronnang, B. 1982, ApJLet, 252, L29

Güsten, R., 1995, in IAU Symposium 136, The Galactic Center of the Galaxy, ed. M. Morris, Kluwer Academic Publisher, p. 89

Güsten, R., Walmsley, C.M. & Pauls, T.A. 1981, A & A, 103, 197

Güsten, R., Walmsley, C. M., Ungerechts, H. & Churchwell, E. 1985, A & A, 142, 381

Hasegawa, T., Sato, F., Whiteoak, J.B., Miyawaki, R. 1994, ApJ, 429, L77

Ho, P. T. P., Ho, L. C., Szczepanski, J. C., Jackson, J. M. & Armstrong, J. T. 1991, Nature 350, 309

Hüttemeister, S., Wilson, T. L., Mauersberger, R., Lemme, C., Dahmen, G. & Henkel, C. 1995, A & A 294, 667

Hüttemeister, S., Wilson, T. L., Bania, T. M. & Martín-Pintado, J. 1993, A & A 280, 255

Langer, W.D. & Glassgold, A.E. 1990, ApJ, 352, 121

Lindqvist, M., Sandqvist, A., Winnberg, A., Johansson, L. E. B. & Nyman, L. A. 1995, A & AS113, 257

Mackay, D.D.S. 1995, MNRAS, 274, 694

Martín-Pintado, J., Bachiller, R. &Fuente, A. 1992, A & A, 54, 315

Martín-Pintado, J., de Vicente, P., Fuente, A. & Planesas, P. 1996 , ApJ, in press

Martín-Pintado, J., de Vicente, P., Wilson, T.L., Gaume, R., 1996, in ESO-IRAM-NFRA-Osala Workshop on "Science with Large Millimeter Arrays", in press

Mezger, P. G., Zylka, R., Chini, R., Salter, C. J. & Wink, J. E. 1989, A & A, 209, 337

Morris, M., Polish, N., Zuckerman, B., Kaifu, N. 1983, AJ, 88, 1228

Odenwald, S.F. & Fazio, G.G. 1984, ApJ, 283, 601

Poglitsch, A., Stacey, G. J., Geis, N., Haggerty, M., Jackson, J.,Rumitz, M. Genzel, R. & Townes, C. H. 1991, ApJLet, 374, 33

Peng, Y., Vogel, S.N.& Carlstrom, J.E. 1996, ApJ, in press

Seab, C. G. &Shull, J. M. 1983, ApJ113, 257

Serabyn, E. & Güsten, R. 1987, A & A 184, 133

Serabyn, E. & Güsten, R. 1991, A & A, 242, 376

Serabyn, E. & Morris, M., 1994, ApJLet, 424, L91

Soufe, Y. 1990, PASJ, 42, 827

Tielens, A. G. G. M., McKee, C. F.,Seab, C. G. & Hollenbach, D. J. 1994, ApJ, 431, 321

Turner, B.E. 1992, ApJ, 388, L35

Ziurys, L.M., Fribeg, P. & Irvine, W.M. 1989b, ApJ, 343, 201

Wilson, T.L., Ruf, K., Walmsley, C.M., Martin, R.N., Pauls, T.A. & Bartla, W. 1982, A & A115, 185

The Molecular Gas in the Galactic Center Region based on $C^{18}O$ Measurements

G. Dahmen[1], S. Hüttemeister[2], T.L. Wilson, R. Mauersberger[3], and A. Linhart

MPIfR, Auf dem Hügel 69, 53121 Bonn, Germany

L. Bronfman

Dept. de Astronomía, U. de Chile, Casilla 36-D, Santiago, Chile

A.R. Tieftrunk, K. Meyer, and W. Wiedenhöver

MPIfR, Auf dem Hügel 69, 53121 Bonn, Germany

T.M. Dame and E.S. Palmer

CfA, 60 Garden Street, Cambridge, MA 02138, U.S.A.

J. May, J. Aparici, and F. Mac- Auliffe

Dept. de Astronomía, U. de Chile, Casilla 36-D, Santiago, Chile

Abstract. A large scale $C^{18}O(J = 1-0)$ survey of the central few hundred pc ($-1.05° \leq l \leq +3.6°$, $-0.9° \leq b \leq +0.75°$) of the Galaxy is presented. These $9'$ resolution data were obtained with the 1.2 m Southern Millimeter-Wave Telescope (SMWT) at CTIO and compared to $^{12}CO(1-0)$ data also obtained with this telescope. In addition, $HNCO(5_{0,5}-4_{0,4})$ line data included in the spectrometer passband are also presented. Although both ^{12}CO and $C^{18}O$ are thought to be tracers of the molecular mass, their distribution and line shapes differ significantly. The $C^{18}O(1-0)$ line is mucher weaker than expected from the "Standard $I_{^{12}CO(1-0)}/\mathcal{N}_{H_2}$ Conversion Formula" obtained by Strong et al. (1988). LVG calculations suggest that on large scales the $^{12}CO(1-0)$ emission is only of intermediate ($\tau = 1-5$) or low optical depth ($\tau < 1$). In this case, the "Standard Conversion Formula" overestimates $\mathcal{N}_{H_2}$. Our $C^{18}O(1-0)$ data combined with other H_2 tracers indicate a total molecular mass of $(3^{+2}_{-1}) \cdot 10^7$ $M_\odot$. The existence of a widespread component of molecular gas with low density (thin gas) is found to be very likely.

---

[1] Physics Department, Queen Mary & Westfield College, University of London, Mile End Road, London E1 4NS, U.K.

[2] CfA, 60 Garden Street, Cambridge, MA 02138, U.S.A.

[3] Steward Observatory, The University of Arizona, Tucson, AZ 85716, U.S.A.

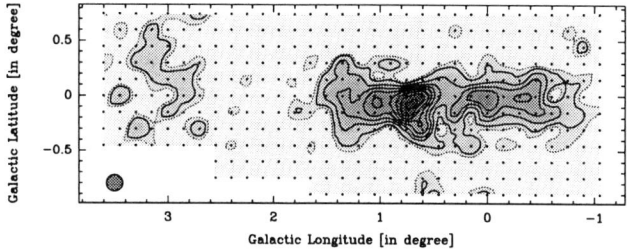

Figure 1. The integrated intensity of the Galactic Center region in $C^{18}O(1-0)$. The velocity over which the intensity is integrated ranges from -225.0 to $+225.0$ km s^{-1}. The solid contour levels range from 3.9 to 28.05 in steps of 3.45 K km s^{-1} where the lowest level is the 3σ-value. The dashed contour is at 2.6 K km s^{-1} which is the 2σ-value. The circle in the lower left corner of the plot indicates the beam size of 9.'2.

1. The Instrumentation of the 1.2 m SMWT

The primary antenna of the SMWT is a 1.2 meter parabolic aluminium dish. The telescope has a 3 mm liquid nitrogen cooled superheterodyne receiver and a main-beam size of 8.'8 at 115.3 GHz. A detailed description of the telescope system can be found in Bronfman et al. (1988, 1989).

For the $C^{18}O(1-0)$ survey (line frequency 109.782160 GHz), the telescope control computer system and software as well as the data reduction facilities were upgraded (Dahmen 1995). The Max-Planck-Institut für Radioastronomie (MPIfR) in Bonn, Germany, made available a broadband (795 MHz) AOS (Linhart 1994).

2. Observations and Data Reduction

The observations were carried out between the beginning of August 1993 and the end of August 1994. A detailed description of the observing procedure, the observations, and the data reduction is given in Dahmen (1995). The calibration stability was excellent over the course of the survey. The data were scaled to T_{MB} using comparison measurements done with the 1.2 m NMWT, where the scaling is known with a high accuracy (Cohen et al. 1986).

3. Results of the $C^{18}O(1-0)$ Galactic Center Survey

The survey presented here covers the area of $-1.05° \leq l \leq +3.6°$ and $-0.9° \leq b \leq +0.75°$. In Figure 1, we show a contour map of the integrated intensity of the $C^{18}O(1-0)$ line, covering the complete emission range of the Galactic center region from -225.0 to $+225.0$ km s^{-1}. The main $C^{18}O$ emission regions coincide with the known continuum and CO peaks Sgr A (with the extension to Sgr B1), Sgr B2, Sgr C and Sgr D. In addition, Clump 2 is weak but visible. In the middle

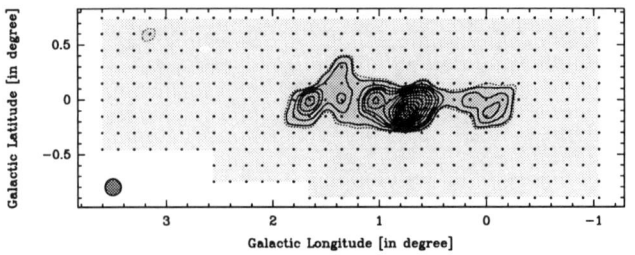

Figure 2. The integrated intensity of the Galactic Center region in HNCO($5_{0,5}$–$4_{0,4}$). The velocity over which the intensity is integrated ranges from -37.5 to $+137.5$ km s^{-1}. The solid contour levels are 2.4, 5.0, 7.5, and from 10.0 to 42.0 in steps of 4.0 K km s^{-1} where the lowest level is the 3σ-value. The dashed contour is at 1.6 K km s^{-1} which is the 2σ-value. The circle in the lower left corner of the plot indicates the beam size of 9$.\!'$2.

panel of Figure 3, a channel map of 50 km s^{-1} width centered at $+50$ km s^{-1} is shown. A detailed discussion of the data can be found in Dahmen (1995) and will be published soon (Dahmen et al. 1996).

4. The HNCO($5_{0,5}$–$4_{0,4}$) Line in the Survey

Because of the large bandwidth of the AOS the emission of the $5_{0,5}$–$4_{0,4}$ transition of HNCO at 109.905573 GHz fell mostly (depending on $v_{\rm LSR}$) into the range of the spectrometer. In Figure 2, the integrated intensity is plotted as a contour map, covering the velocity range from -37.5 to $+137.5$ km s^{-1}. This is the complete emission range of the HNCO($5_{0,5}$–$4_{0,4}$) line which is covered by the spectra. Most notably the HNCO($5_{0,5}$–$4_{0,4}$) emission is much more restricted to the Galactic plane than the C^{18}O(1–0) emission.

5. ^{12}CO(1–0) Measurements

To determine the comparability of our C^{18}O data we have also taken some ^{12}CO(1–0) data (using the same system) in March 1994 and August 1994. Our ^{12}CO map covers the inner region of the C^{18}O map (see Figure 3). As in the case of C^{18}O, the data were scaled to $T_{\rm MB}$. In the top panel of Figure 3, a channel map of 50 km s^{-1} width centered at $+50$ km s^{-1} is exemplarily shown. A detailed discussion of the ^{12}CO(1–0) data can be found in Dahmen (1995). As also shown there, our ^{12}CO data are in an excellent agreement with the ^{12}CO(1–0) survey obtained by Bitran (1987) in 1984 and with the ^{12}CO(1–0) data being obtained with the 7 m Bell-Labs telescope since 1986.

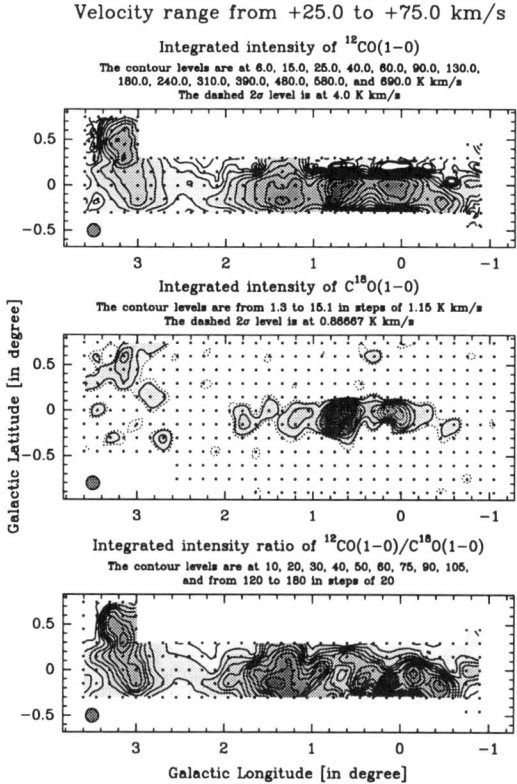

Figure 3. The integrated intensity of the Galactic Center region of ^{12}CO(1–0), C^{18}O(1–0), and their ratio in the velocity interval from +25 to +75.0 km s^{-1} (50 km s^{-1} width). The circle in the lower left corner of the plots indicates the beam size of 9′. The ratio was calculated with an 1-σ-threshold, thus, if the content of a channel either in ^{12}CO or in C^{18}O was below 1σ r.m.s. the 1σ value was taken instead for the calculation of the ratio.

6. C^{18}O and ^{12}CO in Comparison

The global differences between the C^{18}O data and the ^{12}CO data are:

1. The ^{12}CO emission at Sgr B2 is not as strikingly strong compared to the other emission maxima.

2. The ^{12}CO emission is much more widespread than the C^{18}O emission. In fact there is no position inside the map where no ^{12}CO emission is visible. However, the smaller latitude extent of the C^{18}O emission might be influenced by the detection limit of the rather weak C^{18}O(1–0) line.

3. The ^{12}CO emission from Clump 2 is much stronger compared to the other features than in C^{18}O.

A comparison of the line shapes of C^{18}O and ^{12}CO toward selected positions shows that they often differ significantly. A detailed analysis (see Dahmen 1995) shows that the standard conversion formula for the determination of the ^{12}CO column density does not work in the Galactic center region and that the assumption of high optical depth for the ^{12}CO emission is not generally valid there.

7. Integrated Intensity Ratios

An analysis of the integrated intensity ratio (see, e.g., the bottom panel of Figure 3) shows that the ^{12}CO/C^{18}O ratio in the Galactic Center region is generally higher than the value of about 15 which is expected from the compilation of the "standard" conversion factors. Wherever C^{18}O is above the detection limit this ratio is at least of the order 40, mostly of the order of 60 to 80, in several areas of the order 90 to 120, and toward a few positions even as high as 190.

¿From LVG calculations (see Dahmen 1995), it was confirmed that the large scale ^{12}CO emission in the Galactic Center region is dominated by emission with intermediate ($\tau = 1 - 5$) or low optical depths ($\tau < 1$). High optical depth emission ($\tau \geq 10$) is restricted to very limited areas such as Sgr B2. For the 4 intensity ratio ranges, the following conditions were found:

1. For low intensity ratios of about 40: $n_{H_2} \sim 10^{3.5}$ cm^{-3}, $T_{kin} \sim 50$ K, and $\tau \sim 3.0$.

2. For intermediate ratios of 60 to 80 which are the most common: $n_{H_2} \sim 10^{3.0}$ cm^{-3}, $T_{kin} \sim 50$ K, and $\tau < 2.0$.

3. For high ratios of 90 to 120 as found in several areas, in particular in the Sgr D region and in Clump 2: $n_{H_2} \sim 10^{3.0}$ cm^{-3}, $T_{kin} \sim 100$ K, and $\tau < 1.0$.

4. For very high ratios up to 190 which are present toward a few positions: $n_{H_2} \sim 10^{2.0}$ cm^{-3}, $T_{kin} \sim 150$ K, and $\tau \sim 2.0$.

8. The Molecular Mass in the Galactic Bulge

Using the standard conversion formula (Strong et al. 1988), the total molecular mass is found to be $2.8 \cdot 10^8$ M$_\odot$ from the ^{12}CO(1–0) emissivity found by Bitran (1987). On the contrary, from the emissivity of C^{18}O(1–0), the total molecular mass is found to be $1.7 \cdot 10^7$ M$_\odot$, taking the average excitation conditions into account (see Dahmen 1995). Because ^{12}CO emission of intermediate and low optical depth is dominant in the Galactic center region the existence of a widespread component of molecular gas with low density (thin gas) is found to be very likely. Invisible in C^{18}O due to subthermal excitation conditions, its additional mass contribution is estimated to be about $0.7 \cdot 10^7$ M$_\odot$. Comparing this to other tracers of molecular mass (see Table 1 and Dahmen 1995), we obtain a weighted best estimate of $\mathcal{M}_{mol} = (3^{+2}_{-1}) \cdot 10^7$ M$_\odot$ by ignoring the result from the standard conversion formula for ^{12}CO.

Table 1. The gas mass in the central 600 pc of the Galactic Center from different tracers

Tracer	M_{mol}	Reference
$^{12}CO(1-0)$	$2.8 \cdot 10^8 \, M_\odot$	Dahmen (1995) / Bitran (1987)
$C^{18}O(1-0)$	$1.7 \cdot 10^7 \, M_\odot$	Dahmen (1995)
$^{12}CO(1-0)$ (thin gas)	$0.7 \cdot 10^7 \, M_\odot$	Dahmen (1995)
Dust / IRAS	$3.6 \cdot 10^7 \, M_\odot$	Cox & Laureijs (1989)
Dust 800 μm	$> 0.4 \cdot 10^7 \, M_\odot$	Lis & Carlstrom (1994)
Dust / COBE	$3.1 - 7.0 \cdot 10^7 \, M_\odot$	Sodroski et al. (1994) data of Bitran (1987)
0.1–1.0 GeV γ-rays	$< 5.8 \cdot 10^7 \, M_\odot$	Blitz et al. (1985)

Probably, the high ^{12}CO emissivity is caused by the large scale emission being dominated by a rather small fraction of gas at moderate or low τ and the non-virialisation of a considerable fraction of the molecular gas in the gravitational potential of the Galactic bulge. This results in an overestimate of the molecular gas mass when applying the standard conversion factor. The main cause of the difference in mass estimates from the two CO isotopes is, therefore, not the emission from the peaks but the extended emission from ^{12}CO of rather low optical depth.

References

Bitran, M.E. 1987, Ph.D. Thesis, University of Florida
Blitz, L., Bloemen, J., Hermsen, W., Bania, T.M. 1985, A&A 143, 267
Bronfman, L., Cohen, R.S., Alvarez, H., May, J., Thaddeus, P. 1988, ApJ 324, 248
Bronfman, L., Alvarez, H., Cohen, R.S., Thaddeus, P. 1989, ApJS 71, 481
Cohen, R.S., Dame, T.M., Thaddeus, P. 1986, ApJS 60, 695
Cox, P., Laureijs, R. 1989, *IRAS Observations of the Galactic Center*. In M. Morris (ed.), *The Center of the Galaxy*, IAU Symp. 136, 121
Dahmen, G. 1995, Ph.D. Thesis, Universität Bonn
Dahmen, G., Hüttemeister, S., Wilson, T.L., Mauersberger, R., Linhart, A., Bronfman, L., Tieftrunk, A.R., Meyer, K., Wiedenhöver, W., Dame, T.M., Palmer, E.S., May, J., Aparici, J., Mac-Auliffe, F., to be submitted to A&AS
Linhart, A. 1994, Diploma Thesis, Universität Bonn
Lis, D.C., Carlstrom, J.E. 1994, ApJ 424, 189
Sodroski, T.J., Bennett, C., Boggess, N., Dwek, E., Franz, B.A., Hauser, M.G., Kelsall, T., Moseley, S.H., Odegard, N., Silverberg, R.V., Weiland, J.L. 1994, ApJ 428, 638
Strong, A.W., Bloemen, J.B.G.M., Dame, T.M., Grenier, I.A., Hermsen, W., Lebrun, F., Nyman, L.-Å., Pollock, A.M.T., Thaddeus, P. 1988, A&A 207, 1

CS (J=2-1) Observations of the Sgr B Complex and FIR 21

F. Yusef–Zadeh

Dept. Physics and Astronomy, Northwestern University, Evanston, IL 60208

K.I. Uchida

Max Planck Institut für Radioastronomie, Bonn, Germany

D. M. Mehringer and D. Roberts

Department of Astronomy, University of Illinois at Urbana, Champaign

L.-A. Nyman

European Science Observatory, La Silla, Chile

S. Casement

Dept of Astronomy, UCLA, Los Angeles, CA 90095

M. Lindqvist

Leiden Observatory, Postbus 9513, Leiden, the Netherlands

Abstract. Preliminary results on the velocity structure of CS (J=2-1) gas toward Sgr B1, Sgr B2 and the Galactic source FIR 21 (G0.28-0.47) are presented based on observations made wth the SEST. The general distribution of dense gas is compared with radio continuum VLA images. We present a number of new velocity features in the Sgr B region including a spectacular bright core at V_{LSR}=42 km s^{-1} situated within 2' of Sgr B2 Main. The association of molecular gas with FIR 21 is also discussed.

1. Introduction

The Sgr B complex is located about 100 pc from the Galactic center and is a region of considerable interest because of its ongoing star formation. This region consists of Sgr B1 (G0.5-0.0), G0.6-0.0 and Sgr B2 (G0.7-0.0) all of which extend over 20' along the Galactic plane. Recent radio continuum observations of this region suggest that while these features are physically related to each other (Mehringer et al. 1992), they are at different phase of their evolution. Motivated by this suggestion, the three components of the Sgr B complex were mapped in the CS (2-1) line in order to investigate the spatial and velocity distributions of dense molecular gas in this region. We have also searched for CS emission from

FIR 21 (Odenwald 1984), a star forming region possibly outside the Galactic center region.

2. CS (2–1) Observations

Emission from the CS $J = 2 - 1$ transition (97.98 GHz) was observed with the 15 m Swedish-ESO Submillimeter Telescope (SEST). The observations were distributed in a 45″ spaced grid ($0.3^0 \times 0.2^0$ in size) in the central region of Sgr B and 1′ spaced grid, (4′ × 4′) in size, centered at the radio continuum peak of FIR21. The beam size is about 45″. A 4-minute integration was done per position, resulting in an RMS noise level of about 0.15 K. The backend was the SEST "High resolution" AOS with 100 MHz bandwidth and 2048 channels. The velocity coverage was between $V_{LSR} = 200$ and -75 km s^{-1} with the central channel corresponding to $V = +55$ km s^{-1}. The velocity resolution was originally 0.13 km s^{-1} per channel, but was convolved to 1 km s^{-1}. The processing of the data was done with COMB, a spectral line reduction developed at AT&T Bell Labs.

3. Preliminary Results: Velocity Structure

3.1. Sgr B2 and Sgr B1

A bright CS core about 2′ SE of Sgr B2 Main with the peak velocity about +42 km s^{-1} is found. This source, which we call Sgr B2 South-East (SE), shows a remarkable resemblence to the velocity structure of the Sgr B2 Main, although high-resolution observations of Sgr B2 Main indicate that it is source of bipolar outflow and of maser activity (Mehringer 1995). Unlike Sgr B2 Main, North and South, the new feature is not associated with any strong compact ionized sources. Figure 1 show contours of CS emission at 45 km s^{-1} superimposed on the continuum gray-scale image at λ20cm. The Sgr B2 Main peak is displaced by ≈ 15 − 20″ with respect to its ionized counterpart, whereas Sgr B2 SE shows no ionized counterpart. The CS spectrum of Sgr B2 SE consists of two components both of which are blue-shifted by about 9 km s^{-1} with respect to the velocity components of Sgr B2 Main and North. In fact, the low-velocity component of this new feature is considered to be the most prominent molecular CS (2-1) feature in the Sgr B complex other than Sgr B2 Main.

There appears to be a shell-like molecular feature at $V_{LSR} = 35 - 45$ km s^{-1} surrounding the ionized gas associated with Sgr B1 and Sgr B2. The anti-corelation between ionized and molecular gas at this velocity suggests that these two HII complexes may be associated with the same parent molecular cloud. This kinematic cavity has been noted previously in large-scale CO survey of the Galactic center (Bally et al. 1988) and has been discussed previously in terms of a hole created in Sgr B2 due to cloud-cloud collision (Hasegawa et al. 1994).

Two high-velocity CS features are found at $V_{LSR} \approx$113-120 km s^{-1}. One of these sources coincides with a prominent HII region located about 2.2′ west of Sgr B2 Main at l=37′ 36.6″, b=-1′ 45.7″. The other molecular peak lies about 1′ north of Sgr B2 Main where there is lack of continuum emission. The two CS peaks extend over 5′ and show velocities ranging between 110 to 150 km

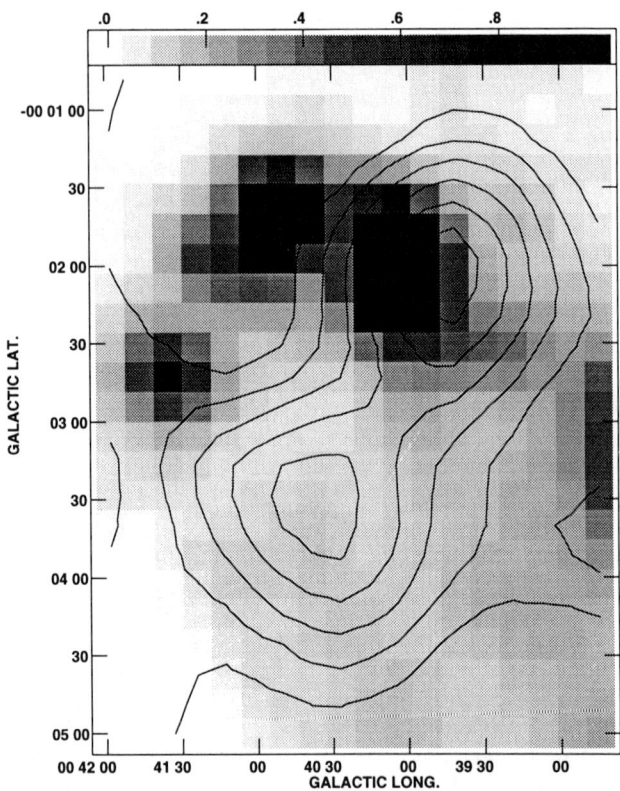

Figure 1. Contours of CS (2–1) emission at a velocity of 45 km s^{-1} are set at 1.2, 1.4, 1.6, 1.8, 2, and 2.2 K- km s^{-1} and are superimposed on the grayscale continuum image of Sgr B2 at λ20cm (Mehringer et al. 1993). The grayscale ranges between 0 and 1 Jy/beam. The two molecular peaks correspond to Sgr B2 Main and Sgr B2 South-East.

s^{-1}. The HII region coincident with one of the high-velocity CS peaks exhibits a recombination line velocity of about +100 km s^{-1} (Mehringer et al. 1993), thus suggesting that it is associated with the high-velocity molecular feature. High-resolution imaging of the continuum feature shows a face-on ionized shell, source V in Mehringer et al. (1993). These new high-velocity star forming molecular clumps avoid the region where a hole in the distribution of molecular gas is noted at a velocity of about 40 km s^{-1}.

3.2. FIR 21

Emission from the CS J = 2 - 1 transition has been detected towards G0.28−0.47 or FIR 21. Radio continuum observations of G0.28−0.47 at 4.8 GHz using the VLA shows an extended 3' long crescent-shaped feature with a bright compact core at its apex (Odenwald 1989). The CS emission is seen at velocities between 12 and 24 km s^{-1} and peaks about 12'' to the east of the radio continuum Core. Perhaps the most remarkable aspect of the CS emission, however, is its correspondence to the radio continuum Crescent. The circular contours of the CS emission, near the emission peak, closely follows the curvature of the arch-shaped Crescent, thus suggesting that the CS and the continuum features are physically associated with each other.

Analysis of the individual spectra show that the highest CS line velocities (18 km s^{-1}) are seen at, and slightly to the west of the CS peak. The strongest component is seen at a velocity of 18 km s^{-1} and displays a FWHM linewidth of $\sigma = 4.5$ km s^{-1}. The small FWHM linewidth of this component preclude it as having originated form the Galactic center region.

A second, somewhat weaker, velocity component is seen at 3.2 km s^{-1}. The FWHM linewidth of this component, $\sigma = 2.6$ km s^{-1}, also implies a source foreground to the Galactic center. However, this component's relationship to G0.28−0.47 is, at best, doubtful, since it shows no strong correlation in morphology or position to the radio continuum emission. The peak of the +3.2 km s^{-1} CS component is centered to the northwest of the Core and Crescent features.

References

Bally et al. 1988, ApJ, 324, 223

Hasegawa, T., Sato, F. Whiteoak, J.B. and Miyawaki, R. 1994, ApJ, 429, L27

Mehringer, D.M., Palmer, P., Goss, W.M. and Yusef-Zadeh, F. 1993, ApJ, 412, 684

Mehringer, D.M. 1995, ApJ, 454, 782

Odenwald, S.F. and Fazio, G.G. 1984, ApJ. 283, 601

A Hot Ring in the Sgr B2 Molecular Cloud

Pablo de Vicente and Jesús Martín-Pintado
Observatorio Astronómico Nacional (IGN), Campus Universitario, Apartado 1143, E-28800 Alcalá de Henares, Spain

Tom L. Wilson
Max-Planck Institut für Radioastronomie, Auf dem Hügel 69, D53121 Bonn, Germany

Abstract.
Single dish mapping of several lines of CH_3CN toward the Sgr B2 molecular cloud are presented. These data have been used to derive the kinetic temperature and density structure of this prototypical giant molecular cloud in the Galactic Center. The data reveal the presence of four different components in the Sgr B2 molecular complex: the hot cores, the warm envelope, the hot ring and the very hot diffuse envelope. The most remarkable feature is a hot (120 K) ring of gas surrounding the ultracompact HII regions. The gas in the hot ring is probably related to shocks. The high temperature of the hot cores (300 K) and the kinetic temperature distribution for short distances can be accounted by gas-dust collisions. The heating of the warm envelope is probably due to the dissipation of turbulences.

1. Introduction

The Sagittarius B2 molecular cloud is one of the most active regions of star formation in the Galaxy. Massive star formation is occurring in two regions, Sgr B2M and Sgr B2N, which are very strong IR emitters (Thronson & Harper 1986) and contain all signposts of this activity like ultracompact HII regions (Martin & Downes 1972; Gaume & Claussen 1990), hot cores (Vogel, Genzel & Palmer 1987) and maser emission in OH, H_2O, H_2CO and SiO (Gaume & Claussen 1990; Kobayashi et al. 1989; Gaume & Mutel 1987). The star-forming regions, which contain 10^5 $M_\odot$, are embedded in a giant molecular cloud with a total mass of $7 \cdot 10^6$ $M_\odot$ (Lis & Goldsmith 1989).

In spite of the large mass of the envelope, little is known about its physical properties. Large scale studies of this remarkable molecular clouds have only been made in ^{12}CO and $C^{18}O$ (Scoville et al. 1975; Lis & Goldsmith 1989). Absorption line studies for the line of sight toward the continuum sources (Wilson et al. 1982, Henkel et al. 1983) indicate the presence of a warm ($T_k \geq 100$ K) (Wilson et al. 1982; Hüttemeister et al. 1993) and moderate density envelope ($10^3 - 10^4$ cm^{-3}). Unlike the molecular clouds in the disk, the gas kinetic temperatures in the envelope are above the dust temperatures (~ 20K), and cloud-

cloud collisions have been proposed to heat the gas in the molecular clouds in the galactic center (Wilson et al. 1982, Hasegawa et al. 1994).

2. Observations and data analysis

Using the unmatched sensitivity and angular resolution of the IRAM 30-m telescope we have mapped with an angular resolution of 12-24″ the J=5–4, J=8–7 and J=12–11 lines of the symmetric top rotors CH_3CN and $CH_3^{13}CN$, and the 11-10 line of HC_3N. These data have been combined with an LVG analysis to determine the kinetic temperature and the H_2 density structure of the molecular gas in the Sgr B2 cloud at scales from 0.03 pc to 20 pc.

The LVG analysis adjusts several velocity components, derived from the HC_3N profiles by fitting gaussians. Due to the broad lines in Sgr B2, we have also considered in our model, overlapping between the K=0 and K=1 components. Our LVG analysis model fits, simmultaneously, the main isotope and $CH_3^{13}CN$ for all K components. The free parameters used in the analysis were the hydrogen density, the kinetic temperature, the column density and the A/E ratio, which was allowed to vary from 1 to 3, although in almost all cases was 1. The $^{13}C/^{12}C$ ratio was 0.05.

3. The kinetic temperature and hydrogen density distribution

The kinetic temperature distribution in Sgr B2 is shown in Fig. 1. The temperature varies from 300 K towards the dense hot cores to 40 K for the warm envelope up to distances of 8 pc from Sgr B2M. Four hot cores have been detected. Two of them are associated with the star forming regions in Sgr B2. The kinetic temperarure for theses cores is $\sim$ 300 K and the H_2 density $\simeq 3-10 \cdot 10^6$ cm^{-3}. The other two cores, near Sgr B2N, are newly found hot spots, with $T_K \simeq 200$ K and $n(H_2) \sim 10^6$ cm^{-3}.

Surrounding the cores there is an envelope of warm (40-80 K) and dense gas ($\sim 2 \cdot 10^5$ cm^{-3}) which contains $\sim 2 \cdot 10^6$ $M_\odot$. The density decreases with distance (for distances between 1 and 8 pc) from Sgr B2M as $n(H_2) \sim 2.96 \cdot 10^5$ cm$^{-3}(r/\text{pc})^{-0.87}$.

The most remarkable feature in the kinetic temperature map is the hot ring. This feature appears in the T_K map as a ring like structure of higher temperature (120 K versus 60-80 K) surrounding the star forming region Sgr B2M and Sgr B2N. The radius of the ring is $\sim$ 2 pc and its thickness is 1.4 pc. Fig.1 shows four spectra taken towards the hot ring and four outside it. The spectra from the hot ring clearly cannot be explained by lower kinetic temperatures while those taken out of the ring cannot be fitted by higher T_k.

There is an additional component not seen in the kinetic tempertaure maps. This component is seen towards the continuum source Sgr B2M from CH_3CN absorption lines. The absorption arises from a very hot diffuse envelope which contains gas at very high temperature (500 K) and low density (10^3 cm^{-3}). Additional observations are needed to determine the extent of this envelope.

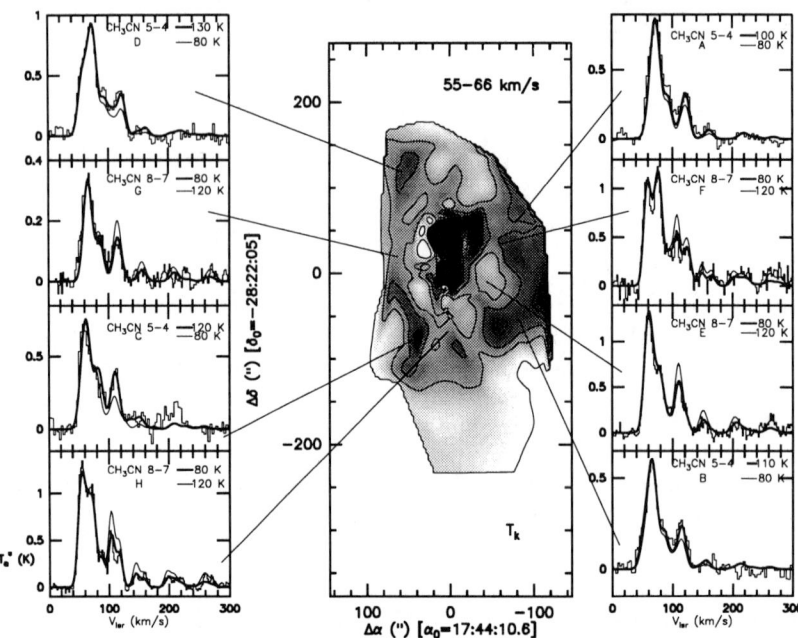

Figure 1. Kinetic temperature structure in Sgr B2. The contour levels are 200, 120, 80 and 40 K. At both sides, CH$_3$CN spectra with the predicted profiles for 80 and 120 K towards selected positions in the warm envelope

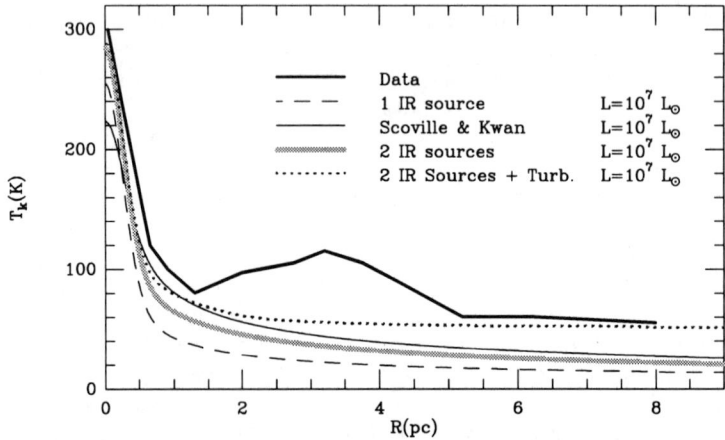

Figure 2. Kinetic temperature as a function of distance to Sgr B2M as derived from CH$_3$CN and predicted by models. The thick solid line represents the averaged T$_K$ obtained from our maps.

4. Heating mechanisms

The analysis of the thermal balance of the molecular cloud in Sgr B2 (de Vicente et al. 1996) shows that the temperature of the hot cores and the kinetic temperature distribution for distances to the IR sources smaller than 1 pc can be accounted for by gas-dust collisions (Fig. 2). However, the dust temperatures in the warm envelope are too low (10–20 K) to heat the molecular gas by this mechanism. Heating by dissipation of turbulent motions in the envelope of Sgr B2 can explain the gas kinetic temperature derived from the CH_3CN data.

The presence of the hot ring requires the existence of another heating mechanism. It has been proposed (de Vicente et al. 1996) that the hot ring might be associated to the interface between the warm envelope and the ionized bubble created by the OB stars recently formed in the Sgr B2 core. In this interface, heating by UV photons and/or shock fronts produced by the expansion of the ionized gas could explain the hot ring. The morphogy of the hot ring resembles that of the "hole" of ^{13}CO emission at 40-50 km s^{-1} reported by Hasegawa et al. (1994). These authors proposed that cloud-cloud collisions are responsible of this morphology. This possibility cannot be ruled out to explain the origin of the hot ring.

References

de Vicente, P., Martın-Pintado, J., Wilson, T.L. 1996, A&A in press.
Flower D.R., Pineau des Fôrets G. Walmsley C.M. 1995, A&A 294, 815.
Gaume R.A., Mutel R.L., 1987, ApJS 65, 193.
Gaume R.A., Claussen M.J., 1990, ApJ 351, 538
Hasegawa T., Sato F., Whiteoak J.B, Miyawaki R., 1994, ApJ 429, L77
Huettemeister S., Wilson T.L., Henkel C., Mauersberger R., 1993, A&A 276, 445
Kobayashi H. et al., IAU 136, *The Center of the Galaxy*, 181-187, Ed.: Morris M., Kluwer Academic Publishers
Lis D.C., Goldsmith P.F., 1989, ApJ 337, 704
Martin A.H.M., Downes D., 1972, Astrophys. Letters 11, 219
Scoville N.Z., Solomon P.M., Penzias A&A 1975, ApJ 201, 352
Thronson H.A., Harper D.A., 1986, ApJ 300, 396
Vogel S.N., Genzel R., Palmer P., 1987, ApJ 316, 243
Wilson T.L., Ruf K., Walmsley C.M., Martin R.N., Pauls T.A., Batrla W., 1982, A&A 115, 185

Kinematical, Physical and Evolutionary Status of the Galactic Center Giant Molecular Cloud G1.6-0.025

Andrej M. Sobolev

Astronomical Observatory, Ural State University, Lenin ave. 51, Ekaterinburg 620083, Russia

Abstract. The giant molecular cloud G1.6–0.025 is of particular interest because it shows evidence for a collision between the main cloud and a cloudlet with relative velocity $\sim$ 100 km/s. The G1.6 main cloud consists of clumps with sizes varying from $3''$ (0.1 pc) to $2'$ (4 pc). The hydrogen number densities in the large clumps vary from 10^2 to $5 \cdot 10^4$ cm^{-3} and in the small clumps reach the values greater than 10^6 cm^{-3}. The kinetic temperature in the G1.6 main cloud exceeds 50–100 K. The high values are partially a consequence of the collision with the cloudlet. The region influenced by cloud-cloud collision covers an area of $\sim 4'$ in diameter. It is traced by spike features the in $4_{-1} - 3_0$ CH$_3$OH-E line, which arises from methanol that is enhanced by the passage of shock waves. A probable protostar cluster is present in the same direction.

1. Introduction

The giant molecular cloud G1.6-0.025 (hereafter G1.6) was first observed during the 5 GHz H$_2$CO survey by Whiteoak & Gardner (1979). Later (Gardner et al. 1985, hereafter GWFPK) obtained maps in NH$_3$(1,1), (2,2) and (3,3). On the basis of a kinematical analysis GWFPK concluded that G1.6 consists of a number of clumps with dimensions d = $2' - 5'$, velocities V$_{lsr}$ = 50 $-$ 75 km/s, and line widths ΔV= 15 $-$ 20 km/s. G1.6 also contains a compact region ($<$ $81''$) toward ($17^h46^m12^s$, $-27°33'55''$), which is characterized by relatively narrow (about 6 km/s) emission peak at V_{lsr} = 43 km/s. The ammonia rotational temperature of the cloud is rather high and exceeds T$_{rot}$ > 50 K, the hydrogen volume density is n$_H$ = 10^3–10^4 cm^{-3} and the hydrogen column density is N$_H$ = $2 \cdot 10^{23}$ cm^{-2} (with adopted [NH$_3$/H$_2$] = $5 \cdot 10^{-7}$). In a following paper, Gardner & Boes (1987), hereafter GB) reported the detection of a clump with V$_{lsr}$ = 165 km/s, ΔV $\sim$ 25 km/s, d $\sim$ $5'$, N$_H$ = $3.5 \cdot 10^{22}$ cm^{-2} (with [NH$_3$/H$_2$] = 10^{-7}), T$_{rot}$ $\sim$ 120 K and n$_H$ $<$ $5 \cdot 10^3 cm^{-3}$, centered in ($17^h46^m12^s$, $-27°33'12''$).

The G1.6-0.025 cloud and the 165 km/s clump were also detected in the ^{13}CO(1-0) and CS(2-1) line surveys of Galactic center by Bally et al. (1987). Later, Whiteoak & Peng (1989), hereafter WP), mapped G1.6 in the $2_0 - 3_{-1}E$ methanol line. They detected absorption with maximal T$_d$ = -2 K, V$_{lsr}$ = 45 $-$ 75 km/s, ΔV= 10 $-$ 30 km/s in in an area covering $18' \cdot 12'$. They also found clumps near V$_{lsr}$= 165 km/s, and near V$_{lsr}$ = 105 km/s, centered at ($17^h46^m48^s$, $-27°47'30''$). The clump towards ($17^h46^m12^s$, $-27°33'15''$) showed evidence for multiple velocity structure, with $T_d = -1.2$ and $T_d = -0.8$ K, near velocities V_{lsr} = 46 km/s and 57 km/s, respectively, and corresponding line widths of ΔV = 9.5 and 17.5 km/s.

On the basis of the observational data from WP and model brightness temperature calculations for a number of CH_3OH–E lines, Sobolev (1990) inferred a hydrogen density of $n_H < 10^4$ cm^{-3} and a methanol column density of $N_m = 10^{15.5} - 10^{16.8}$ cm^{-2} for the main cloud, and a methanol column density of $N_m = 10^{14.5} - 10^{14.9}$ cm^{-2} for the high-velocity clumps.

2. $4_{-1} - 3_0 E$ line observations and large-scale structure of G1.6 giant molecular cloud

The methanol line $4_{-1} - 3_0 E$ from giant molecular clouds (GMC) is regarded to be a maser line since its first detection by Turner et al. (1972). Recent observations by e.g. Haschick et al. (1990), Berulis et al. (1991) (hereafter BKSS) confirmed this hypothesis. Results from excitation calculations have shown that the $4_{-1} - 3_0 E$ transition should be inverted in the whole range of GMC physical parameters (Sobolev 1990 & BKSS). The $4_{-1} - 3_0 E$ maser itself is energetically weak, but the information one can derive from it is very important. The masing line intensity depends critically on the column density variations. It provides a unique tool to study the cloud's structure.

The large-scale structure of the G1.6 in $4_{-1} - 3_0 E$ line was studied by Haschick & Baan (1993), hereafter HB) employing the Haystack 36.7 m antenna in Westford, MA. HB have shown that the emission in the 50 km/s component occurs in three distinct and aligned knots, and a knot located north of the central knot covering an area of 12' by 9'. The sizes of the central and the eastern $4_{-1} - 3_0 E$ methanol components are about 2', while the northern knot and the western component cover about 1'. The triple structure has the same alignment as $2_0 - 3_{-1} E$ methanol absorption structure obtained by WP. The central component of the $4_{-1} - 3_0 E$ emission coincides with the peak of the $2_0 - 3_{-1} E$ map. Regarding the NH_3 emission, the central and the eastern $4_{-1} - 3_0 E$ components roughly align with two peaks of NH_3 (integrated emission) while the western component has no counterpart in NH_3. The difference of the G1.6 $4_{-1} - 3_0 E$ line map, compared to the maps in the $2_0 - 3_{-1} E$ and the NH_3 lines, origin from the high sensitivity of the $4_{-1} - 3_0 E$ line intensity to variations of column density, to the chemical evolution of the source and to the decrease of the hydrogen volume density n_H to the periphery of the cloud.

Emission from the $4_{-1} - 3_0 E$ line near 165 km/s was detected by HB as well. The high velocity emission shows an elongated structure and is located north of the central knot of the low-velocity emission. A prominent knot in the 165 km/s emission coincides with the northern low-velocity component. The condensation Ea of the $NH_3(1,1)$ map by GWFPK corresponds to the peak of the $NH_3(3,3)$ map for the high-velocity cloud (GB). The upper limit for the brightness temperature near V_{lsr} = 105 km/s was determined to be $T_d < 0.3$ K (HB).

3. $4_{-1} - 3_0 E$ spectrum in the vicinity of $(17^h 46^m 12^s, -27°33'15'')$ and structure and evolution of G1.6-0.025 cloud

3.1. Phenomenological description of spectrum : summary emission of clumps

Observations of BKSS conducted with the RT-22 facility of the Lebedev Physical Institute show that the $4_{-1} - 3_0 E$ spectrum toward $(17^h 46^m 12^s, -27°33'15'')$ differs from the other spectra by the presence of two narrow features with V_{lsr} = 54 and 58 km/s. We refer to these two lines as the 'spikes', following Morimoto et al. (1985). The spike features were detected by HB as well. Following WP, we refer to the latter position as the center position (0,0).

With the aim to determine the sizes of spike sources and study the wide component's behaviour, additional observations were conducted at positions $(0, 1')$, $(-1', 0), (0, -1'), (1', -1')$ and $(1', 0)$. Spikes near 54 and 58 km/s are present in 3 of them. While the line widths and the central velocities of the narrow peaks remain the same within the observational accuracy, the spike intensities show substantial variations from point to point. These pronounced variations demonstrate that the line arises from compact sources located to south-west from the center position (0,0). Sources with similar phenomenological characteristics in the $4_{-1} - 3_0 E$ line are not unique and were observed in many different objects (see, e.g., Haschick et al., 1990 and BKSS). Usually they appear as spike features also in the spectra of $7_0 - 6_1 A^+$ and other methanol lines, which characterize Class I methanol masers (Batrla et al. 1987). The nature of spike sources was discussed in BKSS and will be further discussed below.

To separate the spectra of the compact spike sources from the spectra of the ambient cloud, we have decomposed the profiles into Gaussian components and a linear continuum. The residuals of the resultant fit and the observed spectra lie well within the observational errors. Statistical equilibrium calculations on the basis of $4_{-1} - 3_0 E$ (BKSS), $1_0 - 0_0 A^+$ (Slysh et al. 1994, Kalenskii & Sobolev 1994) and $2_0 - 3_{-1} E$ (WP) were performed to estimate methanol column densities in these sources of $N_m \sim (1-2) \cdot 10^{15} cm^{-2}$ and hydrogen densities of $n_H \sim 10^3 - 10^5 cm^{-3}$.

The additional information on the clump characteristics can be obtained from the comparison of the $2_0 - 3_{-1} E$ profile towards the (0,0) position (WP data) and the NH_3 line profiles in the nearby position $(0,-40'')$ (GWFPK data). Because class I methanol maser sources are not seen in $2_0 - 3_{-1} E$ and NH_3 lines, their widths must be compared with those of the $4_{-1} - 3_0 E$ line, obtained after the subtraction of spikes. The $4_{-1} - 3_0 E$ profiles are about twice as narrow than the $2_0 - 3_{-1} E$ line profiles and about thrice as narrow as the NH_3 line profiles. This phenomenon can not be the consequence of beam effects, because the HPBW of WP was equal to $2'$, and the HPBW of GWFPK (equal to $1\rlap{.}'35$) differs not much from that of the RT-22 facility. The difference in line widths can neither be explained by maser narrowing of the $4_{-1} - 3_0 E$ line; its moderate brightness temperature and the statistical equilibrium calculations demonstrate that the optical depth is relatively low ($|\tau| < 1$). At the same time the optical depth of the clumps in $2_0 - 3_{-1} E$ and in the NH_3 lines exceeds a value of 1. Hence, the difference in the line widths may be explained by the hypothesis of a clump with velocity dispersion increasing to the periphery.

The phenomenon may be a consequence of the presence of clumps with V_{lsr} = 45–50 km/s and V_{lsr} = 60–65 km/s, which contribute little to the $4_{-1} - 3_0 E$ emission but which clearly show up as $2_0 - 3_{-1} E$ absorption and NH_3 emission. Our statistical equilibrium calculations indicate this case might occur if the hydrogen volume density of the clump is low. This will be further discussed in the next section.

3.2. On the nature of compact region with V_{lsr} = 45 km/s : a probable protostar cluster?

The observations show that the $4_{-1} - 3_0 E$ emission in the range of V_{lsr} = 40 – 50 km/s is weak, while the corresponding NH_3 components (45 km/s emission according to GWFPK and GB) and $2_0 - 3_{-1} E$ (46 km/s component according to WP) are very strong and dominate the line profiles. It can be shown that the NH_3 emission at 45 km/s and the $2_0 - 3_{-1} E$ absorption at 46 km/s most probably originate in the different parts of the same clump. Indeed, the similarity of the radial velocities suggests a spatial connection of the line forming regions, whereas the estimates of the hydrogen densities for the sources suggest that they differ by more than by 1.5 - 2.5 orders of magnitude. The density estimate of the methanol source was obtained from statistical equilibrium calculations (BKSS) : strong $2_0 - 3_{-1} E$ absorption with $T_d < -1$ K and simultaneous weak $4_{-1} - 3_0 E$ emission with $T_d < 1$ K can occur only if $n_H < 10^{3.5}$ cm^{-3}. In the NH_3 emission region the density seems to be significantly higher, since the ratio of NH_3 line intensities observed toward (0,0) at 45 km/s is very high: (2,2)/(1,1) = 1,4. Calculations of Stutzki & Winnewisser (1985) and Sobolev et al. (1989) indicate a hydrogen volume density of $n_H > 10^6 cm^{-3}$ and a kinetic temperature $T_k > 50$ K. Note that GWFPK remarked that high rotational NH_3 temperatures together with low antenna temperatures of NH_3 indicate a low beam filling factor of $\phi_f < 0,1$.

We conclude that the NH_3 emission towards the (0,0) direction, at $V_{lsr} \sim$ 45 km/s, arises from relatively hot dense clumps ($n_H > 10^6 cm^{-3}$, $T_k > 50$ K). These clumps are surrounded by material at significantly lower densities of $n_H < 10^{3.5}$ cm^{-3}, absorbing in the $2_0 - 3_{-1} E$ line. Note, that collisions thermalize the NH_3 $4_{-1} - 3_0 E$ transition and the corresponding line does not become not maser. Together with the low filling factor, this explains the absence of methanol emission at V$lsr \sim$ 45 km/s. It can be easily shown that gas pressure and magnetic fields can not provide the dynamical equilibrium of the noted components. Thus, the NH_3 compact sources are most probably gravitationally bound objects of the protostellar nature.

It can be noted that GWFPK, GB and WP observations combined with statistical equilibrium results in a minimal possible value of the $2_0-3_{-1} E$ brightness temperature of $T_d > -2.3$ K and an estimate of the source sizes of

$$d(NH_3) \sim 60'' \text{ and } d(CH_3OH) > 80''.$$

We suggest to perform observations with an angular resolution better than 30–40″ as an observational test of the conclusions derived here.

3.3. Parameters of the sources of $V_{lsr} = 54$ and 58 km/s spikes under the hypothesis of collision–radiative pumping in rotational transitions

As noted above, the spectra with spikes are well fitted by the sum of Gaussians and a linear background. Our least squares fit provides Gaussian parameters for the spike profiles of $T_a - T_c = 2.7 - 5.1$ K, $\Delta V = 0.9 - 1.8$ km/s, $V_{lsr} = 54$ km/s and $T_a - T_c = 1.8 - 4.7$, $\Delta V = 3.5 - 4.5$ km/s, $V_{lsr} = 57 - 58$ km/s, where T_a is the antenna temperature of the source and T_c is the antenna temperature of the ambient cloud. Note that the 58 km/s component may be fitted by similar Gaussians, but the observational accuracy is poor and no attempt was made to determine the corresponding parameters.

It was shown in BKSS that the available data on spike features can be explained by the hypothesis of collision-radiative pumping in the rotational transitions (CRr). IRAS sources and other strong factors which lead to a redistribution of the level population seem to be absent in the region under study, we propose that CRr pumping is the most probable mechanism in our case. We use the CRr hypothesis to obtain a lower limit of the source sizes and to estimate the values of other physical parameters.

The brightness temperature under CRr pumping has the absolute maximum value $T_{s,max} = 7\,000$ K. Hence, the lower limits of the sizes are

$$d > D_L[(T_a - T_c)/(T_{s.max} - T_c)]^{1/2} = 3''\!.2 \text{ and } 3''\!.1$$

for 54 and 58 km/s spikes, correspondingly. At a distance to the objects of $R_0 \sim 8.5$ kpc, these values correspond to linear sizes of about $4 \cdot 10^{17}$ cm. Clumps with such sizes are typical for GMC cores (see observations of Stutzki & Güsten (1990) and the theory of Kolesnik & Ohul'chansky (1990), and for boundaries of young HII regions (observational data of Mangum et.al. 1992). This may indicate that the clumps in the speike sources are quasi-stable features or protostars at the earliest stages of their collapse. Following Stutzki & Güsten (1990), we estimate their physical parameters on the basis of the virial equation for a uniform ball of molecular hydrogen :

$$V_d^2 = 0.6\,G\,M/(d/2) = 0.4\,p\,G\,m_H\,n_H\,d^2, \qquad (1)$$

where $V_d = 0.5\,\Delta V\,(\ln 2)^{-0.5}$ is the Doppler line width and M is the mass of the ball.

Combined with the above estimates of the sizes and the Gaussian fits results, the latter equation allows to estimate masses of the clumps to be $> 70\,M_\odot$ and $> 300\,M_\odot$ for the 54 and 58 km/s spike sources, correspondingly. A more detailed elaboration on the parameters for the spike sources is given in Sobolev (1992).

A comparison of the obtained results with the data of Stutzki & Güsten (1990) shows that a large number of clumps in GMC core may have similar parameters. At the same time actual number of spikes is not large and rarely exceeds 2 – 3 in the whole bulk of the cloud. If we assume that the methanol column densities in the clumps are not high but that the spike sources have exceedingly high abundances of methanol relative to hydrogen, with the values $[CH_3\,OH/H_2\,] \sim 10^{-7} - 10^{-6}$, this contradiction disappears.

Recent observations by Saito et al. (1993) have shown that the grain mantle evaporation is the main process producing the large enhancement of the methanol abundance relative to molecular hydrogen (see Plambeck & Menten (1990) for discussion of various possibilities). This fact is strongly supported by recent work of Hartquist et al.(1995) which show that considerable amount of methanol can be sputtered from grain mantles in the shocked regions. Observational data on class I methanol masers demonstrates a close connection between spike sources and high-speed outflows (e.g., Plambeck & Menten 1990). A non-stationarity of chemistry in G1.6 is supported by the detection of H_2 emission by Pak et al.(1996). The realization of this scenario in G1.6 needs the presence of a high-speed flow. Such a flow may exist near the (0,0) position where the 165 km/s clump is observed. The velocity of this clump with respect to the G1.6 main source exceeds 100 km/s. So, collision of these objects should give rise to sufficiently strong shock waves. The same shocks may induce star formation in the 45 km/s NH_3 emission region.

4. Concluding remarks on the physical and evolutionary status of G1.6-0.025

We have demonstrated that the G1.6 giant molecular cloud consists of clumps with sizes varying from $3''$ (0,1 pc) to $2'$ (4 pc). Column densities of hydrogen and methanol in the large clumps are about $(1-4) \cdot 10^{23}$ and $(1-2) \cdot 10^{15}$ cm^{-2}, correspondingly. Hydrogen densities in the large clumps vary from 10^2 to $5 \cdot 10^{-3}$ cm^{-3} and in the small clumps reach values exceeding $> 10^6$ cm^{-3}. The high kinetic temperature in the cloud of $T_{kin} > 50 - 100$ K is a consequence of heating by radiation from Galactic Center (see, e.g., Krugel & Tutukov (1978) and collisions with individual clumps. Possibly, a collision occurs toward $(17^h46^m12^s, -27°33'15'')$ where the 165 km/s clump is observed. Shocks connected with this collision may lead to enhancement of the methanol abundance in some small-scale clumps with masses $> 70 - 300 M_\odot$. These clumps are the origin of the spike features in $4_{-1} - 3_0 E$ line. A possible protostar cluster in that direction may be the result of that collision.

References

Batrla, W., Matthews, Menten, K.M., and Walmsley, C.M., 1987, Nature, 326, 49
Bally, J., Stark, A.A., Wilson, R.W., and Henkel, C. 1987, ApJS, 65, 13
Berulis, I.I., Kalenskii, S.V., Sobolev, A.M., Strelnitskij, V.S. 1991, As. Ap. Trans., 1, 1 (BKSS)
Gardner, F.F., Boes, F. 1987 Proc. astr. Soc. Austr., 7, 185 (GB)
Gardner, F.F., Whiteoak, J.B., Forster, J.R., et.al. 1985 Proc. astr. Soc. Austr., 6, 176 (GWFPK)
Hartquist, T.W., Menten, K.M., Lepp, S., Dalgarno, A., MNRAS, 272, 184
Haschick, A.D., Baan, W.A. 1993, ApJ, 410, 663
Haschick, A.D., Menten, K.M., Baan, W. 1990, ApJ, 354, 556

Kalenskii, S.V., Sobolev, A.M., Pis'ma a Astron. Zhurn., 1994, 20, 113
Kolesnik, I.G., Ohul'chansky, Ya.Yu. 1990, In : Physical Processes in Fragmentation and Star Formation. R.Capuzzo-Dolcetta et.al., eds., Kluwer Academic Publishers, 81
Krugel, E., Tutukov, A.V. 1978, A&A, 63, 375
Mangum, J.G., Wooten, A. and Mundy, L.G. 1992, ApJ, 370
Morimoto, M., Ohishi, M., Kanzawa, T. 1985, ApJ, 288, 556
Pak, S., Jaffe, D.T., Keller, L.D. 1996, ApJ, 457, L43
Plambeck, R.L., Menten, K.M. 1990, ApJ, 364, 555
Saito, S., Mikami, H., Yamamoto, S., Murata, Y., Kawabe, R., 1993, in IAU Coll. 140
Slysh, V.I., Bachiller, R., Berulis, I.I., et al., Astron. Zhurn., 1994, 71, 37
Sobolev, A.M. 1990, Astron. Tsirk., N1543, 7
Sobolev, A.M. 1992, Soviet Ast., 36, 590
Sobolev, A.M., Sumina, S.V., Karamysheva, S.Yu., 1989, Astronomo-geodezicheskie issledovaniya, Sverdlovsk, 150
Stutzki, J., Güsten, R. 1990, ApJ, 356, 513
Stutzki, J., Winnewisser, G. 1985, A&A, 144, 13
Turner, B.E., Gordon, M.A., Wrixon, G.T. 1972, ApJ, 177, 609
Whiteoak, J.B., Gardner, F.F. 1979 MNRAS, 188, 445
Whiteoak, J.B., Peng, R.-S. 1989, MNRAS, 239, 677 (WP)

Part 1. Kinematics of the Gas and Dust
Section C. Sgr A and the Radio Arc

The Galactic Center
ASP Conference Series, Vol. 102, 1996
Roland Gredel, ed.

Kinematics of the Sgr A Cloud Complex

Robert Zylka

*Institut für Theoretische Astrophysik, Universität Heidelberg,
Tiergartenstr. 15, 69121 Heidelberg*

Abstract. I discuss the kinematics of the neutral gas within a projected distance of $\sim 10'$ from Sgr A* and speculate on the influence the explosion that created the synchrotron source Sgr A East may have had on the distribution of this material. I also point out some links between the main clouds in this region and argue that Sgr A East Core, M−0.02−0.07, M−0.13−0.08 and gas westwards of Sgr A East form one coherent cloud complex.

1. Introduction

The molecular clouds in the Sgr A Complex were among the first detected by radio astronomers in OH absorption (Bolton et al. 1964, Robinson et al. 1964 and Goldstein et al. 1964). But even now, more than three decades later and with spatial resolution two orders of magnitude higher, we still cannot give an authoritative explanation for the kinematics of the neutral gas in this region, i.e., we cannot define the coherent kinematic features and explain their dynamics. Sub-arcminute studies of the neutral gas within the Sgr A Complex were published by Serabyn and Güsten (1987, CS(2−1), HPBW 25"), Mezger et al., (1989, 1.3mm-continuum, HPBW 11"), Zylka et al., (1990, 1.3mm-continuum, ^{13}CO(2−1), C^{18}O(2−1), C^{34}S(2−1), HPBW 24-11"), Serabyn et al., (1992, CS(5−4), CS(7−6), HPBW 29-20"), Tsuboi et al., (1994, CS(1−0), HPBW 34"), Lis and Carlstrom (1994, 800μ-continuum, HPBW 25"), Lindqvist et al., (1995, C^{18}O(1−0), HNCO(5−4), HPBW 46"), Hasegawa, (CO(1−0), HPBW 17", this volume). The interferometric studies of Sandqvist et al., (1987), Okumura et al., (1989, 1991), Ho et al., (1991) and Ho (1994) have much higher resolution, but miss a large fraction of the molecular emission.
Here I present spectroscopic data from the Sgr A Complex obtained with the IRAM 30-m Telescope, compare it with high-resolution observations made elsewhere and discuss the kinematics and the possible connections between the different components of this complex.

2. The main components of the Sgr A Radio and Cloud Complex

Within the central 50 pc the volume and column density of the ISM attains its highest values. Very compact GMCs with kinetic gas temperatures ≥ 70 K are embedded in a predominantly molecular intercloud gas. The outstanding features seen in this region are:

- **Sgr A West** — the central HII region within which is embedded the point-like source **Sgr A***. Most of the gas within Sgr A West is ionized. The ionized features show the well-known spiral structure, which can also be recognized in Fig. 1 in the 1.3 mm continuum map (HPBW $\sim 11''$).

- **Circumnuclear Disk** (CND) — a torus of neutral gas surrounding Sgr A West (see Fig. 2) containing a few $10^4\,M_\odot$. The motion of the CND is mainly rotation around the Galactic Centre at velocities $\sim 110\,{\rm km\cdot s^{-1}}$ with the same sence as the galactic rotation (Güsten et al. 1987). The distribution of material within the CND is highly assymetric - the southern lobe is roughly three times more massive than the northern lobe. The estimated age of the CND is $\sim 10^4$ years.

- **High Negative Velocity Gas** (HNVG) — detected in H_2CO absorption by Güsten and Downes (1981). This blueshifted gas maybe associated with the ionized gas of Sgr A West (Zhao et al., 1995).

- **Sgr A East** — a synchrotron shell-like source created by an explosion a few 10^4 years ago (Mezger et al., 1989). Its extent is marked in Fig. 1 with thin solid contours. The explosion that created Sgr A East took place inside a dense molecular cloud refered to by Zylka et al. (1990) as

- **Sgr A East Core** — a GMC with mass of few $10^5\,M_\odot$. Its western part is disrupted and its eastern part is compressed by the explosion, forming the **M−0.02−0.07** cloud (see Fig. 3).

- **M−0.13−0.08** — a GMC with mass of few $10^5\,M_\odot$ located to the south of Sgr A West. It shows two very compact condensations $-3.3'$ and $-5.2'$ south of Sgr A* (see Fig. 1) but its centre is at $\sim -4.5'$.

In the following I argue that the Sgr A East Core, the M−0.02−0.07 and the M−0.13−0.08 clouds form one coherent cloud complex and refer to it as **Sgr A Cloud Complex**. Note however that interpretating spectral line data, one should keep in mind the *caveat*: The combination of two-dimensional projected positions with radial velocity as a third dimension does not contain sufficient information to allow for unique deprojection into three-dimensional space. Such a determination of the third spatial coordinate always carries the risk of being misled by a coincidence of features which are projected onto the same position in two-dimensional space and which, by chance, have the same radial velocities without being physically connected. This necessarily introduces a certain ambiguity in the following interpretation.

A map showing the distribution of the neutral gas within the central few arcminutes, together with some representative $^{13}CO(2-1)$ spectra, is presented in Fig. 1. For reference the extent of 6 cm continuum emission showing Sgr A East and the fountain-like thermal emission (Yusef-Zadeh, Goss, this volume) extending to the west are also marked. The coordinates are offset relative to Sgr A* in seconds of arc in (R.A.,DEC.). Note that due to the observing method used for the mm-continuum observations (on-the-fly double-beam mapping), the extended emission associated with the Sgr A East Core (size $> 5'$ in east-west direction) has been suppressed. This makes the dust shell surrounding Sgr A

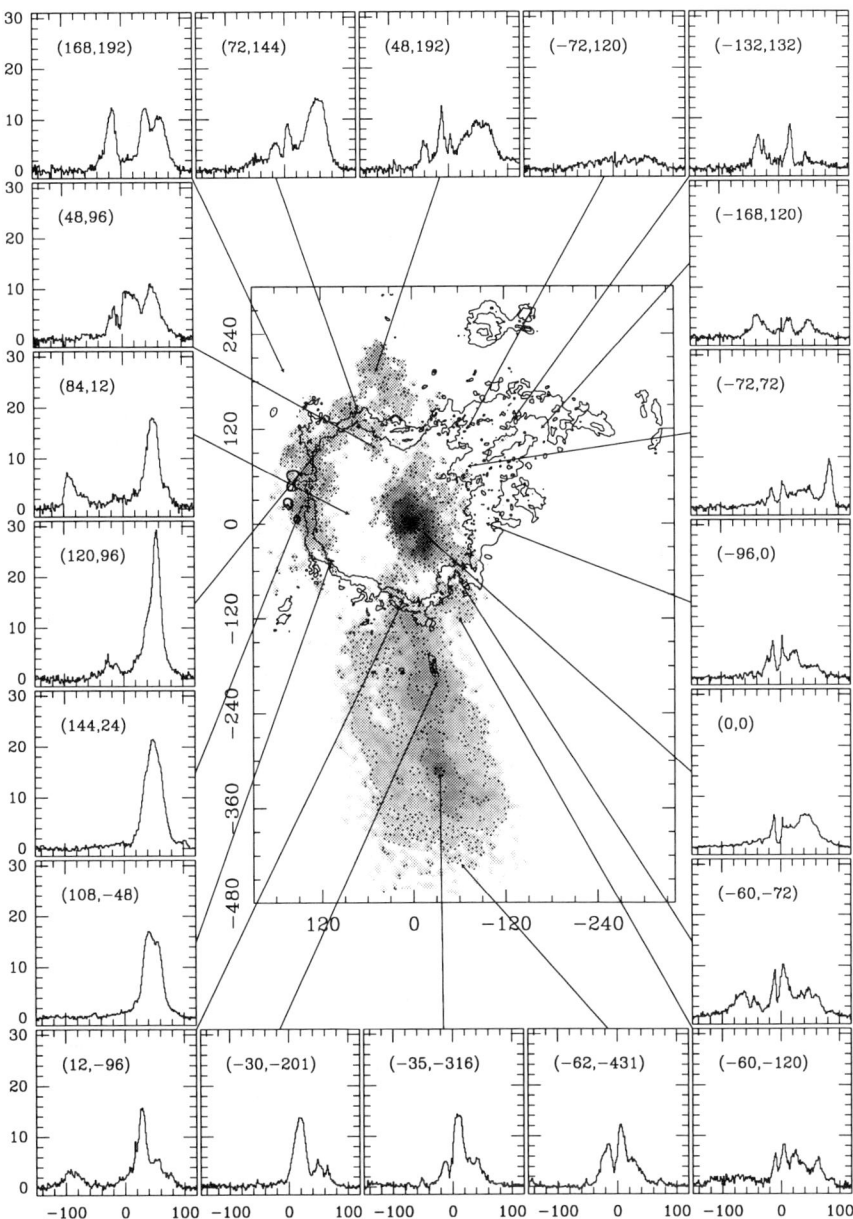

Figure 1. Overlay of the 1.3 mm dust and free-free emission (Mezger et al., 1989, Zylka et al., 1990, grey scale) on 6 cm continuum radio image (Yusef-Zadeh & Morris, 1987, thin contours). Also shown are a number of representative ^{13}CO(2–1) spectra. The coordinates are offset relative to Sgr A* in (R.A.,DEC.) in arcseconds. The scale of the x- and y-axis for the spectra is km·s^{-1} and K of T_{MB}, respectively.

East clearly visible in the 1.3 mm map, although it accounts only for $\sim 10\%$ of the mass of Sgr A East Core.

3. Kinematics of the components of the Sgr A Complex

The ^{13}CO(2−1) spectra presented in Fig. 1 illustrate the kinematics of the Sgr A Cloud Complex. All ^{13}CO(2−1) lines toward the central region are affected by strong absorption by gas in the 3 kpc Arm (v ~ -53 km·s^{-1}) and two other local features at v ~ -31 km·s^{-1} and v ~ -5 km·s^{-1} (see Zylka et al., 1994). These features are apparent throughout the entire field.
In the M−0.13−0.08 cloud the spectra show a dominant velocity feature, whose radial velocities increase linearly with galactic longitude from v ~ 0 km·s^{-1} in the south to v ~ 30 km·s^{-1} in the north. Perpendicular to the galactic plane there exist much higher and variable velocity gradients within this cloud (see Fig. 2). The spectra observed at the southern edge of Sgr A East show additional velocity components at v ~ -150 to -50 km·s^{-1}, arising from the gas in the southern lobe of the CND. The M−0.13−0.08 cloud extends much farther to the north than reported so far. Towards and north-westwards of Sgr A* this cloud splits into two velocity features: one bends towards negative velocities (v ~ -20 km·s^{-1} and possible reaching even more negative values), the other bends towards higher positive velocities (as high as v ~ 80-90 km·s^{-1}). Towards the eastern edge of Sgr A East the gas of M−0.13−0.08 attains radial velocities of v ~ 40-70 km·s^{-1} (see Fig. 3, bottom left) and connects with the estern part of the Sgr A Core, i.e the M−0.02−0.07 cloud. M−0.13−0.08 and M−0.02−0.07 thus form one coherent cloud complex. The kinematics becomes more complicated north- and westwards of the center of Sgr A East. In this region the gas of the Sgr A East Core separates into two main velocity components which connect with the v ~ -20 km·s^{-1} and v ~ 80-90 km·s^{-1} components of M−0.13−0.08. Just before these components appear, i.e. at $\Delta b \sim 0''$, the line widths reach the highest values (see the spectrum at $(\Delta \alpha, \Delta \delta) = (48'', 96'')$). Further out, towards positive galactic latitude offsets, up to 6-7 velocity components are found. Inside of Sgr A East there exist highly blue-shifted gas with v ~ -110 to -50 km·s^{-1}.

3.1. Gas affected by the explosion of Sgr A East

Mezger et al. (1989) have shown that the relativistic plasma of the synchrotron source Sgr A East is expanding into the dense molecular gas of the Sgr A East Core. They estimated the energy input of the explosion that created Sgr A East to be roughly 40 supernova explosions and its age to be $\sim 5 \cdot 10^4$ yr. The interferometer data of Ho et al. (1991, 1994) lead to similar results, although they show only part of the Sgr A East Core. The influence of Sgr A East on the kinematics of the neutral gas within the central $\sim 10'$ is well visible in the ^{13}CO data presented here. The explosion that created Sgr A East has apparently accelerated molecular gas to radial velocities in the range -110 to 110 km·s^{-1} (Fig. 2 and 3). The explosion disrupted the western part of the Sgr A East Core. The line widths here are by a factor of 2-3 larger than elsewhere and reach values as high as ~ 50 km·s^{-1}. At the western edge of Sgr A East there is a drastic change in the complexity of the line profiles mentioned above: to the south-east, i.e. at negative galactic lattitudes, one observes (at most) two velocity

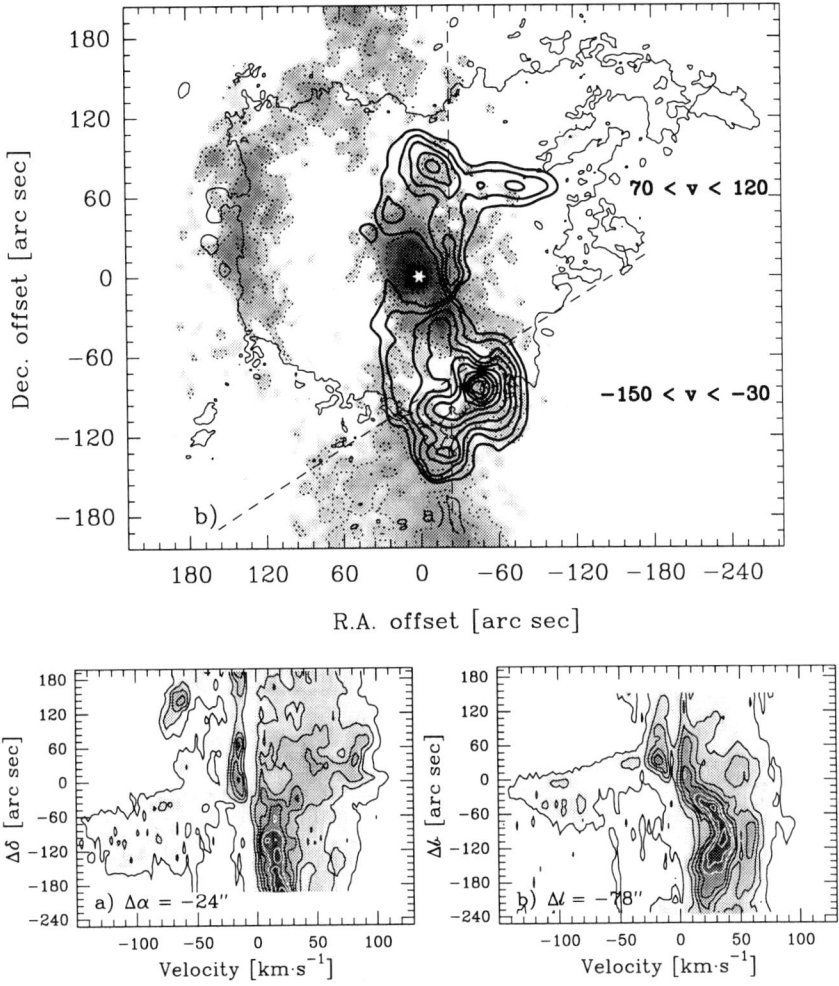

Figure 2. Upper panel: Overlay of the 1.3 mm (Mezger et al., 1989, grey scale) and 6 cm (Yusef-Zadeh, & Morris, 1987, thin contours) continuum data with ^{13}CO(2−1) emission integrated over two velocity intervals. While in the velocity range $-150 < v < -30$ the entire emission shown arises from the CND, in the velocity range $70 < v < 120\,\mathrm{km\cdot s^{-1}}$ only a small fraction of the emission can be associated with the northern lobe of the CND (at position $(\Delta\alpha, \Delta\delta) = (-12'', 84'')$). The north-south and east-west elongated clouds are affected by the explosion that created Sgr A East. Lower panels: Two position-velocity cuts through the ^{13}CO(2−1) data cube showing (left) the disruption (at $\Delta\delta \sim -90''$) of the northern part of the M−0.13−0.08 cloud due to the explosion into components down to $v \sim -15\,\mathrm{km\cdot s^{-1}}$ and up to $v \sim 80\,\mathrm{km\cdot s^{-1}}$ (at $\Delta\delta \sim 40''$) and (right) the connection in velocity-space of the gas from the northern part of M−0.13−0.08 with the gas in the CND.

components, whereas to the north-west of Sgr A East up to 6-7 components can be observed. At the south-western egde of Sgr A East, the M−0.13−0.08 cloud also shows signs of being affected by the explosion (Fig. 2). The separation into the two velocity componets described above can be interpreted as the near (at $\sim -15\,\mathrm{km\cdot s^{-1}}$) and the far (up to $\sim 90\,\mathrm{km\cdot s^{-1}}$) side of the disrupted cloud.
Zhao et al. (1995) investigated the high negative velocity gas (HNVG) observed in OH absorption at $v \sim -180\,\mathrm{km\cdot s^{-1}}$, and found it to form a coherent layer in front of Sgr A West and East. They suggested a close relation between the HNVG and the ionized gas within Sgr A West. It is not clear, however, if the HNVG has been expelled by the explosion that created Sgr A East.

3.2. Feeding the CND

The dust continuum emission observed at 1.3 mm correlates very well with the boundaries of Sgr A East, forming a shell-like structure around this source (Fig. 1). In the southern part of Sgr A East the morphology of the dust shell agrees with that of the southern lobe of the CND (see Fig. 1 and 2). In addition there appears to be a connection in velocity space between the gas in the northern part of M−0.13−0.08 and the southern lobe of the CND. The Δb-v plot at $\Delta l = -78''$, shown as a dashed line labeled b) in Fig. 2, indicates a continuous transition in velocity and velocity gradient from the gas in the northern part of M−0.13−0.08 (which seems to be disrupted by the explosion that created Sgr A East, see section 3.1) to the gas forming the southern lobe of the CND. This suggests that the CND could have been formed out of explosion debris which have been captured by by the gravitational potential of the central mass. Note that the estimates of the age of the explosion and of the CND do agree (Mezger et al., 1989). Such an origin of the CND could explain its nonequilibrium state, its "warp" and "tilt" (Güsten et. al., 1987). It could also explain the fact that its southern "lobe" is much more massive than its northen "lobe", as well as the existence of many kinematic streamers within the CND. Ho (1994) suggested another possibility of feeding the CND by an east-west streamer in the velocity range 50-70 km·s^{-1}. The ^{13}CO data presented here show that this streamer could also be associated with the front part of the Sgr A East Core disrupted by the explosion.

4. Line-of-sight distribution within the Sgr A Complex

The exact location of Sgr A East and of the molecular material along the line-of-sight as well as its association with the GC is still not clear. Pedlar et al, (1989) have shown that Sgr A East is located (at least partly) behind Sgr A West and thus behind the Galactic Centre, because the 90 cm synchrotron emission from Sgr A East is absorbed by Sgr A West. Therefore the bulk of Sgr A East Core must also be located behind the Galactic Centre, while the HNVG must lie in front of the Galactic Centre, since it is (partly) seen in absorption against Sgr A West and East (Zhao et al, 1995). In a similar way, i.e. by comparing NIR extinction with submm dust emission one can locate — at least in a rough way — the components of the Sgr A Cloud Complex along the line-of-sight (Zylka et al., 1990). A further example is the determination of the inclination of the CND relative to the line-of-sight. NIR extinction shows that the south-eastern

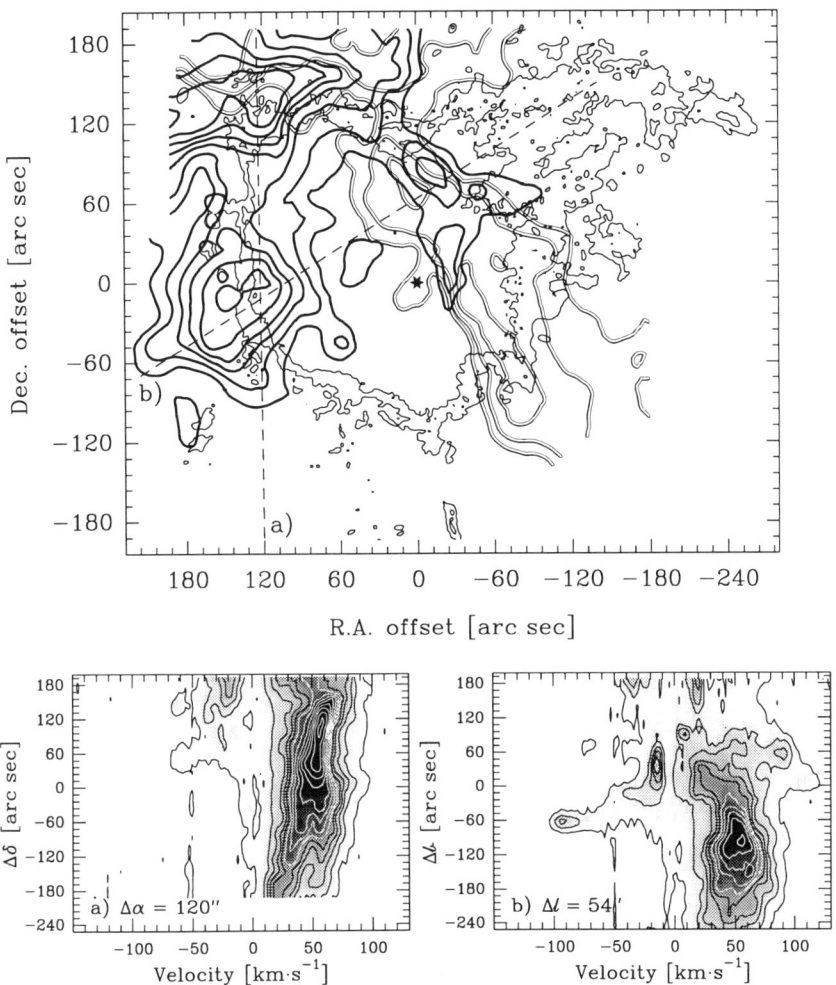

Figure 3. Upper panel: ^{13}CO(2−1) emission integrated over the velocity interval $-20 < v < -10 \, \text{km} \cdot \text{s}^{-1}$ (double-lined contours) and $60 < v < 100 \, \text{km} \cdot \text{s}^{-1}$ (thick solid contours). The solid contours alone outline the eastern part of the Sgr A East Core (i.e. M−0.02−0.07), while solid and double-lined contours outline the disrupted northern and western part of this cloud. The 6 cm continuum emission is shown for reference (thin contours). Lower panels: Two position-velocity cuts through the ^{13}CO(2−1) data cube showing (left) the transition from M−0.13−0.08 to M−0.02−0.07 along the eastern side of Sgr A East and (right) the bending of the gas from M−0.02−0.07 ($v \sim 50 \, \text{km} \cdot \text{s}^{-1}$) up to $v \sim -20 \, \text{km} \cdot \text{s}^{-1}$. This negative velocity feature represents the front side of the disrupted material. At $\Delta b > 60''$ this Δb-v plot shows the disrupted gas which spans the velocity range from ~ -50 up to $\sim 120 \, \text{km} \cdot \text{s}^{-1}$. The "bullet" at $(\Delta l, \Delta b, v) = (54'', -60'', -90 \, \text{km} \cdot \text{s}^{-1})$ is also gas affected by the explosion.

part of the CND is in front of the central star cluster, while its north-western part is behind the cluster (see Mezger, this volume). It is noteworthy that the Sgr A Cloud Complex alltogether exhibits a similar inclination angle.

Acknowledgments. I thank P.G. Mezger and K. Uchida for helpful discussion and critical reading of the manuscript

References

Bolton, J.G., van Damme, K.J., Gardner, F.F., Robinson, B.J. 1964, Nature, 201, 279

Goldstein, S.J., Gundermann, E.J., Penzias, A.A., Lilley, A.E. 1964, Nature, 203, 65

Güsten, R., Downes, D. 1981, A&A, 99, 27

Güsten, R., Genzel, R., Wright, C.M.H., Jaffe, D.T., Stutzki, J., Harris, A.I. 1987, ApJ, 318, 124

Ho, P.T.P., Ho, L.C., Szczepanski, J.C., Jackson, J.M., Armstrong, J.T., Barrett, A.H. 1991, Nature, 350, 309

Ho, P.T.P 1994, in Genzel, R., Harris, A.I. (eds), NATO ASI Series, Vol. 445, 149

Lis, D.C., Carlstrom, J.E. 1994, ApJ, 424, 189

Lindqvist, M., Sandqvist, Aa., Winnberg, A., Johansson, L.E.B., Nyman, L.-Å. 1995, A&AS, 113, 257

Mezger, P.G., Zylka, R., Salter, C.J., Wink, J.-E., Chini, R., Kreysa, E., Tuffs, R. 1989, A&A, 209, 337

Okumura, S.K., Ishiguro, M., Fomalont, E.B., Chikada, Y., Kasuga, T., Morita, K.-I., Kawabe, R., Kobayashi, H., Kanzawa, T. 1989, ApJ, 347, 240

Okumura, S.K., Ishiguro, M., Fomalont, E.B., Hasegawa, T., Kasuga, T., Morita, K.-I., Kawabe, R., Kobayashi, H. 1991, ApJ, 378, 127

Pedlar, A., Anantharamaiah, K.R., Ekers, R.D., Goss, W.M., van Gorkom, J.H., Schwarz, U.J., Jun-Hui Zhao 1989, ApJ, 342, 796

Robinson, B.J., Gardner, F.F., van Damme, K.J., Bolton, J.G., 1964, Nature, 202, 989

Sandqvist, Aa., Karlsson, R., Whiteoak, J.B., Gardner, F.F. 1987, in Backer (ed), AIP Conference Proceedings, 155, 95

Serabyn, E., Güsten, R. 1987, A&A, 184, 133

Serabyn, E., Lacy, J.H., Achtermann, J.M. 1992, ApJ, 395, 166

Tsuboi, M., Handa, T., Ukita, N. 1994, in Genzel, R., Harris, A.I. (eds), NATO ASI Series, Vol. 445, 95

Yusef-Zadeh F., Morris, M. 1987, ApJ, 320, 545

Zhao, J.-H., Goss, W.M., Ho, P.T.P. 1995, ApJ, 450, 122

Zylka, R., Mezger, P.G., Wink, J.-E. 1990, A&A, 234, 133

Zylka, R., Mezger, P.G., Wilson, T.L., Mauersberger, R. 1994, in Genzel, R., Harris, A.I. (eds), NATO ASI Series, Vol. 445, 161

Millimetre Spectral Line Observations of the Galactic Centre

Aa. Sandqvist

Stockholm Obervatory, S-133 36 Saltsjobaden, Sweden. e-mail: sandqvis@astro.su.se

Abstract. We have surveyed the Sgr A region (roughly $-12'$ to $+18'$ in l and $-12'$ to $+3'$ in b), including the Radio Arc, in the 110-GHz lines of $C^{18}O$ ($J = 1-0$) and HNCO ($J_{kk'} = 5_{05} - 4_{04}$) using the SEST. We have observed 554 positions with a grid spacing of $45''$. The observational data are presented in form of $l-b$, $l-v$ and $b-v$ maps, recently published in A&AS 113, 257. Observations of selected regions in the Galactic Centre in the 220-GHz $C^{18}O$ ($J = 2-1$) line have also begun and the Circumnuclear Disk has been mapped with the SEST in the 268-GHz HCO^+ ($J = 3-2$) line with a grid spacing of $10''$. Future mapping programs are being planned in the 119-GHz O_2 line, as well as in submillimetre lines of O_2 (487-GHz) and H_2O (557-GHz), using the 1.1-m Odin satellite telescope, which has a launch date in the fall of 1997. The Odin satellite is a Swedish project in collaboration with Canada, Finland and France.

1. The Galactic Centre and $C^{18}O$

The central regions of our Galaxy have been extensively studied at wavelengths between the near infrared and radio portions of the spectrum (see e.g. reviews by Genzel & Townes 1987, and Sandqvist & Genzel 1993). A remarkable variety of structure on scales ranging from kiloparsecs down to the subparsec level has been observed, among them some of the most mysterious and bizarre astronomical structures known. In the innermost hundred parsecs of the Galactic Centre region there is first of all the so-called 'Arc' with its long (≈ 60 pc) and narrow (≈ 1 pc), parallell filaments of synchrotron radio emission perpendicular to the Galactic plane (Yusef-Zadeh 1989). These nonthermal filaments appear to be connected to the Sgr A Complex by a filamentary thermal 'Bridge'. The Sgr A Complex consists of a nonthermal shell component, Sgr A East, and a thermal component, Sgr A West. The source Sgr A West with its 'spiral arms' consisting of infalling ionized gas (Killeen & Lo 1989), contains in its innermost regions the unique nonthermal radio source, Sgr A*, which is believed by many to be the manifestation of a black hole (albeit 'starving') in the centre of the Milky Way system (Lo 1989). All these structures together with many more, are probably parts of a complex mechanism involving magnetic fields and an understanding of these processes may improve our understanding of much more energetic phenomena occurring in the centres of active galaxies. Our Galactic Centre (GC) after all is the closest galactic nucleus!

Observations obtained during the early 70s demonstrated that molecular clouds dominate the interstellar medium in the inner 500 pc of the Galaxy and that the density of molecular clouds is far higher in this region than in any other part of the Galaxy. Although it is less than 0.2% of the Galactic disk by volume, nearly 10% of the Galactic mass is found there. Many of these molecular clouds, such as the $+20$ and $+50\,\mathrm{km\,s^{-1}}$ clouds and the Circumnuclear Disk, are intimately entwined and interacting with the continuum complex described above (Sandqvist 1989a; Zylka et al. 1990; Serabyn et al. 1994).

The common isotopic specie of carbon monoxide, $^{12}C^{16}O$ (hereafter called ^{12}CO), is the most often used tracer of molecular gas. Comparisons of spectra taken in the $J = 1 - 0$ 115-GHz transition of this specie with the $J = 1 - 0$ transition of the rarer ^{13}CO or the much rarer $^{12}C^{18}O$ (hereafter called $C^{18}O$) species demonstrate that the ^{12}CO emission line is highly saturated, having optical depths of several hundred in some directions toward the GC region. The various isotopic forms of the CO molecule are excited into emission in regions having molecular hydrogen densities $n(H_2) > 200\,\mathrm{cm^{-3}}$. Most of the gas in the GC has much higher densities than this and can be detected in molecules of high dipole moments such as NH_3 (Güsten et al. 1981), H_2CO (Sandqvist 1989a), and CS (Bally et al. 1987). Because the ^{12}CO line is so saturated, it is not a good tracer of the cloud distribution, or of the column density along the line of sight. Therefore, it is necessary to observe the molecular gas in less opaque lines to measure accurately the distribution of the column density and the kinematics of this gas. However, even the optical depth in the ^{13}CO $J = 1 - 0$ line sometimes exceeds unity in the GC clouds (Bally et al. 1988). Therefore, observations of $C^{18}O$ would yield the most reliable results as far as gas distribution and kinematics go, since the ratio $[^{16}O]/[^{18}O]$ is of the order of $175 - 400$ (Wannier 1989). Still this isotopic specie is as easily excited as $C^{16}O$ and the GC clouds are so dense that the lines of $J = 1 - 0$ and $2 - 1$ $C^{18}O$ are easily detected (Bally et al. 1987; Sandqvist 1989b; Lindqvist et al. 1995).

2. SEST Observations in the $C^{18}O$ ($J = 1 - 0$) and HNCO ($J_{kk'} = 5_{05} - 4_{04}$) Lines

We have had seven observing sessions between 1990 and 1993 using the 15-m Swedish-ESO Submillimetre Telescope (SEST) and the 115-GHz receiver for this project. A total number of 554 positions have been observed with a spacing of $45''$ in the $C^{18}O$ ($J = 1 - 0$) and HNCO ($J_{kk'} = 5_{05} - 4_{04}$) lines. The data cover most of the so-called Sgr A Complex. The extent of the survey is not rectangular but covers roughly the region $-12' \leq l \leq +18'$ and $-12' \leq b \leq +3'$. The average 1σ rms of the survey is $\approx 0.02 - 0.03$ K. An observational paper has recently been published in *Astronomy and Astrophysics Supplement Series* (Lindqvist et al. 1995). It presents velocity-integrated maps using a velocity interval of $5\,\mathrm{km\,s^{-1}}$. The $C^{18}O$ maps (71 in total) cover the velocity range $V_{LSR} = -165\,\mathrm{km\,s^{-1}}$ to $V_{LSR} = +190\,\mathrm{km\,s^{-1}}$. The HNCO maps (35 in total) cover the velocity range $V_{LSR} = -60\,\mathrm{km\,s^{-1}}$ to $V_{LSR} = 115\,\mathrm{km\,s^{-1}}$. Longitude-velocity and latitude-velocity maps of $C^{18}O$ and HNCO are also presented in that paper.

The difference in distribution between the two molecules is striking and must be due to the different excitation conditions. The great adantage with the HNCO molecule is that it is easily excited (rotationally) by IR radiation (Churchwell et al. 1986). This enables us to locate regions of enhanced IR radiation which can be compared with direct IR observations (Odenwald & Fazio 1984).

A summary of the observational parameters for the SEST observations is presented in Table 1.

Table 1. SEST Observations of the Galactic Centre Region.

Frequency (GHz)	Molecule	Transition J	Beamwidth FWHM	Grid Spacing	Number of Positions	η_{mb}
109.8	$C^{18}O$	$1-0$	$46''$	$45''$	554	0.72
109.9	HNCO	$5_{05} - 4_{04}$	$46''$	$45''$	554	0.72
219.6	$C^{18}O$	$2-1$	$24''$	$20''$	28	0.55
267.6	HCO^+	$3-2$	$20''$	$10''$	75	0.45

3. Observations of the $J = 2 - 1$ $C^{18}O$ Line in the Immediate Environment of Sgr A

The largest molecular clouds near the Sgr A Complex are the so-called +20 and +50 km s^{-1} clouds which appear to be condensations in a continuous molecular belt lying more or less parallell to the Galactic plane (see figure 4 - from Sandqvist 1989a). The outer edges of the radio continuum shell Sgr A East fit well against the edges of the +20 and +50 km s^{-1} clouds, suggesting physical contact and interaction between the continuum and molecular components of the complex. One important question is whether these molecular clouds can be considered to be reservoirs feeding molecular gas into the Circumnuclear Disk which is rotating about Sgr A West with velocities of the order of 100 km s^{-1}. Our previous $J = 1 - 0$ $C^{18}O$ observations had insufficient resolution to clearly resolve the Circumnuclear Disk and we therefore need the $J = 2 - 1$ $C^{18}O$ line to address this question.

The value of $C^{18}O$ as a probe of the molecular gas has been illustrated by Genzel et al. (1990) who detected blue-shifted, high-velocity gas towards Sgr A East. This gave for the first time direct kinematic evidence for a recent high energy explosion at the centre. Zylka et al. (1995) have also shown some evidence for a continuous connection between the Circumnuclear Disc and the +20 km s^{-1} cloud. Some preliminary $J = 2 - 1$ $C^{18}O$ line observations, performed with SEST a few years ago, also illustrate the promise of this isotopic line. A two-directional scan was performed across the Sgr A region, starting at the +20 km s^{-1} cloud, going north in Declination (DEC) across Sgr A* and the Circumnuclear Disk (CND), and then east in Right Ascension (RA) to the +50 km s^{-1} cloud. Figure 1 shows a profile map of the observations, and also includes a small map of the Circumnuclear Disk. The resulting two-directional position-velocity diagram is seen in the lower part of Figure 1. The signatures of the Circumnuclear Disk as well as the +20 and +50 km s^{-1} clouds stand out markedly. In addition, there is a growing negative-velocity wing clearly discernable in going from the +20 km s^{-1} to the +50 km s^{-1} cloud across the face of the Sgr A East continuum shell. We obviously need to perform a complete mapping

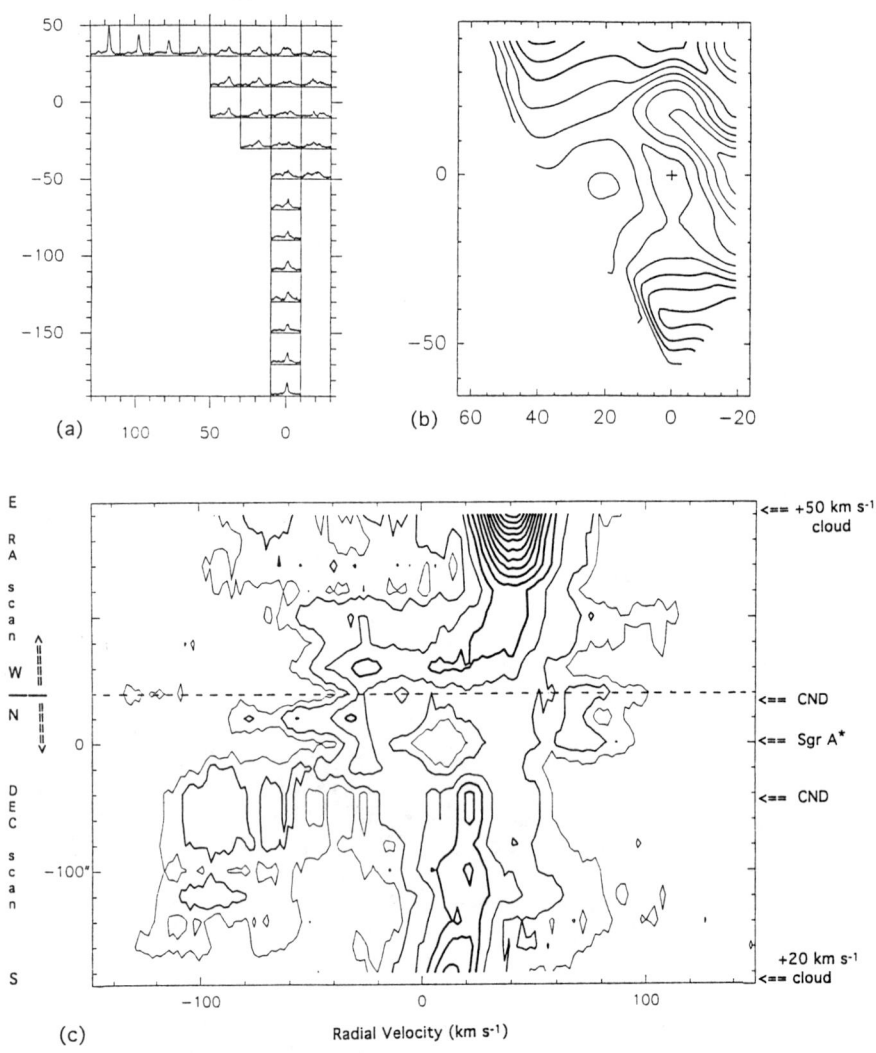

Figure 1. $J = 2-1$ $C^{18}O$ line observations towards Sgr A; the equatorial offsets, in arcseconds, are centered on Sgr A*. (a) Profile map, the velocity range is -150 to $+150$ km s^{-1}, the main beam brightness temperature ($T_{\rm mb}$) range is -0.2 to $+2.5$ K. (b) The total integrated line $T_{\rm mb}$-intensity (-150 to $+150$ km s^{-1}) in the CND; the lowest contour and the contour interval are 9 K km s^{-1}. (c) The two-directional $T_{\rm mb}$ position-velocity diagram; the lowest contour (thin line) value is 0.09 K, the next contour value and the contour interval thereafter are 0.18 K.

of the Sgr A region in the $J = 2 - 1$ $C^{18}O$ line to properly interpret these preliminary, enticing results.

Another important question is that of the excitation mechanism of the CO molecule in these clouds. There is considerable evidence that the interstellar dust is colder than the gas in the GC region (Güsten et al., 1981). Due to the huge potential well in the GC there could be tidal forces that disrupt the molecular clouds, leading to turbulent shocks which will heat the medium. There is observational evidence that the $+20$ km s^{-1} cloud is a debris of a tidal disruption which is falling towards the GC, dissipating a lot of gravitational energy and therefore heating the medium (e.g. Okumura et al., 1991). We hope that the future planned observations will yield indications of varying temperature in the immediate neighbourhood of Sgr A using the ratio between the $J = 1 - 0$ and $J = 2 - 1$ $C^{18}O$ line intensities after convolution of the $J = 2 - 1$ data to the resolution of the $J = 1 - 0$ data.

Several of the dynamical phenomena described above give rise to broad and weak line wings, e.g. explosions, turbulence, collisions, and tidal disruptions. It is therefore important that these $J = 2 - 1$ $C^{18}O$ line observations be sufficiently sensitive and stable to give line profiles with high S/N and linear baselines, where weak line wings can be well identified.

4. SEST Mapping of the Circumnuclear Disk in the 268-GHz HCO$^+$ ($J = 3 - 2$) Line

The advantage of using the 268-GHz HCO$^+$ ($J = 3 - 2$) emission line to probe the structure of the CND was demonstrated by Sandqvist, Wootten and Loren (1985) when the CND's "missing" southeastern part was detected and mapped using the NRAO 12-m telescope. Due to its formation process, HCO$^+$ can be used as a tracer of enhanced cosmic ray ionization and shock regions. A new map of the CND has now been made using the SEST, which has an angular resolution of 20″ at this frequency (as compared with the 28″-resolution of the 12-m telescope). The SEST map of the HCO$^+$ ($J = 3 - 2$) total integrated line T_{mb}-intensity (-125 to $+125$ km s^{-1}), with a grid spacing of 10″, is shown in Figure 2.

The general shape of the CND is clearly revealed in the HCO$^+$ ($J = 3 - 2$) line, although the HCO$^+$-distribution is very clumpy throughout the CND, with a significant concentration in the southwestern part. Just to the west of Sgr A*, there seems to be a streamer component in towards Sgr A* from the CND. This streamer stands out clearer in velocity maps around $+30$ to $+50$ km^{-1} and may have some relation to the Sgr A* OH streamer found by Sandqvist et al. (1989). A corresponding dust streamer may possibly be seen in the 800 μm JCMT maps of Zylka, Ward-Thompson & Mezger. Further investigation of this matter is taking place.

5. The Odin Satellite

The Odin satellite is a combined aeronomy and astronomy mission for millimetre and submillimetre spectral line observations of the Earth's upper atmosphere and the interstellar medium. The aeronomical scientific motivation is to study

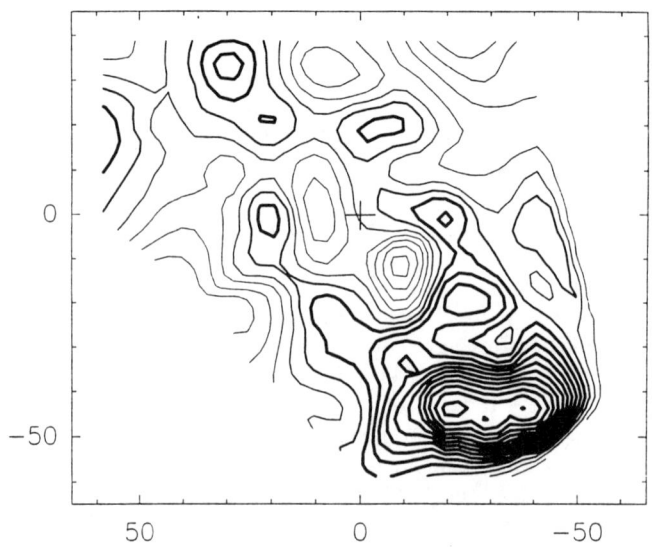

Figure 2. SEST map of the HCO$^+$ ($J = 3 - 2$) total integrated line T_{mb}-intensity (−125 to +125 km s^{-1}) in the CND; the lowest contour level is 22 and the contour interval is 11 K km s^{-1}, the thickest contour lines begin at a value of 89 K km s^{-1}. The equatorial offsets from Sgr A* (cross) are in arcseconds.

ozone-related chemistry and its connection with the destruction of ozone in the polar regions of the stratosphere. The astronomical scientific motivation is mainly the detection and mapping of molecular oxygen and water vapour in different regions of the Galaxy. These molecules, if abundant, are important coolants. Other molecules will also be observed, and objects in the solar system, as well as relatively nearby galaxies, will be probed.

Odin has a 1.1 meter reflector, which results in angular resolutions of 2 arcmin at submillimetre and 9 arcmin at millimetre wavelenghts. Five receivers are planned, tuned to 119, 494, 549, 555 and 571 GHz. The four submillimetre receivers are tunable over a 17-GHz band, centred on the above frequencies. There will be two 1000-channel hybrid auto-correlators with channel resolutions of 1, 0.5, 0.25 and 0.15 MHz, as well as an acousto-optical spectrometer with a channel resolution of 1 MHz. This permits simultaneous observations with three receivers. The important species to be observed for astronomy are O$_2$, H$_2$O, CO (5–4), NH$_3$, H$_2^{18}$O, ^{13}CO (5–4) and C I, while for aeronomy O$_3$, NO, NO$_2$ and ClO are important. In addition to the above-mentioned receivers, Odin will also carry a UV/optical slit spectrograph and IR imager (OSIRIS) for studies of aerosols and additional molecules by the aeronomers. A schematic diagram of the satellite and receiver system is shown in Figure 3.

The exact Odin astronomical observation program is of course strongly dependent on the detectability of the O$_2$ and H$_2$O lines. However, regions in the Galactic Centre, especially near the Sgr A Complex with its shock regions and high-temperature molecular clouds, are strong contenders for observing time. In

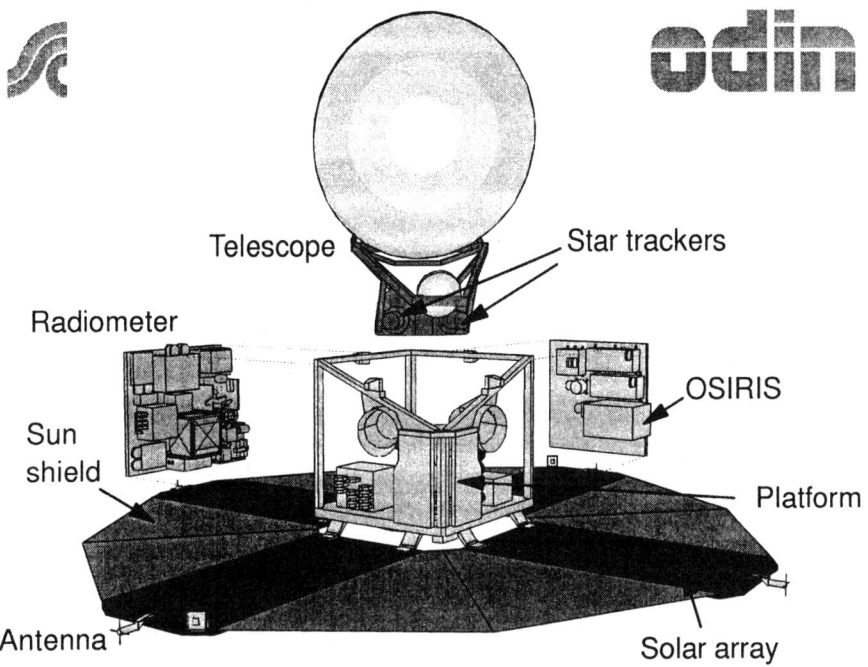

Block diagram of radiometer

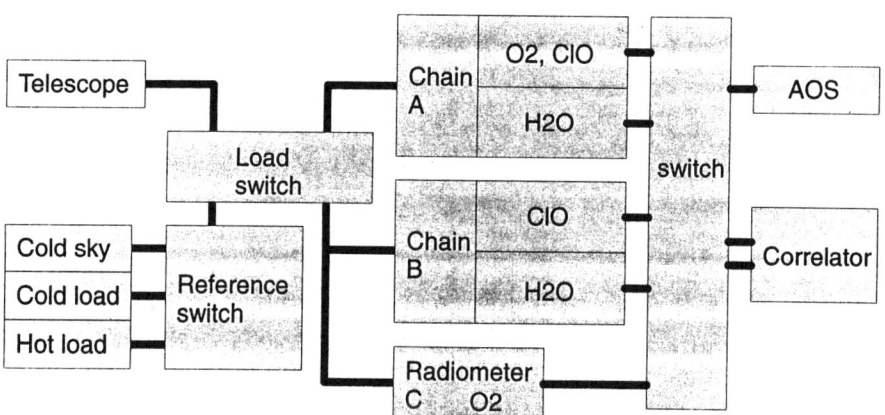

Figure 3. Schematic drawings of the components of the Odin satellite for millimetre and submillimetre spectral line observations of the Earth's atmosphere and the interstellar medium. The diameter of the telescope reflector is 1.1 m.

addition, spectral scan observations, utilizing the tunability of the submillimetre receivers, are being planned for the giant molecular complex in Sgr B2.

Odin will be launched in the fall of 1997 and has an expected lifetime of 2 years. It is a Swedish project in collaboration with Canada, Finland and France.

References

Bally, J., Stark, A.A., Wilson, R. W., & Henkel, C. 1987, ApJS, 65, 13
Bally, J., Stark, A.A., Wilson, R.W., & Henkel, C. 1988, ApJ, 324, 223
Churchwell, E., Wood, D., & Myers, R.V. 1986, ApJ, 305, 405
Genzel, R., Stacey, G.J., Harris, A.I., Townes, C.H., Geis, N., Graf, U.U., Poglitsch, A., & Stutzki, J. 1990, ApJ, 356, 160
Genzel, R., & Townes, C.H. 1987, ARA&A, 25, 377
Güsten, R., Walmsley, C.M., & Pauls, T. 1981, A&A, 103, 197
Killeen, N.E.B., & Lo, K.Y. 1989, in The Center of the Galaxy, M. Morris, Dordrecht: Kluwer, 453
Lindqvist, M., Sandqvist, Aa., Winnberg, A., Johansson, L.E.B., & Nyman, L.-A. 1995, A&AS, 113, 257
Lo, K. Y. 1989, in The Center of the Galaxy, M. Morris, Dordrecht: Kluwer, 527
Odenwald, S.F., & Fazio G.G. 1984, ApJ, 283, 601
Okumura, S.K., Ishiguro, M., Fomalont, E.B., Hasegawa, T., Kasuga, T., Morita, K.-I., Kawabe, R., & Kobayashi, H. 1991, ApJ, 378, 127
Sandqvist, Aa. 1989a, A&A, 223, 293
Sandqvist, Aa. 1989b, The Messenger 57, 15
Sandqvist, Aa., & Genzel, R. 1993, in Central Activity in Galaxies, Aa. Sandqvist & T.P. Ray, Springer, 1
Sandqvist, Aa., Karlsson, R., & Whiteoak, J.B. 1989, in The Center of the Galaxy, M. Morris, Dordrecht: Kluwer, 421
Sandqvist, Aa., Wootten, A., & Loren, R.B. 1985, A&A, 152, L25
Serabyn, E., Keene, J., Lis, D.C., & Phillips, T.G. 1994, ApJ, 424, L95
Townes, C. H. 1989, in The Center of the Galaxy, M. Morris, Dordrecht: Kluwer, 1
Wannier, P. G. 1989, in The Center of the Galaxy, M. Morris, Dordrecht: Kluwer, 107
Yusef-Zadeh, F. 1986, PhD Thesis, Columbia University
Yusef-Zadeh, F. 1989, in The Center of the Galaxy, M. Morris, Dordrecht: Kluwer, 243
Zylka, R., Mezger, P.G., Wilson, T.L., & Mauersberger, R. 1995, in Nuclei of Normal Galaxies: Lessons from the Galactic Centre, R. Genzel, & A. Harris, Dordrecht: Kluwer, (in press)
Zylka, R., Mezger, P.G., & Wink, J.E., 1990, A&A, 234, 133

Detection of OH (1720 MHz) Masers in Sgr A East

A. J. Green

School of Physics, University of Sydney, Sydney, NSW 2006, Australia

D. A. Frail and W. M. Goss

NRAO, P.O. Box O, Socorro, NM 87801, USA

F. Yusef-Zadeh

Northwestern University, Dept of Physics and Astronomy, Evanston, IL 60208, USA

D. A. Roberts

NCSA, 405 N. Mathews Ave., Urbana, IL 61801, USA

Abstract.
Australia Telescope Compact Array (ATCA) and Very Large Array (VLA) observations of the Sgr A region have been made in the satellite line of the OH molecule at 1720 MHz. Narrow line emission, which we believe to be caused by maser action, has been detected at six positions in the non-thermal source Sgr A East, all close to the interface of Sgr A East with the molecular cloud M−0.02−0.07. The velocities of all the proposed maser features are similar to the systemic velocity of +50 km s^{-1} of this giant molecular cloud. Significant circular polarization is observed from seven spectral components; if this is interpreted as Zeeman splitting, there is good evidence for the existence of strong magnetic fields, ≤3.5 mG. We propose that these masers are produced by shock excitation of molecular gas, similar to those which have been recently detected in Galactic supernova remnants (Frail, Goss & Slysh 1994, Frail et al. 1996).

1. Introduction

OH maser emission has long been regarded as a probe of dynamical and physical conditions in molecular gas. However, it has been traditionally used as a diagnostic for HII regions and evolved stars (Reid & Moran 1988, Cohen 1989). Recent research by Frail, Goss & Slysh (1994) has revealed that maser emission in the OH satellite line at 1720 MHz (generally weaker or absent for the traditional sources) can be an effective tracer of shock interaction with molecular clouds, particularly for supernova remnants (SNRs).

The central square degree of the Galactic center is a complex region of co-existing thermal and non-thermal gas. 1720 MHz OH maser emission has been used as a probe to investigate the shocked molecular gas and the magnetic fields in the Sgr A region (Yusef-Zadeh et al. 1996). It has been anticipated that strong magnetic fields exist near the Galactic center (Genzel, Hollenbach & Townes 1994, Morris 1994), and there are a few measurements of the Zeeman effect (Killeen, Lo & Crutcher 1992, Plante, Lo & Crutcher 1995, Marshall, Lasenby & Yusef-Zadeh 1995). However, it has been difficult to find direct evidence for shock interactions in the molecular clouds (Burton & Allen 1992; Pak et al. 1996).

The direct measurement of the shocked gas and magnetic field has been difficult for several reasons. The linewidths of molecules observed near the Galactic center are typically broad (Bally et al. 1988), and there are numerous unrelated velocity features along the line of sight to confuse the picture. Furthermore, the high density of H_2 and an intense UV radiation field result in a photodissociation region overwhelming the shocked infrared lines (Sternberg & Dalgarno 1989, Pak, Jaffe & Keller 1996).

Elitzur (1976) proposed that the satellite line of the OH molecule at 1720 MHz is preferentially populated by strong collisional pumping from H_2 molecules which have been excited by a shock front. The resulting maser lines are bright and narrow and easily identified as tracers of the interaction of shocks with adjacent molecular gas clouds. Zeeman splitting can also be readily measured to give an estimate of the magnetic field. Elitzur suggested that the conditions to produce these masers be restricted to gas densities of $10^3 - 10^5$ cm^{-3} and kinetic temperatures in the range 25−200 K. In this paper we present evidence for masers delineating the interaction of the non-thermal source Sgr A East and the molecular cloud M−0.02−0.07. An accompanying paper by Yusef-Zadeh et al. describes similar maser detections for the circumnuclear disk (CND) and the SNR G359.1−0.5. A description of these results is also given by Yusef-Zadeh et al. (1996).

2. Observations

The observations presented here were made in three separate runs with the VLA of the National Radio Astronomy Observatory[1] and the ATCA[2]. The first measurements with the VLA were made in February 1995 using the compact DnC configuration of the telescope. Details are given in Yusef-Zadeh et al. (1995). In summary, right and left circular polarization was measured over 128 channels with a total bandwidth in two IF pairs of 432 km s^{-1}. After Hanning smoothing the velocity resolution of the data was 4.25 km s^{-1} and the synthesized beam was $\sim 38''$. The ATCA observations were carried out in August 1995 using a 6 km baseline configuration, as outlined in Frail et al. (1996). Orthogonal components of linear polarization were recorded for the velocity range −430 km s^{-1} to +230 km s^{-1} over 1024 channels, giving a

---

[1] The National Radio Astronomy Observatory is a facility of the National Science Foundation, operated under a cooperative agreement by Associated Universities, Inc.

[2] The Australia Telescope National Facility is operated in association with the Division of Radioastrophysics by CSIRO

velocity resolution of 1.36 km s^{-1} and a synthesized beam of $\sim 8''$ after Hanning smoothing.

Confirmation of the preliminary detections of masers features was carried out early in 1996 with the VLA in the CnB and C configurations (see Frail et al. 1996). Again, left and right circular polarization was measured over 128 channels with a total bandwidth of 34 km s^{-1}. After Hanning smoothing of the data, a velocity resolution of 0.27 km s^{-1} was achieved. The synthesized beam is $\sim 15''$ and the rms noise in the spectra is about 15 mJy beam^{-1}.

3. Results

The locations of the six maser candidates are shown in Figure 1, superimposed on the 5 GHz continuum image of the Sgr A region from Pedlar et al. (1989). The strongest continuum emission is from Sgr A*, the compact non-thermal radio source at the Galactic center. The spiral-shaped thermal feature Sgr A West is also shown, surrounded by the non-thermal partial shell source Sgr A East. All the 1720 MHz OH detections are distributed along the rim of Sgr A East and we propose that they are physically associated with the source.

Spectra of Stokes I ($=[RCP + LCP]/2$) and V ($=[RCP - LCP]/2$) were produced for each of the maser candidates, where RCP and LCP are the measured hands of right and left circular polarization, respectively. Figure 2 shows these spectra, while Table 1 contains the parameters calculated from Gaussian fitting to the line profiles of the Sgr A East masers. The velocities of the line peaks are confined to a narrow range, even though the maser positions are distributed widely around the boundary of the source. All the spectra show narrow emission lines with widths <2 km s^{-1}, significantly less than the 10–50 km s^{-1} linewidths of CO and CS molecular clouds found in the Galactic center region (Bally et al. 1988). Previous OH measurements were seen as absorption features (Killeen et al. 1992). This evidence strongly suggests that the sources are masers, however, not from compact HII regions since the characteristic tracers (the OH transitions at 1665 and 1667 MHz) were not detected above 20 mJy (1σ level).

If the Stokes V profiles are due to the Zeeman effect then an estimate of the local line-of-sight magnetic field, B_{los}, can be made from the scaling of the derivative of the Stokes I profile ($dI/d\nu$) to fit the Stokes V spectrum. From Roberts et al. (1993), the magnetic field is derived numerically using $V = C\ dI/d\nu$ where $C = 0.6536\ B_{los}\ Hz\ \mu G^{-1}$. The results of this analysis (shown as dashed lines in the Stokes V profiles in Figure 2) are likely to be an overestimation of the actual B_{los}, following recent work on maser polarization by Elitzur (1996a). A further refinement of this procedure may be necessary since the asymmetry in the V profiles has been recently proposed as as indication of maser saturation (Elitzur 1996b). Higher angular resolution studies should clarify the situation.

4. Discussion

The non-thermal shell source, Sgr A East, is thought to be an SNR located just behind the Galactic Center (Yusef-Zadeh and Morris 1987, Pedlar et al.

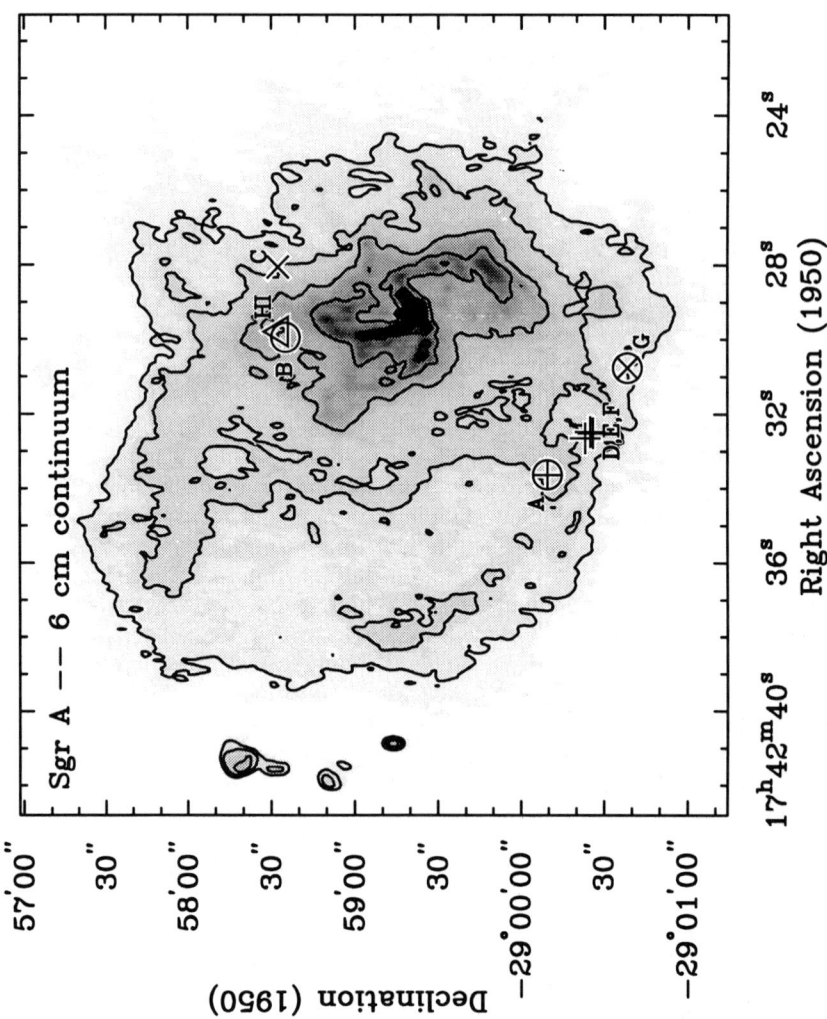

Figure 1. A continuum image of the Sgr A region at 5 GHz from Pedlar et al. (1989) with a spatial resolution of $3.5'' \times 1.8''$ showing the location of the OH(1720 MHz) maser features. The masers are named as listed in Table 1, following the terminology of the results presented in Yusef-Zadeh et al. (1996). The position of the HI Zeeman splitting measurements at $+130$ km s^{-1} by Plante et al. (1995) is also shown (triangle).

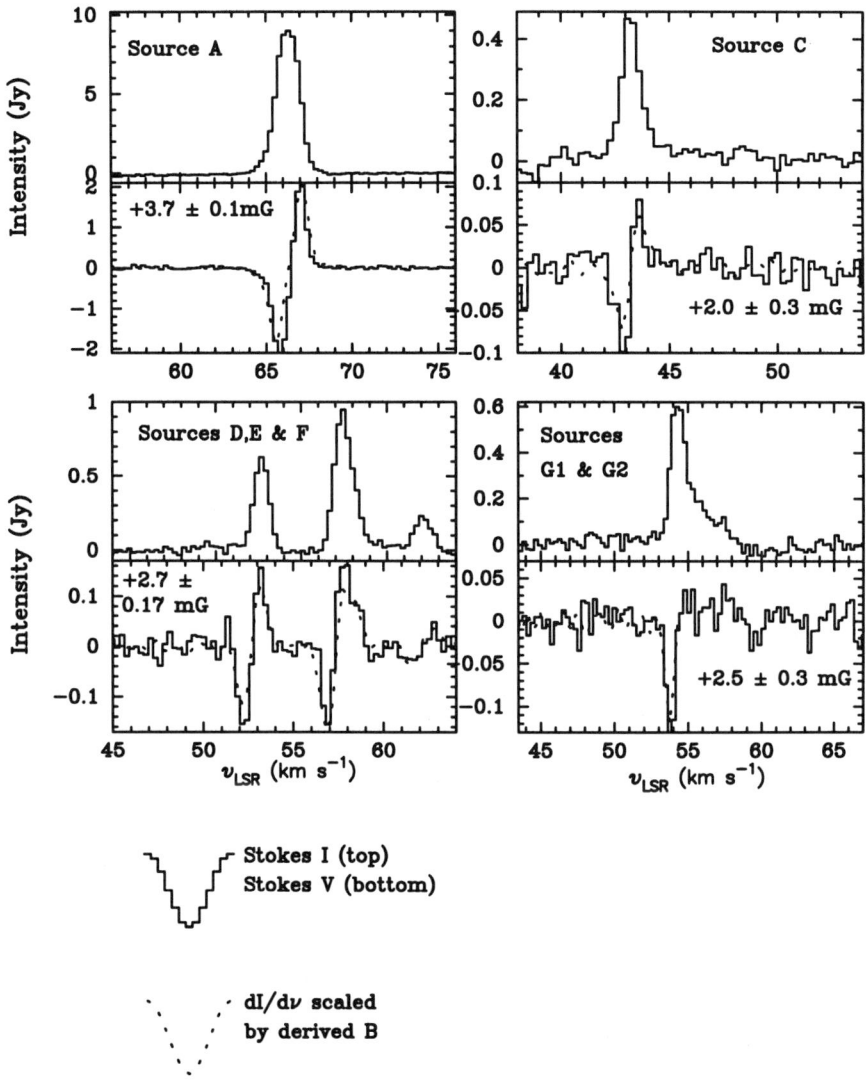

Figure 2. Profiles of Stokes I (top histogram) and V (bottom histogram) in the OH(1720 MHz) line for the maser detections in Sgr A East. The dotted line superimposed on the V spectrum is the derivative of the I spectrum scaled by magnetic fields given in Table 1. For spectra containing more than one feature, the I spectrum is decomposed into several Gaussian components and independent fields were fitted to each component.

Table 1. Gaussian Fits for OH(1720 MHz) Maser Features

Maser Designation	α_{1950} (h m s)	δ_{1950} (o ′ ″)	S_p (Jy)	V_{LSR} (km s^{-1})	ΔV (km s^{-1})	B_{los} (mG)
Sgr A OH1720:A (66)	17 42 33.63	−29 00 09.6	9.55	+66.3	1.57	+3.7±0.1
Sgr A OH1720:C (43)	17 42 28.07	−28 58 32.3	0.46	+43.3	1.1	+2.0±0.3
Sgr A OH1720:D (53)	17 42 32.51	−29 00 26.1	0.63	+52.7	1.1	+2.7±0.2^a
Sgr A OH1720:E (56)	17 42 32.65	−29 00 23.3	0.91	+57.5	1.3	+2.7±0.2^a
Sgr A OH1720:F (62)	17 42 32.48	−29 00 25.1	0.22	+62.1	1.1	+2.7±0.2^a
Sgr A OH1720:G1 (54)	17 42 30.77	−29 00 38.5	0.52	+54.3	1.0	+2.5±0.3
Sgr A OH1720:G2 (56)			0.18	+55.5	3.2	(1.0)^b

^{a}Three components fitted simultaneously.
^{b}3 σ upper limit.

1989). However, it would have had to be an abnormally energetic explosion ($\geq 10^{52}$ erg) to produce the observed shell. One possible explanation is that the SNR exploded into a wind bubble created by the progenitor star (Mezger et al. 1989). Another model invokes some other kind of violent event, such as the tidal disruption of a star by a massive black hole (Khokhlov & Melia 1996). The 1720 MHz masers which are located on the south-eastern interface of Sgr A East with the molecular cloud M−0.02−0.07 have radial velocities between +52.7 and +66.3 km s^{-1} (see Figure 1). These velocities are consistent with the kinematics of the molecular gas in the region as seen in HI absorption and CS emission (Serabyn, Lacy & Achtermann 1992, Lasenby, Yusef-Zadeh & Lasenby 1989). The maser on the NW side of Sgr A East has a comparable velocity of V_{LSR}=+43 km s^{-1} and a similar magnetic field strength. The orientation of the magnetic field is positive for all of these maser features. Such similarity in physical properties strongly suggests that the masers are all produced as the expanding shocks from the Sgr A East shell collide with the surrounding molecular cloud, whose systemic velocity is +50 km s^{-1}.

This giant cloud (M−0.02−0.07) has been previously interpreted as interacting principally with the eastern part of Sgr A East (Zylka, Mezger & Wink 1990, Lasenby et al. 1989, Mezger et al. 1989). The maser in the NW sector of the source, with a velocity of +43 km s^{-1} is significant in that it confirms that molecular gas surrounds the western part of Sgr A East. In this picture the radial velocities of the masers to the SE and NW of Sgr A East are redshifted and blueshifted, respectively, with respect to the systemic velocity of +50 km s^{-1} of the M−0.02−0.07 molecular cloud. The mechanism for the interaction process may be similar to the SNR W28, where a shock is driven into the surrounding molecular cloud, compressing the gas and stimulating the OH maser emission (Frail et al. 1994). Sheet-like geometry results and it is thought that a C-type shock excites the H$_2$ molecules (with density $10^3 - 10^5$ cm^{-3} and kinetic temperature 25−200 K), which collisionally pump the 1720 MHz transition of the OH molecule (Elitzur, Hollenbach & McKee 1989). The substantial magnetic field strengths, ≤ 3.5 mG, indicate that the gas and the field lines have been compressed by the shock.

field strengths, ≤ 3.5 mG, indicate that the gas and the field lines have been compressed by the shock.

5. Summary

We have detected OH (1720 MHz) masers near the boundary of the shell source Sgr A East. In a companion paper Yusef-Zadeh et al. report the detection of masers, with kinematic velocities of +134 km s^{-1}, located close to the rotating disk of gas known as the circumnuclear disk (CND). HI measurements in the same region show similar velocities and Zeeman splitting. It is clear from their kinematic, morphological and magnetic field characteristics that these two groups of masers are physically associated with Sgr A East and the CND. The conditions required for this maser action to occur place tight constraints on the environment existing at these locations. Evidence is presented for molecular gas surrounding the western half of the Sgr A East shell. Magnetic fields of ≤ 3.5 mG are inferred from the Zeeman splitting seen in the Stokes V profiles. The detection of these masers demonstrate directly for the first time the presence of shocked gas at the Galactic center, located in the M$-0.02-0.07$ giant molecular cloud. Future high resolution studies will clarify the question of the magnetic field strengths. If proper motion of the maser spots can be detected, then an accurate picture of the dynamics of the gas clouds at the Galactic center will emerge.

Acknowledgments

We thank M. Elitzur for useful discussions. F. Yusef-Zadeh's work was supported in part by NASA grant NAGW-2518. D. Roberts acknowledges support from the NSF grant AST94-19227.

References

Bally, J., Stark, A. A., Wilson, R. W. & Henkel, C. 1988, ApJ, 324, 223
Burton, M. & Allen, D. A. 1992, Proc. Astr. Soc. Australia, 10, 55
Cohen, R. J. 1989, Rep. Prog. Phys., 52,881
Elitzur, M. 1976, ApJ, 203, 124
Elitzur, M. 1996a, ApJ, 457, 415
Elitzur, M. 1996b, ApJ, submitted
Elitzur, M., Hollenbach, D. & McKee, C. F. 1989, ApJ, 346, 983
Frail, D. A., Goss, W. M. & Slysh, V. I. 1994, ApJ, 424, L111
Frail, D. A., Goss, W. M., Reynoso, E. M., Giacani, E. B., Green, A. J. & Otrupcek, R. 1996, AJ, 111, 1651
Genzel, R., Hollenbach, D. & Townes, C. H. 1994, Progress in Physics, 57, 41
Khokhlov, A. & Melia, F. 1996, ApJ, 457, L61.
Killeen, N. E. B., Lo, K. Y. & Crutcher, R. 1992, ApJ, 385, 585
Lasenby, J., Yusef-Zadeh, F., & Lasenby, A. N. 1989, in I.A.U. Symposium No. 136, The Centre of the Galaxy, ed. M. Morris (Dordrecht: Kluwer), 365
Marshall, J., Lasenby, A. & Yusef-Zadeh, F. 1995, MNRAS, 274, 519
Mezger, P. G., Zylka, R., Salter, C. J., Wink, J. E., Chini, R., Kreysa, E. & Tuffs, R. 1989, A&A, 209, 337
Morris, M. 1994, in Nuclei of Normal Galaxies: Lessons from the Galactic Centre, eds: R. Genzel and A. I. Harris, (Kluwer), 185
Pak, S., Jaffe, D. T. & Keller, L. D. 1996, ApJ, 457, L43
Plante, R. L., Lo, K. Y., & Crutcher, R. M. 1995, ApJ, 445, L113
Pedlar, A., Anantharamaiah, K. R., Ekers, R. D., Goss, W. M., & van Gorkom J. H. 1989, ApJ, 342, 769
Reid, M. J. & Moran, J. M. 1988, in Galactic and Extragalactic Radio Astronomy, ed. G. L. Verschuur & K. I. Kellermann (Springer, NY), 255
Roberts, D. A., Crutcher, R. M., Troland, T. H. & Goss, W. M. 1993, ApJ, 412, 675
Serabyn, E., Lacy, J. H., & Achtermann, J. M. 1992, ApJ, 395, 166
Sternberg A. & Dalgarno, A. 1989, ApJ, 338, 197
Yusef-Zadeh, F. & Morris, M. 1987, ApJ, 322, 721
Yusef-Zadeh, F., Roberts, D. A., Goss, W. M., Frail, D. A. & Green, A. J. 1996 Ap. J. Lett., accepted for publication
Yusef-Zadeh, F., Uchida, K. & Roberts, D. A. 1995, Science, 270, 1801
Zylka, R., Mezger, P. G., & Wink, J.E. 1990, A&A, 234, 133

The Galactic Center
ASP Conference Series, Vol. 102, 1996
Roland Gredel, ed.

Physical Condition of UV Heated Warm Gas around the Galactic Center

Yukari Y. Yui[1]

Comminications Research Laboratory, Nukuikitamachi 4-2-1, Koganei, 184 Japan

Takao Nakagawa, Haruyuki Okuda[2], Hiroshi Shibai[2]

The Institute of Space and Astronautical Science, Yoshinodai 3-1-1, Sagamihara, 229 Japan

Norihisa Hiromoto, Yasuo Doi[1]

Comminications Research Laboratory, Nukuikitamachi 4-2-1, Koganei, 184 Japan

Abstract. We made large-scale [CII] 158 μm line survey observations with a balloon-borne telescope incorporated with a stressed Ge:Ga detector developed at CRL/ISAS. The system NEP during the observation was 6×10^{-16} W/Hz$^{-1/2}$, and the detection limit of the final smoothed map is 1.5×10^{-5} ergs s^{-1} cm^{-2} sr^{-1} (3σ). The spatial resolution is 15'. We observed the Galactic Plane (including the G.C.) and the ρ Oph cloud, and compared the results. The observed [CII] line emission is relatively weak toward the Galactic Center, and the physical condition is similar to that of the ρ Oph cloud core region. We show a quantitative estimate of physical conditions in the Galactic Center as an ensemble of the ρ Oph core region.

1. Introduction

The far-infrared [CII] fine structure 158μmline is the dominant coolant of neutral interstellar gas (Tielens and Hollenbach 1985), and one of the brightest emission lines id the Galaxy (Shibai et al. 1991; Wright et al. 1991). Hence the line is generally expected to be a powerful probe for studying physical conditions in neutral clouds in the Galaxy. Moreover, since the ionization potential of neutral carbon (11.3 eV) is lower than that of neutral hydrogen (13.6 eV), even B-type stars, which cannot form extensive HII regions, can form C$^+$ regions around them. Thus, [CII] emission is expected to trace the distribution of B-type stars as well as that of O-type stars.

---

[1]The Institute of Space and Astronautical Science, Yoshinodai 3-1-1, Sagamihara, 229 Japan

[2]Comminications Research Laboratory, Nukuikitamachi 4-2-1, Koganei, 184 Japan

There are thermal componet in the Galactic Center (Reich et al. 1987). If it is due to photoionization, $\sim 10^{52}$ Lyman continuum photons are required (Güsten 1989). Sources for such a flux have not yet been identified, however, and still controversial. The most possible explanation for the thermal flux is photoionization by hot, young stars. From far-infrared survey of the Galactic Center region, Odenwald and Fazio (1984) calculated cluster models which consist of O and B-type stars. They concluded that (3-5) $\times 10^4$ B0 stars embedded in an interstellar medium can explain the the thermal emission from the Galactic Center, and that there are fewer early O-type stars relative to B-type stars than are predicted by the Salpeter ILF.

2. Observations and Results

We made large-scale [CII] 158 μm line survey observations (Yui et al. 1993; Mochizuki et al. 1994; Nakagawa et al. 1995) with a balloon-borne telescope incorporated with a stressed Ge:Ga detector developed in CRL/ISAS (Hiromoto et al. 1989). The system NEP during the observation was 6×10^{-16} W/Hz$^{-1/2}$, and the detection limit of the final smoothed map is 1.5×10^{-5} ergs s^{-1} cm^{-2} sr^{-1} (3σ). The spatial resolution is 15'.
We observed the Galactic plane (including the Galactic Center) and the ρ Ophiuchi cloud. The observed [CII] line emission is relatively weak toward the Galactic Center, and the physical condition of the Galactic Center is quite different from that of the Galactic plane (Nakagawa et al. 1995).
On the other hand, there are three gas components in the observed region of the ρ Oph cloud (Yui et al. 1996). The physical condition of these three components are different from each other, and that of the ρ Oph cloud core region is very similar to the Galactic Center (Figure 1 and 2).

3. Discussion

In this section, we discuss a possibility of early B-type stars to be sources of thermal flux in the Galactic Center, comparing the observed parameters of the Galactic Center with the ρ Oph cloud core. The energy source of the ρ Oph cloud core is known, the B2V star HD147889. Hence we discuss if the Galactic Center can be explained as an ensemble of ρ Oph cloud core.

3.1. Galactic Center ($r \leq 40$ pc)

There are non-thermal and thermal flux in the Galactic Center region, and the heating sources of them have not yet been identified. The observed parameters within the central 40 pc of the Galaxy are as followings.

$$IRE \sim 30$$

$$I^{GC}_{[CII]}/I^{GC}_{FIR} \sim 1.8 \times 10^{-3}$$

$$L^{GC}_{bol} = 2 \times 10^8 \ L_\odot \quad (IRAS \ 60\mu m, \ 100\mu m)$$

$$L^{GC}_{[CII]} = 9 \times 10^4 \ L_\odot \quad (\text{Nakagawa et al. 1995})$$

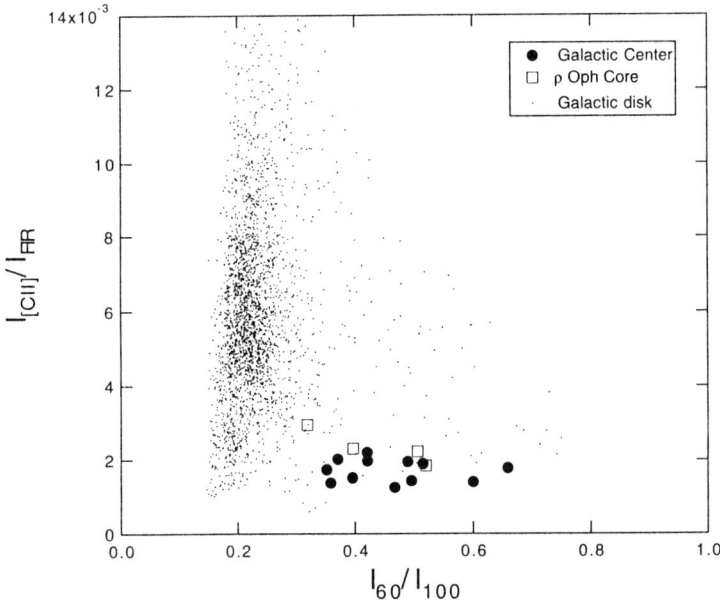

Figure 1. [CII]/FIR intensities versus $60\mu m/100\mu m$ intensities for Galactic Center and ρ Oph cloud core.

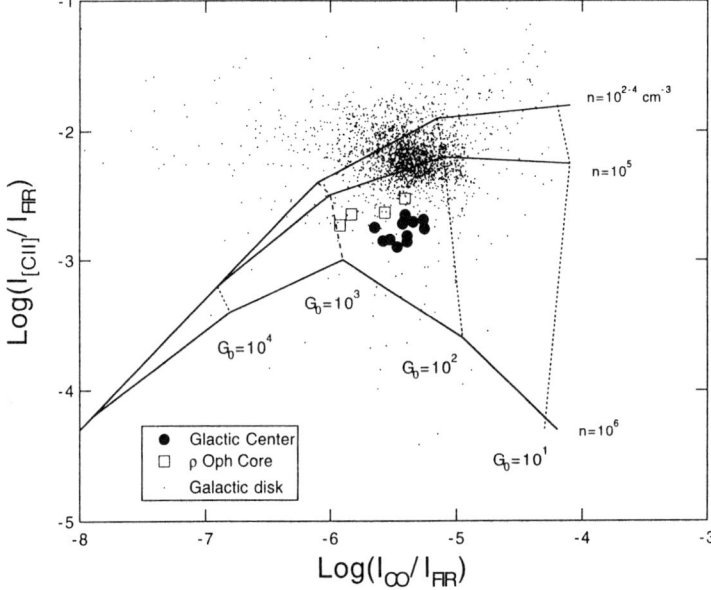

Figure 2. CO/FIR intensities versus [CII]/FIR intensities for Galactic Center and ρOph cloud core.

$$N_{\text{Lyc}}^{\text{GC}} \sim 8 \times 10^{50} \text{s}^{-1} \quad \text{(Odenwald and Fazio 1984)}$$

The UV radiation field of the Galactic Center is relatively softer than the Galactic disk of $IRE \sim 10$. The intensity ratio of [CII] to far-infrared of the Galactic Center is much smaller than those of the Galactic disk (4-8 $\times 10^{-3}$). If the thermal flux is caused by photoionization, the Lyman continuum photons required for it is $\sim 8 \times 10^{50} \text{s}^{-1}$.

3.2. ρ Ophiuchi Cloud Core ($r \leq 0.6$ pc)

The main energy source of the ρ Oph Cloud core is the B2V star HD147889. The physical parameters of the core region are as followings.

$$N_{\text{Lyc}}^{\text{B2}} \sim 10^{46} \text{s}^{-1} \quad \text{(Puxley, Hawarden, and Mountain 1990)}$$

$$L_{\text{bol}}^{\rho\text{Oph}} = 7500 \ L_\odot$$

$$L_{\text{[CII]}}^{\rho\text{Oph}} = 3 \ L_\odot$$

$$I_{\text{[CII]}}^{\rho\text{Oph}} / I_{\text{FIR}}^{\rho\text{Oph}} \sim 2 \times 10^{-3}$$

The intensity ratio of [CII] to far-infrared is smaller than those of the Galactic disk, and very close to that of the Galactic Center.

3.3. Comparison of the Galactic Center with the ρ Oph Core

If the thermal flux from the Galactic Center is due to hot young stars, how many early B type stars are required ? In the followings, we devide observed parameters of the Galactic Center by those of the ρ Oph core.

$$L_{\text{bol}}^{\text{GC}} / L_{\text{bol}}^{\rho\text{Oph}} \sim 2.7 \times 10^4$$

$$L_{\text{[CII]}}^{\text{GC}} / L_{\text{[CII]}}^{\rho\text{Oph}} \sim 3 \times 10^4$$

$$N_{\text{Lyc}}^{\text{GC}} / N_{\text{Lyc}}^{\text{B2}} \sim 8 \times 10^4$$

Those numbers are almost the same, and shows that there are $\sim 10^4$ B2 and earlier type stars in the central 40 pc around the Galactic Center. The result thus derived by the [CII] observation is consistent with that by Odenwald and Fazio (3-5 $\times 10^4$).

4. Conclusion

The far-infrared [CII] line is a powerful probe to trace embedded early B-type stars. The [CII] observations of the Galactic Center and the ρ Ophiuchi cloud core suggests that (3-8) $\times 10^4$ early B-type stars are embedded in the central 40 pc region of the Galaxy, and that they are the dominant source of the thermal emission of the region.

A H$_2$ fluorescence

There is another approach to explain the energy source of the thermal flux from the Galactic Center.
If there are $\sim 10^4$ early B-type stars in the central 40 pc of the Galactic Center, infrared H$_2$ fluorescence can also be observed.
The heating source of the NGC2023 is also known, and it is a B1.5 star. The total H$_2$ fluorescence luminosity from the NGC2023 is,

$$L^{\text{NGC2023}}_{\text{H}_2 v=1-0S(1)} = 0.5 \ L_\odot \qquad \text{(Gatley et al. 1987)}.$$

If $(3-8) \times 10^4$ early B-type stars are embedded in the Galactic Center,

$$L^{\text{GC}}_{\text{H}_2 v=1-0S(1)} / L^{\text{NGC2023}}_{\text{H}_2 v=1-0S(1)} = (3-8) \times 10^4$$

then the expected luminosity of the H$_2$ fluorescence from the Galactic Center is,

$$L^{\text{GC}}_{\text{H}_2 v=1-0S(1)} = (1.5 - 4) \times 10^4 \ L_\odot.$$

On the other hand, the observed value is,

$$L^{\text{GC}}_{\text{H}_2 v=1-0S(1)} = 2.5 \times 10^4 \ L_\odot \qquad \text{(Pak, Jaffe, and Keller 1995)}.$$

This is also consistent with our conclusion.

References

Gatley, I., et al. 1987, ApJ, 318, L73

Güsten, R. 1988, in The Center of the Galaxy, M. Morris, Dordrecht: Kruwer, 1989, 89

Mochizuki, K., Nakagawa, T., Doi, Y., Yui, Y. Y., Okuda, H., Shibai, H., Yui, M., Nishimura, T., & Low, F. J. 1994, ApJ, 430, L37

Nakagawa, T., Doi, Y., Yui, Y. Y., Okuda, H., Mochizuki, K., Shibai, H., Nishimura, T., & Low, F. J. 1995, ApJ, 455, L35

Odenwald, S. F. & Fazio, G. G. 1984, ApJ, 283, 601

Pak, S., Jaffe, D., & Keller, L. D. 1996, ApJ, 457, L43

Puxley, P. J., Hawarden, T., G. & Mountain, C. M. 1990, ApJ, 364, 77

Reich, W., Sofue, Y. & Fürst, E. 1987, PASJ, 39, 573

Shibai, H., et al. 1991, ApJ, 374, 522

Tielens, a. g. g. M, & Hollenbach, D. 1985, ApJ, 291, 722

Wright, E. L., et al. 1991, ApJ, 381, 200

Yui, Y. Y., Nakagawa T., Doi, Y., Okuda, H., Shibai, H., Nishimura, T., & Low, F. J. 1993, ApJ, 419, L37

Yui, Y. Y., Nakagawa T., Doi, Y., Okuda, H., & Shibai, H. 1996, in preparation

KWIC Imaging of the Galactic Center

H. M. Latvakoski, G. J. Stacey, T. L. Hayward, and G. E. Gull

Cornell University, Ithaca, NY 14853 USA

Abstract. We present images of the Galactic Center at 31.5 and 37.7μm obtained with the Kuiper Widefield Infrared Camera (KWIC) on the KAO. These images cover a $10' \times 24'$ region that includes the entire circumnuclear disk, mini-spiral, thermal arches, and pistol/sickle region. Each of these features is clearly visible in the images. The spatial resolution of the raw images is $8''$, and a maximum likelihood deconvolution reduces this to less then $5''$. We have calculated color temperatures and optical depths from these images and present a physical model to explain the observed far-infrared dust distribution.

1. Introduction

The Galactic Center contains many unique and interesting features including the radio point source Sgr A*, the associated mini-spiral and circumnuclear ring and the thermal and non-thermal arches $20'$ to the north. The thermal arches are long, thin structures (20-30 parsecs long and less than 2 wide) that are roughly parallel to the galactic plane.

There has been debate about the source of ionization of the thermal arches since their discovery with the VLA (Yusef-Zadeh, Morris and Chance 1984). The structure and uniform brightness and ionization have generally been taken as evidence that the ionization source is not recently formed hot stars, as they would have to be embedded or near the arches at fairly uniform intervals (Serabyn and Guesten 1987). Also, dust temperature maps do not show peaks at the positions of the arches, which would be expected for embedded sources (Davidson et al. 1994). These reasons and the proximity of strong magnetic fields (in the non-thermal arches) have lead many to suggest magnetic ionization phenomena are involved (Yusef-Zadeh, Morris and Chance 1984; Serabyn and Guesten 1987). However, the emission produced by these processes will be considerably different from that from photoionization (Hollenbach, Chernoff, and McKee, 1989). Line and continuum observations of the arches match predicted ionization by star formation much better than ionization by magnetic phenomena (Genzel et al. 1990; Erickson et al. 1991; Poglitsch, et al. 1994).

These observations also require a low ionization level (approximately O9 stars) to match the data. Since the arches must be very young structures (the circular rotation periods here are about 2Myr), hotter more luminous stars would normally be expected if the stars formed in the arches (Erickson, et al. 1991). A cluster of Brγ and HeI emission line stars near an arch may be the source of ionization, especially if they are associated with other OB stars (Cotera et al.

1995). These stars appear to be type WN (a short post main sequence phase) which would make them some 10Myr old. A possible explanation is that the filaments have recently moved near groups of older stars that are now ionizing them on their edges (Genzel and Poglitsch, 1995).

In the central few parsecs, molecular tracers such as HCN (Guesten, et al, 1987) show a ring like structure surrounding a nearly empty cavity. This circumnuclear disk of material with an inner radius of about 1.5pc and tilted 70° to our line of sight appears to consist of multiple clouds in orbit about the Galactic Center (Genzel and Poglitsch, 1995). The radio continuum and other tracers of ionized material show an unusual group of filaments located inside the cavity and collectively named the mini-spiral. The western arc of the mini-spiral coincides with the circumnuclear ring for most of its western half and is likely the ionized edge of this portion of the ring. The remaining mini-spiral features are believed to be streamers of material close to the Galactic Center. In the case of the northern arm, the ionized material is probably just the edge of the much larger feature seen in the 50 and 90μm infrared continuum and in [OI] (Davidson et al, 1992; Jackson, et al, 1993). These streamers are likely to be tidally stretched molecular clouds with low angular momentum that are ionized as they pass the Galactic Center on a close orbit (Genzel and Poglitsch, 1995). Recent H92α data taken with high velocity resolution suggests the eastern arm and bar are really pieces of one streamer (Roberts and Goss, 1993), so only two streamers are needed to explain the mini-spiral.

The source of ionizing radiation for the mini-spiral and circumnuclear ring has long been a mystery (Genzel and Poglitsch 1995), but recent work indicates that it may be a cluster of luminous stars (Krabbe et al. 1991; Krabbe et al. 1995). Dust temperature maps of the inner parsec from the mid infrared continuum (4.8-20μm) show correlations with the positions of these stars which suggests they are the source of the ionization (Gezari, Dwek, and Varosi 1994). These stars are sufficient to produce the observed luminosity of the inner parsec.

2. Observations

We observed the Galactic Center at 31.5 and 37.7μm on May 15, 18, and 20, 1995 using KWIC on the KAO. A complete description of KWIC is given elsewhere (Stacey et al, 1993; Latvakoski et al, 1995). Briefly, KWIC is an infrared spectrometer/ spectrophotometer that is sensitive from 18-40μm. The detector is a SIRFT funded, Rockwell International 128X128 Si:Sb BIB array, permitting both a large field of view (5.8' × 5.8') and oversampling of the diffraction limited beam: 2$''$73 pixels, 8$''$5 beam at 38μm. Two modes of operation are available, a high resolution mode for observing spectral lines and a low resolution mode for observing the far-infrared continuum. The resolution used for our Galactic Center images was about 35.

Images were taken in a 10' × 24' region which includes the entire Sgr A region as well as the thermal arches and the Sickle/Pistol. The individual images were sky subtracted (i.e. chopping and nodding), flat fielded using images of a heated blackbody internal to the instrument, and added to make a large mosaic. Mars was used as our primary calibrator; it was taken to be a 212°K blackbody as is appropriate at the time of our observations (Wright, 1976).

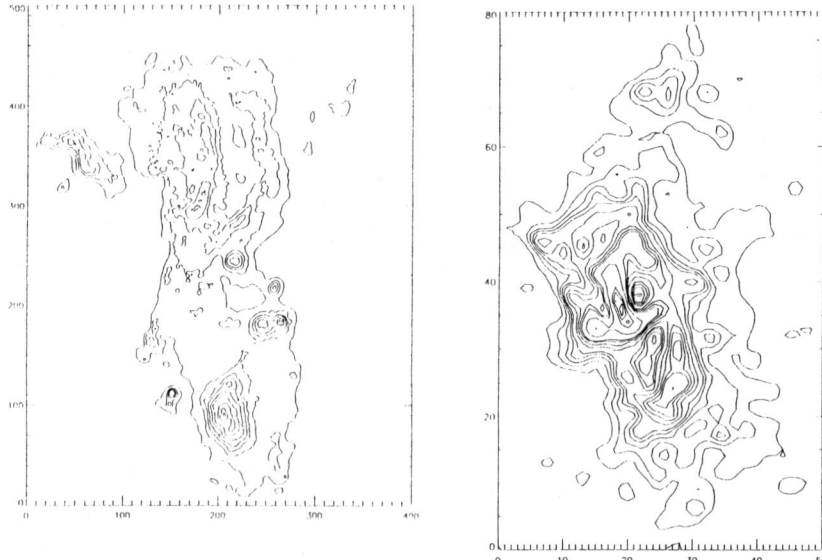

Figure 1. The large scale 37.7μm map of the Galactic Center region. The contour levels are 1, 2, 3, 4, 6, 10, 20, 30 Jy/pixel. The beam is 8$''$.5.

Figure 2. The 31.5μm map of Sgr A west. The contours levels are 2, 4, 6, 8, 10, 15, 20, 25, 30, 50, 70, 90, 110, 130, 150 Jy/pixel. Sgr A* is marked by the cross. The beam size is about 3$''$.

In the Sgr A west region, the signal to noise ratio is very high ($\geq$140 over the entire circumnuclear ring), which allows for the use of Maximum Likelihood Deconvolution to remove irregularities in the point spread function and to reduce the spot size of the features. We used the Lucy deconvolution package available for IRAF. The deconvolved Galactic Center images are significantly sharper than the originals, but all the features seen in the deconvolved images may also be found in the originals. The effective beam sizes in the deconvolved images are about 3$''$ and 5$''$.5 (FWHM) for the 31.5 and the 37.7μm images, respectively.

3. Results and Discussion

3.1. Arches

Our 37.7μm map of the arches (Figure 1) clearly shows all the thermal features that are observed in the radio continuum, but, as expected, none of the non-thermal arches are seen. The correlation between the thermal features in the far-infrared and the radio is extremely good. As the 37μm continuum traces mainly photodissociated material (Stacey et al, 1995), this overlap implies that the ionized hydrogen and photodissociated material are mixed which is what one would expect for embedded sources of luminosity.

It appears less likely that Cotera et al.'s (1995) cluster is the luminosity source since if it were one would expect the radio continuum emission to be offset toward the cluster relative to the far-infrared. The cluster could still be the source of ionization if the stars are a significant distance in front of or behind the arch. However, the location of the stars and a comparison of the morphology of the molecular clouds (Serabyn and Guesten, 1987) with the arches rules this out.

A dust temperature map created from the 31.5 and 37.7μm images (and assuming dust emissivity is inversely proportional to wavelength) shows the arches have a fairly uniform temperature to within our resolution. This, implies uniformly distributed sources. Davidson, et al (1992) obtained a similar result from 50 and 90μm data with a much larger beam. Over most of the arches we observe 30-60 Jy per 8$''$5 beam and a temperature of about 75 K. Dust which radiates this much flux at this temperature must have a total luminosity of a few by $10^4 L_{sun}$, which is the luminosity of a B0 star. In order to have the temperature map appear uniform, there must be roughly one star per beam. Thus, a uniform distribution of B0 stars could reproduce the observed results. This estimate compares well with the O9 stars estimated by Erickson et al (1991) based on the ionization state of the ambient gas.

3.2. Sgr A West

Our 31.5 and 37.7μm images (Figures 2 and 3) show for the first time both the mini-spiral and the entire circumnuclear ring. The data also shows the ring to be quite clumpy which is particularly obvious in the 31.5μm map. Prominent clumps can be observed in the northwestern and southwestern portions of the ring and a group of clumps is seen to the northeast, one of which (outside the ring) aligns with an H_2O maser (Yusef-Zadeh and Mehringer, 1995). The maps also show deep holes between the ring and mini-spiral where the cavity is fairly empty; the contrast between the peak and the hole to its northwest is 40:1 in the space of 9$''$. The ring is weaker relative to the northern arm in the 31.5μm map than in the 37.7μm map, which shows the ring has a lower color temperature. Both images show several elongated streamers just outside the ring. One of these looks like a natural extension of the bar and another is possibly an extension of the eastern arm. In addition, there is clearly a considerable amount of flux visible outside both the ring and these streamers. This background flux accounts for about 1/3 of the total flux in the 3.5$'$ square around the Galactic Center.

A comparison of our data with radio continuum (Lo and Claussen, 1993) and HCN observations (Guesten et al, 1987) shows that the far-infrared western arc is located outside the radio arc but inside the HCN ring. As the radio continuum traces ionized gas, the far-infrared traces photodissociated material, and the HCN traces molecular gas, these observations show a clear progression of HII region, PDR, and molecular cloud out from the center. This confirms that the western arc is a component of a centrally heated ring. Comparing the radio and far-infrared mini-spiral shows that the radio arms are located closer to the center than the far-infrared arms, which demonstrates that these features are also centrally heated ionization fronts. This characterization of all mini-spiral features as ionization fronts has been discussed extensively by Telesco, Davidson, and Werner (1996), and need not be repeated here. Note that our

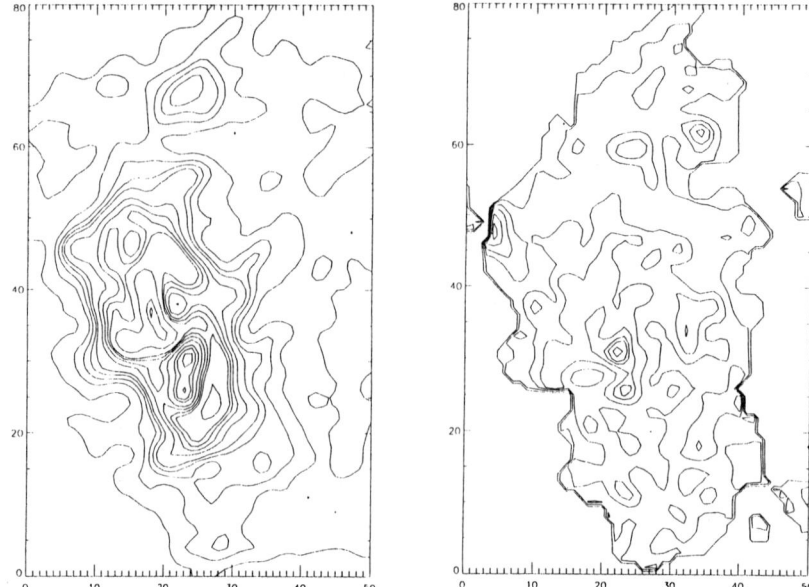

Figure 3. The 37.7μm map of Sgr A west. The contour levels are 2, 4, 6, 8, 10, 15, 20, 25, 30, 50, 70, 90, 110 Jy/pixel. The beam size is about 5″.5.

Figure 4. The color temperature map. The contour levels are 30, 50, 70, 90, 120, 150, 180, 240, 300°K. The beam size is 7″.1

31.5μ map agrees in detail with the 30m map of Telesco, Davidson, and Werner (1996) which covers the inner 1.5 arcminutes of this region at about the same spatial resolution.

In order to calculate color temperatures and optical depths from our data, the interstellar extinction to the Galactic Center must be taken into account. The maps shown here were made by using a constant extinction correction of $\tau(30)=0.8$ (McCarthy et al, 1980; Telesco, Davidson and Werner, 1996) and assuming a $1/\lambda$ emissivity law.

Our color temperature map (Figure 4) shows an obvious gradient from the center toward the edges indicating a central heating source. The various warm spots scattered throughout the map may correspond to additional luminosity sources. On the northern arm and bar, the positions of the warm spots agree well with those observed in the mid- infrared (Gezari, Dwek, and Varosi, 1993; Telesco, Davidson, and Werner, 1996). The temperatures of these hot spots are about what one would expect given the positions of the known Helium line stars.

Our optical depth map (Figure 5) shows cavities of very low optical depth to the south of Sgr A* as well as to its northwest. All the mini-spiral features are seen to have a higher optical depth. The ring is of very high optical depth particularly along the major axis of the ellipse. Note that the optical depth observed on the ring is about as high on the east as on the west side; the ring is equally massive on both sides, the eastern half is fainter only because it is cooler.

From the optical depth of the northern arm and the grain model of Drain and Lee (1984) we estimate the mass of the northern arm below where it crosses the ring is about 130 solar masses or half that derived from the [OI] emission. This may just reflect the large beam size (22″) at [OI] which may include some flux from the ring as well. The optical depth map clearly shows clumps along the ring, which correlate well with observed HCN clumps (Guesten et al, 1987).

Overlaying our 37.7μm map on the [OI] map demonstrates excellent correspondence between the two tracers at the northern arm. The portion of the [OI] arm extending beyond the ring is well matched by a streamer in our images. Of the remaining streamers outside the ring, one (or possibly two) streamers seem to be natural extensions of the bar/eastern arm, and the rest are probably additional tidally stretched molecular clouds headed toward the Galactic Center.

4. Computer modeling

In order to further study this interesting region we constructed a computer model. The model calculates expected values for the luminosity, infrared continuum emission, observed dust temperature, and radio continuum emission. It requires as inputs the locations and luminosities of the sources of radiation and the location and density (UV/optical extinction) of dust in the region. A dust grain emissivity model is also needed.

The density ($n=10^4$), thickness (0.5pc) and flare angle (32°) of the ring are based on previous work (Jackson, et al, 1993), while the inner radius, tilt angle, and shape were based on best matches to our data. We find the ring has an inner radius of 1.58pc, a 65° tilt from face on, and it is a slight ellipse with Sgr A* at one focus and the other .1pc to the north-northeast. These results agree with previous work (Guesten et al, 1987; Jackson, et al, 1993). The west side of the ring is assumed to be tilted toward us as is suggested by Roberts and Goss, (1993). The northern arm is modeled as a long thin streamer of dust nearly in the plane of the disk. The bar and eastern arm are modeled as a single parabolic feature, centered on Sgr A*, that is well out of the plane of the disk. The vertex is roughly toward us and south of Sgr A*. The eastern arm is then the limb brightened edge of this parabolic streamer. The model shows the central cavity to be fairly empty; a uniform density of about 50 cm^{-3} reproduces the observed flux. The source of luminosity is assumed to be the cluster of stars; the six brightest are included in the model.

A total source luminosity was calculated by matching the fluxes on the southwestern portion of the ring. Since the ring is optically thick in the UV, and the distance from the ring to the sources is known, the observed flux values on the southwestern ring will depend ONLY on the source luminosity and the properties of the grains. A source luminosity of about $2.5*10^7 L_{sun}$ is estimated, and the grain properties seem better matched by a $1/\lambda$ emissivity law than a $1/\lambda^2$ emissivity law.

In general, the model reproduces the data quite well. Figure 6 shows the predicted far-infrared continuum. All the observed features are reproduced in the proper locations and at about the proper intensities. The modeled total luminosity and fluxes match the observed results if the 1/3 contributed by the background (which is not modeled) is considered. The model cannot reproduce

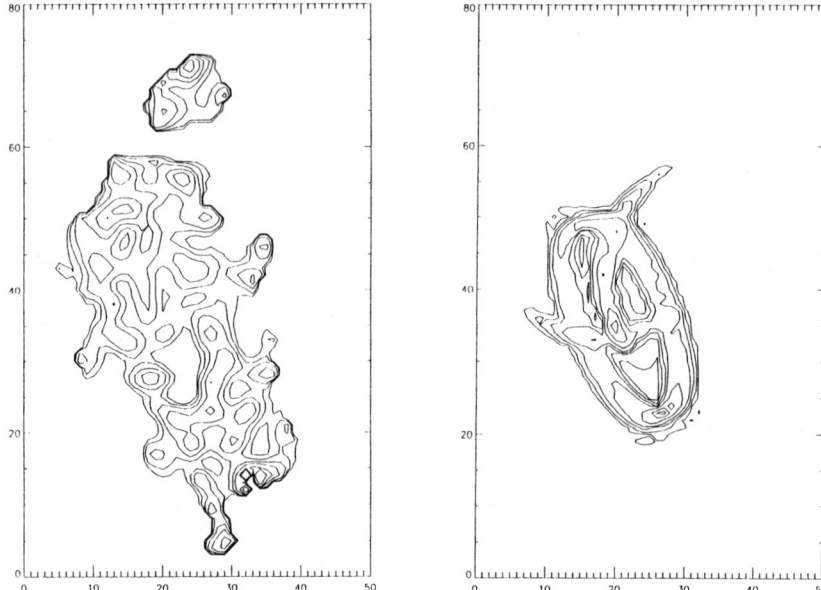

Figure 5. The optical depth map. The contour levels are 0.0125, 0.025, 0.05, 0.1, 0.2, 0.4.

Figure 6. The predicted 37.7μm flux from the model. The contour levels are 4, 8, 15, 25, 50, 90, 130 Jy/pixel.

the clumpy structure of the data since a smooth distribution of dust was used. The entire ring is reproduced by the model, with the eastern portion dimmer than the western because the northern arm is partially in the way. The modeled radio continuum does not show the eastern half of the ring. Lyman continuum from the center is absorbed by the northern arm but enough lower energy radiation gets through to heat the dust in the eastern portion of the ring so that the ring radiates in the far-infrared but not in the radio.

References

Cotera, A.S., Erickson, E.F., Allen, D.A., Colgan, S.W.J., Simpson, J.P., Burton, M.G. 1995, Airborne Astronomy Symposium on the Galactic Ecosystem, 73, 511.

Davidson, J.A., Morris, M., Harvey, P.M., Lester, D.F., Smith, B., and Werner, M.W. 1994, Nuclei of Normal Galaxies: Lessons from the Galactic Center, eds. R. Genzel, and A.I. Harris, (Dordrecht: Kluwer).

Davidson, J.A., Werner, M.W., Wu, X, Lester, D.F., Harey, P.M., Joy, M., and Morris, M. 1992 Ap. J., 387, 189.

Drain B.T., and Lee, H.M., 1984, Ap. J., 285, 89.

Erickson, E.F., Colgan, S.W., Simpson, J.P., Rubin, R.H., Morris, M., and Haas, M.R. 1991 Ap. J., 370, L69.

Genzel, R., and Poglitsch, A., 1995, Airborne Astronomy Symposium on the Galactic Ecosystem, 73, 447.

Genzel, R., Stacey, G.J, Harris, A.I., Townes, C.H., Geis, N., Graf, U.U., Poglitsch, A., and Stutzki, J., 1990, Ap. J., 356, 160.

Gezari, D.Y, Dwek, E., and Varosi, F. 1994, Nuclei of Normal Galaxies: Lessons from the Galactic Center, eds. R. Genzel and A. Harris, (Dordrecht: Kluwer).

Guesten R., Genzel R., Wright, M.C.H., Jaffe D.T., Stutzki J., and Harris A.I., 1987, Ap. J., 318, 124.

Hollenbach, D.J., Chernoff, D.F. and McKee, C.F. 1988, Proceedings of the 22nd Eslab Symposium on Infrared Spectroscopy in Astronomy. Krabbe, A., Genzel, R., Drapatz, S., and Rotaciuc, V. 1991, Ap. J., 382, L19.

Jackson J.M., Geis N., Harris A.I., Madden S., Poglitsch A., Stacey G.J., and Townes C.H., 1993, Ap. J. 385, 585.

Krabbe, A., Genzel, R., Drapatz, S., Rotaciuc, V., 1991, Ap. J., 382, L19.

Krabbe, A., Genzel, R., Eckart, A., Najarro, F., Lutz, D., Cameron, M., Kroker, H., Tacconi-German, L.E., Thatte, N., Weitzel, L., Drapatz, S., Geballe, T., Sternberg, A., and Kudritki, R., 1995, Ap. J., 447, L95.

Latvakoski, H.M., Stacey, G.J., Hayward, T.L., Gull, G.E., and Peng, L., 1995, Airborne Astronomy Symposium on the Galactic Ecosystem, 73, 447.

Lo, K.Y. and Claussen, M., 1983, Nature, 306, 647.

McCarthy, J.F., Forrest, W.J., Briotta, D.A., and Houck, J.R., 1980,Ap. J., 242, 965.

Poglitsch, A., Geis, N., Genzel, R., Herrmann, F., Jackson, J.M., Madden, S.C., Nikola, T., Stacey, G.J., and Townes, C.H. 1994, Nuclei of Normal Galaxies: Lessons from the Galactic Center, eds. R. Genzel and A.I. Harris, (Kluwer: Dordrecht).

Roberts D.A. and Gross W.M., 1993, Ap. J. Supp, 86, 133.

Serabyn, E. and Guesten, R. 1987, A. A., 184, 133.

Stacey, G.J., Hayward, T.L., Latvakoski, H.M., and Gull, G.E., 1993, Infrared Detectors and Instrumentation", SPIE Proceedings, 1946, 238.

Stacey, G.J., Gull, G.E., Hayward, T.L., Latvakoski, H.M., and Peng, L., 1995, Airborne Astronomy Symposium on the Galactic Ecosystem, 73, 215.

Yusef-Zadeh, F., Moris, M. and Chance, D. 1984, Nature, 310, 557.

Yusef-Zadeh, F. and Mehringer, D., 1995, Ap. J., 452, L37.

Telesco, C.M., Davidson, J.A., and Werner, M.W., 1996, Ap. J. 456, 541.

Wright E.L., 1976, Ap J, 210, 250.

The Galactic Center
ASP Conference Series, Vol. 102, 1996
Roland Gredel, ed.

Far-Infrared Fine Structure Line Observations of the Galactic Center Radio Arc

A. Poglitsch, R. Genzel, S.C. Madden[1], T. Nikola, R. Timmermann

Max-Planck-Institut für extraterrestrische Physik, Postfach 1603, 85740 Garching, Germany

N. Geis, C.H. Townes

University of California, Physics Department, Berkeley, CA 94720

Abstract. We report observations of the [O III] 88 μm and 52 μm, the [N III] 57 μm, and the [O I] 63 μm fine structure lines toward the "sickle" feature (G0.18–0.04) in the Galactic Center Radio Arc. We also mapped the eastern region of the arched thermal filaments, including the "banana" (G0.10+0.02), in the [O III] 88 μm line. These observations strongly support the concept of combined H II/PDR regions. G0.18–0.04 is a low-density H II region with $n_e \approx 600$ cm^{-3} which is most likely excited by stars of type earlier than O8 located in the Quintuplet cluster. Ionization of the arched filaments and the "sickle" by shock-like mechanisms is ruled out by the large observed O^{++} column densities. The broad line width of the [O III] 88 μm line, particularly toward the "sickle", may indicate an interaction of the plasma with the strong magnetic fields associated with the nonthermal filaments.

1. Introduction

The Radio Arc, located $\sim 13'$ north of Sgr A, has been a challenge to our imagination both by its unique morphology and the question of the source of excitation. The radio continuum observations of Yusef-Zadeh, Morris & Chance (1984) revealed the highly filamentary structure of the Arc. The straight filaments perpendicular to the Galactic plane have a non-thermal radio spectrum, while the arched filaments (E1, E2, western filaments) show a thermal radio spectrum. A schematic view of the structures in the Radio Arc is shown in Figure 1. The "sickle" (G0.18–0.04) at the center of the non-thermal filaments has a thermal spectrum. Both the "sickle" and the thermal arches are likely to be the ionized surfaces of dense (n $\approx 10^5$ cm^{-3}) molecular material as traced by CS emission (Serabyn & Güsten 1987, 1991).

A number of mechanisms have been discussed to explain the observed structures. Relativistic electrons accelerated due to magnetic reconnection may be the cause for the non-thermal emission in the straight filaments. A number

---

[1]Present address: CE Saclay, Service d'Astrophysique, 91191 Gif-sur-Yvette, France

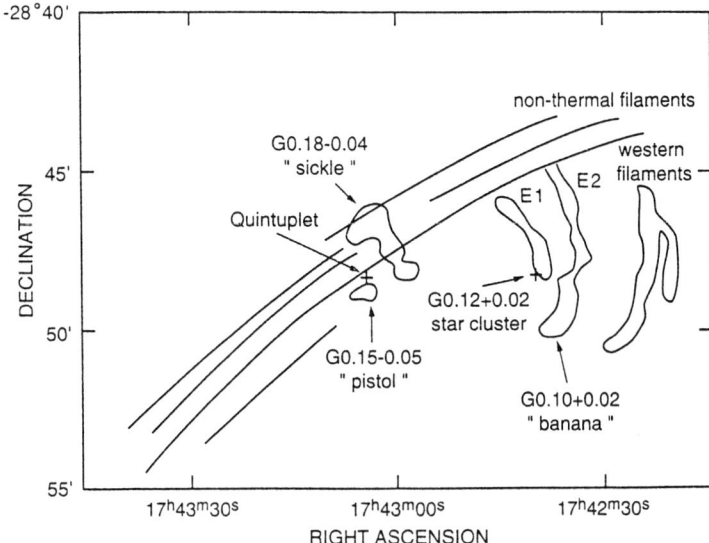

Figure 1. Schematic view of structures in the Radio Arc

of mechanisms have been suggested for the excitation of the thermal arches, ranging from various shock models (Bally et al. 1988; Heyvaerts, Norman & Pudritz 1988; Sofue & Fujimoto 1987) over magnetohydrodynamic interaction of molecular clouds running into the (stiff) poloidal magnetic field (Morris & Yusef-Zadeh 1989) to UV excitation from nearby OB stars (Genzel et al. 1990; Davidson et al. 1994; Morris, Davidson & Werner 1995).

In this paper we present observations of the "sickle" region in the [O III] 88 μm and 52 μm, the [N III] 57 μm, and the [O I] 63 μm fine structure lines, and an [O III] 88 μm map of the eastern arched filament. We will show that these observation strongly support the concept that these regions are ionized and heated by UV radiation.

2. Observations

All observations were carried out aboard NASA's Kuiper Airborne Observatory (KAO) with the MPE/UCB Far-infrared Imaging Fabry-Pérot Interferometer (FIFI; Poglitsch et al. 1991) at an altitude of 12.5 km in several flights between 1992 and 1994. The beam shape was approximately Gaussian with a FWHM of $\approx 22''$. The plate scale on the 5×5 element Ge:Ga detector array (Stacey et al. 1992) was $20''$/pixel. The spectral resolution was 50 km/s at 88 μm, 340 km/s at 52 μm, 250 km/s at 57 μm, and 50 km/s and at 63 μm. The [O III] 88 μm observations were taken fully sampled (half-beam spacing); the other observations are single array pointings.

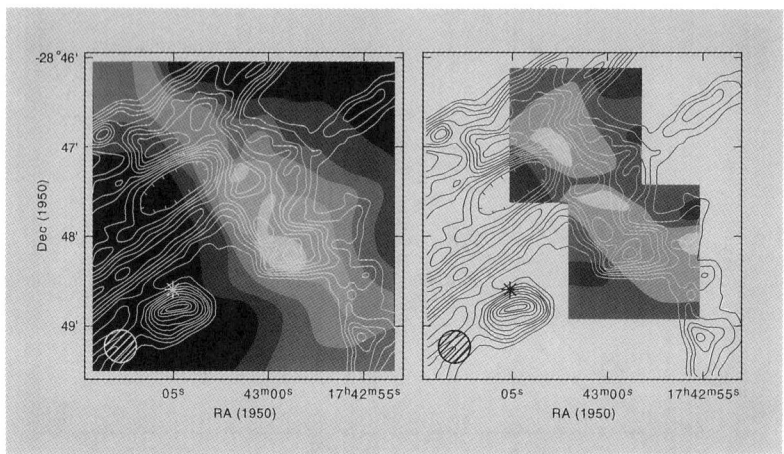

Figure 2. a) [O III] 88 µm map of the "sickle/pistol" region (left panel). The [O III] 88 µm integrated line intensity is represented by the gray scale map; contours are 5, 10, 15, 20, 25 × 10^{-4} erg $s^{-1}cm^{-2}sr^{-1}$. b) [N III] 57 µm map of the "sickle/pistol" region (right panel). The [N III] 57 µm integrated line intensity is represented by the gray scale map; contours are 10, 20, 30 × 10^{-4} erg $s^{-1}cm^{-2}sr^{-1}$. The beam size (22″ FHWM) is shown as the hatched circle. Overlayed as black/white contours is the 6 cm radio map (Yusef-Zadeh, Morris & Chance 1984). The star marks the position of the Quintuplet cluster.

3. Results

3.1. The Sickle/Pistol Region

Figure 2a shows the map of the integrated [O III] 88 µm line emission in G0.18–0.04 (gray scale) overlayed with the 6 cm radio continuum map (white contours) of Yusef-Zadeh, Morris & Chance (1984). The fine structure line emission is well correlated with that of the radio continuum. The positions of the [O III] 88 µm peaks correspond to peaks in the 6 cm map and are also coincident with the peaks of CS $J = 3 - 2$ emission in the map of Serabyn & Güsten (1991). However, no [O III] emission was detected from the "pistol" (G0.15–0.04). The line width ranges from 70 to 110 km/s over the mapped area; the widest line profiles were found where the non-thermal filaments cross the thermal ridge. The velocity centroid shifts by ∼ 70 km/s in east-west direction over the [O III] map.

Using the peak value of $I_{52} = 5.8 \times 10^{-3}$ erg $s^{-1}cm^{-2}sr^{-1}$ for the [O III] 52 µm line toward the southern 88 µm peak position we can derive density and column density of the [O III] emitting region. For an electron temperature of $T_e = 6200$ K (Lang, Goss & Wood 1995), we find an electron density of

$n_e = 600$ cm^{-3} from the [O III] line pair ratio (Dinerstein, Lester & Werner 1985), and a column density of $N(\mathrm{O}^{++}) = 1.1 \times 10^{17}$ cm^{-2}. From the electron density and the emission measure ($EM \sim 2 \times 10^5$ cm^{-6}pc; Sofue, Murata & Reich 1992) we get the column density of ionized hydrogen, $N(\mathrm{H}^+) = 1.0 \times 10^{21}$ cm^{-2}. With $N(\mathrm{O}^{++})/N(\mathrm{H}^+) = 1.1 \times 10^{-4}$ and assuming an (enhanced) relative oxygen abundance of $[\mathrm{O}]/[\mathrm{H}] = 1.2 \times 10^{-3}$ near the Galactic Center (Erickson et al. 1991), we obtain $N(\mathrm{O}^{++})/N(\mathrm{O}) = 0.09$, a ratio which is also found for Sgr A West (Watson et al. 1980).

Figure 2b shows the map of the integrated [N III] 57 μm line emission in G0.18-0.04 (gray scale) overlayed with the 6 cm radio continuum (contours). The emission generally follows the thermal ridge, but the peaks appear somewhat offset from those in the radio continuum and the [O III] 88 μm line. The peak integrated line intensity is $I_{57} = 3.5 \times 10^{-3}$ erg s^{-1}cm^{-2}sr^{-1}. This implies a ratio of $N(\mathrm{N}^{++})/N(\mathrm{O}^{++}) \approx 0.65$ when the emissivity ratio for the two species is taken into account (Lester et al. 1987; Rubin et al. 1994). For a high-excitation source one would expect this ratio to reflect the elemental abundances, $[\mathrm{N}^{++}]/[\mathrm{O}^{++}] \approx [\mathrm{N}]/[\mathrm{O}]$. For a low-excitation source, and particularly for high metallicity, the N^{++} and O^{++} Strömgren radii can decouple due to the different ionization potentials of N$^+$ (29.5 eV) and O$^+$ (35.2 eV), and the ratios will differ: $[\mathrm{N}^{++}]/[\mathrm{O}^{++}] > [\mathrm{N}]/[\mathrm{O}]$. For example, if we apply Rubin's (1985) H II region model U with a metal abundance of $[\mathrm{O}]/[\mathrm{H}] = 1.2 \times 10^{-3}$ and a ratio of $[\mathrm{O}^{++}]/[\mathrm{O}] = 0.09$ as derived above, we deduce $[\mathrm{N}]/[\mathrm{O}] \approx 0.33$. This value is more likely, and is consistent with measurements of other H II regions in the Galactic Center (Simpson et al. 1995). The effective temperature was determined from the fractional O^{++} abundance to be $T_{\mathrm{eff}} \approx 36000$ K, close to the value found for G0.10+0.02 (Simpson et al. 1995).

One frame was recorded in the [O I] 63 μm line toward G0.18-0.04. The peak integrated line intensity of $I_{63} = 7 \times 10^{-4}$ erg s^{-1}cm^{-2}sr^{-1} was found at R.A. = $17^\mathrm{h}42^\mathrm{m}59\overset{\mathrm{s}}{.}7$, decl. = $-28°47'36''$ (1950) which is roughly between the two [O III] peaks. The emission line is centered at $v_\mathrm{LSR} \approx 40$ km/s and thus falls in between the two velocity components of 25 and 80 km/s that were found in the CS emission (Serabyn & Güsten 1991).

3.2. The Eastern Arched Filaments

In Figure 3 we present a map of the eastern arched filaments in the integrated [O III] 88 μm line emission overlayed with the 20 cm continuum contour map (Morris & Yusef-Zadeh 1989). All [O III] features have radio counterparts, which are slightly offset, though. For example, the [O III] image of the "banana" (G0.10+0.02) is well correlated with the 20 cm emission but offset by about $25''$ to the northeast. In general, the [O III] emission appears more extended compared to the radio continuum. Since we have not observed this region in the [O III] 52 μm line we adopt the values for the electron density from Erickson et al. (1991). For the "banana" and their positions 2 and 4 in the E1 filament we then derive O^{++} column densities of 1.7, 1.2, and 1.4×10^{17} cm^{-2}, respectively – comparable to what we have found in the "sickle" region.

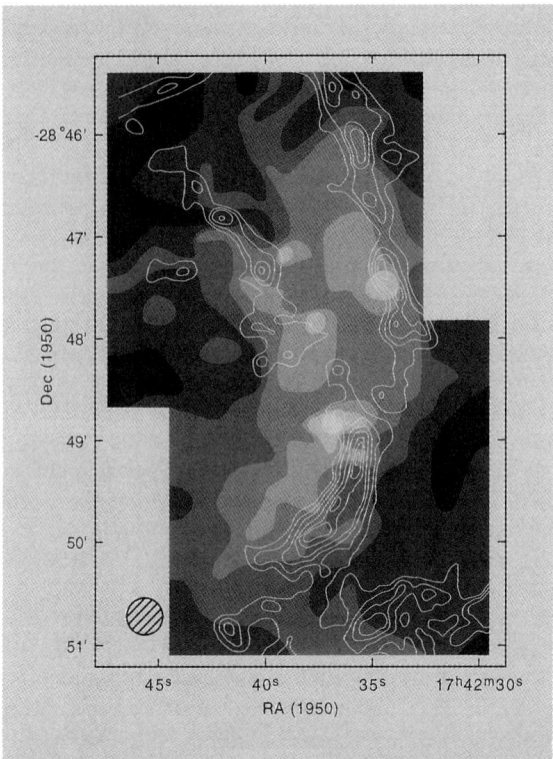

Figure 3. [O III] 88 μm map of the eastern arched filaments. The [O III] 88 μm integrated line intensity is represented by the gray scale map; contours are 5, 10, 15, 20, 25, 30 × 10^{-4} erg $s^{-1}cm^{-2}sr^{-1}$. The beam size (22″ FHWM) is shown as the hatched circle. Overlayed in white is the 20 cm radio continuum (Morris & Yusef-Zadeh 1989).

4. Discussion

4.1. Shocks vs. Photoionization

Shocks and related mechanisms. In order to ionize a large fraction of the interstellar gas, J-shocks are required. Shock velocities in J-shocks are ≥ 40 km/s. J-shocks could, in principle, explain the observed emission measure, $EM \sim 2 \times 10^5$ $cm^{-6}pc$, toward the "sickle" region (McKee & Hollenbach 1987). However, J-shocks fail to produce the observed amounts of N^{++} and O^{++} *by three orders of magnitude* (Shull & McKee 1979). At the same time, the [O I] 63 μm line which would be a typical shock tracer is relatively weak. Therefore, shocks can be excluded as the main source of ionization for the "sickle" and "banana".

An argument against the "critical ionization mechanism" is provided by observations of the HCN and HCO^+ $J = 3 - 2$ lines in G0.18–0.04 (Harris et al. 1994). The basic concept of this ionizing mechanism is similar to MHD-shocks. If an MHD-shock mechanism was indeed responsible for the ionization, one would expect a difference in velocity between the ions and the neutrals. However, the HCN and HCO$^+$ $J = 3 - 2$ observations show no measurable difference between those lines.

This is further support for the rather general finding that the kinetic power of the molecular clouds in the Radio Arc region simply is too small to explain the total FIR luminosity (Davidson et al. 1994; Morris, Davidson & Werner 1995).

Photoionization. While shocks or shock-like mechanisms are clearly ruled out we will show in the following that UV radiation from hot stars is the likely source of ionization for the "sickle" and the thermal arches. The number of stars needed to maintain the observed ionization equilibrium can be estimated in a fairly direct way. The ratio of $N(O^{++})/N(O)$ is determined by the intensity of the UV radiation (Osterbrock 1989) with photon energies above 35.2 eV, the ionization energy of O^+. If we assume that the Quintuplet cluster is the location of the stars which ionize the "sickle" and if we take its projected distance ($R \approx 4.8$ pc) then we can use model calculations for hot stars (Kunze 1994) to determine the minimum number of stars needed to produce the ionizing UV intensity prevalent in the "sickle" region. We find that ~ 1 O6 star or ~ 35 to ~ 60 O8 stars (depending on metallicity) are required to excite the "sickle" region as far as O^{++} is concerned. This may be compared with the Lyman continuum flux emerging from the Quintuplet region which can be estimated from the (thermal) radio continuum flux (Mezger et al. 1979). Integration of the 32 GHz map (Lesch & Reich 1992) over an area within 2.5' from the Quintuplet gives a flux corresponding to ~ 14 O6 stars or ~ 100 O8 stars. Compared with what we found from the O^{++} above, this favours a larger number of late-type O stars over a smaller number of earlier O stars.

In an equivalent way, we derive that about 3 to 10 O8 stars in the star cluster G0.12+0.02 recently found by Cotera et al. (1995) could explain the [O III] emission from the "banana".

4.2. The Stellar Content in the Radio Arc

The Quintuplet Cluster. In the previous section we have shown that the ionizing UV flux from O stars could excite the "sickle" region, and young stars have, indeed, been found in the Quintuplet cluster (Harris et al. 1994; Figer, McLean & Morris 1995). The existence of He I stars implies that massive star formation in the Galactic Center has taken place less than 10^7 yr ago. Taking a standard IMF, one may extrapolate that ≤ 15 stars of type earlier than O8 could be present which would provide the UV flux required to produce the observed amounts of O^{++} amd N^{++}.

The fact that we did not detect any [O III] emission from the "pistol" can only mean that this structure is at a distance much greater than its projected distance from the Quintuplet, and that the exciting stars, e.g. the Pistol Star (Cotera et al. 1996; Figer, McLean & Morris 1995), cannot be hotter than ~ 35000 K. Our finding is supported by the non-detection of [S IV] while [S III] is present (S. Stolovy, private communication).

G0.12+0.02. Recently a cluster of ~ 12 Brγ emission line stars has been detected (Cotera et al. 1996) toward the base of the E1-filament. The discovered stars which are probably high mass loss, blue supergiants are in a relatively short-lived phase. Therefore, it seems very likely that these stars represent only a fraction of hundreds of stars in that area. In fact, Cotera (1995) has discovered additional emission line stars located between Sgr A and the arched filaments.

The observed offset between some of the [O III] emission peaks and the corresponding radio peaks may help to localize the sources of excitation. For the "banana" the [O III] emission is offset to the northeast which would indicate that the UV photons are coming from the the northeast, probably from G0.12+0.02. For the E2-filament the offset is small; in the E1-filament the [O III] emission is slighly offset to the nortwest. This may indicate the presence of an ionizing source between the E1- and E2-filaments, or stars that are embedded in the filaments (Erickson et al. 1995).

5. Conclusions

We have observed the "sickle" region in [O III] 88 μm and 52 μm, [N III] 57 μm, and [O I] 63 μm and mapped the eastern arched filaments in the [O III] 88 μm line. The thermal filaments, including the "sickle", are almost certainly photoionized H II/PDR-regions at the surfaces of dense molecular clouds. Excitation by J-shocks or the "critical ionization mechanism" is ruled out by the observations. The excitation of these regions is due to UV illumination from nearby star clusters, such as the Quintuplet cluster or the G0.12+0.02 cluster at the base of the E1-filament. The "sickle" is excited by the equivalent of $\sim$ 50 O8 stars. Due to its different excitation, the "pistol" cannot be directly related to the "sickle" region. The [N III] observation indicates an enhanced [N]/[O] ratio typical of H II regions near the Galactic Center. The offset between the [O III] peak positions and the radio peaks in the thermal arches indicates that these structures are — at least in part — externally excited.

References

Bally, J., Stark, A.A., Wilson, R.W., & Henkel, C. 1988, ApJ, 324, 223
Cotera, A. 1995, PhD thesis, Stanford University
Cotera, A., Erickson, E.F., Colgan, S.W., Simpson, J.P., Allen, D.A., & Burton, M.W. 1996, ApJ, 461, 750
Davidson, J.A., Morris, M., Harvey, P.M., Lester, D.F., Smith, B., & Werner, M.W. 1994, in Nuclei of Normal Galaxies: Lessons from the Galactic Center, ed. R. Genzel & A.I. Harris (Dordrecht: Kluwer), 231
Dinerstein, H.L., Lester, D.F., & Werner, M.W. 1985, ApJ, 291, 561
Erickson, E.F., Colgan, S.W.J., Simpson, J.P., Rubin, R.H., Morris, M., & Haas, M.R. 1991, ApJ, 370, L69
Erickson, E.F., et al. 1995, in Airborne Astronomy Symposium on the Galactic Ecosystem: From Gas to Stars to Dust, ed. M.R. Haas, J.A. Davidson & E.F. Erickson (San Francisco: ASP), 477
Figer, D.F., McLean, I.S., & Morris, M. 1995, ApJ, 447, L29

Genzel, R., Stacey, G.J., Harris, A.I., Townes, C.H., Geis, N., Graf, U.U., Poglitsch, A., & Stutzki, J. 1990, ApJ, 356, 160
Harris, A.I., Krenz, T., Genzel, R., Krabbe, A., Lutz, D., Poglitsch, A., Townes, C.H., & Geballe, T.R. 1994, in Nuclei of Normal Galaxies: Lessons from the Galactic Center, ed. R. Genzel & A.I. Harris (Dordrecht: Kluwer), 223
Heyvaerts, J., Norman, C., & Pudritz, R.E. 1988, ApJ, 330, 178
Kunze, D. 1994, PhD thesis, Ludwig-Maximilians-Universität, Munich
Lang, C.C., Goss, W.M., & Wood, D.O.S. 1995, ApJ(submitted)
Lesch, H., & Reich, W. 1992, A&A, 264, 493
Lester, D.F., Dinerstein, H.L., Werner, M.W., Watson, D.M., Genzel, R., & Storey, J.W.V. 1987, ApJ, 320, 573
McKee, C.F., & Hollenbach, D.J. 1987, ApJ, 322, 275
Mezger, P.G., Pankonin, V., Schmidt-Burgk, J., Thum, C., & Wink, J. 1979, A&A, 80, L3
Morris, M., Davidson, J.A., & Werner, M.W. 1995, in Airborne Astronomy Symposium on the Galactic Ecosystem: From Gas to Stars to Dust, ed. M.R. Haas, J.A. Davidson & E.F. Erickson (San Francisco: ASP), 477
Morris, M., & Yusef-Zadeh, F. 1989, ApJ, 343, 703
Osterbrock, D.E. 1989, Astrophysics of Gaseous Nebulae and Active Galactic Nuclei (Mill Valley: University Science Books)
Poglitsch, A., et al. 1991, Int. J. Infrared Millimeter Waves, 12, 895
Rubin, R.H. 1985, ApJS, 57, 349
Rubin, R.H., Simpson, J.P., Lord, S.D., Colgan, S.W.J., Erickson, E.F., & Haas, M.R. 1994, ApJ,420, 772
Serabyn, E., & Güsten, R. 1987, A&A, 184, 133
Serabyn, E., & Güsten, R. 1991, A&A, 242, 376
Shull, J.M., & McKee, C.F. 1979, ApJ, 227, 131
Simpson, J.P., Colgan, S.W.J., Rubin, R.H., Erickson, E.F., & Haas, R.M. 1995, ApJ, 444, 721
Sofue, Y., & Fujimoto, M. 1987, ApJ, 319, 73
Sofue, Y., Murata, Y., & Reich, W. 1992, PASJ, 44, 367
Stacey, G.J., Beeman, J.W., Haller, E.E., Geis, N., Poglitsch, A., & Rumitz, M. 1992, Int. J. Infrared Millimeter Waves, 13, 1689
Watson, D.M., Storey, J.W.V., Townes, C.H., & Haller, E.E. 1980, ApJ, 241, L43
Yusef-Zadeh, F., Morris, M., & Chance, D. 1984, Nature, 310, 557

Interstellar Extinction in the Vicinity of Galactic Center Thermal Radio Emission Regions

Angela S. Cotera

NASA/Jet Propulsion Laboratory 100-22, 4800 Oak Grove, Pasadena, CA 91109

Janet P. Simpson [1], Edwin F. Erickson, Sean W. J. Colgan [1]

NASA Ames Research Center, Moffett Field, CA 94035

Michael G. Burton

University of New South Wales, School of Physics, P.O. Box 1, 2033 Kensington, NSW, Australia

Abstract. Broad band J, H and K' and spectral Brγ 2.166 μm images have been obtained in four areas surrounding regions of thermal radio emission. Localized variations in the extinction are examined by means of color and comparison of the observed Brγ flux with the Brγ emission predicted from radio continuum data. Results show few spatial variations and little correlation with the position of the radio emission features.

1. Introduction

The extinction towards the Galactic Center (GC) is usually assumed to be $A_V \sim 30$, but is known to vary between $20 \lesssim A_V \lesssim 35$ (Catchpole, Whitelock & Glass 1990). With the recent discoveries of hot young stars at several locations within the central ~ 40 pc (e.g. Cotera et al. 1994; Figer, McLean & Morris 1995; Blum, DePoy & Sellgren 1995), there is a renewed interest in better characterization of the stellar populations in the GC. The ability to do this will require a good understanding of the extinction in the region and whether or not that extinction is associated with the known regions of radio emission. Statistical studies of NIR broad band images permit an estimate of the extinction towards the stellar populations.

Spectral images of Brγ 2.166 μm show a strong morphological correspondence with the 6 cm radio images and the diffuse Brγ emission (see Figure 3, Cotera et al. 1996). By comparing the theoretical Brγ flux from radio recombination theory with our measured Brγ emission, we can estimate $A_{Br\gamma}$. By comparing the extinction values obtained from our broad band studies of the stellar population with those we derive from the radio emission regions, we will be able to address the relationship betwed these two separate populations.

---

[1] The SETI Institute

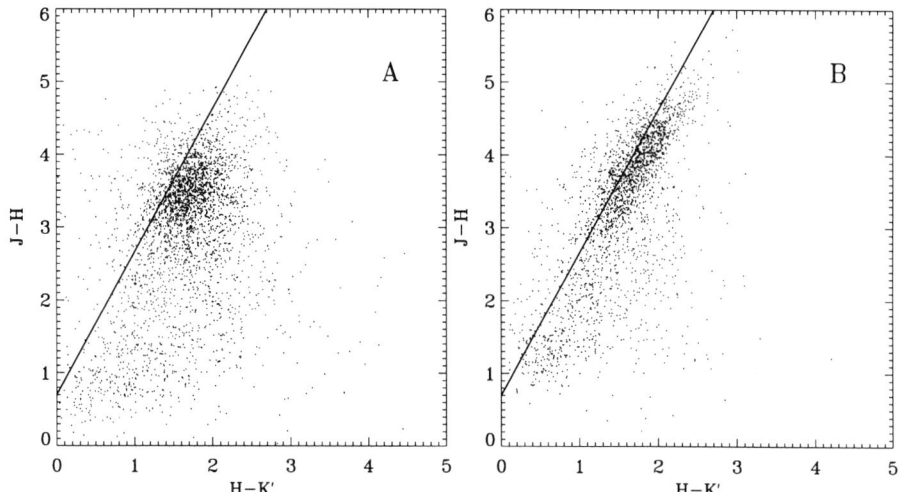

Figure 1. J–H vs H–K' for the stars in A) the field surrounding the Sickle and Pistol and B) the E1 & E2 filaments. The straight line is the extinction curve of He et al. (1995) adjusted to H-K' (Wainscoat & Cowie 1992), and offset by the intrinsic J–H magnitude of 0.7 as found for M giants (Frogel & Whitford 1987).

We have obtained broad band and Brγ emission images in four regions: 1) the Arched Filaments including the E1 & E2 Filaments and G0.10+0.02 (nomenclature Morris & Yusef-Zadeh 1989), 2) the Straight Filaments including G0.18-0.04 (the 'Sickle') and G0.15-0.04 (the 'Pistol ', Yusef-Zadeh & Morris 1987b) , 3) the compact radio emission regions south of the Arched Filaments: H1-H8, and 4) The region including the radio emission regions A-D and Sgr A East (Yusef-Zadeh & Morris 1987a). Due to space constraints this paper will concentrate on the Sickle and Pistol.

2. Observations and Data Reduction

Observations were made at the Anglo-Australian Observatory on 1992 July 13-14 and 1993 June 13-15 with the facility infrared imaging spectrometer IRIS. Broad band J (1.25 μm) H (1.65 μm) and K'(2.1 μm) filters were used for all photometry (K' $\sim$K+0.18(H−K), Wainscoat & Cowie 1992). Broad band figures have been made for each of the four regions (Cotera 1995) which are mosaics of smaller images, 70" × 70" each. There are $\sim$400 stars in each 70"× 70" image at K' which creates problems in determining precise magnitudes due to crowding. The images are also undersampled, making good point spread functions difficult to obtain. The preliminary estimate of the error in the derived magnitudes is $\sim$0.25.

Spectroscopic observations utilized data cubes with axes α, δ and λ. Spectral images of each source were extracted from the data cubes by coadding the

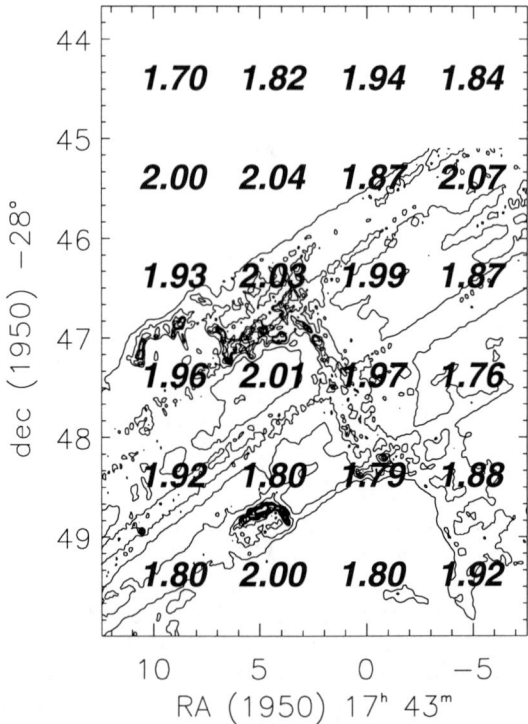

Figure 2. Average H–K' by location overlayed on the 6 cm radio map of the Sickle and Pistol (Yusef-Zadeh & Morris 1987b). Although there are slight variations, the values are the same within the preliminary estimates of the errors. There is no notable variation with location of the Sickle or Pistol.

desired α, δ planes and using adjacent λ planes, containing no emission line features, for sky continuum subtraction. The Brγ flux is then determined by summing over the regions of interest.

3. Results and Disscussion

In Figure 1a the J–H vs. H–K' data is presented for the Sickle and Pistol and in Figure 1b, for the E1 & E2 filaments. The stars in Figure 1b seem to be reasonably fit by the extinction curve of He et al. (1995) based on NIR observations of OB stars. The offset of J–H=0.7 for M giants (Frogel & Whitford 1987) also seems to closely approximate the data. In Figure 1a, the region around the Sickle and Pistol, we see many stars with infrared excess in comparison to those in the E1 and E2 filament data and the extinction curve. The are several possiblities. 1) The seeing for the Sickle and Pistol images was $\lesssim 0.9''$, while for the E1 and E2 Filament the seeing was $\sim 1.2''$. Thus we observe more stars as point sources in the Sickle and Pistol images and the problems inherent in photometry in crowded fields are exacerbated. 2) The bulk of stars that we are seeing may not be M giants and the extinction curve intercept would therefore be incorrect: however, the stellar populations in the Sickle and Pistol region would have to be different from those in the E1 and E2 filaments, and this seems unlikely. 3) The extinction curve, although in reasonable agreement with the curve of Rieke & Lebovfsky (1985), may not be correct for this region.

In Figure 2 we present the average H–K' values for the 24 individual frames taken in the Sickle and Pistol region as a function of position. The average value of H–K' for the entire region is ~ 2.0. Our current estimated errors in H–K' are ~ 0.3 and all values in the image are the same within the error bars. The small variation of extinction with position across the sky does not follow a regular pattern and is not well correlated with position on and off the Straight (nonthermal) Filaments which pass through the region, nor with location of the Sickle or Pistol. A similar pattern is seen within the other three regions. The largest variation in the value of H–K' is seen in the region around the E1 and E2 filaments with $1.5 \lesssim H-K' \lesssim 2.3$, which is still less than a 3σ variation in the H–K' extinction.

In Table 1 we present the Brγ flux observed in the Pistol and the eastern most part of the Sickle. Our observed values are presented with the radio data for the same regions. As the VLA images were available in fits format, we were able to sum over nearly identical regions as those measured in our Brγ images. Assuming Case B recombination, we estimate the expected Brγ emission from the 6 cm radio data, which is unaffected by extinction, and from comparison with our observed Brγ flux derive a value of $A_{Br\gamma}$. In order to examine how the results from the radio emission regions compare to the stellar population, we take a value of H–K'=2.0, and calculate an expected $A_{Br\gamma} \sim 3.6$ using the extinction curve of He et al. (1995).

As can be seen in Table 1, our values for $A_{Br\gamma}$ are considerably less than the predicted value. Possible explanations for the discrepancy in $A_{Br\gamma}$ are: 1) there are Brγ emission stars within the diffuse gas which contribute additional flux. 2) The values of the radio emission may be $\sim 10\%$ low due to losses from the interferometric configuration, however there should also be a correction due to the nonthermal emission which would most likely offset the radio value. 3) As seen in Figure 1, the extinction curve does not agree well with the observed data so a different extinction curve may be needed in this region.

Table 1. Extinction derived from Observations of Brγ Flux

Region	Area arcsec^2	Observed Brγ flux 10^{-19} W/cm^2	Observed[a] 6 cm flux mJy	Predicted Brγ flux 10^{-19} W/cm^2	$A_{Br\gamma}$
Pistol	390	0.62	624	8.4	2.8
Sickle	1280	1.2	1130	15.2	2.7

[a]Data from VLA images (Yusef-Zadeh & Morris 1987b).

References

Blum, R. D., DePoy, D. L. & Sellgren, K. 1995, ApJ, 440, L17

Catchpole, R. M., Whitelock, P. A., & Glass, I. 1990, MNRAS, 247, 479

Cotera, A.S., Erickson, E.F., Allen, D.A., Simpson, J.P., Colgan, S.W.J. & Burton, M.G., 1994 NATO Conference on Nuclei of Galaxies: Lessons Learned from the Galactic Center, ed. R. Genzel & A. I. Harris, (Dordrecht: Kluwer Academic Publishers), p.217

Cotera, A.S., Erickson, E.F., Allen, D.A., Simpson, J.P., Colgan, S.W.J., Burton, M.G., 1996, ApJ, 461, 750

Cotera, A.S. 1995, Ph.D. Thesis, Stanford University

Figer, D.F., McLean, I.S. & Morris, M. 1995, ApJ, 447, L29

Frogel, J. A., & Whitford, A. E., 1987, ApJ, 320 199

He, L., Whitter, D. B. C., Kilkenny, D., & Spencer Jones, J. H. 1995, ApJS, 101, 335

Morris, M., & Yusef-Zadeh, F. 1989, ApJ, 343, 703

Rieke, G. H. & Lebofsky. M. J. 1985, ApJ, 288, 618

Wainscoat, R. J. & Cowie, L. L. 1992, AJ, 103, 332

Yusef-Zadeh, F. & Morris, M. 1987, ApJ, 320, 545

Yusef-Zadeh, F. & Morris, M. 1987, AJ, 94, 1178

The Turbulent Galactic Center: Extremely Strong Scattering near Sgr A*

T. Joseph W. Lazio and James M. Cordes

Dept. of Astronomy, Cornell University and NAIC, Ithaca, NY 14853-6801 USA

Abstract. Angular broadening measurements of Sgr A* and nearby OH/IR stars reveal a region of enhanced, anisotropic broadening approximately centered on Sgr A*, but do not constrain the *radial* distance to the region. If this region is an unrelated superposition $\gtrsim 1$ kpc from Sgr A*, AGNs seen through the region should be $\sim 1''$, roughly double the OH/IR source sizes. If the region is a consequence of the extreme properties of the Galactic center and ~ 100 pc from Sgr A*, AGNs should be *much* larger, ~ 2 *arcmin*. Using the VLA we find a paucity of AGNs within $\sim 0.5°$ of Sgr A* that is due to strong angular broadening of the sources, indicating a scattering region local to the Galactic center. We discuss likely agents for this enhanced scattering and the implications for the detection of pulsars toward the Galactic center.

1. Introduction

Davies et al. (1976) established that the observed size of Sgr A* scaled as λ^2, exactly as expected if angular broadening determines the observed size. The observed size of Sgr A* is now known to scale as λ^2 from 30 cm to 3 mm (Rogers et al. 1994) and to be anisotropic at least over the wavelength range 21 cm to 7 mm (Yusef-Zadeh et al. 1994; Backer et al. 1993; Krichbaum et al. 1993). Maser spots in OH/IR stars within $25'$ of Sgr A* also show enhanced, anisotropic angular broadening (Frail et al. 1994; van Langevelde et al. 1992). These observations indicate that a region of enhanced scattering with an angular size of at least $25'$ is along the line of sight to Sgr A*.

Angular broadening results from density fluctuations in the interstellar plasma. The density fluctuations in turn likely result from velocity or magnetic field fluctuations or both. Thus, identifying the scattering region may provide important clues about the origin of the scattering as well as provide a probe of the velocity or magnetic field in the scattering region.

The aforementioned observations do not constrain the *radial* location of this scattering region, however. The region could be located in or near the Galactic center ($\lesssim 100$ pc) and caused by processes occurring there (GC model). Alternately, the region could be located far from the GC ($\gtrsim 1$ kpc) and be only a random superposition with the GC (RS model). Although the GC model is attractive, other sites of enhanced interstellar scattering are found throughout the Galaxy (e.g., NGC 6634, Moran et al. 1990; Cyg X-3, Molnar et al. 1995)

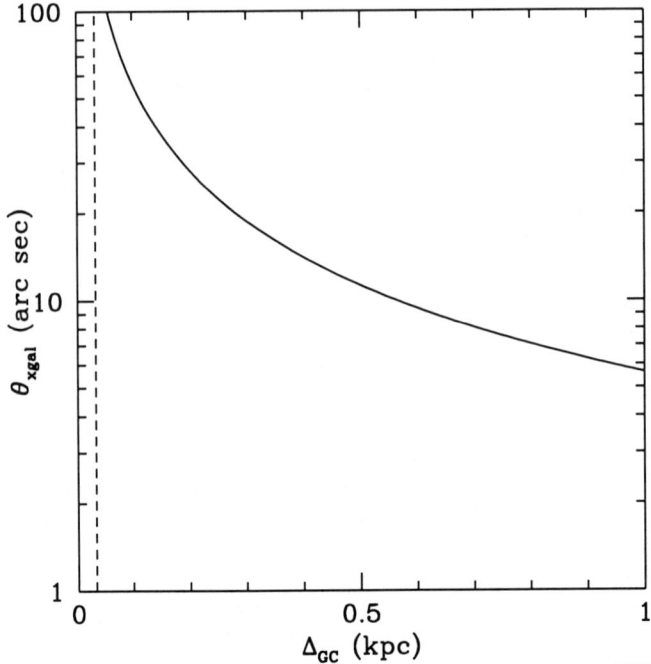

Figure 1. The size of an extragalactic source, at 1.4 GHz, seen through the scattering region in front of Sgr A* as a function of the Galactic center-scattering screen distance, Δ_{GC}. The dashed line indicates the lower limit to Δ_{GC} as derived from the absence of free-free absorption at centimeter wavelengths toward Sgr A*.

and the mean free path for encountering such a region is approximately 8 kpc (Cordes et al. 1991).

An observable consequence of the GC-scattering region distance, Δ_{GC}, is the scattering size of an extragalactic source, θ_{xgal}, seen through the region (van Langevelde et al. 1992). If θ_{GC} is the characteristic size of a GC source seen through the scattering region, an extragalactic source will have a size

$$\theta_{xgal} = \frac{D_{GC}}{\Delta_{GC}} \theta_{GC}. \qquad (1)$$

Here D_{GC} is the GC-Sun distance, which we take to be 8.5 kpc, and, for reference, the size of Sgr A* at 1.4 GHz is 0″.7 (Yusef-Zadeh et al. 1994). Figure 1 shows θ_{xgal} as a function of Δ_{GC}. If the RS model is correct and $\Delta_{GC} \gtrsim 1$ kpc, we expect extragalactic source sizes to be a few arcseconds; if the GC model is correct and $\Delta_{GC} \approx 100$ pc, source sizes could exceed 1 *arcmin*. The lack of free-free absorption toward Sgr A* at centimeter wavelengths provides the constraint $\Delta_{GC} \gtrsim 30$ pc.

2. Observations & Analysis

We undertook a program to identify extragalactic sources near Sgr A* and determine their angular sizes. With the aid of other observers of the GC, we compiled a list of candidate extragalactic sources, ranging from 15' to 2.5° from Sgr A*. Using the VLA in the BnA configuration, we observed these sources at 1.3, 1.7, and 5 GHz; we report on only the 1.3 and 1.7 GHz observations here. Two hour angles were obtained for each pointing center for a total observing time per pointing center of 40 min. In order to provide a large field of view, observations were conducted in spectral-line mode with 3 MHz channels; our total bandwidth is 18 MHz.

We enlarged our sample of candidate extragalactic sources from our initial list of 15 sources to more than 150 sources by adapting pdfCLEAN (Zepka et al. 1994): Deviations from the expected intensity distribution of a noise-only image indicate the contributions of sources. For the VLA, the expected intensity distribution of a noise-only image is Gaussian (Fomalont 1988). The utility of this method is that a source whose *brightness* is below a nominal signal-to-noise threshold (e.g., 5σ) can still be detected. An example is presented in Fig. 2.

None of the sources we have detected show the characteristic λ^2 scaling for their sizes. We do not expect to detect the broadening of Galactic sources (unless they are well beyond the scattering region). The size of Sgr A* at 1.4 GHz is much smaller than our typical synthesized beam of 5". However, if the GC model is appropriate and Δ_{GC} is sufficiently small, θ_{xgal} could be large enough that extragalactic sources are *resolved out* by our observations. This appears to be the case.

In order to quantify our results, we utilize a likelihood method. For the jth field of view, the probability of finding M_j extragalactic sources is

$$p_j(M_j|N_j) = \frac{N_j^{M_j}}{M_j!} e^{-N_j} \qquad (2)$$

given N_j expected extragalactic sources.

The number of extragalactic sources, M_j, is smaller than the actual number of sources found since some sources are Galactic. Zoonematkermani et al. (1990) and Helfand et al. (1992) surveyed this region of the Galactic plane, using the VLA in a similar configuration at 1.4 GHz. A population of Galactic sources, with a *density* of approximately $7.5 \deg^{-2} \exp[-(b/0.3°)^2]$, can be identified in their results (Becker et al. 1992); the number of Galactic sources can then be estimated from the area of a typical field of view ($\sim 1 \deg^2$).

The expected number of extragalactic sources within a field of radius θ_j we estimate from the $\log N$-$\log S$ relation of Katgert et al. (1988) as

$$N_j = \int_0^{\theta_j} d\theta' \, 2\pi\theta' \int_{S_{\min}(\theta')}^{\infty} dS \, f(\phi_{\max,j}|S, \Psi_{GC}, \Delta_{GC}) \frac{dn}{dS}. \qquad (3)$$

Due to the VLA's primary beam attenuation, the minimum detectable flux density for a point source, $S_{\min}(\theta)$, increases with distance from the phase center. For the various fields, $S_{\min}(0)$ is 2 to 8 mJy and is estimated from the inner 2% of a residual image of the primary beam, i.e., after having subtracted all sources

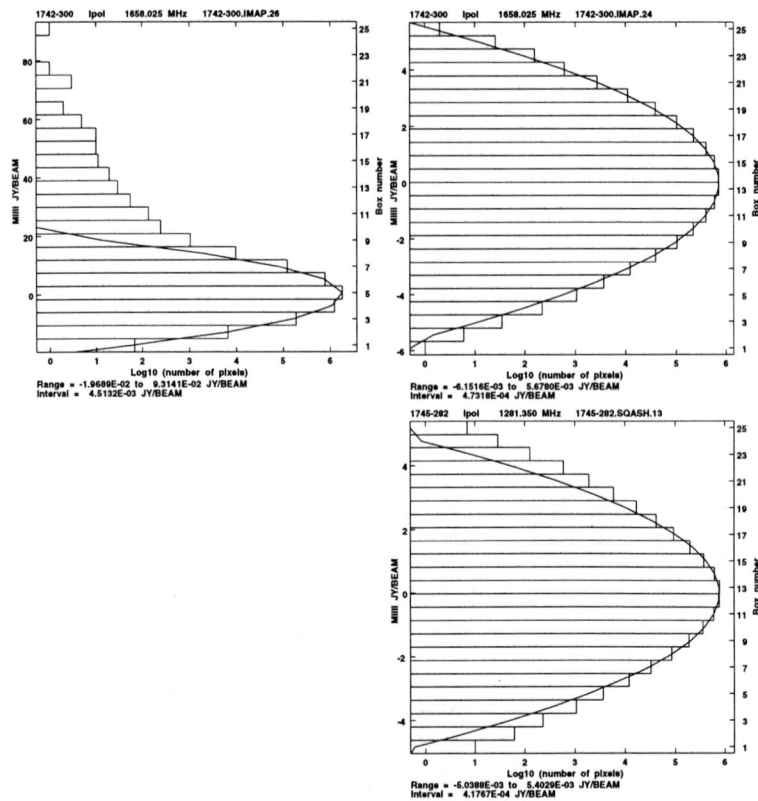

Figure 2. An example of pdfCLEAN. The intensity histogram for the field 358.7−0.0 is shown before (*left*) and after (*right*) the removal of 22 sources. The solid line is a Gaussian with the mean and variance derived from the image, but is *not* a fit to the histogram; note the *logarithmic* abscissa. The lower right panel illustrates that a source with a *brightness* below a nominal signal-to-noise threshold is still detectable with pdfCLEAN. The deviation from a Gaussian is significant, yet only the bin with the largest brightness exceeds the 5σ brightness threshold.

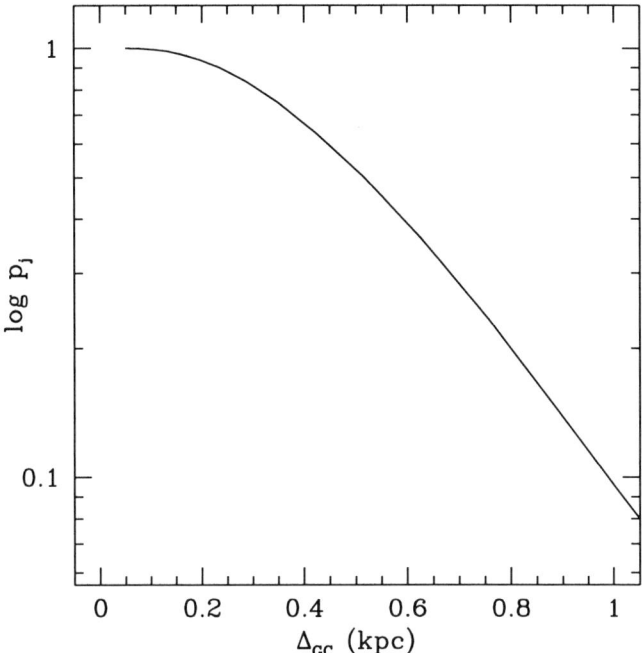

Figure 3. The log likelihood as a function of the Galactic center-scattering screen distance, Δ_{GC}, for the field 359.9+0.2, 15' from Sgr A*.

detected. Over this portion of the residual image the primary beam attenuation is negligible. Finally, our survey is *brightness*-limited. Only a fraction, f, of all sources will have a sufficient brightness to be detected. The maximum *apparent* size for a marginally-detectable source, $\phi_{max,j}$, depends upon S, VLA configuration, and the scattering region. We obtain the apparent size by adding the intrinsic and scattering sizes in quadrature. The distribution of intrinsic sources sizes as a function of S is given by Windhorst et al. (1990). The scattering size depends upon Δ_{GC} (cf. Fig. 1) and the angular extent of the scattering region, Ψ_{GC}. The OH/IR maser observations require $\Psi_{GC} \gtrsim 25'$ (§1.) while observations of H$_2$O masers in Sgr B2(N) require $\Psi_{GC} \lesssim 40'$ (Gwinn et al. 1988). Although the scattering region may have some complex angular shape, we have adopted a disk of radius $\Psi_{GC} = 30'$.

Figure 3 shows p_j as a function of Δ_{GC} for the field closest to Sgr A*, 359.9+0.2. This field is 15' from Sgr A* and contains no extragalactic sources consistent with a scattering region close to Sgr A* ($\lesssim 500$ pc). All other fields are sufficiently far from Sgr A* that they cannot constrain Δ_{GC}, given our assumed size and shape of the region.

Our current analysis favors the GC model and constrains $\Delta_{GC} \lesssim 500$ pc. Work in progress utilizes a more generalized likelihood function, one which also considers as parameters the size and shape of the scattering region and the outer scale within the region (§3.).

3. Sources of Scattering

The density fluctuations which give rise to angular broadening also contribute to free-free absorption and emission. Comparison of the emission measure EM expected from scattering observations with that derived from observations of free-free absorption and emission may indicate the source of the scattering.

Cordes et al. (1991) and van Langevelde et al. (1992) give expressions for estimating the EM from the scattering size. Our data suggest that extragalactic sources are so large as to be resolved out, so we adopt the size of Sgr A* as a characteristic scattering size. We take $\Delta_{\rm GC} = 200$ pc, intermediate between our likelihood results (~ 500 pc) and the size derived from the angular extent of the region (~ 75 pc). The resulting EM is

$$\mathrm{EM} = 10^8 \, \mathrm{pc} \, \mathrm{cm}^{-6} \left(\frac{\Delta_{\rm GC}}{200 \, \mathrm{pc}}\right)^{-2} \left(\frac{l_0}{1 \, \mathrm{pc}}\right)^{2/3} \left(\frac{l_1}{100 \, \mathrm{km}}\right)^{1/3}. \quad (4)$$

This result assumes that the density fluctuations have a Kolmogorov spectrum over a range of spatial scales from an inner scale, l_1, to an outer scale, l_0.

The free-free optical depth in front of Sgr A* is certainly less than unity at 1 GHz (Davies et al. 1976). Sgr A* is not observed below 410 MHz (Pedlar et al. 1989; Davies et al. 1976; but see Duschl et al. 1996). The EM required to produce an optical depth greater than unity between 0.41–1 GHz is EM $\sim 10^{5.5}$–$10^{6.3}$ pc cm^{-6}.

An extended low-density H II region with EM $\approx 10^4$ pc cm^{-6} covering the inner degree is observed in both emission and absorption (Alvarez et al. 1996; Mezger & Pauls 1979). Within this region are arcmin-sized H II regions with EM $\approx 10^{5.5}$ pc cm^{-6} (Pauls et al. 1976).

We require that the EM derived from scattering observations not exceed those derived from free-free absorption and emission measurements. A key free parameter in eqn. (4) is the size of the outer scale, l_0. An outer scale $l_0 \lesssim 10^{-2}$ pc has been suggested for other heavily scattered lines of sight (Frail et al. 1993). Adopting EM $= 10^{5.5}$ pc cm^{-6} as characteristic of that derived from free-free observations, we find an outer scale length of $l_0 \approx 2 \times 10^{-4}$ pc.

Yusef-Zadeh et al. (1994) derive a similar value for the thickness of the ionized layer of molecular clouds using estimates of the ionizing radiation field in the GC and associate these layers with the source of the enhanced scattering. The enhanced scattering could result from the large velocity dispersions of GC molecular clouds, both internal and cloud-to-cloud (e.g., Bally et al. 1988) as well as from the impact of stellar winds and supernova remnants on the molecular clouds.

A final comment on the geometry of the scattering region is warranted. The size given in eqn. (1) is derived assuming the scattering is concentrated in a single region. If the scattering instead arises in N layers, as could be the case if it arises from a collection of molecular cloud layers, the size of an extragalactic source could be as much as N times larger than that shown in Fig. 1. Yusef-Zadeh et al. (1994) estimate $N \lesssim 10$.

4. Implications for Pulsar Searches in the Galactic Center

Only 15 of the approximately 700 known pulsars are within 5° of Sgr A* and none are within 1° (Taylor et al. 1993). In addition to providing powerful probes of the magnetoionic material in the GC, pulsars in or near the GC could be used to constrain the star formation history of the GC. Counterparts to GC X- and γ-ray sources may also involve radio pulsars or their progenitors.

The measured angular broadening of GC sources can be used to predict the *pulse broadening* expected from pulsars in the GC,

$$\tau_{\rm GC} \sim f\left(\frac{\Delta_{\rm GC}}{D_{\rm GC}}\right)\left(\frac{D_{\rm GC}\theta_{\rm GC}^2}{8c\ln 2}\right) \sim \frac{250\,{\rm s}}{\nu_{\rm GHz}^4}. \quad (5)$$

where the scattering region's location along the line of sight is represented by the geometric factor $f(x) \equiv x^{-1}(1-x)$ and we have taken $\Delta_{\rm GC}/D_{\rm GC} = 0.02$.

This pulse broadening is so large that, at conventional frequencies used for pulsar searches, 0.4–1.4 GHz, the pulsed flux is orders of magnitude too small to be detected (Cordes & Lazio 1996). Alternate strategies for detecting GC pulsars are:

High-frequency periodicity searches: The strong frequency dependence of pulse broadening ($\propto \nu^{-4}$) suggests that searches at 5–20 GHz will mitigate the effects of scattering. Pulsar spectra generally decline at higher frequencies, sometimes precipitously, making this option unlikely to find many pulsars. However, recent detections of a few pulsars at frequencies as high as 30 GHz (e.g., Kramer et al. 1996) suggest that some pulsars may show sufficient spectral flattening to be detectable even if at the GC.

Interferometric surveys: Scattering preserves the total flux density while attenuating the pulsed flux. Pulsars may therefore appear as point sources, with sizes comparable to those of Sgr A* and the OH/IR masers, with pulsar-like spectra and polarization in aperture synthesis surveys.

Acknowledgments. The National Radio Astronomy Observatory is operated by Associated Universities, Inc., under cooperative agreement with the National Science Foundation. Travel support was provided by the Cornell Graduate School and an American Astronomical Society International Travel Grant.

References

Alvarez, H., Aparici, J., May, J., & Reich, P. 1996, this volume

Backer, D. C., Zensus, J. A., Kellermann, K. I., Reid, M., Moran, J. M., & Lo, K. Y. 1993, Science, 262, 1414

Bally, J., Stark, A. A., Wilson, R. W., & Henkel, C. 1988, ApJ, 324, 223

Becker, R. H., White, R. L., McLean, B. J., Helfand, D. J., & Zoonematkermani, S. 1992, ApJ, 358, 485

Cordes, J. M. & Lazio, T. J. W. 1996, ApJ, submitted

Cordes, J. M., Weisberg, J. M., Frail, D. A., Spangler, S. R., & Ryan, M. 1991, Nature, 354, 121

Davies, R. D., Walsh, D., & Booth, R. S. 1976, MNRAS, 177, 319

Duschl, W. J., Beckert, T., Mezger, P. G., & Zylka, R. 1996, this volume

Fomalont, E. 1988, in Synthesis Imaging in Radio Astronomy, eds. R. A. Perley, F. R. Schwab, & A. H. Bridle (San Francisco: ASP) p. 213

Frail, D. A., Diamond, P. J., Cordes, J. M., & van Langevelde, H. J. 1994, ApJ, 427, L43

Frail, D. A., Kulkarni, S. R., & Vasisht, G. 1993, Nature, 365, 136

Gwinn, C. R., Moran, J. M., Reid, M. J., & Schneps, M. H. 1988, ApJ, 330, 817

Helfand, D. J., Zoonematkermani, S., Becker, R. H., & White, R. L. 1992, ApJS, 80, 211

Katgert, P., Oort, M. J. A., & Windhorst, R. A. 1988, A&A, 195, 21

Kramer, M., Xilouris, K. M., Jessner, A., Wielebinski, R., & Timofeev, M. 1996, A&A, 306, 867

Krichbaum, T. P. et al. 1993, A&A, 274, L37

Molnar, L. A., Mutel, R. L., Reid, M. J., & Johnston, K. J. 1995, ApJ, 438, 708

Moran, J. M., Greene, B., Rodríguez, L. F., & Backer, D. C. 1990, ApJ, 348, 147

Mezger, P. G. & Pauls, T. 1979, in The Large-Scale Characteristics of the Galaxy, ed. W. B. Burton (Dordrecht: Reidel) p. 357

Pauls, T., Downes, D., Mezger, P. G., & Churchwell, E. 1976, A&A, 46, 407

Pedlar, A., Anantharamaiah, K. R., Ekers, R. D., Goss, W. M., van Gorkom, J. H., Schwarz, U. J., & Zhao, J.-H. 1989, ApJ, 342, 769

Rogers, A. E. E. et al. 1994, ApJ, 434, L59

Taylor, J. H., Manchester, R. N, & Lyne, A. 1993, ApJS, 88, 529

van Langevelde, H. J., Frail, D. A., Cordes, J. M., & Diamond, P. J. 1992, ApJ, 396, 686

Windhorst, R., Mathis, D., & Neuschaefer, L. 1990, in Evolution of the Universe of Galaxies, ed. R. G. Kron (San Francisco: ASP) p. 389

Yusef-Zadeh, F., Cotton, W., Wardle, M., Melia, F., & Roberts, D. A. 1994, ApJ, 434, L63

Zepka, A. F., Cordes, J. M., & Wassermann, I. 1994, ApJ, 427, 438

Zoonematkermani, S., Helfand, D. J., Becker, R. H., White, R. L., & Perley, R. A. 1990, ApJS, 74, 181

Part 1. Kinematics of the Gas and Dust
Section D. The Central Parsec and the CND

Kinematics of the Ionized Gas in Sgr A West: The Minicavity and the Western Arc

D.A. Roberts
University of Illinois, NCSA, 405 N. Mathews Ave, Urbana, IL 61801

F. Yusef-Zadeh
Northwestern University, Department of Physics and Astronomy, Evanston, IL 60208

W.M. Goss
NRAO, P.O. Box 0, Socorro, NM 87801

Abstract. The "western arc" and "minicavity"(approximately $3''$ southwest of Sgr A*) of the Galactic center H II region Sgr A West have been imaged in the H 92α line at 8.3 GHz with the VLA. The kinematics of the ionized gas in the inner 1.5 parsecs of the Galactic center have been determined. The western arc, with an inner radius of 1 pc, appears to be in circular rotation about a point near Sgr A*, with an enclosed mass of $\sim 3.5 \times 10^6\ M_\odot$. Toward the minicavity, a large velocity gradient (>600 km s^{-1} pc^{-1}) is observed along the eastern rim and uniformly large velocities are observed along the western rim. The velocity field of the minicavity is interpreted as gas in a hyperbolic orbit about Sgr A*. Based on this model, the gas is orbiting a point at the position of Sgr A* with a mass of $\sim 3.0 \times 10^6\ M_\odot$.

1. Introduction

One of the important questions concerning the ionized gas in the Galactic center is whether the kinematics of the orbiting gas might constrain the mass of Sgr A*, testing the hypothesis that a black hole resides in the center of the Galaxy (Lynden-Bell & Rees 1971). The major components of Sgr A West include the "southern arm," the "western arc," the "northern arm," the "bar," and the "eastern arm". The "circumnuclear disk" (CND) has been identified by observations H I (Liszt et al. 1983), dust (Becklin, Gatley, & Werner 1982) and molecules (Güsten et al. 1987). From these observations, it appears that the CND surrounds the ionized gas at a radius of 1.7 pc and extends to an outer radius of greater than 8 pc, ($R_o = 8.5$ kpc).

One conspicuous feature in the inner parsec of the Galaxy is the "minicavity," which appears as a hole in the distribution of ionized streamers. At high resolution, the minicavity is observed as a nearly-complete circular feature with a diameter of $\sim 2''$ (0.1 pc) (Yusef-Zadeh, Morris, & Ekers 1990), the center of which is about $3''$ (0.15 pc) southwest of the non-thermal point source Sgr A*.

The minicavity is unique because of its position close to Sgr A* and its unusual morphology and kinematics. Additionally, the large column density near the minicavity shields molecular gas from radiation field and molecular emission and absorption is observed in the area (Marr et al. 1992; Zhao, Goss, & Ho 1995).

Eckart et al. (1992) have observed infrared lines of [Fe III] near the western side of the minicavity and derive a kinetic gas temperature between 4×10^4 and 10^5 K. Lutz, Krabbe, & Genzel (1993) have confirmed the earlier identification of the [Fe III] line by Eckart et al. and propose a model to explain the strong [Fe III] emission from the region. In this model, the minicavity is the result of the interaction of a fast wind with the partially neutral gas orbiting the dynamical center of the Galaxy. In another model, Wardle & Yusef-Zadeh (1992) suggest that the minicavity is formed as a result of the interaction of stellar winds from IRS 16 and the strong gravitational potential of Sgr A*.

The results of two separate observations are presented. In the first project, the Very Large Array of the National Radio Astronomy Observatory [1] was used to determine the global kinematics of the ionized gas of the Galactic center, using the H 92α line, at 2″ resolution. The velocity range of this experiment included the ionized gas of the major features, but did not include gas with $|V_{LSR}| > 200$ km s^{-1}. Globally, kinematic features of the western arc, the northern arm, and the extended bar have been identified. The details of the 2″ H 92α line observations are given in Roberts & Goss (1993). In a second project, ~1″ resolution H 92α line observations of the minicavity of Sgr A West were undertaken, centered on the extreme negative velocities that were previously detected in ~2″ resolution observations (Serabyn et al. 1988; Herbst et al. 1993; and Roberts & Goss 1993). The observational parameters of the high resolution VLA observations are presented by Roberts et al. (1996), who provide a detailed account of these results.

2. Results

2.1. Western Arc Kinematics

H 92α observations of Sgr A West have been carried out with improved resolution and sensitivity over the previous 2 cm H 76α line observations of Schwarz, Bregman, & van Gorkom (1989). These observations have a factor of two greater angular and velocity resolution than recent [Ne II] observations (Serabyn et al. 1988). Kinematic features of the western arc, the northern arm and the extended bar have been identified. The western arc has been modeled as a half ring in circular rotation about a point near the non-thermal point source, Sgr A*. The parameters of the circular orbit model of the western arc are presented in Table 1.

The kinematics of the western arc and northern arm are identified as features in continuous space and radial velocity at the position on where they appear to cross (7″ E, 16″ N of Sgr A*), the profiles have two peaks (see figure

---

[1] The National Radio Astronomy Observatory is a facility of the National Science Foundation operated under a cooperative agreement by Associated Universities, Inc.

11 of Roberts & Goss 1996). The velocity resolution of 14 km s^{-1} of these observations is barely sufficient to resolve the two components. The northern arm and western arc are believed to be unrelated kinematically. This interpretation of the global structure, consisting of distinct features, differs from the interpretation suggested by the recent [Ne II] observations (Lacy, Achtermann, & Serabyn 1991) in which all of the gas in Sgr A West is part of a single Keplerian disk. However, in the observations of Lacy et al. with a velocity resolution twice as coarse of the current data, the two spectral features would be blended together into a single component.

The H 92α line emission has been compared to the Brγ emission (Burton & Allen 1992) in the H II region. This comparison shows the striking result that the Brγ emission is absorbed at the outer edge of the western arc. The extinction is strongly correlated with presence of HCN emission from the CND. Given the previous result that the western part of the CND is inclined out of the plane of the sky along with the suggestion that Brγ emission from the western arc is absorbed by the CND, it is likely that the western arc has the same orientation as the CND, and is in fact related to the CND.

2.2. Minicavity Kinematics

Figure 1 shows a gray scale representation of the peak line intensity (a) and center velocity (b) in the inner parsec of the Galactic center. The contours represent the 8.3 GHz continuum emission from the minicavity. The velocities shown in Figure 1 vary continuously from $\sim$+100 km s^{-1} 0.6 pc (15″) NE of the Sgr A* in the northern arm to -280 km s^{-1} at the position of IRS 2. These velocity and spatial distributions correspond to the "extended northern arm" observed at $\sim$2″ resolution in [Ne II] (Serabyn et al. 1988), in Brγ (Herbst et al. 1993), and in the 2″ H 92α observations (Roberts & Goss 1993).

The central panel of Figure 2 shows a gray-scale and contour representation of the 8.3 GHz continuum emission in the inner 0.8 pc (20″) of the Galactic center; the surrounding panels show spectra taken at the positions marked in the continuum image. In cases where the profiles were taken at the position of infrared sources, the name is shown at the top of the profile (e.g. IRS 2 and IRS 13). The profiles generally have single peaks and a large velocity gradient (>600 km s^{-1} pc^{-1}) is observed in the eastern rim of the minicavity. The velocities in the western edge are extreme; velocities of -250 and -280 km s^{-1} are observed toward IRS 13 and IRS 2, respectively. A second weaker velocity component near 0 km s^{-1}, which has been previously detected in H 92α (Roberts & Goss 1993) and [Ne II] (Serabyn et al. 1988), is observed in some positions on the western side of the minicavity. (Note that the strong line at 0 km s^{-1} in the upper left profile arises from the northern arm gas with the strong velocity gradient and is not related to the 0 km s^{-1} component observed on the western edge.) The crosses in Figure 2 show the center velocity, amplitude, and FWHM velocity width of the Gaussians fitted to the data.

Attempts were made to model the velocity field in the inner 0.8 pc (20″) area of the Galactic center as circular rotation; no satisfactory fits were possible. However, the observed radial velocities of only the eastern side could be modeled as a circular orbit with velocities from +100 to -200 km s^{-1}. The best fit circular orbit model to the observed velocity field on the eastern side of the minicavity

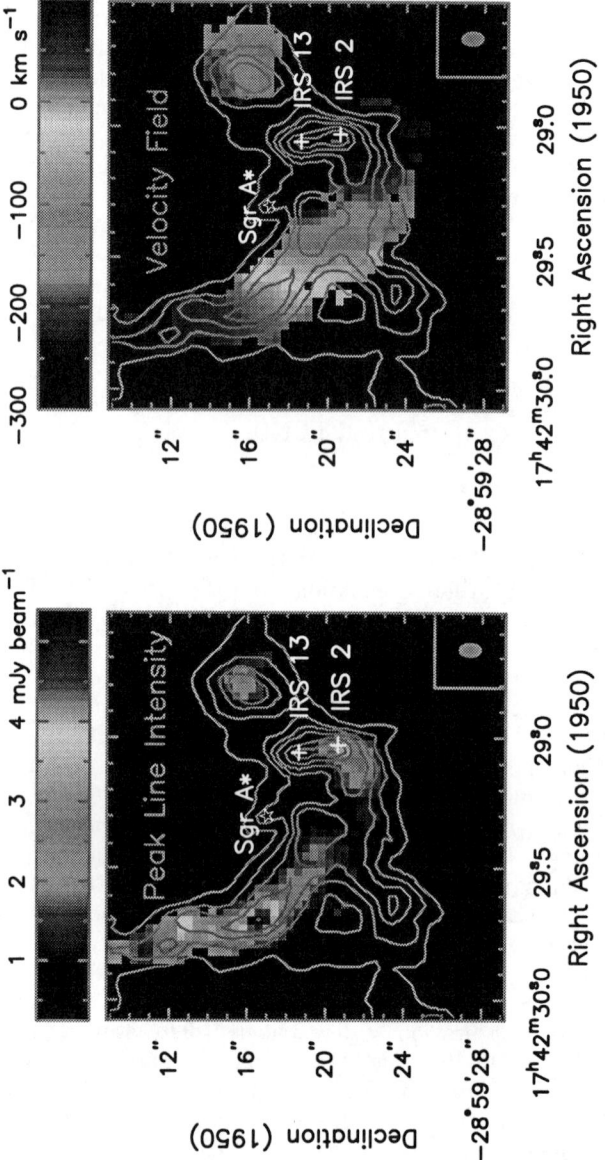

Figure 1. Grey-scale representation of Gaussian parameters of intensity and central velocity from H 92α line emission from the minicavity and northern arm of Sgr A West at a resolution ($\alpha \times \delta$) of $0''\!.75 \times 1''\!.18$ (*lower right corner*). The 8.3 GHz continuum emission is overlaid with the same contour levels as Figure 2.

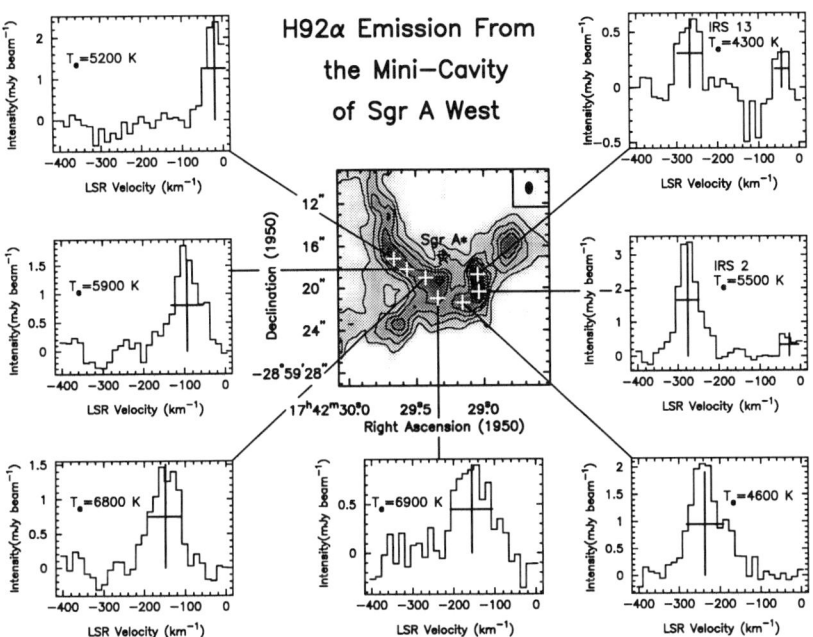

Figure 2. Gray-scale and contour representation of the 8.3 GHz emission from the inner parsec of Sgr A West from the VLA BnA-configuration. The contour levels are at 10, 20, 30, 40, 60, 75, and 90 mJy beam^{-1}. Note that 90% of the continuum emission Sgr A* has been removed. The asterisk marks the position of Sgr A* and crosses mark the positions of the associated line profiles. The cross in each profile shows the peak intensity, center velocity and FWHM velocity width. The electron temperatures are shown for each profile. For IRS 2 and IRS 13, T_e^* is calculated using continuum opacities of 0.8 and 1.1, respectively.

Table 1. Orbital parameters of western arc and minicavity. Parameters for the western arc were determined using a circular orbit model (see Roberts & Goss 1993). The modeling of the minicavity used a hyperbolic model (see Roberts, Yusef-Zadeh, & Goss 1996).

Model parameter	Derived value
Western Arc (circular)	
Inner radius	25″ (1 pc)
Outer radius	35″ (1.5 pc)
Position angle	22°± 5°
Inclination	56°± 5°
Rotational velocity	105 km s^{-1} ± 12 km s^{-1}
Dynamical center[a]	1″± 4″ W, 7″± 4″ S
Enclosed mass	3.5 × 10^6 $M_\odot$
Minicavity (hyperbolic)	
Impact parameters	4″ − 15″ (0.17 − 0.65 pc)
$v_{initial}$	200 km s^{-1}
v_{system}	0
Position Angle (PA)	+27°± 7°
Inclination (incl.)	−56°± 5°
Impact Angle (IA)	−25°± 5°
Dynamical center[a]	0″± 0.″6 E, 0.″4 ± 0.″6 N
Enclosed mass	3.0 ± 0.5 × 10^6 $M_\odot$

[a] relative to Sgr A*

suggests an enclosed mass of 6 ± 2 × 10^6 $M_\odot$. Zhao, Goss, & Ho (1995) have observed OH absorption against Sgr A using the VLA with a resolution ($\delta \times \alpha$) of 2″× 1″. The kinematics of the OH gas in the so-called "high-negative-velocity gas" (HNVG) at −180 km s^{-1} were modeled as gas in a hyperbolic orbit about Sgr A*. The hyperbolic orbit of the OH gas near the minicavity proposed by Zhao et al. motivated an examination of hyperbolic orbits to explain the velocity field of the ionized gas observed in the current observations. Zhao, Goss & Ho (this volume) also discuss the HNVG model based on the OH data.

The velocity field including all of the minicavity (0.8 × 0.8 pc [20″ × 20″]) was modeled as gas in a hyperbolic orbit around a point mass at the position of Sgr A*. The parameters of the best fit model are presented in Table 1.

In the model, an initial velocity of 200 km s^{-1} and a systemic velocity of 0 km s^{-1} were assumed. The orientation of the orbital plane with respect to the plane of the sky is determined by the position angle and the inclination. In the orbital plane, the impact angle (the angle between the initial velocity vector and the line of nodes), a mass of Sgr A*, a dynamical center, and a range of impact parameters are used to predict the observed radial velocity field. The position angle, inclination, impact angle, dynamical center and mass have been varied to minimize the residual difference between the model and observed velocity fields. For the hyperbolic orbital model, the residual is about a factor of two lower than the residuals obtained using the circular orbital model. In the best fit model, the trajectory of the gas originates to the southwest of Sgr A* and passes in front of Sgr A*. Given the model geometry, the closest approach of the ionized gas to

IONIZED GAS IN SGR A WEST   143

Sgr A* is 0.13 pc. The point mass at the position of Sgr A* is predicted to be $3.0 \pm 0.5 \times 10^6$ $M_\odot$.

3. Discussion

3.1. Modeling The Minicavity Kinematics

Using gas dynamics to probe the mass distribution of a system relies upon the understanding of the relative importance of both gravitational and non-gravitational effects. The two most significant non-gravitational effects at the Galactic center are those due to strong magnetic fields and winds from the central stellar cluster. Milligauss strength fields are suggested by modeling of the far infrared polarization (Wardle & Königl 1993) and direct observations of the H I Zeeman effect by Plante, Lo, & Crutcher (1995), in which fields $\lesssim 5$ mG are detected. In addition 1720 MHz OH maser observations in Sgr A West by Yusef-Zadeh et al. 1996 (and Green et al. this volume) also suggest line of sight magnetic fields (assuming negligible saturations effects) of 2-4 mG. However, the pressure from a field even as large as 5 mG is only 1% of the pressure from the bulk motion of a gas ($n_{HI} = 10^5$ cm^{-3}) at a velocity of 300 km s^{-1}. Additionally, the winds from the central stellar cluster are dynamically important but only at very close distances. Given a large mass loss rate 3×10^{-3} $M_\odot$ yr^{-1}, Yusef-Zadeh & Wardle (1993) argue that the winds are dynamically important only at distances from the wind source (IRS 16) less than 0.1 pc.

With a magnetic field of a few milligauss, polarization of the near IR emission at 12 microns is expected (Aitken et al 1991 and this volume). In the northern arm values up to 6.5% are observed. However, Aitken et al (1991) show that a low degree of polarization ($\leq 0.5\%$) is observed near the minicavity. If the grain alignment by the magnetic field is saturated in this region, a linear relationship between the degree of polarization of aligned dust grains and the square of radial velocity is expected. The observations presented by Aitken in this volume are consistent with this expectation. This agreement suggests that the ionized flow associated with the region near IRS 2 may be directed primarily along the line of sight.

3.2. Implications of the Mass Estimate of Sgr A*

The close proximity of the ionized gas of the minicavity (0.13 pc) to Sgr A* provides the strongest evidence for a highly concentrated mass at the position of Sgr A*. The value of the enclosed mass derived for the hyperbolic orbit is similar to the derived value for the much larger structure of the western arc (see Table 1). The fact that similar masses are derived for r < 1 pc and r < 0.13 pc suggests that the highest concentration of mass lies within the inner 0.13 pc.

Recent estimates of the mass distribution based on the velocity dispersions of hot massive stars and evolved M supergiants with different core radii give an upper limit of about 2×10^6 $M_\odot$ (Haller et al. 1996; Krabbe et al. 1995). Our estimate from minicavity kinematics of 3×10^6 $M_\odot$ appears to be consistent to within 2 σ with the determinations from stellar velocity dispersions.

The determination of the mass based on both gas and stellar dynamics provides strong evidence for a large mass within a distance of 0.13 pc that cannot be explained by luminous stars. The most likely candidate for such a concentrated mass is Sgr A*. Additionally the similarity between the mass derived from stellar and gas motions suggests that non-gravitational effects such as winds or magnetic fields are not important in determining the motion of high velocity gas near the dynamical center.

4. Summary

Kinematic features of the western arc, the northern arm, and the extended bar have been identified. The western arc has been modeled as a half ring in circular rotation about a point near Sgr A*; the mass interior to the modeled ring at a radius of 1 pc is $\sim 3.5 \times 10^6 \ M_\odot$. The global kinematics have been determined and suggest that the three major kinematic features are all unrelated. This interpretation of the global structure consisting of distinct features differs from the interpretation suggested by the [Ne II] observations in which all of the gas in Sgr A West is part of a single Keplerian disk.

The minicavity has been observed in the H 92α line at high spatial and spectral resolution. The velocity field of the minicavity is modeled by gas in a hyperbolic orbit about a point at the position of Sgr A* with a mass $\approx 3.0 \pm 0.5 \times 10^6 \ M_\odot$, consistent with previous mass estimates at larger radii. Based on this model, the closest approach of the gas to Sgr A* is 0.13 pc. The kinematics of the ionized gas determined are identical to the kinematics suggested for molecular gas of the high-negative-velocity gas from recent high resolution OH absorption observations by Zhao et al. 1995. The gas of the minicavity is likely to be the ionized part of a molecular cloud impacting close to the Galactic center. Since stars cannot explain the estimated mass, a large fraction of the mass in the inner parsec must exist in a non-stellar component in the inner 0.13 pc.

References

Aitken, D.K., Smith, C.H., Gezari, D., McCaughrean, M., & Roche, P.F. ApJ, 380, 419

Becklin, E.E., Gatley, I., & Werner, M.W. 1982, ApJ, 258, 134

Burton, M., & Allen, D. 1992, PASAu, 10, 55B

Eckart, A., Genzel, R., Krabbe, A., Hofmann, R., van der Werf, P.P. & Drapatz, S. 1992, Nature, 355, 526

Güsten, R., Genzel, R., Wright, M.C.H., Jaffe, D.T., Stutzki, J., & Harris, A .I. 1987, ApJ, 318, 124

Haller, J.W., Rieke, M.J., Rieke, G.H., Tamblyn, P., Close, L., & Melia, F. 1996, ApJ, 456, 194

Herbst, T.M., Beckwith, S.V.M., Forrest, W.J. & Pipher, J.L. 1993, AJ, V105, N3, 956

Krabbe, A., Genzel, R., Eckart, A., Najarro, F., Lutz, D., Cameron, M., Kroker, H., Tacconi-Garman, L.E., Thatte, N., Weitzel, L., Drapatz, S., Geballe, T., Sternberg, A., & Kudritzki, R. 1995, ApJ, 447, L95

Lacy, J.H., Achtermann, J.M., & Serabyn, E. 1991, ApJ, 380, L71

Liszt, H.S., van der Hulst, J.M., Burton, W.B., & Ondrechen, M.P. 1983, A&A, 126, 341
Lutz, D., Krabbe, A., & Genzel, R. 1993, ApJ, 418, 244
Lynden-Bell, D., & Rees, M.J., 1971 MNRAS, 152,461
Marr, J.M., Rudolf, A.L., Pauls, T.A., Wright, M.C.H., & Backer, D.C. 1992, ApJ, 400, L29
Plante, R.L., Lo, K.Y., & Crutcher, R.M., 1995, ApJ, 445, L113
Roberts, D.A. & Goss, W.M. 1993, ApJS, 86, 133
Roberts, D.A. ,Yusef-Zadeh,F., & Goss, W.M. 1996, ApJ,459,627
Serabyn, E., Lacy, J.H., Townes, C.H., & Bharat R. 1988, ApJ, 326,171
Schwarz, U.J., Bregman, J.D., & van Gorkom, J.H. 1989 A&A, 215, 33
Wardle, M. & Yusef-Zadeh, F. 1992, Nature, 357, 308
Wardle, M. & Königl, A. 1993 ApJ. 410, 218
Yusef-Zadeh, F. & Wardle, M. 1993, ApJ, 405, 584
Yusef-Zadeh, F., Morris, M., & Ekers, R. 1990, Nature, 348, 45
Yusef-Zadeh, F., Roberts, D.A., Goss, W.M., Frail, D.A., & Green, A.J. 1996, ApJ, in press
Zhao, J.-H., Goss, W.M., & Ho, T.P. 1995 , ApJ, 450,122

The −185 km s^{-1} Cloud Interacting with the Galactic Center

Jun-Hui Zhao[1]

ASIAA, P. O. Box 1-87, Taipei, Taiwan

W. M. Goss

NRAO, P. O. Box O, Socorro, NM 87801, USA

P. T. P. Ho

SAO, 60 Garden st, MS 78, Cambridge, MA 02138

Abstract. We present VLA observations of the −185 km s^{-1} cloud towards Sgr A. This cloud is tidally disrupted and moves towards the Galactic center under the influence of the central gravitational field. The mini-cavity 3″ SE of Sgr A* could have resulted from the impact of a high-negative-velocity streamer which may be associated with the high velocity cloud.

1. Introduction

The absorbing gas at −185 km s^{-1} in the direction of Sgr A* was discussed by Güsten & Downes (1981). The origin of this high negative velocity gas (or HNVG) was proposed by the authors to be an ejection from the Galactic nucleus. Since then numerous observations at radio wavelengths were carried out to determine the properties of this component (e.g. Marr et.al. 1992; Yusef-Zadeh et.al. 1993; Liszt & Burton 1993; and Zhao, Goss & Ho 1995). The nature of this feature was uncertain due to the lack of high angular resolution observations.

2. High-Resolution VLA OH Observations

Recently we have carried out VLA observations of this feature in the OH lines at V_{LSR}=−185 km s^{-1} with an angular resolution of several arc seconds. Fig. 1 shows the spectra of the $OH_{1667MHz}$ and $OH_{1665MHz}$ line observed with an angular resolution of 2″×1″. The peak optical depths of the absorbing $OH_{1667MHz}$ and $OH_{1665MHz}$ lines at −185 km s^{-1} are 0.07±0.01 and 0.04±0.01, respectively. The ratio of $\tau_{OH1667MHz}/\tau_{OH1665MHz}$ ≈1.8 agrees with that of optically thin OH gas in LTE. The $OH_{1667MHz}$ absorption is the best tracer of the gas in front of

---

[1]SAO, 60 Garden st, MS 78, Cambridge, MA 02138

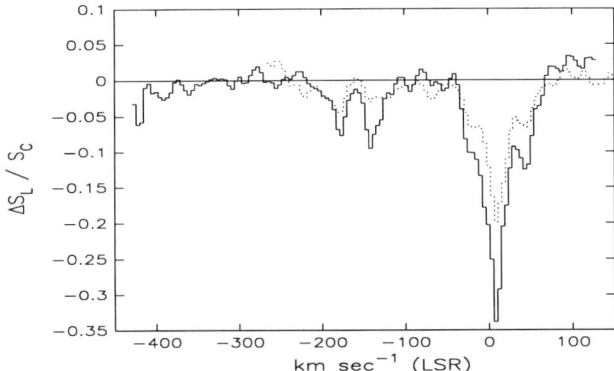

Figure 1. Spectra of the OH lines at 1667 MHz (solid) & 1665 MHz (dashed).

the continuum sources in the Galactic center. Fig. 2 shows the resulting image of the $OH_{1667MHz}$ absorption flux density averaged between −168 to −200 km s^{-1} in the central 4 arcmin of Sgr A West with a resolution of 7″. The overall morphology of the OH absorption is elongated EW with a size of 3′×1′. This elongation is consistent with HI observations carried out by Yusef-Zadeh et. al. (1993). The morphology of the OH gas in absorption is also characterized by a Z-shaped feature centered on Sgr A*.

Fig. 3 shows both the integrated optical depth and optical depth weighted velocity field images of the $OH_{1667MHz}$. The distribution of the OH optical depth appears to be non-uniform over Sgr A West. In particular, a noticeable region with lower optical depth (or OH cavity) is centered NE of but including Sgr A*. A few obvious filaments of OH are observed. In addition, Fig. 3a shows a clump of enhanced optical depth in the region 25″ SW of Sgr A*. Finally, a large change in velocity with higher negative velocities (−220 km s^{-1}) to the south and lower negative velocities (−160 km s^{-1}) to the north is observed. The velocity gradient appears to be consistent with the sense of rotation of the circum-nuclear disk on a similar angular scale (2 min of arc) but the magnitude of the gradient is smaller by at least a factor of six.

3. The Nature of HNVG

Both the structure and the kinematics of the HNVG cloud have been determined. The high-negative-velocity absorbing gas is well mixed with the ionized gas in the center of the Galaxy. The mass of neutral gas seen in absorption at −185 km s^{-1} is about 5×10^3 $M_\odot$, only a small fraction of the entire cloud observed in CO by Liszt and Burton (1993). The observed kinematics are not consistent with ejection, outflow, or a possible association with the expanding molecular ring (EMR) at V=−135 km s^{-1}. The location and morphology suggest that this high velocity cloud may be tidally disrupted by the central gravitation potential. The tidal instability may accelerate the debris of the disrupted cloud.

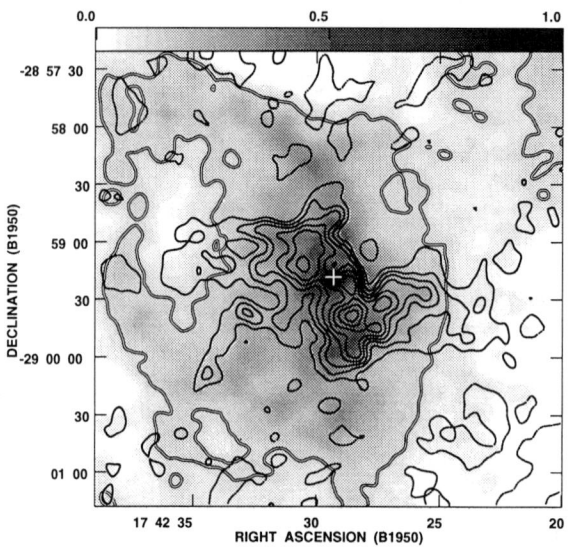

Figure 2. The OH 1667 MHz absorption line image (solid contours) of the HNVG averaged within a velocity range between −168 and −200 km s^{-1}. The double-line contour outlines the Sgr A East, possible SNR, and the greyscale representation is the overall radio continuum emission at 18 cm. The cross indicates the position of Sgr A*. The angular resolution is 7″.

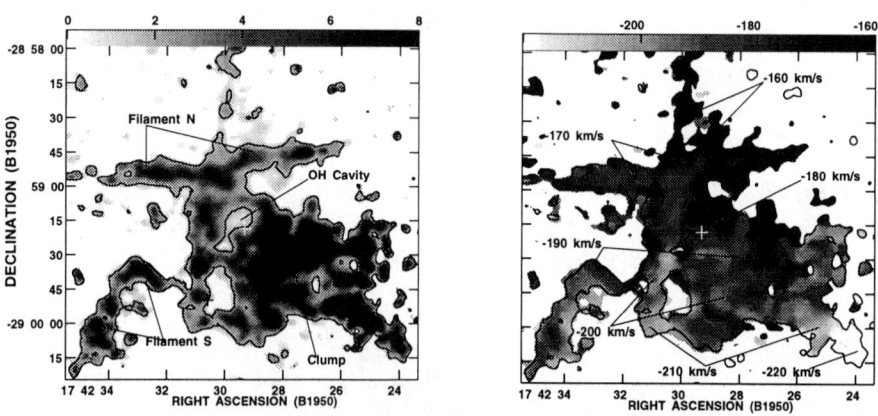

Figure 3. Left: Optical depth integrated over a velocity range between −157 and −230 km s^{-1}. Right: Velocity field weighted by the optical depth. The angular resolution is 5″.

Based on a numerical model (Zhao et. al. 1995), we have demonstrated that the disrupted high velocity gas can move towards the Galactic center and interact with the central gravitational potential, thereby distorting the kinematics of the high velocity gas projected in front of Sgr A West. The gas near Sgr A* may well be heated or ionized by the central energy sources. The model is based on the assumption that a high velocity cloud impacts the Galactic center interacting with the central gravitational potential with a point mass of 4×10^6 $M_\odot$. Fig.4 shows a comparison between observed and calculated velocity-declination diagrams. In general, the central gravitation force produces a velocity gradient for the interacting molecular gas from an average radial velocity of $V_r = -165$ km s^{-1} at 2 pc north of Sgr A* to $V_r = -195$ km s^{-1} at 2 pc south of Sgr A* (left panel). This velocity gradient is consistent with the observed pattern (right panel). In addition, a high-negative-velocity wing which is located at 0.5 pc south of the center and which extends up to –250 km s^{-1} can be produced by this model. This high-negative-velocity wing at this particular location along with the velocity gradient is probably the unique signature of the kinematics of the gas distorted by the central gravitational force.

4. Mini-Cavity

The impacting model we proposed also suggests that the mini-cavity in radio continuum 3″ SE of Sgr A* could have resulted from the impact of a high-negative-velocity streamer. The detailed kinematics observed in H92α (Roberts et. al. 1996) are difficult to explain with the wind model. Both the morphology and the velocity suggest that the central H92α emitting gas is an ionized counterpart of the HNVG which impacts the medium in the center of the Galaxy. The mini-cavity may be filled with a clump of the compressed streamer gas. The density of the gas ($n_H > 10^5$ cm^{-3}) determined from the surrounding ionized medium is large enough for a gas clump (>1″) within the mini-cavity to remain neutral. A column density of shielding material of 10^{22} cm^{-2} is estimated, which could protect the molecular gas from destruction by the central UV radiation field while the less dense, orbiting gas remains photoionized by the same UV radiation field.

Acknowledgments. The NRAO is a facility of the NSF operated under cooperative agreement by Associated Universities, Inc. JHZ was supported in parts by NSC 85-2816-M001-006L of Taiwan.

References

Güsten, R. & Downes, D., 1981, A&A 99, 27.
Liszt, H. S. & Burton, W. B., 1993, ApJL, 407, L25.
Marr, J.M., Rudolph, A.L., Pauls, T.A., Wright, M.C., Backer, D.C., 1992, ApJ, 400, L29.
Roberts, D. A., Yusef-Zadeh, F, & Goss, W. M., 1996, ApJ, 459, 627
Yusef-Zadeh, F., Lasenby, A., & Marshall, J., 1993, ApJ, 410, L27
Zhao, J.-H., Goss, W. M., & Ho, P.T. P., 1995, ApJ, 450, 122.

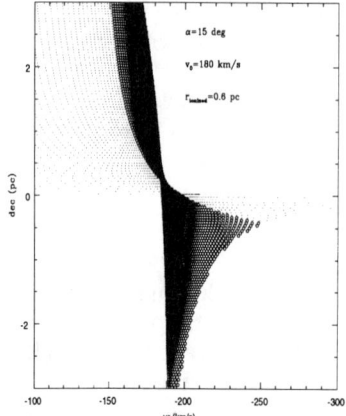 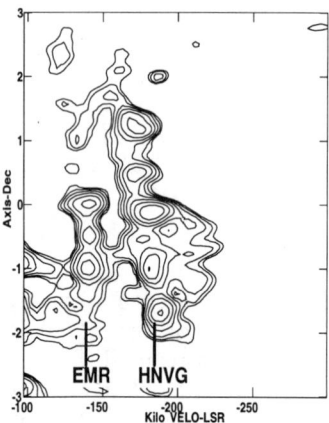

Figure 4. The velocity-declination diagrams. Left: Results are calculated from the impacting model. The pentagon symbols represent the absorbing molecular gas and the dots indicate the ionized gas. Right: The kinematic pattern of HNVG is observed at the right ascension of Sgr A*. Both the HNVG at −185 km s^{-1} and the EMR −135 km s^{-1} are labeled.

The 1720 MHz OH Emission from G359.1–0.5 and the CND

F. Yusef-Zadeh and B.T. Robinson

Northwestern University, Dept Physics and Astronomy, Evanston, IL 60208, US

D.A. Roberts

NCSA, 405 N. Mathews Ave, Urbana, IL 61801

W.M. Goss and D.A. Frail

NRAO, P.O. Box 0, Socorro, NM 87801

A. Green

University of Sydney, School of Physics, Sydney, NSW 2006

Abstract. VLA observations of the CND, which surrounds Sgr A West, and G359.1–0.5 have been carried out at the 1720 MHz transition of the OH molecule. 1720 MHz OH maser emission has been detected from the CND and G359.1-0.5 at velocities near $V_{LSR} = +132$ and -5 km s^{-1}, respectively. The OH maser features toward G359.1–0.5 have been detected along the interface between a large-scale non-thermal continuum shell G359.1–0.5 and its surrounding ring of high velocity molecular CO gas. The V spectra of a number of bright maser sources associated with the CND and G359.1–0.5 suggest strong magnetic fields between 0.4 and 4 mG. We argue that these masers are produced at the boundaries of G359.1–0.5 and the CND and that the maser features signify regions of shocked gas resulting from the supernova expansion and from cloud-cloud collisions at the Galactic center.

1. Introduction

A new and powerful probe to find evidence of shock activity was recently discussed by Frail, Goss & Slysh (1994). They observed numerous distinct 1720 MHz OH maser spots along the interface between the supernova remnant W28 and an adjacent molecular cloud. They suggest that the masers are being pumped collisionally behind the shock where H_2 molecules with densities and temperatures limited to $10^3 - 10^5$ cm^{-3} and 25-200 °K cause population inversion in the OH molecules (Elitzur 1976). More recent cross section calculations (Offer *et al.* 1994; Lockett 1995) show a significant difference between para and ortho-H_2 rates, suggesting that the ortho/para H_2 ratio, as well as the temperature and density of H_2, can be important in collisional pumping of the 1720 MHz OH transition.

Motivated by the 1720 MHz maser observations, a search was conducted to find examples such as the W28 SNR and its associated molecular complex in the Galactic center region. In this paper, we review recent detections of shocked maser emission from G359.1-0.5 and the circumnuclear disk (CND) in Sgr A West surrounding the inner 1' of the Galactic center (Yusef-Zadeh, Uchida and Roberts 1995; Yusef-Zadeh et al. 1996). We argue that the expansion of a supernova into an adjacent molecular cloud and cloud-cloud collisions are responsible for producing shocked OH maser emission. In the accompanying paper by Green et al., the results of maser activity associated with Sgr A East are described (see also Yusef-Zadeh et al. 1996).

1.1. G359.1–0.5

The shell-type SNR G359.1-0.5 was first discovered as a complete shell of polarized emission at λ6cm with a spectral index of $\approx$-0.4 (Reich & Fürst 1984). A ^{12}CO survey of the Galactic center has revealed a nearly continuous ring of molecular gas concentric with this prominent non-thermal radio continuum source (Uchida et al. 1992a). The cloud has a radius of about 12', a radial velocity which ranges between -60 and -190 km s^{-1} and is characterized as a Galactic center cloud because of its large linewidth. This morphological correlation of the non-thermal and the CO shells (Uchida et al. 1992a) as well as an HI absorption line study (Uchida et al. 1992b) suggest that both the non-thermal continuum and CO shells are physically associated with each other and that they both lie near the Galactic center.

The detection of a number of compact 1720 MHz OH maser features along the interface between the SNR shell G359.1-0.5 and its surrounding ring of high-velocity molecular gas was recently reported by Yusef-Zadeh, Uchida & Roberts (1995). An extended and weakly emitting OH feature was also noted along the brightest side of the non-thermal shell. The extended feature is considered to be a weak maser because of its small linewidth and its similarity to the velocity structure of the bright and compact maser sources. The morphological correlation between the neutral gas, the non-thermal shell and the maser features provide strong support for the hypothesis that the 1720 MHz maser line of OH arises from shocked gas by the impact of the expanding SNR into the molecular material. The boundary of the SNR and molecular clouds should be a natural place for population inversion because of the energy flow between the two components of the ISM with very different temperatures.

Follow-up VLA observations with a spatial and spectral resolution of $20.7'' \times 12''$ and 0.27 km s^{-1} were carried out in early 1996 in the 1720 MHz line and in the main lines of the OH molecule at 1665 and 1667 MHz. No main line emission from G359.1-0.5 was detected above $5\sigma = 75$ mJy level, as theoretical considerations had predicted. In the collisional pump model, only the maser amplification of the 1720 MHz transition of OH is expected to be observed (Elitzur 1976). Figure 1 shows the maser spots based on these new observations superimposed on the grayscale continuum image of the western edge of the G359.1-0.5 shell. The brightest of the OH masers appears at the location where a non-thermal filament crosses the western edge of the SNR shell. This filament

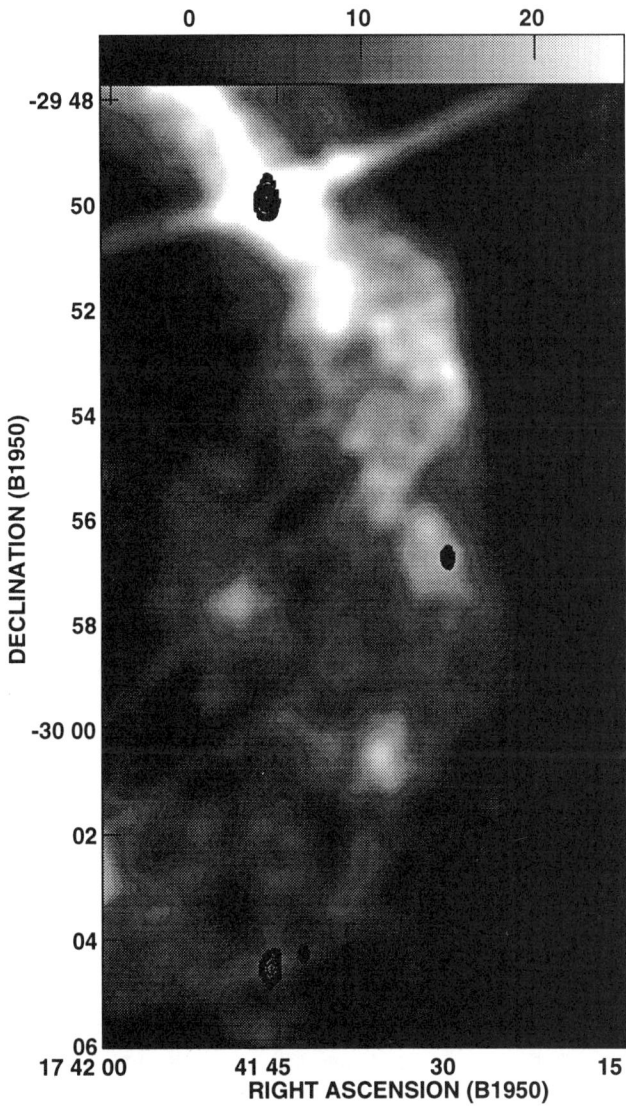

Figure 1. The black spots show the positions of the 1720 MHz OH masers A, B, C1 and C2 integrated in velocity between -7.7 and -3.7 km s^{-1} at a spatial resolution of $21'' \times 12''$. They are superimposed on the gray scale λ20cm continuum image of the western half of G359.1–0.5 SNR shell at a resolution of $33'' \times 31''$.

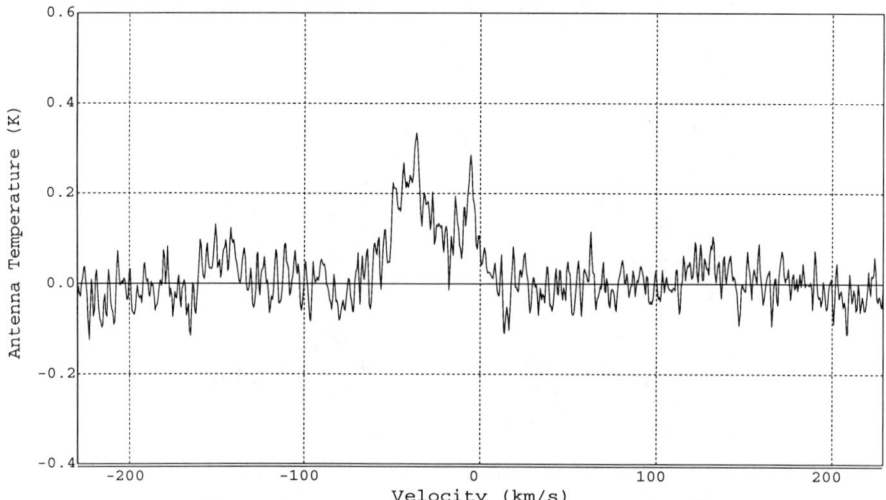

Figure 2. The CS(2–1) spectrum toward the 1720 MHz OH maser source B of G359.1-0.5 showing a peak near -6 km s^{-1}. This peak velocity is close to the velocity of the OH masers toward G359.1-0.5. This spectrum is based on observations made with the 12m Kitt Peak NRAO telescope.

is known as the "Snake" and its morphology is distinguished somewhat from the other Galactic center filaments by its long (20′) and narrow (10″) extent, and by two uncharacteristic kinks along its length (Gray et al. 1991). If there is an interaction between the non-thermal filament and the non-thermal continuum shell, the "Snake" may be responsible for an enhanced shock activity at the location where the bright maser A is noted in Figure 1. We note, however, that Zeeman measurements of the V spectrum of this maser are not unusual as they show similar polarization characteristics to other maser components in G359.1–0.5. Estimates of the line of sight component of the magnetic field based on Zeeman measurements range between 0.4 and 0.6 mG (Robinson et al. 1996).

The extended 1720 MHz OH feature that had been detected in the low-resolution (71″ × 33″) observations (Yusef-Zadeh et al. 1995) was resolved out in the new high-resolution observations as shown in Figure 1. This weak and extended maser may be the site where population inversion is achieved at low OH densities by either radiative or collisional pumps (Elitzur 1976). Similar extended and weak OH features at 1720 MHz have been found at low galactic latitudes based on single-dish surveys of the Galactic plane (Haynes and Caswell 1977; Turner 1982). These extended maser features are argued to be associated with giant molecular clouds confined to the spiral arms of the Galaxy and their excitation is considered to be due to collisional pumping at low temperatures resulting in an enhanced T_{ex} at 1720 MHz (Turner 1982).

A strong argument in favor of the physical association of 1720 MHz OH masers and the surrounding molecular cloud near SNR is the similarity of the velocity of the maser and the systemic velocity of nearby thermal gas (Frail et al. 1994; Frail et al. 1996). However, in the case of G359.1-0.5, a puzzling aspect of the nature of the association between the masers, the molecular CO, and the non-thermal shell is the strong discrepancy that is noted between the ring of molecular gas with velocities in the range between -60 and -90 km s^{-1} and the velocity of 1720 MHz masers near -5 km s^{-1}. The possibility that all three features of G359.1-0.5 are associated with each other and lie near the Galactic center has been discussed by Yusef-Zadeh et al. (1995). Here, we explore the possibilities that these rare masers arise in low-velocity foreground gas which happens to be inverted. In this picture, the source of population inversion is unclear. Furthermore, it is difficult to imagine that a ring of molecular cloud lying in the foreground matches exactly the size of the supernova shell; there is no evidence for maser emission toward other Galactic center continuum sources that lie in the field centered on G359.1-0.5. Finally, the detection of CS (2-1) emission at low negative velocities near the maser positions is evidence that the gas cloud can be associated with the SNR and with the population of Galactic center molecular clouds. Figure 2 shows the CS spectrum at $V_{LSR} = -6$ km s^{-1} centered on one of the bright masers (position B of Yusef-Zadeh, Uchida & Roberts 1995). This spectrum, which is based on recent observation of G359.1-0.5 made with the NRAO 12m telescope (Robinson et al. 1996), shows clearly two velocity components with similar peak antenna temperature. The -6 km s^{-1} velocity feature of the CS gas toward G359.1-0.5 has a similar velocity to the radial velocity of the shocked OH masers. Alternatively, the alignment of the CO ring and the SNR is a coincidence. In this picture, it is conceivable that the molecular ring is not associated with the shell but still lies near the Galactic center because of its large linewidth. One argument against this model is that the largest concentration of the CO ring is concentrated on the brightest side of the SNR. This side of G359.1-0.5 faces the Galactic plane and most of the masers, including the extended masing feature, are found here. This morphology suggests that the CO ring and the shell may be interacting.

1.2. The Circumnuclear Disk

Near IR observations of shocked gas in the CND surrounding Sgr A West have been carried out for more than a decade by imaging the distribution of H_2 emitting gas (Gatley et al. 1984; Pak et al. 1996). The ratio of the V=2-1 and 1-0 S(1) lines of H_2 have been used to probe the shocked region. However, because of the dense environment of the Galactic center surrounded by an intense UV radiation field, it has not been possible to distinguish between shocked and UV heated gas (Gatley et al. 1984; Burton & Allen 1992; Pak et al. 1996). In addition, because of the large linewidths of molecular and atomic clouds near the Galactic center, the broad CO lines accompanying shocked sites (e.g. HH objects, W28) are not good diagnostics of shocked gas in this region.

1720 MHz OH maser emission from Sgr A West has been detected at the velocity of $+134$ km s^{-1}. Figure 3 shows the position of the 1720 MHz line emission drawn as a circle. The maser source observed at a resolution of $\approx 15''$ is superimposed on a gray-scale radio continuum image of Sgr A West at

λ6cm with a resolution of $3.8'' \times 3.2''$ (PA$=-65°$). The high-resolution radio continuum image shows the peak $+132$ km s^{-1} maser line feature arising from a region of weakly emitting ionized gas sandwiched between the Eastern and Northern Arms of Sgr A West. In fact, a radio continuum radiograph of this region, as seen in Figure 4 of Yusef-Zadeh & Morris (1987), reveals a limb-brightened hole in the distribution of continuum emission at the location of the northwest Streamers. This morphology is consistent with the presence of neutral gas surrounded by ionized gas (Yusef-Zadeh, Zhao & Goss 1994). An HI absorption study of this region confirms the presence of the 130 km s^{-1} HI gas at the location of the peak maser emission (Plante, Lo & Crutcher 1995). The position of the HI Zeeman splitting measurements at $+130$ km s^{-1} is shown as a triangle. Remarkably, the strength and the orientation of the magnetic field based on Zeeman measurements of HI gas and the 1720 MHz OH masers agree (see Plante et al. 1995; Yusef-Zadeh et al. 1996). However, if we apply the new method describing the polarization of masers (Elitzur 1996a), the line of sight magnetic field estimates will be lower by a factor of about five; Elitzur (1996b) suggests that the 1720 MHz OH masers in the Galactic center are saturated.

Figure 4 displays contours of the integrated 1720 MHz emission feature based on low-resolution data integrated between $+128$ and $+138$ km s^{-1} which are superposed on the distribution of HCN (1-0) and [OI] line emission from the CND taken from Jackson et al. (1993). The maser feature is located where there is a gap in the distribution of the HCN emission molecular gas (dark-shaded areas) as traced by the HCN emission. The CND is known to orbit the Galactic center with a circular velocity of about 110 km s^{-1}; at this location the kinematics of the HCN and [OI] gas deviate from circular geometry (Jackson et al. 1993). H110α observations also indicate a feature at a velocity up to $+144$ km s^{-1} near this location. The kinematics of the ionized gas at the region of the gap are also known to be inconsistent with circular motion of ionized gas around the Galactic center (Yusef-Zadeh, Zhao & Goss 1994).

A "tongue" of molecular gas at the location where the 135 km s^{-1} OH(1720) maser is located is distinguished from the gas in the CND by its magnetic field properties. The line of sight magnetic field based on HI measurements is estimated to be about -2 mG at the location of the 1720 MHz OH maser (Plante et al. 1995). The far-IR polarization observations of this region are also noted for the unusual geometry of the magnetic fields (Hildebrand & Davidson 1994), showing a distribution which is quite similar to mid-IR polarization measurements along the Northern arm (Aitken et al. 1991). This geometry of the field differs from that predicted by an axially symmetric model of the CND which is dominated by circular motion (Hildebrand & Davidson 1994), but argues for the coupling of the ionized gas of the Northern arm to the tongue of molecular gas observed in the gap of the CND. The kinematics as well as morphological and magnetic characteristics suggest that the $+130$ km s^{-1} feature is associated with the neutral gas that is observed between the Northern and Eastern arms (Jackson et al. 1993; Yusef-Zadeh et al. 1994). In this picture, the highly blue shifted molecular and atomic gas clouds noted in HCO$^+$, HI, and OH studies (Marr et al. 1992; Pauls et al. 1993; Yusef-Zadeh, Zhao & Goss 1994; Yusef-Zadeh 1994; Zhao, Goss & Ho 1995), the red shifted [OI] (Jackson et al. 1993), and the 1720 MHz OH gas cloud are likely to be part of a single feature. The molecular species closer to the Galactic center have velocities ranging between

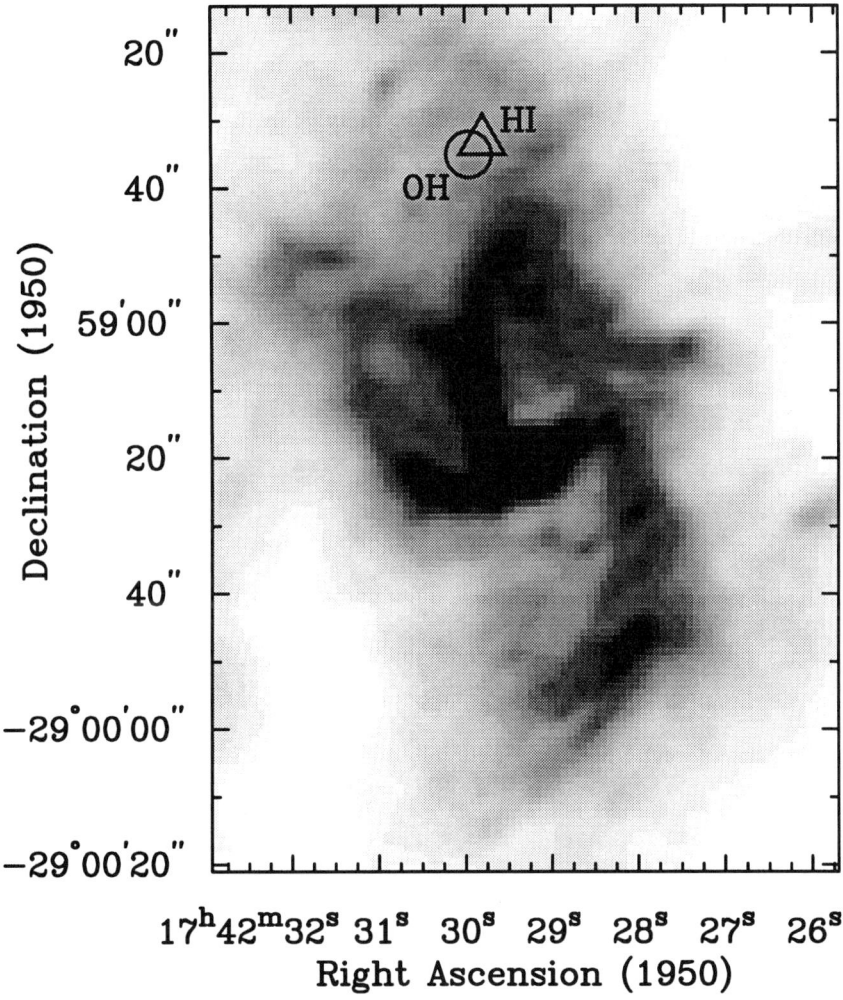

Figure 3. A λ6cm continuum image of Sgr A West with a spatial resolution of $3.7''\times3.2''$ showing the location of the 135 km s^{-1} OH(1720 MHz) maser feature as a circle. The position of the HI Zeeman splitting measurements at +130 km s^{-1} by Plante et al. (1995) is also shown as a triangle.

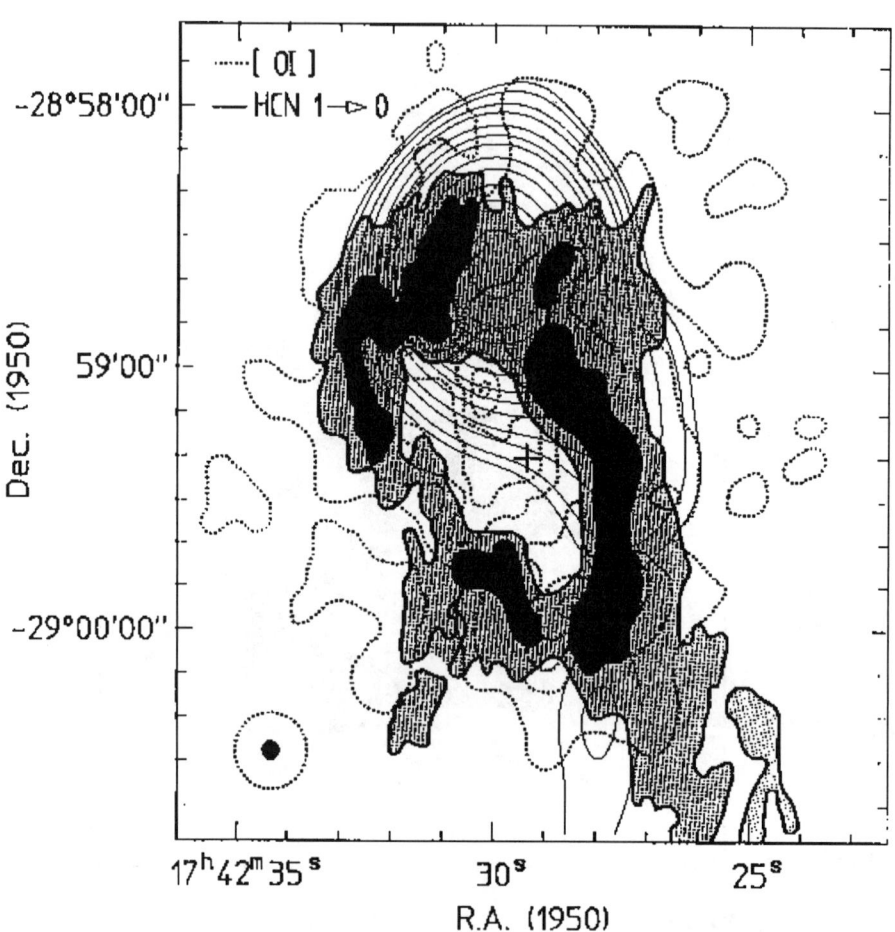

Figure 4. Black, thin, solid contours of 1720 MHz OH emission integrated between +128 and +138 km s^{-1} are displayed at levels (0.75, 1, 1.5, 2, 3, 4, 5, 7, 9, 12, 15, 20)×0.1 Jy km s^{-1} and are superimposed on the [OI] (broken contours) and HCN (dark and shaded areas) distribution of the CND (Jackson et al. 1992). The peak maser emission coincides with a gap in the distribution of the highest density HCN gas displayed as dark areas.

−150 and −180 km s^{-1} whereas the clouds closer to CND have velocities ranging between +70 to +140 km s^{-1}. This feature is considered to be interacting with the circular-moving gas in the CND. This cloud, with an assumed initial velocity of $\approx$ +150 km s^{-1}, could be colliding with the CND gas with a relative velocity of $\approx$ +50 km s^{-1}. The resultant collision would disrupt a segment of the CND near the gap and produce the shocked gas at $\approx$+130 km s^{-1}. Assuming that the area from which the shocked emission arises is $\approx$ 30″ (1.2 pc), the total luminosity that the shock generates is $\approx 10^3$ L$_\odot$ assuming that the gas number density is about 2×10^3 cm^{-3}.

Is the intruding cloud infalling or outflowing from the Galactic center? If the cloud is infalling, a consequence of the interaction is that the red shifted motion of the intruding cloud should become blue shifted as it nears the Galactic center. At the location where the cloud collides with CND, the 130 km s^{-1} OH maser delineates the shocked region. The schematic diagram in Figure 5 presents a picture in which the intruding cloud is stretched along its trajectory as the magnetic field is expected to become uniform, as suggested by mid and far-IR polarization data. The Zeeman splitting of HI and OH suggest a line of sight magnetic field between −4 and −2 mG field at the position of the shocked gas. The large strength of the magnetic field can be explained as a result of the compression of the field as the intruding cloud collides with the the CND. In this picture the Northern and Eastern arms delineate the edges of the intruding cloud photoionized by the UV radiation field at the Galactic center (Jackson *et al.* 1992). The weakly-emitting ionized Streamers beyond the CND may then be due to the low-density ionized rims of the intruding cloud blown by the powerful winds associated with the IRS 16 cluster (Krabbe *et al.* 1991).

The above infall picture does not explain the origin of the intruding high velocity cloud. The expansion of the Sgr A East shell into its molecular counterpart (M–0.02–0.07) can account naturally for the origin of the high velocity neutral clouds toward the Galactic center. However, the red-shifted radial velocity of the infalling cloud is inconsistent with the relative location of Sgr A East with respect to the CND. Low-frequency radio continuum observations have revealed that Sgr A East must lie behind Sgr A West (Yusef-Zadeh and Morris 1987; Pedlar *et al.* 1989). Thus, we consider an alternative schematic model, as presented in Figure 6, in which a high negative velocity cloud is accelerated toward the Galactic center as it follows a parabolic orbit around Sgr A* colliding with the northern part of the CND. This model implies that Sgr A East is behind but in the vicinity of Sgr A West and the CND.

There is no evidence favoring the infall vs the outflow models. In the outflow picture, however, the expansion of Sgr A East into the +50 km s^{-1} molecular cloud can naturally explain the origin of peculiar-moving neutral and ionized clouds (e.g. Yusef-Zadeh, Lasenby & Marshall 1993).

Acknowledgements: We thank Jennifer Reddy for drawing Figures 5 and 6 and Mark Morris for useful discussions. Yusef-Zadeh's work was supported by NASA grant NAGW-2518. The National Radio Astronomy Observatory is a facility of the National Science Foundation, operated under a cooperative agreement by Associated Universities, Inc.

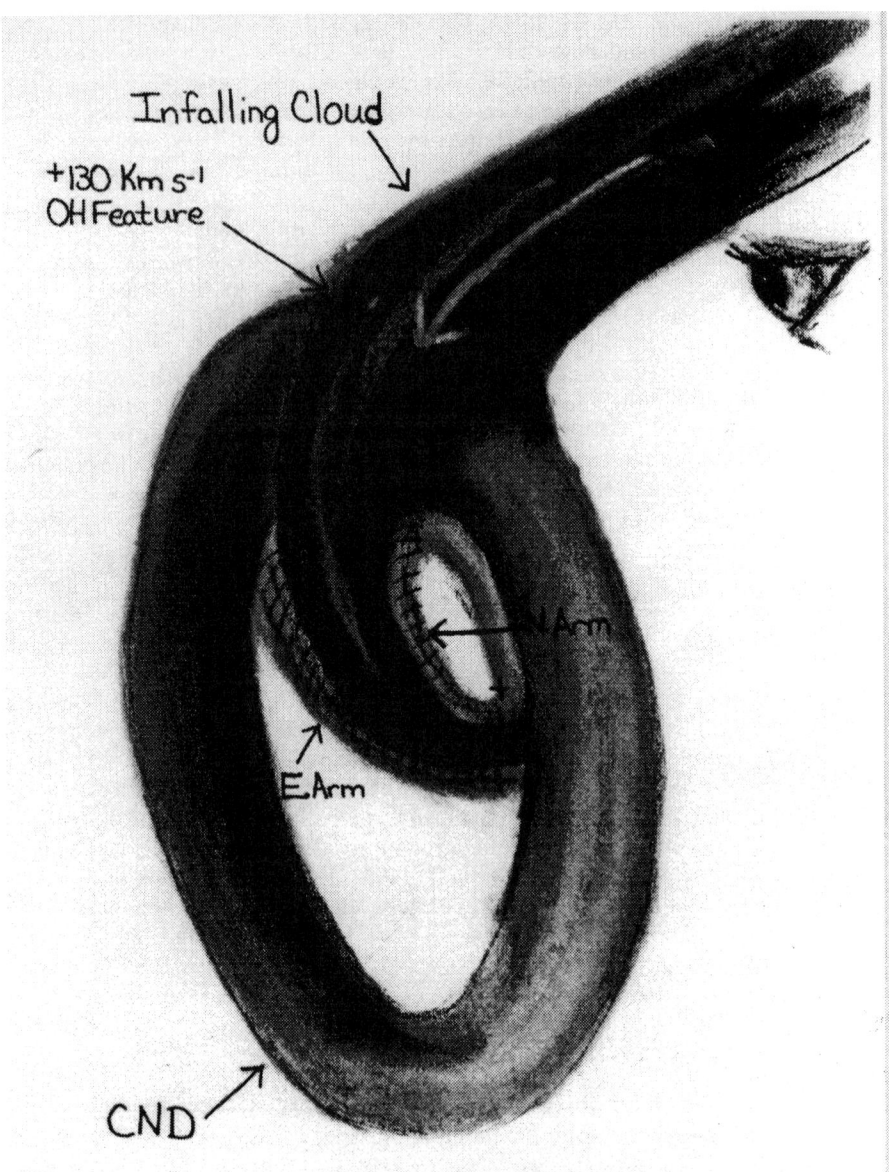

Figure 5. A schematic diagram showing the infall model of gas flow toward the Galactic center. This model represents a tongue of gas infalling toward the center. The cross represents the position of Sgr A*.

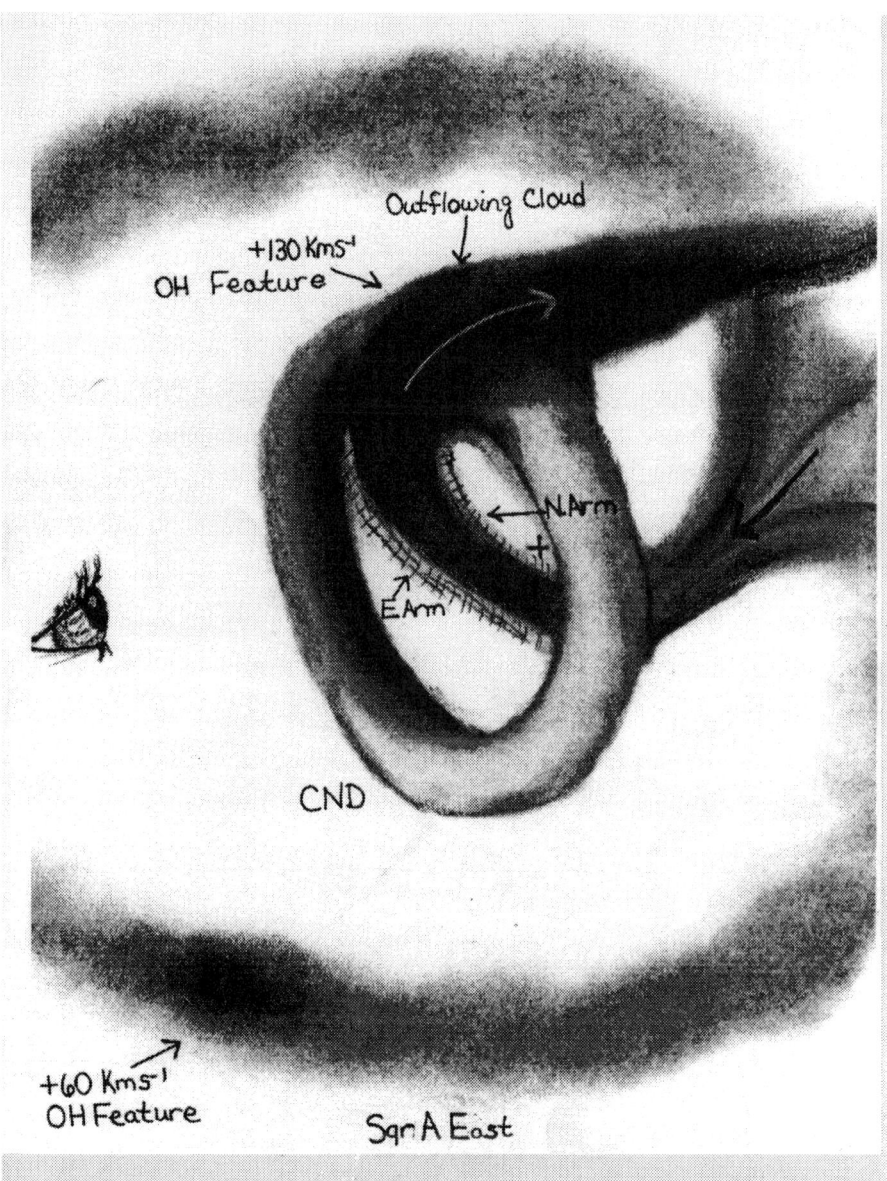

Figure 6. A schematic diagram showing the outflow model of gas flow from the Galactic center. In the outflow picture, the origin of the cloud is associated with the expansion of the Sgr A East SNR which lies behind the Northern and Eastern arms of Sgr A West.

References

Aitken, D.K., Gezari, D., Smith, C.H., McCaughrean, M. & Roche, P.F. 1991, ApJ, 419, 419

Burton, M. & Allen, D.A. 1992, in Proc.Astr.Soc.Australia, 10, 55

Elitzur, M. 1976, ApJ, 203, 124

Elitzur, M. 1996a, ApJ, 457, 415

Elitzur, M. 1996b, ApJ, submitted.

Frail, D.A., Goss, W.M. & Slysh, V.I. 1994, ApJ, 424, L111

Frail, D.A., Goss, W.M., Reynoso, E.M., Giacani, E.B., Green, A.J. & Otrupcek, R. 1996, A.J., 111, 1651

Gatley, I., Jones, T.J., Hyland, A.R., Wade, R., Geballe, T.R. & Krisciunas, K. 1986, MNRAS, 222, 299

Gray, A.D., Cram, L.E., Ekers, R.D. & Goss, W.M. 1991, Nature, 353, 237

Haynes, R.F. & Caswell, J.L. 1977, MNRAS, 178, 219

Hildebrand, R.H. & Davidson, J.A. 1994, in *The Nuclei of Normal Galaxies: Lessons from the Galactic Center*, eds. R. Genzel and A. Harris, Dordrecht: Kluwer, p199

Jackson, J. M. et al. 1993, ApJ, 402, 173

Krabbe, A., Genzel, R., Drapatz, S. & Rotaciuc, V. 1991, ApJ, 382, L19

Lockett, P. 1995, BAAS, 27, 34.01

Marr, J.M., Rudolph, A.L., Pauls, T.A., Wright, M.C., Backer, D.C. ApJ, 1992, 400, L29

Offer, A.R., van Hemert, M.C. & van Dishoeck, E.F. 1994, J.Chem.Phys, 100, 362

Pak, S., Jaffe, D.T. & Keller, L.D. 1996, ApJ, 457, L43

Pedlar et al. 1989, ApJ, 342, 769

Plante, R.L., Lo, K.Y., & Crutcher, R.M. 1995, ApJ, 445, L113

Reich, W. & Fürst, E. 1984, A.A.Suppl. 57, 165

Robinson, B.T. et al. 1996, in preparation.

Turner, B.E. 1982, ApJ. 255, L33

Uchida, K.I., Morris, M., Bally, J., Pound, M. & Yusef-Zadeh, F. 1992a, ApJ, 398, 128

Uchida, K.I., Morris, M. & Yusef-Zadeh, M. 1992b, AJ, 104, 1533

Yusef-Zadeh, F., Lasenby, A. & Marshall, J. 1993, ApJ, 410, L27

Yusef-Zadeh, F. & Morris, M. 1987, ApJ, 322, 721

Yusef-Zadeh, F., Roberts, D.A., Goss, W.M., Frail, D. & Green, A. 1996, ApJ, submitted.

Yusef-Zadeh, F., Uchida, K.I. & Roberts. D.A. 1992, Science, 283, 601

Yusef-Zadeh, F., Zhao, J.-H. & Goss, W.M. 1995, ApJ, 442, 646

H_2 Emission from the Galactic Centre Molecular Ring

S.K. Ramsay Howat

Royal Observatory, Blackford Hill, Edinburgh EH9 3HJ, UK

C.M. Mountain

Gemini 8-m Telescopes Project, 950 N. Cherry Avenue, PO Box 267323, Tucson, Arizona 85726, U.S.A.

T.R. Geballe

Joint Astronomy Centre, 660 N. A'Ohoku Place, University Park, Hilo, Hawaii 96720, U.S.A.

Abstract. Observations of the ring of molecular gas at 2pc radius from Sgr A* have been used to investigate the nature of the central source. The ring was first discovered in the light of the 1-0 S(1) line of molecular hydrogen and a number of groups have since measured rotational-vibrational lines of H_2 emission to determine whether the spectrum is characteristic of gas thermally excited by a shock or radiatively excited by UV photons. We present analysis of new NIR spectra of lines of sight in the molecular ring and show that the derived H_2 column densities are consistent with fluorescent emission from a dense gas. High spatial resolution, velocity-resolved spectra of the 1-0 S(1) line confirm existing models of the rotation of the ring. It is shown that the observed line widths ($\sim$50 kms^{-1}) can be produced without a shock.

1. Introduction

The physical conditions and dynamics of the Galactic centre circumnuclear disk (CND) have been studied in many molecular species (e.g. Jackson et al. 1993) as many authors investigate the nature of the central source through its effect on the CND. The H_2 emission was initially thought to be excited by a shock due to the impact of a wind from a central mass loss source (Gatley et al. 1984), a deduction based on both the observed ratio of the 1-0 S(1)/2-1 S(1) lines and the observed width of the H_2 lines ($\sim$100kms^{-1}). H_2 may also be excited by UV photons from young stars; the fluorescent emission from H_2 has been measured in star-forming regions (e.g. Orion, M17). The motivation for this renewed study of the H_2 emission from the CND, at higher spatial and spectral resolution than achieved previously, can be summarised as follows. Work by Burton, Hollenbach & Tielens (1990) and Sternberg & Dalgarno (1989) on fluorescent emission from high density gas showed that the spectrum of emission from the lowest energy levels does not discriminate between fluorescent emission from dense H_2 and

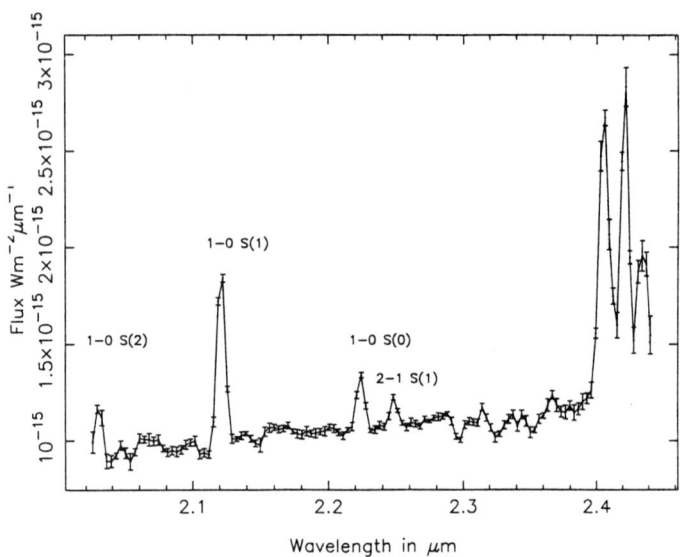

Figure 1. A CGS4 spectrum of the Galactic centre nuclear ring, showing H_2 emission lines. This position is 28.6″E, 40.4″N of SgrA*.

thermal emission from shock excited gas (see below). The high UV flux from the central source and the association of the H_2 emission in the CND with other molecular species excited in star forming regions suggest that the H_2 may be radiatively excited.

UV excitation of H_2 is a non-LTE process. Absorption of a photon in the range 92-108nm excites the molecule into the first electronic band from which it either dissociates or decays into the high rotational-vibrational levels of the ground state. In the latter case, the details of the cascade through those levels define a characteristic emission spectrum in the near-infrared, modelled by Black & van Dishoeck (1987). One result from this modelling is that the 1-0 S(1)/2-1 S(1) line ratio has a value $\sim$3, significantly different the LTE value of $\sim$10 expected for shock excitation. This ratio became the diagnostic for radiative excitation vs. shock excitation, and has been widely used. Gatley et al. measured a value of $\sim$10 for the CND and concluded that the gas was thermally excited. Subsequent modelling of fluorescent emission from gas of higher density illuminated by a stronger UV field by Burton, Hollenbach & Tielens (1990) and Sternberg & Dalgarno (1989) showed that the lower energy levels are affected by collisions in a gas with density greater than the critical density of $\sim 10^5 cm^{-3}$. The column densities in the lower levels become 'thermalised'; the column density approaches the value for LTE. Thus, lines from higher energy levels must be measured to determine the excitation mechanism.

Analysis of new infrared spectra of the H_2 in the CND, both the line ratios from K band spectra and dynamical information from velocity resolved spectra show that the emission is consistent with either shock excitation or excitation of a dense gas by UV photons.

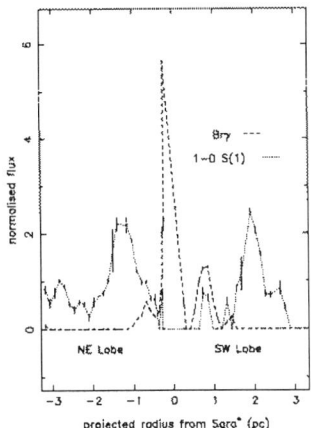

Figure 2. The distribution of ionised and molecular gas in the central few parsecs of the Galactic centre, as traced by the H I Brγ line and the H$_2$ 1-0 S(1) line. Note the existence of molecular emission at a projected distance inside the radius of the ring ($\sim$ 1.5pc).

2. Observations

Using CGS4 on UKIRT two slit positions were measured across the peak H$_2$ emission from the CND, including both the CND and the ionised region surrounding IRS16/SgrA*. Both low and high spectral resolution spectra were taken. The K band spectrum from 2.03μm- 2.44μm was measured with R($= \frac{\lambda}{\delta\lambda}$) $\sim$ 333, a pixel scale of 3″ and a total slit length of 90″. Another K band spectrum was taken with R$\sim$650; two overlapping spectra were taken to cover the wavelength range from 2.006μm-2.399μm. The high resolution, echelle, spectra have R$\sim$8000, or a velocity dispersion per pixel of 37kms^{-1}. The pixel scale was 2.5-3.8 arcsecs per pixel with the echelle grating. The 3″ pixels correspond to 0.14pc at the distance of the Galactic centre (8kpc).

3. Results and Discussion

3.1. The Low Resolution Spectra

A spectrum from the first set of measurements are shown in Figure 1. It is of a position in the molecular ring, offset 28″.6 east, 40″.4 north of Sgr A*. The spectrum is dominated by emission lines from H$_2$ - the 1-0 S(0), S(1), S(2) and 2-1 S(1) lines are seen; the Q branch lines are partially resolved. The 1-0 S(1)/ 2-1 S(1) ratio measured is 9.23$\pm$0.86. The lines detected in the ionised region were the HI Brackett γ line, the 2p-2s HeI line and the 2.217μm line first discovered at the Galactic centre by Allen, Hyland & Hillier (1990).

The distribution of ionised and molecular gas is shown in a plot of Brγ line intensity and 1-0 S(1) line intensity (Figure 2). The Brγ is centrally peaked, falling to zero at radii greater than $\sim$1.5pc, the radius of the CND. H_2 emission peaks in the CND, but some molecular gas is also seen at a projected radius of 0.8pc, implying that it lies within the central cavity or is foreground material. This result is repeated in the echelle spectra discussed below.

Table 1.   Measured strengths of the 1–0 S(0), 1–0 S(2) and 2–1 S(1) lines compared with the predicted values for the shock models

Line	Flux relative to 1-0 S(1) measured	predicted by S&H
1–0 S(0)	0.23±0.04	0.22
1–0 S(2)	0.32±0.04	0.36
2–1 S(1)	0.11±0.01	0.10

From the relative strengths of the H_2 lines measured (Table 1), the excitation temperature of the gas and the ortho/para ratio were determined. The rotational and vibrational excitation temperatures are 1404K and 2220K respectively; the ortho/para ratio is 3.0±0.3. An ortho/para ratio of less than 3 or a vibrational excitation temperature much larger than the rotational temperature indicate fluorescent emission. The values from our observations indicate that the lower energy levels are thermally populated.

The measured line ratios and intensities have been compared with results from two models: the Burton, Hollenbach & Tielens (1990) model of a radiatively excited dense gas and the Shull & Hollenbach (1978) model of excitation by a shock in which the shock speeds are not large enough to dissociate the H_2. The line ratios can be explained by both models.

Comparing the 1-0 S(1)/2-1 S(1) ratio with the models of Burton, Hollenbach & Tielens (1990), a gas density of $5 \cdot 10^5 cm^{-3} - 7 \cdot 10^7 cm^{-3}$ illuminated by a UV field of $3 \cdot 10^4 - 10^5 G_o$ is required. This would give a flux of $2 \cdot 10^{-4}$ergs s^{-1} cm^{-2}, in good agreement with the observed flux which lies in the range $6.6 \cdot 10^{-5}$ to $5.3 \cdot 10^{-4}$ergs s^{-1} cm^{-2}. Evidence from other observations suggest that such conditions do exist within the molecular ring: Davidson et al. (1992) estimate that the central source has a luminosity of $10^7 L_{Solar}$, which would give $I_{uv} = 5 \cdot 10^4 G_o$. Densities of greater than $10^5 cm^{-3}$ are implied from observations of many species in the CND (e.g. Jackson et al. 1993).

The best-fit to a shock model of the H_2 in the ring implies that a gas of density $3 \cdot 10^5 cm^{-3}$ is excited by the passage of a shock of speed v_s=10kms^{-1}. This density is consistent with those seen in the ring by other authors. The intensity of the 1-0 S(1) line predicted by this model is $5.46 \cdot 10^{-3}$ ergs s^{-1}cm^{-2}, considerably brighter than the observed values.

All these results have been corrected for the extinction to the Galactic centre assuming a $\lambda^{-1.75}$ power law for the extinction (Draine 1988) and A_K=3.3 (Wade et al. 1987). The measured 1-0 S(1)/2-1 S(1) ratio is higher than the value of

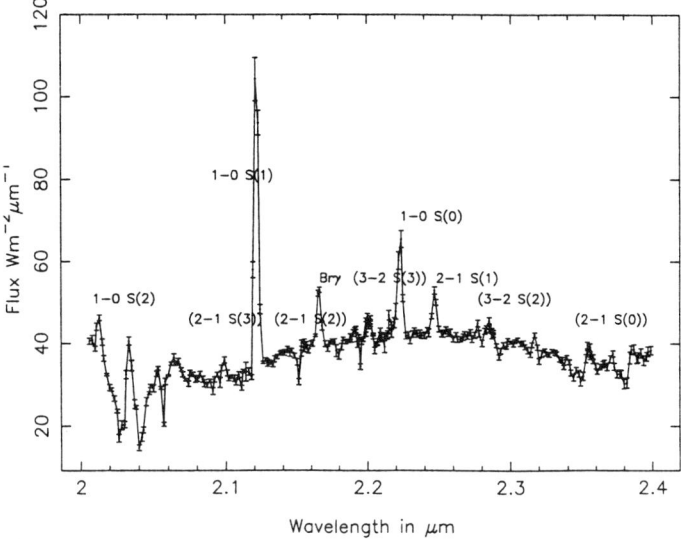

Figure 3. The Galactic centre molecular ring, observed with resolving power R~665. The simulated spectrum, assuming that the emission is fluorescent, has been overlayed. Both the detected and non-detected lines are labelled, with the non-detected lines in parantheses.

~6 determined by Burton & Allen (1993). Burton & Allen used a lower value for the extinction; dereddening our data using their value we find a ratio of 8.4±1.

A limiting factor in making these observations is the presence of telluric features and stellar absorption features from stars in the field of view. The K band spectrum was repeated at higher resolution (Figure 3) to try to resolve the higher excitation lines. The telluric features have been removed by ratioing by an A star observed at a close airmass to the Galactic centre observations. A K star was also observed at the same time. The continuum from this star was subtracted from the Galactic centre observations to correct for the late-type stellar features after the stellar spectrum had been ratioed with the A star. Residuals of both types of feature are still seen in the continuum; these provide the limit to how well the higher excitation lines can be measured at this resolution. No new lines were observed. To test the significance of these non-detections, a spectrum was simulated to include the higher excitation lines at the ratios predicted by Black & van Dishoeck (1987). It is assumed that the column densities of 2-1 and 3-2 branches are those predicted by the fluorescent cascade i.e. that the 2-1 S(1) line strength can be used to predict that strengths of the other lines. To populate these levels collisionally, and invalidate this assumption, requires densities in excess of $10^7 \rm{cm}^{-3}$ and an I_{uv} of 10^5. The simulated spectrum is indistinguishable from the original, within the errors; the measured emission spectrum remains consistent with the conclusions outlined above.

3.2. The Echelle Spectra

Table 2. The parameters for the models for the dynamics of the Galactic centre plotted. The R_o are scaled to the Galactic centre distance of 8kpc.

CND Edge R_o (pc)	Velocity at CND edge V_o kms^{-1}	Radial velocity dependence	Reference
1.54	100	-0.5	Harris et al. 1985
1.30	130	-0.35	Lugten et al. 1986
1.56	110	-	Jackson et al. 1993

The line widths measured for the 1-0 S(1) line at the Galactic centre are typically $\sim$50kms^{-1}, broader than the instrumental resolution of 38kms^{-1}. Many of the observed lines contain two components, such as the spectrum shown in Figure 4.

Since the velocity resolved data consist of only two lines of sight across the center, the dataset has not been used to constrain a new model of the dynamics of the CND, but rather compared with existing models. Figure 5 shows the velocities measured for the H$_2$ gas plotted with three models representative of those to be found in the literature. The parameters of the models are contained in Table 2. The best match to our data is from the Harris et al. model of a ring with velocity 100kms^{-1} at the edge (R=1.54pc) and a $r^{-0.5}$ dependence of the velocity on the radius of the ring.

The bulk of the gas shows velocities consistent with rotation of the ring. However, in common with other investigators, we see evidence for clumps of gas that do not participate in the rotation of the ring. The spectrum in Figure 4 is from a position to the SW of SgrA* where a clump of red-shifted emission (v$\sim$50kms^{-1}) seen extending from 45″ − 35″W and 40″ − 30″S. Features with this velocity that do not fit the rotation of the ring have been seen by Jackson et al. (1993) in their HCN data. This material may be associated with the +50kms^{-1} molecular cloud. Another streamer of gas extends from 3″.9W, 4″.3N to 9″.7W, 13″.3S of SgrA*; it gas a velocity gradient greater than that of the ring and is apparently located within the central cavity.

The broad line widths observed by Gatley et al. (1986) supported the theory that the H$_2$ was excited by a shock. Using the Harris et al. (1985) model for the dynamics of the ring, the velocity was integrated over each pixel of the CGS4 slit to give a value for the line width expected from the rotation alone. The measured dispersions and those calculated are shown in Figure 5. It is seen that the velocity dispersion due to rotation is sufficient to account for the width of the lines.

4. Summary

The molecular hydrogen line ratios measured in the CND can be explained by either radiative excitation of a dense gas ($5 \cdot 10^5 - 10^7$cm^{-3}) by UV photons from a source of strength $10^4 - 10^5 G_o$ or thermal excitation of a gas of density $3 \cdot 10^5$cm^{-3} by shock with $v_s = 10$kms^{-1}. The spatial and velocity distributions

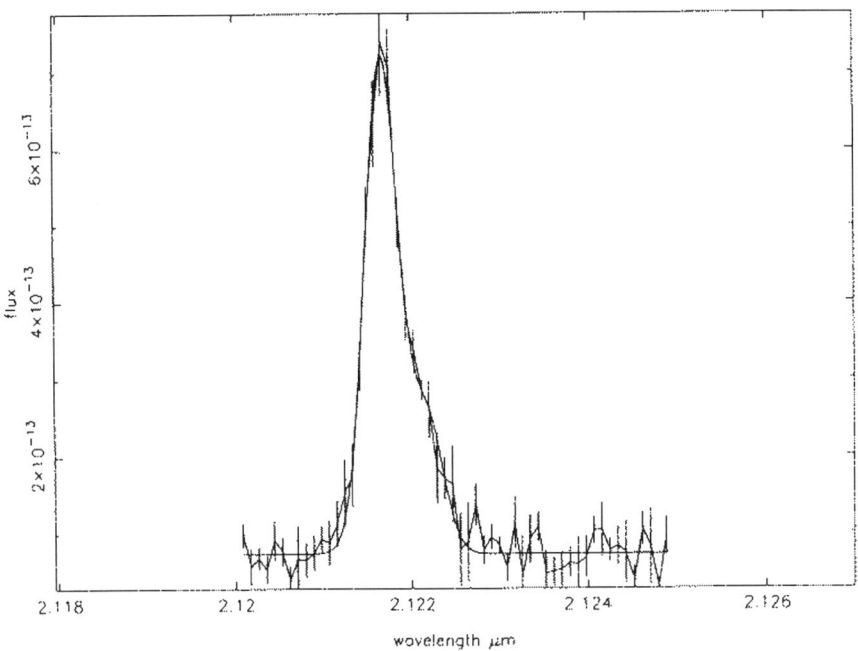

Figure 4. A typical double-peaked spectrum, measured in the southwest of the CND. The brightest emission follows the expected rotation, but a component of gas with a red-shifted velocity of $\sim 50 \mathrm{kms}^{-1}$ is also seen.

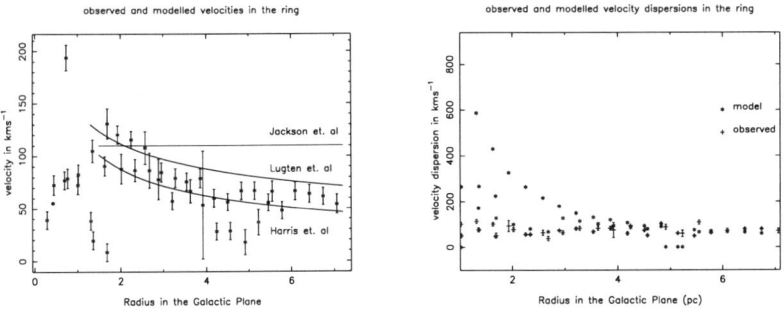

Figure 5. The measured and modelled line widths due to rotation of the ring.

of the emission confirm the Harris et al. (1985) model of the ring, implying a central point like mass source of $4 \cdot 10^6$ MSolar. Taking this model for the velocity field of the ring, the observed line widths can be explained by the rotation of the ring integrated over the 3″ pixels and no thermal broadening is required to explain the line widths. There is some evidence for molecular gas at a projected radius smaller than that of the ring, but the data do not distinguish between foreground gas or gas within the central cavity.

References

Allen, D.A., Hyland, A.R. & Hillier, D.J. 1990, MNRAS, 244, 706
Black, D.H. & van Dishoeck, E.F. 1987, ApJ, 322, 412
Burton, M.G. & Allen, D.H. 1993, in 'Astronomical Infrared Spectroscopy: Future Observational Directions', A.S.P. Conference Series 41, ed. S.Kwok, 289
Burton, M.G., Hollenbach, D.J. & Tielens, A.G.G.M. 1990, ApJ, 365, 620
Davidson, J.A., Werner, M.W., Wu, X., Lester, D.F., Harvey, P.M., Joy, M. & Morris, M. 1992, ApJ, 387, 189
Draine, B.T. 1989, in Proceedings of the 22nd ESLAB Symposium 'Infrared Spectroscopy in Astronomy', 93
Gatley, I., Jones, T.J., Hyland, A.R., Beattie, D.H. & Lee, T.J. 1984, MNRAS, 210, 565
Gatley, I., Jones, T.J., Hyland, A.R., Wade, R., Geballe, T.R. & Krisciunas, K. 1986, MNRAS, 222, 299
Harris, A.I., Jaffe, D.T., Silber, M. & Genzel, R. 1985, ApJ, 294, L93
Jackson, J.M., Geis, N., Genzel, R., Harris, A.I., Madden, S.C., Poglitsch, A., Stacey, G.J. & Townes, C.H. 1993, ApJ, 402, 173-184
Lugten, J.B., Genzel, R., Crawford, M.K. & Townes, C.H. 1986, ApJ, 306, 691
Shull, J.M. & Hollenbach, D.J. 1978, ApJ, 220, 525
Sternberg, A. & Dalgarno, A. 1989, ApJ, 338, 197
Wade, R., Geballe, T.R., Krisciunas, K., Gatley, I. & Bird, M.C. 1987, ApJ, 320, 570

Molecular Line Maps of the Galactic Centre Circumnuclear Disc

Glenn J. White

Physics Department, Queen Mary & Westfield College,
University of London, Mile End Road, London E1 4NS, England.

Abstract. Prelimary results are discussed for a long-term programme carried out with the James Clerk Maxwell 15m Telescope, to map the structure and dynamics of the Circumnuclear Disc at the Galactic Centre in a wide range of millimetre and submilletre wavelength molecular line transitions

1. Observations and Data Reduction

The data were obtained with the 15 metre James Clerk Maxwell Telescope (JCMT) at Mauna Kea, Hawaii at various times between 1987 and 1994. The beam sizes of the observations in this study ($\sim$ 10 - 23 arc seconds) corresponded to linear resolutions of $\sim$ 0.4 - 0.8 pc (15 arc seconds $\approx$ 0.62 pc), and the map centres are located at the position of SgrA*: $\alpha_{1950} = 17^h42^m29^s.3$, $\delta_{1950} = -28°59'19''$.

2. History of the Galactic Plane Mapping Project

The first observations during commissioning of the JCMT in 1987 in the CO J=2-1 and 3-2, CS J= 7-6 , HCN J= 4-3 lines showed the $C^{18}O$ and ^{13}CO isotopes to be surpisingly weak, whereas observations of the HCN J= 4-3 and CS J= 7-6 lines showed them to be relatively intense. Mapping was hard work in those days when system temperatures were typically 3000 - 10,000 K. A more extensive series of maps were followed in 1988 by a spectral scan of the 330 - 357 GHz window towards three bright spots which had been identified on the rim of the CND, supplemented by more speculative targeted searches in the 220 - 270 GHz window for species such as H_2O, O_3 etc. Between 1989 and 1994, fully sampled maps over an area about 2 x 2 arc minutes were made for all the lines seen in the spectral survey which had antenna temperatures greater than about 1 K. These maps were supplemented by deeper observations in key isotopomers between 1990 and 1995, and the programme was completed by April 1996 with the acquisition (after years of being beaten by Mauna Kea weather !) of maps in the CO J=4-3 and $^3P_1$ - $^3P_0$ atomic carbon lines maps, and a larger area (4' x 8' along the galactic plane) in the CO and ^{13}CO J= 3-2 lines. The final key piece of work remaining to be completed is the ISO high resolution spectral survey, which is contained within the SWS and LWS Guaranteed time programmes.

Lines which have been studied (with maps available for the majority) include ^{13}CN N= 2-1 (217.142 GHz), C^{18}O J= 2-1 (219.560 GHz), ^{13}CO J= 2-1 (220.398 GHz), CN N= 2-1 (226.632 GHz), CN N= 2-1 (226.875 GHz), CO J= 2-1 (230.538 GHz), C^{34}S J= 5-4 (241.016 GHz), CH$_3$OH 5_0 - 4_0 E (241.700 GHz), CH$_3$OH 5_{-1} - 4_{-1} E (241.767 GHz), CH$_3$OH 5_0- 4_0 A$^+$ (241.791 GHz), CS J= 5-4 (244.936 GHz), SO 6_6 - 5_5 (258.255 GHz), C$_2$H (260.03 GHz), H$_{13}$CO$^+$ J= 3-2 (260.256 GHz), SiO J= 6-5 (260.518 GHz), HCN J= 3-2 (265.816 GHz), ^{13}CO J= 3-2 (330.588 GHz), SO 3_3 - 3_2 (339.341 GHz), CN (340.035 GHz), CN (340.248 GHz), CH$_3$OH 13_1 - 13_0 A$^-$ (342.730 GHz), CS J= 7-6 (342.883 GHz), SO 8_8 - 7_7 (344.310 GHz), H^{13}CN J= 4-3 (345.340 GHz), CO J= 3-2 (345.796 GHz), H^{13}CO$^+$ J= 4-3 (346.999 GHz), SiO (347.331 GHz), HCN J= 4-3 (354.505 GHz), HCO$^+$ J= 4-3 (356.734 GHz), CO J= 4-3 (461.041 GHz), CI ^{3}P$_1$ - ^{3}P$_0$ (492.161 GHz)

A complete discussion of these data are beyond the scope of this short paper, however, it is possible to summarise some general conclusions.

(1) The chemistry of the CND is on the whole photon dominated excitation - rather than collisionally dominated as in many dense molecular cores.

(2) Some species are very under-abundant. There are however some chemical indicators we have detected, such as SiO, which may suggest high temperature, or shock activity - but recent detection's of this species in absorption against SgrB$_2$ may suggest that our understanding of the relationship of SiO to shocked material should be treated with caution.

(3) Many lines, even dominant isotopomers, are optically thin, and gross variations of line shapes from isotope to isotope as well as species to species can complicate the analysis. Many of the problems encountered in the 3 mm window, where line of sight self absorption in lower J-transitions can have significant effects on the line profiles, are minimised by observing in the higher rotational transitions - which require a higher excitation environment.

(4) Most main isotopes - as well as main lines (with the exception of CO) - are optically thin. The CO isotopes are surprisingly weak - beam-matched data suggest an environment with $n_{H2} \sim 10^5$ - 10^6 cm^{-3} and $T_{kin} \sim 200$ - 300 K.

(5) The ring dynamics are traceable in all molecular species - even CO. The CND's structure may be better described by group of co-rotating filaments within a loose ring structure, than an coherent body. This complicates simple kinematic modelling with inclined rotating rings. Strong non-circular motions are however seen, particularly to the SW of the centre perhaps indicative of an infall - or outflow activity.

(6) There are suggestions of larger scale collapse outside the ring, and strong kinematic activity at the edge of the SW streamer, matching the velocities of material in the ionised fingers - where probably some shock or tidal stripping activity is happening.

(7) The atomic gas CI, has a different distribution to molecular material - although some high velocity gas in associated with the CND.

(8) There is little evidence evidence of a chemically distinct rim along the HII region interface.

(9) The CND's mass depends strongly on the filling factor - but is probably $\sim$ 250 - 500 M$_\odot$, with a filling factor $\sim$ 3 - 10 percent.

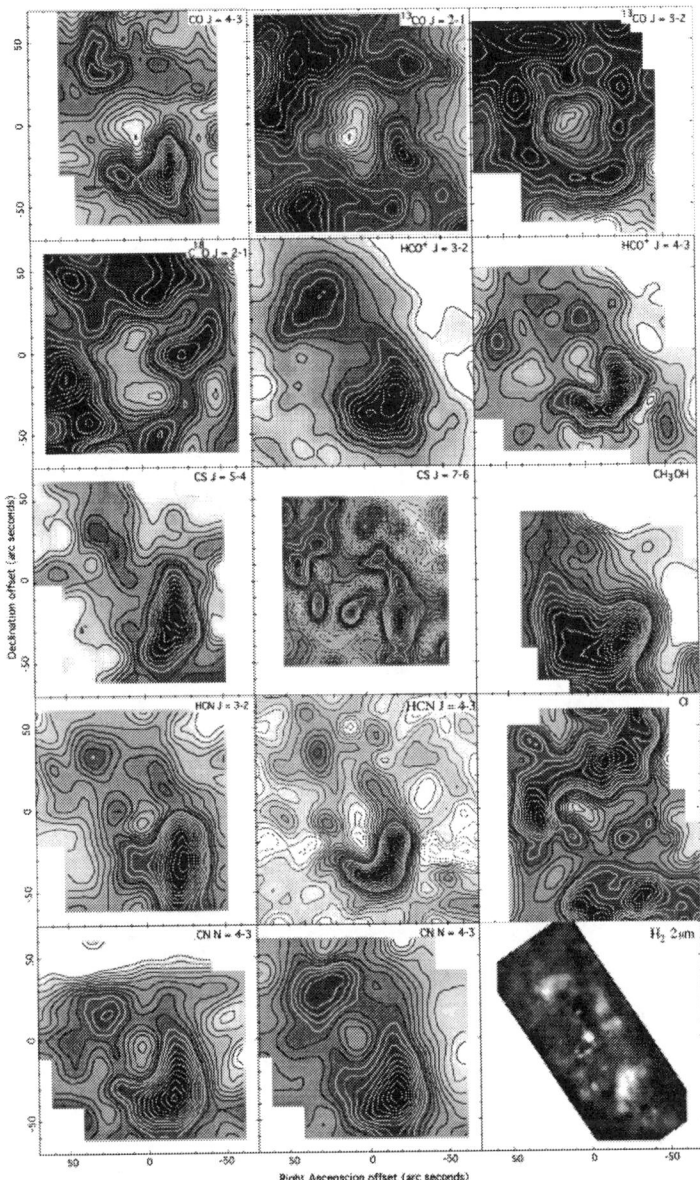

Figure 1. Maps showing the integrated emission for the area around the Galactic Centre, integrated between −120 and +120 km s^{-1}. The beamsize varies between $\approx$ 20 arc seconds at the lower frequencies to $\approx$ 9 arc seconds at the CO 4-3 and atomic carbon frequencies

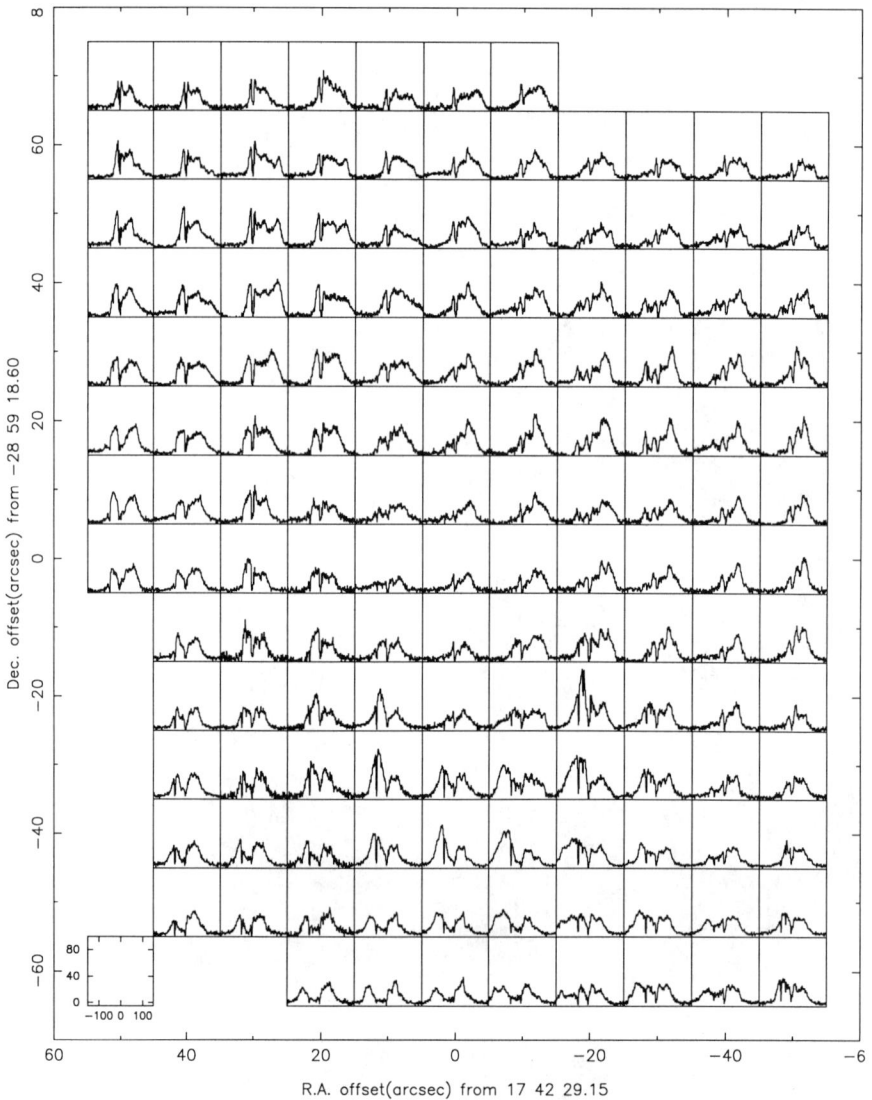

Figure 2. CO $J = 4$-3 map of the region around the Circumnuclear Disc. These spectra are 10 second integrations at each position, with only linear baselines subtracted

(10) A comparison of the line shapes of $C^{18}O$ and ^{12}CO toward selected positions shows that they often differ significantly. A detailed analysis shows that the standard conversion formula for the determination of the ^{12}CO column density does not work in the Galactic center region and that the assumption of high optical depth for the ^{12}CO emission is not generally valid there.

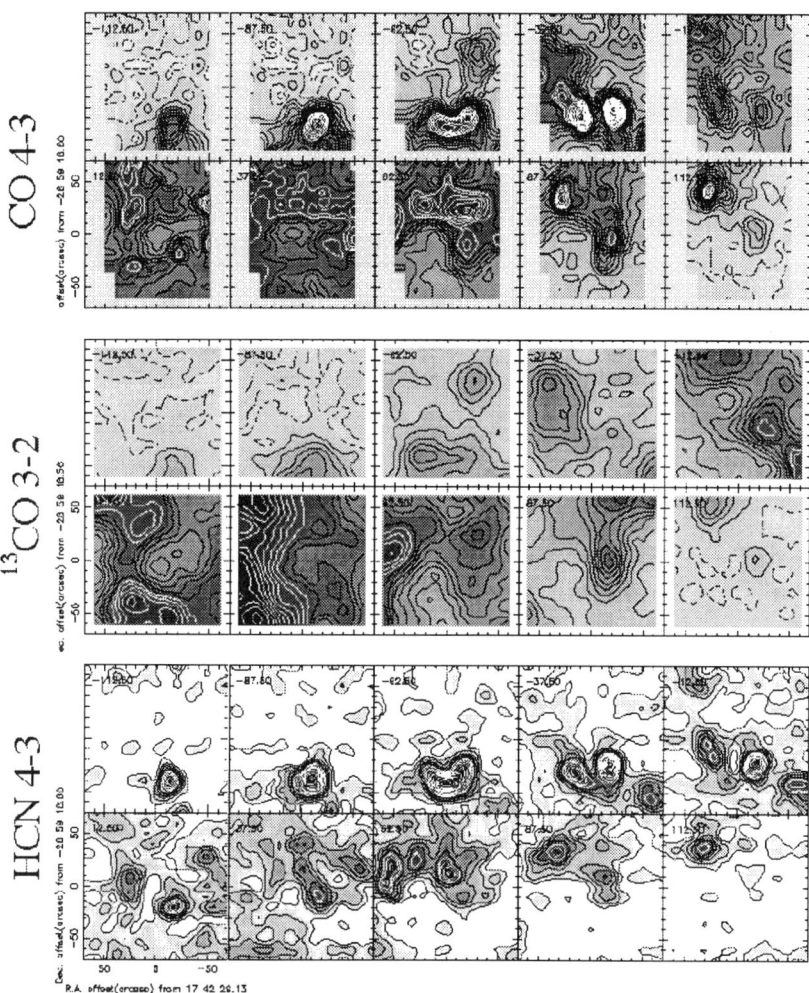

Figure 3. Channel maps for three of the species mapped in this study. It is quite remarkable that the CO line is able to trace the structure of the disk so clearly, despite the fact that it is normally quite saturated in molecular clouds

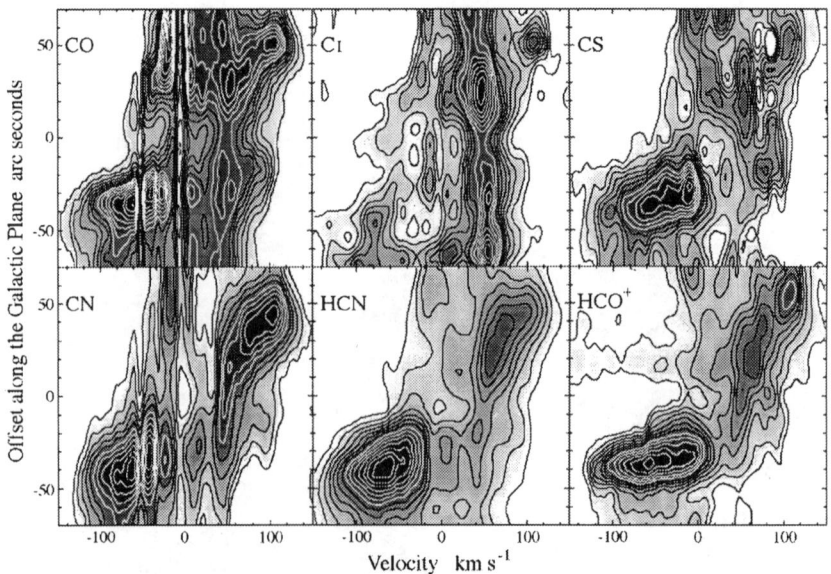

Figure 4. Position velocity maps along galactic longitude, centred on the Galactic Centre. The signature of the ring can be seen ≈ 40 - 50 arc seconds from the central position

3. Acknowledgements

Most of this data was obtained as back-up observations to other high frequency *JCMT* programmes over the last ten years, in collaboration with: Ian Gatley, Rachael Padman, Peter Williams, Henry Matthews, Bill McCutcheon, Norio Kaifu, Saeko Hayashi, Hiroko Suzuki, Simon Pinnock, Jane Greaves, who are all co-owners of the data set, and will form part of the authorship of the formal published papers. Additional data were also obtained by the *JCMT* Service programme, some just by sheer gritty determination to defy the weather, and some at the discretion of the various *JCMT* Directors, who found creative ways for us to finish this large programme. In addition to the normal publishing format, fits cubes, movies, spectral grids and images will be made available shortly on the WWW of this rather unique data set about the molecular environment of the CND, as a result of the provision os specialised image processing equipment by The Royal Society through their Research Grant Scheme.

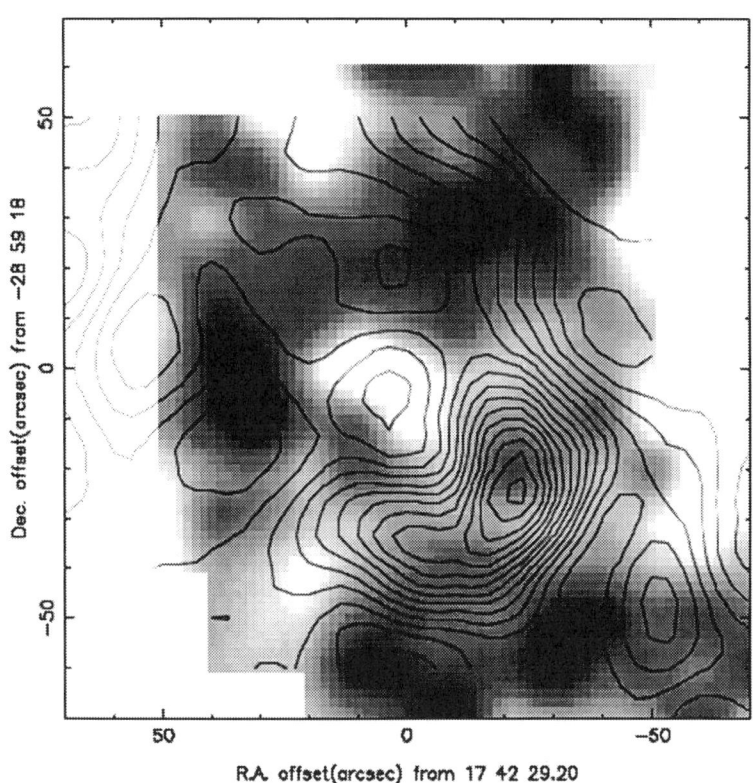

Figure 5. Overlay showing the relative distributions of the HCO$^+$ distribution (as contours) and atomic carbon (as greyscales). It is notable how the CI in the south of the ring sits further out from the edge of the molecular ring, as indicated by most of the tracers shown in Figure 1

References

Becklin, E.E., Gatley, I. and Werner, M.W. 1982, ApJ, 258, 135.
Dent, W.R.F., Matthews, H.E., Wade, R. and Duncan, W.D., 1993, ApJ, 410, 650.
Duschl, W.J. 1989, MNRAS, 240, 219.
Genzel, R. and Townes, C.H. 1987, ARAA, 25, 377.
Genzel, R., Hollenbach, D. and Townes, C.H. 1994, Rep. Prog. in Phys, 57, 417
Harris, A.I., Jaffe, D.T., Silber, M. and Genzel, R. 1985, ApJ Lett, 294, L93.

Mezger, P.G., Zylka, R., Salter, C.J., Wink, J.E., Chini, R., Kreysa, E. and Tuffs, R. 1989, A&A, 209, 337.

Okumura, S., Kawabe, R., Ishiguro, M., Kasuga, T., Morita, K.I. and Ishizuki, S. 1991, *IAU Symposium 146, Dynamics of Galaxies and their Molecular Distributions*,ed F. Combes, Dordrecht, Kluwer Publishers.

Pedlar, A., Anantharamaiah, K.R., Ekers, R.D., Goss, W.M., van Gorkom, J.H., Schwarz, U.J. and Jun-Hui Zhao. 1989, ApJ, 342, 796.

Serabyn, E. and Güsten, R. 1987, A&A, 184, 133

Zylka , R., Mezger, P., Wilson, T.L. and Mauersberger, R. 1994, p 161, in The Nuclei of Normal Galaxies, edited by R. Genzel and A. I. Harris. Kluwer Academic Publishers.

Zylka, R., Mezger, P.G., Ward-Thompson, D., Duschl, W.J. and Lesch, H. 1995, preprint.

10μm Imaging Polarimetry

David Aitken[1,2]
University of Hertfordshire, Division of Physical Sciences, College Lane, Hatfield, Herts AL10 9AB, UK

Craig Smith and Toby Moore
Physics Dept., University College ADFA, Canberra, ACT2600, Australia

Patrick Roche
Physics Dept., Astrophysics, Keble Road, Oxford, OX1 3RH, England

Abstract. Thermal emission from magnetically aligned dust grains produces the observed mid-infrared polarization in the northern arm and east-west bar of SgrA West; recent arcsecond resolution imaging polarimetry at 12.5μm of these ionized filaments is presented. A lower limit $\sim 2mG$ is found for the magnetic field in the northern arm and the IRS16 complex appears to be displaced from the northern arm by $\sim 0.15pc$ along the line of sight. The field directions in the northern arm support the view that this featues delineates the path of orbital motion rather than that it is part of a spiral pattern in circular motion about SgrA*

1. Introduction

Infrared polarization in the Galactic Centre was first reported by Dyck, Capps and Beichman (1974), and subsequent polarimetry through the near and mid-infrared (Knacke and Capps 1977; Lebofsky et al. 1982; Kobeyashi et al. 1980) revealed a complexity of polarizations and position angles among the sources in the central parsec which were variously attributed to absorption or emission by aligned grains and subject to essentially the same interstellar extinction and polarization. Spectropolarimetry from 8 to 13μm (Aitken et al. 1986) was able to separate the interstellar polarization (ISP) from the intrinsic component and show that the latter is due to thermal emission from aligned grains, and that the way in which the emissive position angle varies between the sources and connects them argues strongly for a magnetic alignment mechanism. Further studies of the central 15 arcseconds with the NASA/Goddard array camera (Aitken et al.

---

[1] Visiting Professor, Physics Dept., University College, ADFA
University of New South Wales

[2] mailing address:
5 Dewsbury Cottages, York, YO1 1HB, UK

1991) showed that the polarization is continuous through the ionized filaments and it was considered that the aligning field is amplified by shearing motions within the northern arm and indicates the direction of flow to be along that of the arm. A more recent polarimetric image of the central 30 × 45 arcsecond at 12.5μm with slightly improved resolution obtained with the N-band imaging polarimeter, NIMPOL (Smith, Aitken and Moore 1994) on the AAT in August 1992 and June 1993 is presented here. NIMPOL uses a 128 × 128 SiGa DRO array supplied by Amber Engineering Inc and has a pixel scale on the AAT of 0.25″. At short integrations on bright stars the diffraction limit of 0.7″ (FWHM) was acheived but on the long integrations needed for polarimetry the resolution appears to have been degraded to about 1″.

A necessary condition for Davis-Greenstein alignment (Davis and Greenstein 1951) is that the gas temperature significantly exceeds that of the grains. This condition is generously satisfied in the ionized filaments of the Galactic Centre and grain alignment is feasible for magnetic fields $\mathbf{B} \gtrsim$ mG. Further, it is highly probable that interstellar grains contain the superparamagnetic inclusions (Bradley 1994) which Jones and Spitzer (1967) proposed would enable alignment in relatively weak magnetic fields. Draine (1996) describes the properties of such inclusions, and the increased dissipation within the grains should lead to effective saturation of alignment so that it is independent of, or only weakly dependent on, field and other physical parameters in the filaments provided $\mathbf{B} \gtrsim$ mG. These aspects will be considered in more detail in a later publication, but it is considered established that magnetic alignment occurs in the filaments and that the polarization $\mathbf{E}$ vector from emission is orthogonal to the transverse component of magnetic field.

2. The Observations

The observed polarization is a combination of thermal emission from SgrA overlaid by an ISP contribution of 1.8% at position angle 0° from aligned grains in the interstellar medium (ISM). This value is arrived at from spectropolarimetry of IRS3 which shows only an absorptive interstellar component of this amount at 12.5μm, a value which has been checked for consistency over the central region in two independent ways: (1) the intrinsically unpolarized [NeII] line (12.81μm) in IRS1 has polarization 2.9 ± .9% at 0 ± 9° (Aitken et al. 1986) and (2) separation of the emissive and absorptive components of polarization in a sample of ∼ dozen positions within the region gives 5° with an rms scatter of 5° (an outline of this latter method is given in Aitken (1996)). The adopted value of 0° for the ISP to the Galactic Centre differs from the position angle $\simeq$ 20° found by Kobayashi et al. (1980) at 2.2μm over SgrA and the difference can be interpreted in terms of a twist in the interstellar field direction over a line of sight through which the fraction of silicate material is increasing towards the centre (Roche and Aitken 1985).

Subtraction of 1.8% from the Stokes Q component unfolds the ISP from the image; errors of a few degrees in position angle only affect the result at the tenth percent level. Fig. 1 shows the polarization vectors in SgrA West (rotated through 90°), after correction for the ISP. The amplitude of the vectors is not a measure of field strength because of the unknown susceptibility of the grain material and also because grain alignment is likely to be saturated.

Fig. 1 shows broad agreement with the earlier array results (Aitken et al. 1991, Fig. 3), extends the coverage to a larger area at lower surface brightness and confirms the kink in position angle at IRS10. Noticeable in Fig. 1 is the correspondence between the field directions and the general morphology of the northern arm, being closely north-south in the portion of the arm north of IRS5, following the lateral shift in the arm between IRS5 and IRS10, and smoothly curving through IRS1 to become nearly east-west with almost vanishing polarization south of SgrA*; here the arm appears to lose its identity in the east-west bar. Further west the field directions, indicated by much weaker polarization, roughly follow this feature along its northern edge but with more complex structure elsewhere and in particular at IRS2 where the field is north-south. In the eastern arm the field appears not to be aligned with the filament but instead is mainly north-south.

3. Interaction with the Emission Line Stars

One of the remarkable features of the polarization is its independence of the presence of emission line stars which lie along the line of sight. Fig 2 compares the surface brightness and polarization with stellar positions in the central 15″ at slightly higher spatial resolution than Fig 1.

It is clear that some of the mid-infrared peaks, such as IRS1, 2, 10 and maybe 9, correspond to emission line stars, and these regions also are temperature peaks (Smith, Aitken and Roche 1990; Gezari 1992). They are considered to be either single or clusters of OB stars (Krabbe et al. 1991, 1995) and are suffering mass loss in fast winds. These coincidences imply that the stars are embedded in the ionized streamers and it is evident that both the polarization fraction and position angle, and hence both the field direction and grain alignment, are unperturbed by their presence. Three inferences follow: (1) grain alignment is not affected by the changed physical conditions, including radiation driven streaming of grains through gas, and this implies that magnetic alignment is strong and saturated, (2) if there is a population of non-alignable grains, as has been argued (eg Goodman et al. 1995), their relative contribution remains the same in spite of the temperature change, and (3) the field must be able to constrain the flows. The latter point leads to a lower limit on the field strength by requiring that the ram pressure of the wind at 0.04pc (one arcsecond) is not more than the energy density of the field. For a mass loss rate $10^{-5} M_\odot yr^{-1}$ and an outflow velocity of $\sim$200km s^{-1} this leads to $B \gtrsim 2$mG in the vicinity of IRS1, a value which is similar to estimates (Aitken et al. 1991; Hildebrand 1989) based on the method of Chandrasekhar & Fermi (1953) and is also independant of the magnetic properties of grains and the alignment mechanism. Killeen, Lo, and Crutcher (1992) and Plante et al. (1994) from Zeeman studies find comparable values for the line-of-sight field in the circumnuclear disk (CND).

A number of much hotter and more luminous HeI stars lie projected against some of the ionized filaments yet these are not associated either with surface brightness peaks, or noticeable temperature enhancements. Such is the case for some of the components of the dominant HeI star cluster, IRS16. While much of

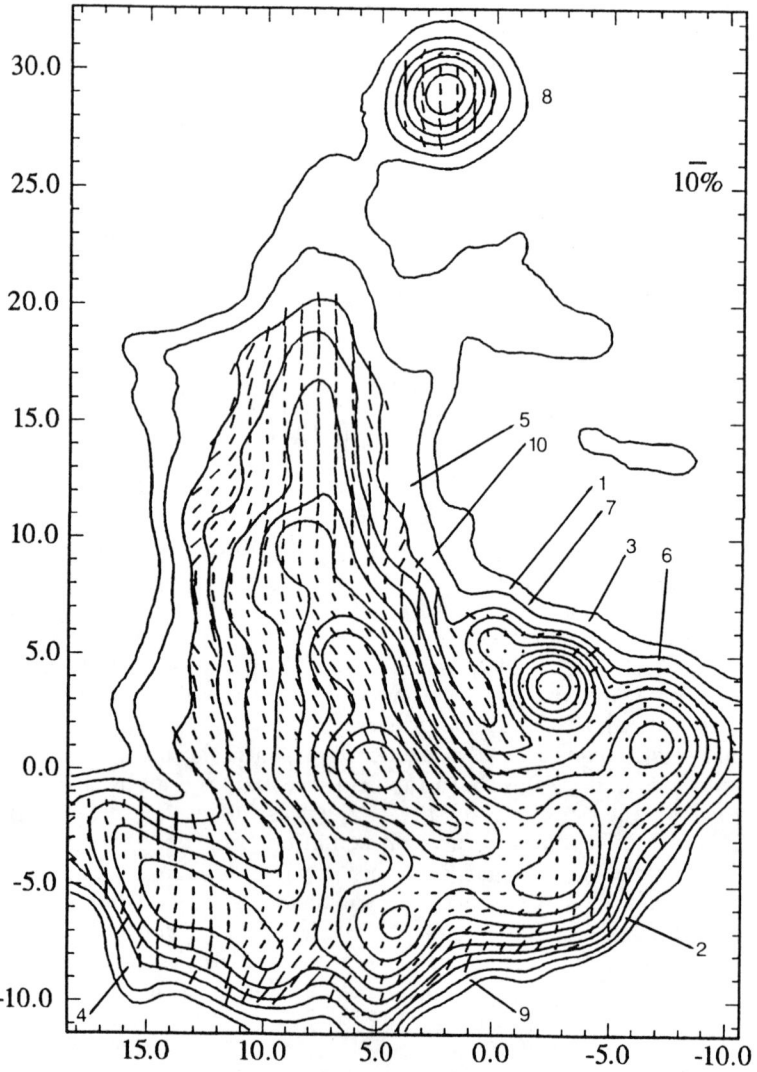

Figure 1. Fig 1 shows surface brightness contours at 12.5μm in the central region of SgrA overlaid with vectors orthogonal to the emissive polarization and thus denoting field directions. The contours are logarithmic with a factor $2^{1/2}$ interval with the lowest contour at 350mJy$''^{-2}$. The data have been smoothed with a truncated gaussian of FWHM 1.5″and outer radius 1.1″and vectors are displayed at 0.75″ intervals. Offset are in arcseconds from SgrA*.

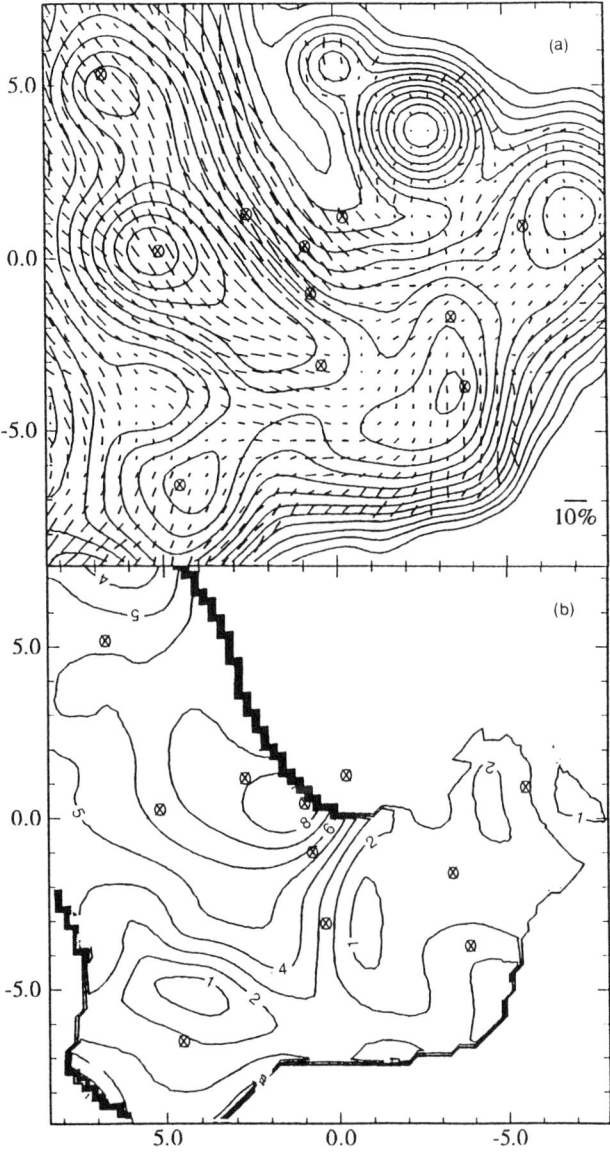

Figure 2. Fig. 2(a) Surface brightness and field directions in the inner 15″: contours are logarithmic with a factor $2^{1/3}$ interval from the lowest contour at $500\text{mJy}''^{-2}$ and vectors presented every 0.5″. The data have been smoothed with a truncated gaussian of 1″ FWHM with outer radius 0.75″. (b) Contours of polarization percentage for surface brightness exceeding $2\text{Jy}''^{-2}$ The peak contour is 8% and the interval is 1%. The polarization data have been smoothed by a truncated gaussian of 2″ FWHM and outer radius 1.5″.

the dust heating in the filaments will be from nebular radiation one would also expect the stellar continuum to heat grains directly and the absence of significant temperature and surface brightness correlations suggests that the components of IRS16 are displaced along the line of sight from the ionized streamers. This would also explain why, with a significantly larger mass loss rate and wind velocity than IRS1, for instance, the polarization directions are undisturbed.

Inspection of Fig 2(b) shows that the polarization amplitude is larger on the west side of the northern arm and peaks close to the centroid of the IRS16 cluster. This does suggest some interaction with a wind from the cluster, or possibly from SgrA*. Compression of the arm and its field by a wind could have the effect not only of increasing the field strength (which would have little effect if alignment is near saturation) but also of increasing the polarization by reducing fluctuations in the line-of-sight component of magnetic field. A separation of IRS16 from the northern arm by the equivalent of several arcseconds ($\sim$.15pc) along the line of sight is suggested.

4. The magnetic fields and the orbital motions

The energy densities of the fields envisaged, $B^2/8\pi \simeq 2 \times 10^{-7}$ ergs cm^{-3}, are larger than the thermal gas kinetic energies and of the same order as the turbulent energies, but small compared with gravitational and tidal forces within the central parsec. The frozen-in fields in the streamers cannot constrain the overall flow but instead will reflect the cumulative strain in the material which amplify and re-orientate initial fields, although instabilities of the same order as the turbulence may arise.

Serabyn & Lacy (1985) have proposed that the filamentary structures represent the path of orbital motion. The strains of both shear and stretch will be in the direction of the orbital motion; this agrees with the observed field directions. Later Lacy, Achtermann & Serabyn (1991) argued that the northern arm and western arc form a continuous one-armed spiral structure in which the material is in circular Keplerian orbits about a point near SgrA*. The shear and stretch strains then have different directions and the field direction will depend on the relative magnitudes of these respective strains, and it is not obvious that they should be oriented as observed along the filaments.

Lacy etal (1991) also interpreted the one-armed spiral as a density wave, and in galaxy spiral arms the magnetic fields do appear to be directed along the arms, presumably the result of non-gravitational stresses at work, and possibly of ambipolar diffusion. To what extent density wave structures are expected in the very different conditions and time scales involved in the central parsec is not clear, although Fridman (1996) has argued on hydrodynamic grounds that density waves are ubiquitous. Roberts & Goss (1993) find, however, that the key components of the one-armed spiral, the northern arm and the western arc, are separate kinematic structures, with the western arc crossing the northern arm about 20″ north of SgrA*. This region is seen in the surface brightness contours in Fig. 1 and the field directions show no tendency to curve towards the west at this position as would be expected if the western arc and northern arm were a continuous spiral structure.

If the alignment is indeed saturated, or nearly so, then although the polarization amplitude gives no information on field strength it will be related to the dip angle the field makes with the plane of the sky. Such an effect is evident just south of SgrA* where the polarization minimum can be identified with the closest approach of the northern arm to the line-of-sight. If ϕ is the angle between the field and the plane of the sky then we expect $p = p_0\cos^2\phi$, where p_0 is the polarization for $\phi = 0$. Taking $p_0 = 6 - 7\%$ and $p \simeq 1\%$ just south of SgrA* we find $\phi \simeq 65°$, close to the inclination of the northern arm orbit proposed by Serabyn and Lacy (1985) from velocity studies of [NeII], and to the tilt of the plane of the CND (eg Lacy 1994).

There is a general correspondence between the polarization fraction and the radial velocity field (Roberts, Yusef-Zadeh & Goss 1996) and but while there is a steep gradient of polarization to the east of the minimum where the velocity change is large, there is not a corresponding rise to the west where the radial velocity is still increasing (negatively). This can be understood if the flow continues close to the line of sight and supports the suggestion (Goss et al. 1996) that the orbit is hyperbolic.

Even if alignment is saturated, the polarization amplitude may suffer dilution, be affected by fluctuations in field directions along the line of sight and, more seriously, by systematic changes along the line of sight, and these factors should be considered if the polarization amplitude is used as a guide to the line-of-sight component of field. The position angle, however, is only affected by the last of these effects.

The waviness noted in the radio images of Yusef-Zadeh and Wardle (1993) is emphasized in the polarization image, but appears to be due more to changes of the pattern in the region of IRS10 rather than near IRS1; this may be an example of magnetic instability. The polarimetry also implies a much larger major axis for the northern arm than that of Herbst et al. (1993) and it is noticeable that at IRS8, where the arm might seem to meet the the circumnuclear disk, the field vectors are north-south and are not directed along the northen arm.

Finally we remark that there is little direct evidence for any field departures from north-south within the central few parsecs except within a few arcseconds of SgrA*. Since the CND is now revealed to be a very fragmented structure this calls into doubt the existence of a poloidal field within the region.

5. Conclusions

A lower limit of 2mG is derived for the magnetic field in the northern arm and grain alignment appears to be saturated in the inner parsec of the Galaxy; this can provide a probe for the line-of-sight component of magnetic field.
The IRS16 complex appears to be displaced from the northern arm along the line-of-sight by $\sim$ .15pc
The polarization pattern in the northern arm suggests that it is not a continuation of the western arc into a spiral structure.

Acknowledgments. We have appreciated useful comments from George Reike and Miller Goss. DKA gratefully acknowledges the support of a Leverhulme Trust Grant and use of facilities at the Royal Observatory Edinburgh and the University of York. As always the support of AAO staff was first rate.

References

Aitken, D.K., Roche, P.F., Bailey, J.A., Briggs, G.P., Hough, J.H., Thomas, J.A., 1986, MNRAS, 218, 363 (ARBBHT)
Aitken, D.K., Gezari, D., Smith, C.H., McCaughrean, M., Roche, P.F., 1991, ApJ, 380, 419 (AGSMR)
Aitken, D.K., 1996, in Polarimetry of the Interstellar Medium, ASP Conf. Series, (in press)
Bradley, J.P., 1994, Science, 265, 925
Chandresekhar, S., Fermi, E., 1953, ApJ, 118, 113
Davis, L., Greenstein, J.L., 1951, ApJ, 114, 206
Draine, B.T., 1996, in Polarimetry of the Interstellar Medium, ASP Conf. Series, (in press)
Dyck, H.M., Capps, R.W., Beichman, C.A., 1974, ApJ, 188, L103
Fridman, A., 1996 this volume
Gezari, D., 1992, in The Center, Bulge and Disk of the Milky Way ed L. Blitz (Kluwer, Dordrecht), 23
Goodman, A.A., Jones, T.J., Lada, E.A., Myers, P.C., 1995, ApJ, 448, 748
Roberts, D.A., Yusef-Zadeh, F., Goss, W.M., 1996 ApJ, 459, 627
Herbst, T.M., Beckwith, S.V.W., Forrest, W.J., Pipher, J.L., 1993, Astron J 105, 956
Hildebrand, R.H., 1989, in Interstellar Dust IAU Symp 135 ed L.J.Allamandola & A.G.G.M.Tielens (Kluwer), 275
Jones, R.V., Spitzer, L., 1967, ApJ, 147, 943
Killeen, N.E.B., Lo, K.Y., Crutcher, R., 1992, ApJ, 385, 585
Knacke, R.F., Capps, R.W., 1977, ApJ, 261, 271
Krabbe, A., Genzel, R., Drapatz, S., Totacuic, V., 1991, ApJ, 382, L19
Krabbe, A., et al. 1995, ApJ in press
Lacy, J.H., 1994, in The Nuclei of Normal Galaxies, ed. R.Genzel and A.I.Harris, Kluwer, 165
Lacy, J.H., Achterman, J.M., Serabyn, E., 1991, ApJ, 380, L71
Lebofsky, M.J., Rieke, G.H., Deshpande, M.R., Kemp, J.C., 1982, ApJ, 236, 672
Kobeyashi, Y., Kawara, K., Kosaza, T. Sato, S. Okuda, H., 1980 PASJ, 32, 291
Plante, R.L., Lo, K.Y., Crutcher, R., Killeen, N.E.B., 1994, in The Nuclei of Normal Galaxies, ed. R.Genzel and A.I.Harris, Kluwer, 205
Roberts, D.A., Goss, W.M., 1993, ApJS, 86, 133
Roche, P.F., Aitken, D.K., 1985, MNRAS, 215, 425
Serabyn, E., Lacy, J.H., 1985, ApJ, 293, 445
Smith, C.H., Aitken, D.K., Roche, P.F., 1990, MNRAS, 246, 1
Smith, C.H., Aitken, D.K., Moore, T.J.T., 1994, in Instrumentation for Astronomy, VIII SPIE, 2198, 736
Yusef-Zadeh, F., Wardle, M., 1993, ApJ, 405, 584

Part 2. The Stellar Population
Section A. The Central Stellar Cluster

Astrometry in the Galactic Center Region

Michael R. Rosa[1]
The Space Telescope European Coordinating Facility, European Southern Observatory, D-80548 Garching, Germany

Abstract. Accurate positional referencing of sources observed at various frequencies in the crowded Galactic Center region is of crucial importance for the inference of physical properties from energy distributions and morphological aspects. The current situation regarding astrometric grids and registration of maps from different wave bands is reviewed. It is shown that various methods to transfer the radio position of SgrA* onto far red and infrared images yield an uncertainty ellipse of characteristic size $1''$.

1. Introduction

The stellar cluster core near the Galactic Center has been the subject of many investigations ever since it was mapped for the first time in the near-infrared by Becklin & Neugebauer (1968). Large numbers of individual sources have been detected in the $30''$ ($1.3 pc$ at an assumed distance of $8.5 kpc$) vicinity of the non-thermal compact radio source SgrA* in the near IR (eg. Eckart etal. 1993) and even in the far red (cf. Rosa etal. 1992). At lower frequencies (FIR, micro, mm, cm) the morphology of the region is dominated by a mixture of diffuse, of filamentary, and of more point like sources. Many of the features seen at a given wavelength may be correlated with or even counterparts of sources detected at other wavelengths. By necessity cross wave band identifications play a key role in the immediate neighborhood of the probable black hole accretion disk powering the non–thermal source SgrA*.

Precise co–registration of maps is a pre–requisite to make inferences from positional coincidences of sources that are detected at different wavelengths. However, the detailed morphology changes quite rapidly even at very small differences in effective wavelength because of the heavy extinction towards the region, because of the sheer number density of objects in the volume of the core region and because of the crowding of foreground and background objects along this particular line of sight through the Galaxy. This becomes particularly acute as the spatial resolution of the contemporary observations is drastically increasing, so that sources that previously were thought to be coincident might now appear to be genuine or apparent pairs of different physical nature (eg. enshrouded O4 stars, BSGs, WR stars, late SGs).

---

[1]Affiliated to the Astrophysics Division of the Space Science Department of the European Space Agency

2. Role of Astrometry

There are four areas in which astrometry enters implicitly or explicitly, namely

1. Individual images or mosaicing: implicit plate scales and orientation, or explicitly verified using astrometric reference stars or grids, plus placement into absolute reference frame.

2. Alignment and co-registration of images from different wave bands: coordinate transformations using source coincidences or explicit plate scales, orientation as under (1) plus reduction to same absolute positional reference frame.

3. Placement of SgrA* relative to IR visible radio sources: requires (1) on both, the IR and the radio maps.

4. Placement of SgrA* using absolute positions: requires (1) on the IR maps and a reduction to the same absolute positional reference frame for both, IR and radio positions respectively.

As shall be discussed in the following, all four areas have been involved in the observational research on the Galactic Center region, but in none of these have all the requirements been met explicitly and full extent.

3. Imaging and Alignment Techniques

For historical reasons, IR "maps" have usually been referenced to two visually and IR bright foreground stars "A" and "B" (cf. Rosa etal. 1992 (RZMM92) for details and references about the astrometry), particularly so because the first "maps" where obtained with 1 D or tiny 2 D detectors in telescope stepping mode. Provided the plate scale and alignment of these maps known, there was no problem to obtain co-registration at the one-to-several arcsecond resolutions available between 1 and 10 μ at the time (but see eg. Allen, Hyland & Jones 1983). At this resolution the SgrA* radio point source (position from eg. Brown 1981) seemed to be located close to the famous IR source "complex" IRS 16 (cf. eg. Forrest, Pipher & Stein 1986 for discussion).

Co-registration of higher resolution maps obtained at 1, 2, 3, 8, and 12 μ could no longer rely solely on these stars, nor on the near-IR bright late type supergiant IRS 7 (cf. Becklin etal. 1987), because these sources are fading away at $2 - 4$ μ. The multistep process necessary to register such maps, partly by shifting contour plots, and the plugging in of the SgrA* absolute position or its relative offset from the radio visible IRS 7, is described in very detail by Tollestrup etal. 1989. The IR to radio alignment even included the discussion of thermal radio sources and their probable IR counterparts.

Such techniques yielded typical uncertainties of order 0.5″ for SgrA* relative to the K band IRS 16 sources. The error is partly due the intrinsic uncertainties of the **absolute** positions for A, B and IRS 7 and the chosen reference frames (Perth70 etc.), partly due to assuming the coincidence of the radio and K band images of IRS 7 (see below), and also due to the fact that none of these maps had truly astrometric plate solutions.

4. Astrometric References and Grids

The earliest positional reference still in use are the two stars A,B (see above). Astrometry of these has been reported by Ricker etal. (1982; source unpublished), Biretta, Lo & Young (1982; source unpublished), Storey etal. (1982; SAO frame), Storey & Allen (1983; Perth70, FK4; hereafter SA83), Vanderspek & Rieker (1983; Taff & Stansfield; hereafter VR83) ref. stars), Forrest, Pipher & Stein (1986; source unpublished); Müller, Rosa & Röser (1994; PPM FK5; hereafter MRR94); Fabricius and J≤rgensen (1995; CAMC with correction to FK5; hereafter FJ95).

Astrometric grids of fainter stars, many of which can actually be detected at wavelength of 900 nm to 2 μ as well, have been provided by VR83 (19 stars, POSS red), MRR94; (55 stars, POSS/ESO red); FJ95 (54 stars on CCD i-band). Because of measuring problems and the use of an obsolete reference catalogue the VR83 are not recommended (cf. RZMM92, FJ95). MRR94 use 86 PPM/FK5 stars on Schmidt plates spanning epochs from 1958 to 1992, hence involving a proper motion analysis. FJ95 obtain absolute positions for a single epoch (1993) CCD image by observations of selected brighter stars with the CAMC meridian circle and correction to the FK5. Currently there is an unresolved discrepancy between the **absolute** J2000 positions of MRR94 and FJ95 of order 0.01 s in RA and 0.18″ in DEC (see also discussion on coordinate transformations), but the internal agreement of the two new grids (22 stars in common) is as good as 0.10″ at the mean MRR94 epoch (1985).

Astrometry for the classical reference object IRS 7, common to the IR and radio regimes, has been provided by Becklin et al (1987; 1.2 μ), MRR94 on RZMM92 data (Gunn z), FJ95 on RZMM92 data (Gunn z), and Yusef-Zadeh & Morris (1991; radio). Another radio measurement can be inferred from the data published by Zhao, Goss, Lo & Ekers (1991). All these measurements are summarized in Table 1.

Table 1. Positions of IRS 7 (17^h 42^m RA(s) -28° 59′ DEC(″); B1950. Estimated uncertainties are of order 0.02^s, 0.2″ (cf. original reference).

Reference	RA (s)	DEC (″)	Remark
Radio (ZGLE91)	29.348	13.08	offset from SgrA*
Radio (YZM91)	29.319	12.66	offset from SgrA*
1.65 μ	29.320	13.00	BDGWJ87 in Perth 70
Gunn Z (RZMM92)	29.330	13.10	in RZMM92 ref. frame
Gunn Z	29.379	12.73	RZMM92 in MRR94 grid
Gunn Z	29.340	12.90	RZMM92 in FJ95 grid

Positions of SgrA* are listed in Table 2. Pre–1992 work has also been compiled by RZMM92. More recent determinations are from Yusef–Zadeh etal. (1990) and Marcaide etal. (1992).

5. Coordinate Transformations

We are now approaching the sub–sub–arcsecond level in cross–correlating sources between the far red, IR and radio regimes, on observational material spanning

Table 2. Positions of SgrA* ($17^h\ 42^m$ RA(s) $-28°\ 59'$ DEC(")); B1950. Estimated uncertainties are of order 0.02^s, $0.2''$ (cf. original reference)

Reference	RA (s)	DEC (")	Remark
Brown 81	29.335	18.60	
Tollestrup etal 89	29.318	18.39	
Yusef-Zadeh etal. 90	29.312	18.38	
Marcaide etal. 92	29.352	17.90	transf. from J2000

years and decades and epoch difference. Most of us here at this conference are not usually confronted with rigorous astrometry in their every days work. It is therefore important to remark upon the following caveats.

1. More recent optical–, and therefore also IR– positions, in the end are referred to one of two fundamental catalogues FK4 B1950 or FK5 J2000.

2. Transition between the two is not merely precession, but also involves a change of equinox, elliptical aberration and correction of zonal errors in the FK4.

3. Tertiary reference grids (see above) have an epoch attached and therefore need to be precessed.

4. While the radio source SgrA* at the very center likely does not have any proper motion (cf. Backer & Sramek 1987), most of the red and IR foreground objects will have, hence to be included when changing epoch.

5. SgrA* itself is measured in the radio J2000/FK5 system against mostly extragalactic references, while the optical reference system is composed of stars taking part in the galactic rotation. Transformation therefore also includes possible secular terms between the radio and the optical frames.

To date none of these items has been comprehensively and thoroughly discussed with its particular relevance for the Galactic Center studies.

6. So Where is SgrA* on IR Frames

Keeping the previous discussion in mind, we may proceed to obtain current best estimates for the positioning of SgrA* in the IR sky. There are three families of procedures, namely

1. Absolute position of SgrA* using the optical faint star grids to obtain an absolute positioning of the IR image.

2. Absolute position of SgrA* using the IR or FR (Gunn z) determined offset of IRS 7 from the optical reference stars to obtain an absolute positioning of the IR image.

3. Relative offset of SgrA* from IRS 7 on IR image showing IRS 7.

The difference between family 1 and 2 is rather subtle, but involves essentially the transfer of several tertiary standards onto a single quarternary one. Family 3 is special in that it claims the identity of the radio and IR image of IRS 7. In fact, IRS 7 shows an extended structure shaped by a possible galactic wind (see YM91), and its is questionable, whether this identity really holds. It might well be that the peak radio emission is due to a build up of density in the shadow of IRS 7 proper (a few .1″ north of IRS 7).

Figure 1 shows the result on a K band map, for convenience the one published by Eckart etal. 1992. It can be seen that the different attempts to position the radio source onto the IR frame yield an elliptical error ellipse of effective radius .7″. The spread along the major axis is largely due to the difference in RA between the Yusef–Zadeh etal. (1990; FK4,B1950) and Marcaide etal. (1992; FK5,J2000) positions of SgrA*, combined with the rather similar RA difference between the SA83 (FK4,B1950), FJ95 (CAMC,J2000) and MRR94 (FK5,J2000) grids. The minor axis reflects mostly the uncertainty in the DEC difference between the radio images of SgrA* and IRS 7. This does not include the possible systematic offset between the radio and IR images of IRS 7, which would move the SgrA* location further south.

7. Conclusion

Currently there is still room to associate the non–thermal radio source SgrA* with everything in a 1″ radius circle located approximately 1″ south of IRS 16 NW and 1″ west of IRS 16 C, depending on which method and which reference data are used to do the positioning. Obviously, cross–identification of OH masers with IR **stars** within the imaged field would greatly improve the situation by allowing for a immediate co–registration of radio and IR maps. However, the process will at require the identification of the IR counter parts, which has to be done on IR images that have been rectified and positioned by proper astrometry in the first place.

Items to be pursued, such as

- Improvement and critical discussion of astrometric grids in the region.

- Transfer of such grids (FK5, J2000) onto IR images.

- Critical discussion of radio–to–optical position transfer.

- Proper motion analyses of the reference grids.

- Establishing radio and mm reference sources with likely IR counter parts within 1 arcmin of the Galactic Center.

should involve experts on IR imaging, on radio mapping and on general astrometry as well. Hopefully the present contribution can stimulate such a an inter–discipline collaboration.

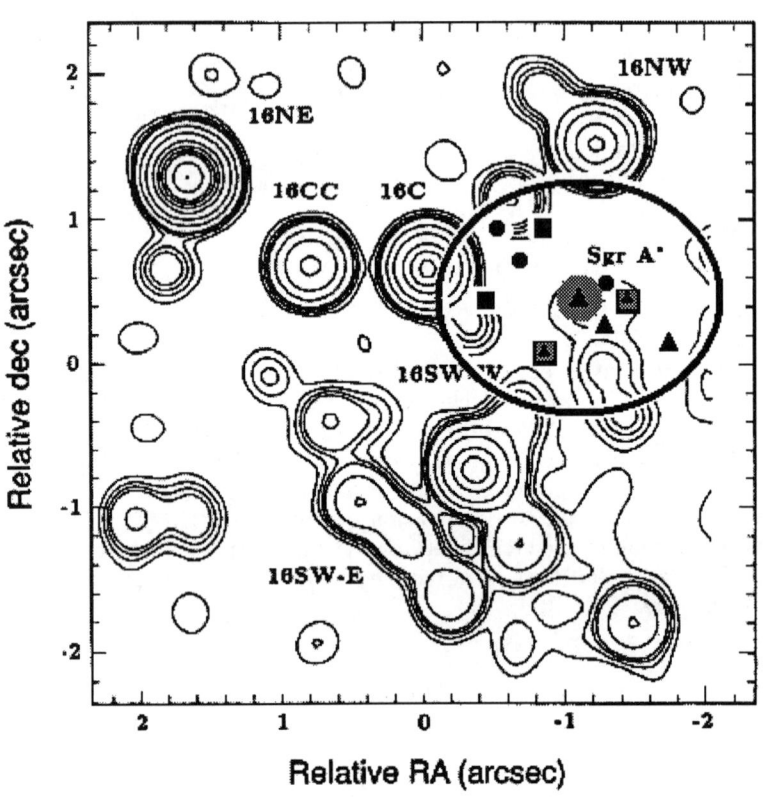

Figure 1. Positioning SgrA* on an IR map. Symbols: Recentering of actual map using optical IRS 7 pos. and Marcaide+ SgrA* (circles); with Yusef-Zadeh+ SgrA* (triangles). Using radio offsets from IRS 7 instead for SgrA*: Boxes. Most probable location hatched.

References

Allen Hyland Jones 1983
Backer, D.C. & Sramek, R.A. 1987, in The Galactic Center, AIP Conf Proc. 155, D.C. Backer, NY: AIP, 1987, 163
Brown, R.L., Johnston, K.J. & Lo, K.Y. 1981, ApJ, 250, 155
Becklin, E.E. & Neugebauer, G. 1968, ApJ, 151, 145
Becklin, E.E., Dinerstein, H., Gatley, I., Werner, M.W. & Jones, B. 1987, in The Galactic Center, AIP Conf Proc. 155, D.C. Backer, NY: AIP, 1987, 162
Biretta, J.A., Lo, K.Y. & Young, P.J. 1982, ApJ, 262, 578
Eckart R., Genzel, R., Krabbe, A., Hofmann, R., van der Werf, P.P., & Drapatz, S. 1992, Nature, 355, 526
Eckart, R., Genzel, R., Hofmann, R., Sams, B.J. & Tacconi–Garman, L.E. 1993, ApJ, 407, L77
Fabricius, C. & J$\leq$rgensen, H.E. 1995, A&A, 296, L1 (FJ95)
Forrest, W.J., Pipher, J.L. & Stein, W.A. 1986, ApJ, 301, L49
Marcaide, J.M., Alberti, A., Bartel, N. etal. 1992, A&A, 258, 295
Müller, Th., Rosa, M.R. & Röser, S. 1994, A&A, 283, L29 (MRR94)
Ricker, G.R., Bautz, M.W., DePoy, D.L. & Meyer, S.S. 1982, ApJ, 260, L59
Rosa, M.R., Zinnecker, H., Moneti, A. & Melnick, J. 1992, A&A, 257, 515 (RZMM92)
Storey, J.W.V., Straede, J.O., Jorden, P.R., Thorne, D.J. & Wall, J.V. 1982, Nature, 296, 333
Storey, J.W.V. & Allen, D.A. 1983, MNRAS, 204, 1153 (SA83)
Tollestrup E.V., Capps R.W. & Becklin E.E. 1989, AJ, 98, 204
Vanderspek, R. & Rieker, G.R. 1983, AJ, 88, 1264 (VR83)
Yusef-Zadeh, F., Morris, M. & Ekers, R.D. 1990, Nature, 348, 45
Yusef-Zadeh, F. & Morris, M. 1991, ApJ, 371, L59 (YZM91)
Zhao, J.-H., Goss, W.M., Lo, K.Y. & Ekers, R.D. 1991, Nature, 354, 46 (ZGLE91)

Observations of the Galactic Center with SHARP: First Stellar Proper Motions

A. Eckart and R. Genzel

Max-Planck Institut für extraterrestrische Physik Garching, FRG

Abstract. Speckle interferometric observations of the Galactic Center in the near-infrared with SHARP at the NTT have lead to the detection of stellar proper motions. The proper motion and radial velocity dispersions indicate a $2.4 \times 10^6 M_\odot$ dark mass with a mass density of at the center of our Galaxy. This mass is either a core collapsed cluster of 10-20 $M_\odot$ stellar black holes or, most likely, a single massive black hole.

1. Introduction

Over the past 5 years we have been conduction a program to study the properties of the central nuclear stellar cluster via near-infrared high spatial resolution measurements using the MPE speckle camera SHARP at the at the 3.5 m New Technology Telescope (NTT) of the European Southern
Observatory (ESO). The SHARP observations resolved the 2.2μm emission in the central parsec into about 600 stars (Eckart et al. 1992, 1993, 1994, 1995) and gave the first evidence for an extended infrared source (Sgr A*(IR)) at the position of the compact radio source Sgr A* (R) which is the most likely candidate for a possible massive black hole at the center of our Galaxy (e.g. Lo 1989). This central source was resolved into a small cluster of at least half a dozen stellar objects (Eckart et al. 1995). Imaging spectroscopy revealed a small cluster of luminous and probably massive, blue supergiants (Allen, Hyland, and Hillier 1990, Krabbe et al 1991). Sellgren et al. (1990) found evidence that the depth of the 2.3μm CO bandhead absorption decreases in the central 10", perhaps indicating a lack of late type stars there.

Over two decades now the observational evidence for a dark central mass concentration at the core of the our Galaxy has been steadily growing via studies of radial velocities of gas and stars (Lacy et al. 1980, Serabyn and Lacy 1985, Genzel et al. 1985, Sellgren 1990, Krabbe et al. 1995, Haller et al. 1986, Genzel 1996). By now for 222 stars within the central parsec (at 8 kpc 1"=3D0.04 pc) radial velocities are known. Genzel et al. 1996 found that a compact (core radius ≤ 0.06pc) central dark mass of 2.2 to 3.2×10^6 $M_\odot$(for a distance of 8.0 kpc) is required if the stellar motions are isotropic. Here we report the first results of a programme to determine the proper motions of stars which directly test the assumption of isotropy.

2. Observations

For this purpose we have been carrying out a program of high resolution 2.2μm (K-band) imaging at the 3.5 m New Technology Telescope (NTT) of the European Southern Observatory (ESO) in La Silla, Chile since 1991. Using a high resolution camera developed specifically for this project (SHARP, Hofmann et al. 1993), we have used speckle imaging techniques to obtain diffraction limited resolution (0.15" FWHM at 2.2.μm, ee Eckart et al. 1992, 1993, 1994 for details). The resulting images reach to K-magnitudes of about 16 and show $\approx$600 stars (dynamic range >8 mag) in the central 25" (1 pc) diameter field centered on or near Sgr A*.

3. Positional Reference Frame

Progress has been made in linking the infrared and radio reference frames for the Galactic Center. We identified 5 H_2O/SiO maser stars within the central 20" of Sgr A* (Menten, Reid et al. 1996). Observing in the near-infrared with SHARP there are 2 stars in the same field of view as Sgr A*. With this information the radio and IR positional reference frame can now be linked to better than one 50mas SHARP pixel. We determined the position of Sgr A* in the NIR to within $\approx$40 mas EW, and $\approx$25 mas NS. This analysis indicates that Sgr A*(R) is not coincident with any of the K=3D15^m members of the stellar cluster in the central 2". However, it clearly demonstrates that Sgr A* is a member of the cusplike cluster in the central 1" of the Galaxy.

4. Stellar Proper Motions

For the present study we analysed $\approx$50 independent images from observing runs in 1992.25, 1992.65, 1993.65, 1994.27 and 1995. 6. For our 50mas pixel scale PSF cross-correlation and Gaussian fitting in the raw SSA images (Christou 1991), the LUCY (Lucy 1974) deconvolved images and the diffraction limited restored maps give the same positions to within 5mas for bright isolated sources and 10mas for close multiple or very faint isolated sources.

To obtain the proper motions we proceeded as follows: Relative pixel offsets from IRS16NE were then determined for each image image from a cross correlation within the central 3×3 pixels around the peak of each star with a mean PSF of the 3 brightest stars in each image. We determined the zero order (centroid position), first order (rotation angle and pixel scale) and the three second order instrumental parameters for each coordinate and image with respect to a reference frame constructed from the 1994.27 epoch data. For this purpose the 2×6 instrumental parameters were obtained by solving an over-determined nonlinear equation for N stars via orthonormalization of the 12×N matrix. The solutions turned out to be very stable in terms of selection of the N stars, as long as N>>20>N_{min}=3D6. The fitted pixel scales differed by less than 0.5% between different images and the second order distortion parameters were of the order of less than 10^{-3}. Therefore the main fit parameters were the position of the centroid of the N stars and the camera rotation angle. After correction we

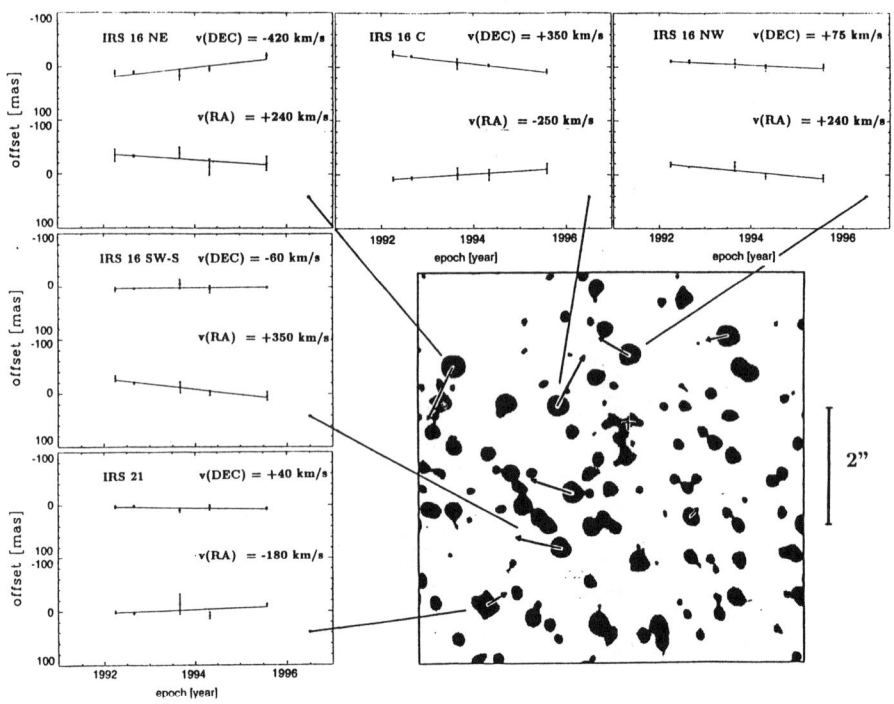

Figure 1. Proper motion measurements of selected stars in the central few arcseconds.

found that the final positional fit errors ranged between 8 and 20mas per images for the brighter, isolated stars.

In Fig.1 we show the RA- and Dec-offsets of 5 selected bright and isolated stars as a function of time, along with the fitted proper motions. The locations of these stars can be seen from the grey-scale 2.2 μm image which also shows the best fit proper motion vectors.

Significant motions ($\geq 4\sigma$) in at least one coordinate are detected for 7 stars. For 35 stars between 0.035 pc and 0.35 pc from the compact radio source Sgr A* the intrinsic proper velocity dispersion per coordinate is 160($\pm$15) km/s at a mean projected radius of 0.12 pc. This value is in excellent agreement with the recent radial velocity dispersion results and indicates that the stellar velocity field is indeed close to isotropic. The full space velocities of early type stars in the central 0.1 pc are of the order of 500 km/s. In addition to the best 7 cases with motions statistically detected at least at the 4σ level, we have included

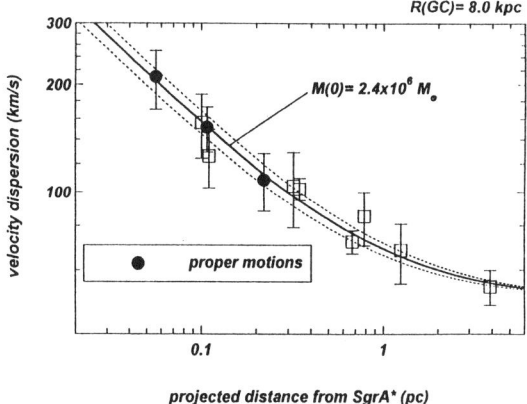

Figure 2. Projected stellar velocity dispersions as a function of projected distance from Sgr A*.

for an un-biased analysis another 28 isolated stars at radii between 0.9 and 8.8 from Sgr A* . Eight of those are HeI emission line stars, five are late type supergiants/AGB stars. Our analysis also shows that the velocity dispersion is very similar in all three coordinates and hence any anisotropy of the stellar motions must be small. For Galactic Center distances between 7 and 9 kpc the proper motion and radial velocity dispersion for the thirteen stars agree very well with 163($\pm$25) km/s and 187($\pm$40) km/s, respectively. In Fig.2 we plot of radial and proper motion velocity dispersion as a function of projected distance from Sgr A*, including 35 stars from this study, and 222 stars for which radial velocity stars are available (Genzel et al. 1996, McGinn et al. 1989, Rieke and Rieke 1988, Lindqvist, Habing and Winnberg 1992).

Inspecting the proper motion vectors within the central few arcseconds it appears that the (early type, HeI) stars in the IRS16 cluster show a coherent streaming pattern that may be interpreted as rotation about a centroid located within 0.5"(0.02 pc) of Sgr A*. In contrast the late type stars near the center (which are likely at much larger true distance; Sellgren et al. 1990, Haller et al. 1986, Genzel et al. 1996) show no such pattern and also much smaller velocities.

5. Evidence for a Massive Black Hole

Fitting to the projected velocity dispersion data in Fig.2 a model with a central point mass plus an extended isothermal cluster of dispersion $\approx$50 km/s, or calculating the enclosed masses from the Bahcall-Tremaine projected mass estimator, the Virial theorem and the Jeans equation results in central masses between 2 and 2.7×10^6 M$_\odot$ within a linear distance of 0.15 pc, and between 1.4 and 3.2×10^6 M$_\odot$ within 0.075 pc of the dynamic center. For central dark mass the combined data indicate a density of at least 6.5×10^9 M$_\odot$pc^{-3} and a core radius less than 0.035pc.

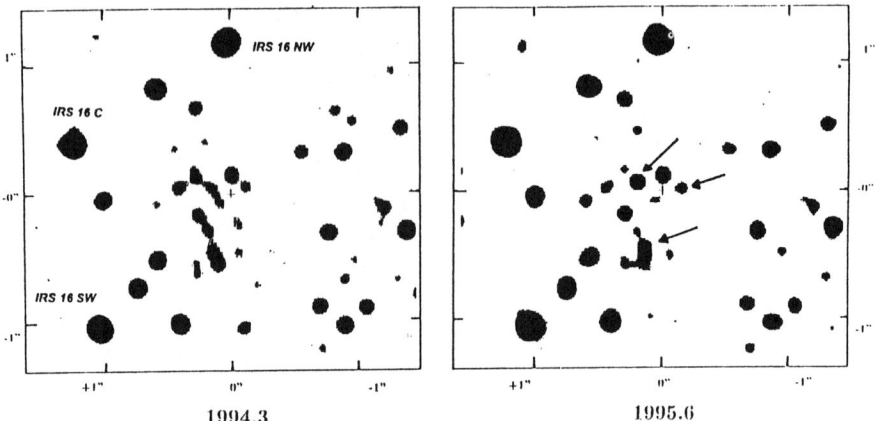

Figure 3. Comparison of the best 2μm maps of the central Sgr A* cluster, taken in 1994 and 1995. Arrows indicate stars with high velocities.

Large proper motions consistent with a large central mass are also indicated by a structural comparison of the images of the stellar cluster in the central 2". Fig.3 shows grey-scale representations of our best images (1994,1995) of the Sgr A*(IR) cluster (Eckart et al. 1994) in the direct vicinity of the compact radio source. There are significant changes of the positions of the stars very close to Sgr A* (radii 0.01 pc), suggesting $>10^3$ km/s motions. This evidence will be tested by future observations, but the data are consistent with a location and concentration of the $\approx 2.4 \times 10^6 M_\odot$ dark mass directly on Sgr A*.

The Galactic Center, along with the mega-maser galaxy NGC 425 (Greenhill et al. 1995, Myoshi et al. 1995) are now known to contain dark masses with densities approaching 10^{10} $M_\odot pc^{-3}$. Given our knowledge about the stellar content of the Galactic Center, it appears unlikely that the mass concentration is a cluster of solar mass remnants (neutron stars or white dwarfs, Genzel et al. 1996). It is either a core collapsed cluster of 10-20 $M_\odot$ stellar black holes (Morris 1993, Lee 1995) or, most likely, a 2.4×10^6 $M_\odot$ single massive black hole.

Acknowledgments. A number of people at MPE and ESO have been involved in making this experiment possible and carry it out. We would like to especially thank N.Ageorges, S.Drapatz, R.Hofmann, A.Krabbe, B.Sams, L.E.Tacconi-Garman, P.M. Duhoux and H.van der Laan. We thank L.Tacconi and N.Thatte for valuable comments.

Discussion

Rene Mendez': In your astrometric solution I noticed that you have not included color and/or magnitude terms. Color terms will arise because of different color refraction, while magnitude terms could arise because of the Cassegrain nature of NTT.

A. Eckart: The sources in the Galactic Center have very similar apparent colors due to the large amount of extinction. Also they have very similar magnitudes over the different epochs. Possible residual effects due to color and magnitude are taken care of by the second order terms in our solution. Furthermore we are currently only interested in the relative proper motions not in the absolute once, so that differential color and magnitude variations will have little effect on our solution.

References

Allen, D.A., Hyland, A.R. and Hillier, D.J., MNRAS 244, 706 (1990)

Christou, J.C. , Experimental Astr. 2, 27 (1991)

Eckart, A., Genzel, R., Krabbe, A., Hofmann, R., van der Werf, P.P. and Drapatz, S. , Nature 355, 526 (1992)

Eckart, A. Genzel, R., Hofmann, R., Sams, B.J. and Tacconi-Garman, L.E. , Ap.J. 407, L77 (1993)

Eckart, A. , Genzel, R., Hofmann, R., Sams, B.J., Tacconi-Garman, L.E. and Cruzalebes, P. , in The Nuclei of Normal Galaxies, eds. R.Genzel and A.I. Harris, Kluwer (Dordrecht), 305 (1994)

Eckart, A. Genzel, R., Hofmann, R., Sams, B.J. and Tacconi-Garman, L.E. , Ap.J. 445, L26 (1995)

Genzel, R., Watson, D.M., Crawford, M.K. and Townes, C.H., Ap.J. 297, 766 (1985)

Genzel, R., Thatte, N., Krabbe, A., Kroker, H. and Tacconi-Garman, L.E. 1996, Ap.J. in press

Greenhill, L.J, Jiang, D.R., Moran, J.M., Reid, M.J., Lo, K.Y. and Claussen, M.J., Ap.J. 440, 619 (1995)

Haller, J.W., Rieke, M.J., Rieke, G.H., Tamblyn, P., Close, L. and Melia, F., Ap.J. 456, 194 (1986)

Hofmann, R., Blietz, M., Duhoux, P. , Eckart, A., Krabbe, A. and Rotaciuc, V. , in Progress in Telescope and Instrumentation Technologies, ed. M.H. Ulrich, ESO Report 42, 617 (1993)

Krabbe, A. et al., Ap.J. 447, L95 (1995)

Krabbe, A., Genzel, R., Drapatz, S. and Rotaciuc, V., Ap.J. 382, L19 (1991)

Lacy, J.H., Townes, C.H., Geballe, T.R. and Hollenbach, D.J., Ap.J. 241, 132 (1980)

Lucy, L.B. , A.J. 79, 745 (1974)

Lee, H.M., MNRAS 272, 605 (1995)

Lindqvist, M., Habing, H. and Winnberg, A. 1992, Astr.Ap. 259, 118 (1992)

Lo, K.Y., "The Center of the Galaxy", (ed.) Morris, M., (Kluwer:Dortrecht), p.527. (1989)

McGinn, M.T., Sellgren, K., Becklin, E.E. and Hall, D.N.B. , Ap.J. 338, 824 (1989)
Morris, M., Ap.J. 408, 496 (1993)
Rieke, G.H. and Rieke, M.J., Ap.J. 330, L33 (1988)
Reid, M., Ann.Rev.Astr.Ap.31,345 (1993)
Sellgren, K., McGinn, M.T., Becklin, E.E. and Hall, D.N.B., Ap.J. 359, 112 (1990)
Serabyn, E. and Lacy, J.H., Ap.J. 293, 445 (1985)
Menten, K.M., Eckart, A., Reid, M.J. and Genzel, R., in prep. (1996)
Myoshi, M., Moran, J.M., Hernstein, J., Greenhill, L., Nakai, N., Diamond, P. and Inoue, M. , Nature 373, 127 (1995)

Quantitative Spectroscopy of the He I Cluster

Francisco Najarro and Rolf P. Kudritzki[1]

Universitäts-Sternwarte München, Scheinerstraße 1, D-81679 München, Germany

Alfred Krabbe, Reinhard Genzel and Dieter Lutz

Max-Planck-Institut für extraterrestrische Physik, 85740 Garching bei München, Germany

D. John Hillier

Department of Physics and Astronomy, University of Pittsburgh, 3941 O'Hara Street, Pittsburgh, PA 15260, USA

Abstract. We present first results on quantitative infrared spectroscopy of the brightest He I emission line stars in the Galactic center. The observed He I and H broad emission lines are caused by extremely strong stellar winds ($\dot{M} \sim 5$ to $80 \times 10^{-5} M_\odot \text{yr}^{-1}$) with relatively small outflow velocities ($V_\infty \sim 300$ to 1000 km s^{-1}). The effective temperatures of the objects range from 17,000 K to 30,000 K with corresponding stellar luminosities of 1 to $30 \times 10^5 L_\odot$. Strongly enhanced helium abundances ($N_{He}/N_H > .5$) are found. These results indicate that the He I emission line stars are evolved blue supergiants close to the evolutionary stage of Wolf-Rayet stars. They power the central parsec and belong to a young stellar cluster of massive stars which formed some 5×10^6 years ago. Statistical evidence for a concentration of dark mass ($\sim 3 \times 10^6 M_\odot$) in the Galactic center is derived from the stellar velocity dispersion.

1. Introduction

The He I emission line cluster in the Galactic center discovered by Krabbe et al. (1991) provides a unique opportunity to test star formation and evolution in connection with the ionization and energetics of the Galactic center. Results of a detailed spectroscopic investigation of the brightest source (AF star) of the cluster in the He I 2.058μm line by Najarro et al. (1994) revealed that the AF star is a helium-rich blue supergiant/Wolf-Rayet star. It is characterized by a strong stellar wind and constitutes a moderate source of Lyman continuum photons. The latter result clearly suggested that the cluster of He I stars can make a significant contribution to the total luminosity and the Lyman continuum flux of the inner parsec. Therefore, it was crucial to analyze the spectra of other He I

---

[1] Max-Planck-Institut für Astrophysik, 85740 Garching bei München, Germany

objects of the cluster and to obtain their stellar parameters to better constrain their role in the energetics of the central parsec. We report here first results of an extensive new study of the Galactic center stellar cluster. Stellar parameters are derived for eight He I objects.

2. Summary of observations and first results

Spectroscopic observations of the He I cluster have been carried out using CSG4, FAST, SHARP and 3D, and are described in detail in Krabbe et al. (1995).

Table 1. Observed properties for the brighter GC He I sources. All fluxes are dereddened and have been scaled to a distance of 1 kpc assuming a distance of 8.5 kpc to the Galactic Center.

Star	A(H-K)	$F_{2.2\mu}$ (Jy)	He I$_{2.06}$	Bγ He I$_{7-4}$	Pα He I$_{4-3}$ (10^{-10} erg s^{-1} cm^{-2})	He I$_{2.112}$ C III+N III	He II$_{2.19}$	FWHM km s^{-1}
7W	2.7	20.1	5.56	1.10	<13	1.75	<.3	1000
13E1	2.7	139.	11.6	5.66	<50	8.47	~.9	1000
16NE	2.0	125.	3.02	1.11	...	1.21	<.3	600
16NW	2.2	72.4	1.16	.421	...	.462	...	1000
16C	2.3	115.	2.69	.894	...	1.98	<.3	600
16SW	2.4	145.	5.04	2.09	...	2.02	<.5	650
15SW	2.3	66.3	7.53	1.04	<7	.909	...	850
15NE	2.3	46.3	6.27	1.30	<12	1.19	<.2	950

Table 1 summarizes the observational results obtained for the He I objects analyzed in this work. Before using the above results as constraints to model the objects some remarks must be made concerning some of the observed values. From Table 1 we see that the extinction shows high variability on small spatial scales, and therefore individual estimates must be made for each object under consideration. For the wavelength range around the K-band we assume $A(K) = 1.53(A(H) - A(K))$ and $A(\lambda) = A(K)(\lambda/2.2\mu)^{-1.75}$ (Krabbe et al. 1995). The measured FWHM of the emission lines were used as an initial guess for V_∞, the terminal velocity of the stellar wind. The final V_∞ value was then obtained by fitting the computed profiles iteratively with the observed ones. Except for IRS 13E, all He II 2.189μm line fluxes are upper limits. All available Pα fluxes were used as checks for the consistency of the results since the measurements suffer from a much higher uncertainty. The observed feature at $\lambda\approx 2.11\mu$m is expected to be a blend of the He I 2.112μm and N III and/or C III lines (Najarro 1995). Indeed, detailed inspection of the observed spectra indicates that in many of the sources the He I 2.112μm we clearly identify a broad blue component centered at $\lambda\approx 2.105\mu$m as well as two red components at $\lambda\approx 2.115\mu$m and $\lambda \approx 2.121\mu$m. Hence, the observed λ 2.112μm fluxes were considered as upper limits, and, if possible, the true He I 2.112μm line flux was estimated by subtracting a $\sim 2V_\infty$ wide region centered at 2.1120μm from the observed 2.11μm feature. As pointed by Najarro et al. (1994) the well separated He I and Pα components constitute the best indicator in the observed wavelength range to determine whether hydrogen is present or highly depleted in the objects. The **CGS4** spectra of IRS7W, 13E1, 15NE and 15SW do not show the hydrogen Pα line after removing the diffuse emission component, which indicates that these objects must be pure He stars (see also Geballe et al. 1994).

To model the objects we proceeded as in Najarro et al. (1994) and used the iterative, non-LTE method presented by Hillier (1987, 1990) which solves the radiative transfer equation in spherical geometry, subject to the constraints of statistical and radiative equilibrium, for the expanding atmospheres of early-type stars. We considered a pure H-He atmosphere with 12 H, 49 He I ($n \leq 10$) and 12 He II levels. The stellar parameters obtained for the above eight objects are presented in Table 2. Agreement between observed and computed values was always better than 10%. Below we disscus our results for the analyzed objects.

Table 2. Derived stellar parameters for GC He I sources. $R_{2/3}$ is the photospheric radius at Rosseland optical depth 2/3, $T_{2/3}$ the effective temperature derived for L_* and $R_{2/3}$ and $\eta = \dot{M}V_\infty/(L_*/c)$ is the "performance" number.

	\multicolumn{8}{c}{IRS Object}							
	16NE	16C	16SW	16NW	7W	13E1	15NE	15SW
R_* ($R_\odot$)	85	85	90	70	23	60	45	62
L_* ($10^5 L_\odot$)	22.0	19.0	25.9	10.3	4.10	22.6	9.10	10.5
$T_{\rm eff}$ (10^4 K)	2.41	2.32	2.44	2.20	3.04	2.89	2.66	2.35
$R_{2/3}$ (R_*)	1.12	1.12	1.14	1.10	1.53	1.95	1.17	1.13
$T_{2/3}$ (10^4 K)	2.27	2.20	2.29	1.77	2.46	2.07	2.46	2.20
He/H	1.0	3.0	1.3	1.3	>500	>500	>100	>100
$\dot{M}$ ($10^{-5} M_\odot$ yr^{-1})	9.50	10.5	15.5	5.30	20.7	79.1	18.0	16.5
V_∞ (km s^{-1})	550	650	650	750	1000	1000	750	700
η	1.17	1.78	1.92	1.91	24.9	17.3	7.33	5.43
V_{rad} (km s^{-1})	-40	200	300	-100	-250	0	-150	-200
Log Q(H$^+$)	49.6	49.5	49.7	49.0	49.2	49.9	49.4	49.3
Log Q(He$^+$)	< 45	< 45	< 45	< 44	< 44	45.0	< 44	< 44
Log Q(He^{++})	< 32	< 32	< 32	< 31	< 33	< 33	< 33	< 32

IRS 7W & IRS 13E. IRS 7W is the hottest ($T_{\rm eff} \approx 30$ kK) but also the least luminous (not considering the AF star) analysed object of the new sample. The stellar parameters derived are similar to those of WN8 and WN9 stars (Hamann et al. 1993, Crowther et al. 1995a). Its observed near-IR spectrum clearly indicates a later spectral type than WN8 (Hillier 1985), and probably corresponds to a WN9 or WN10 type. However, unlike WNL stars, H is much too depleted in IRS 7W, He/H$\geq$500. It is important to note that the hydrogen abundance is strongly constrained by the absence of H Pα. This, together with the high degree of extension of the atmosphere $R_{2/3}/R_* >$1.5 and its high "performance" number ($\eta \approx$25) relate the object to WNE-w stars (Hamann et al. 1993), though the effective temperature of IRS 7W is low compared with other such stars.

IRS 13E is also a rather hot source ($T_* \approx 29$ kK) but because of its enormous degree of extension $R_{2/3}/R_* \approx 2$, $V(R_{2/3}) > V_\infty/3$, it presents a much lower $T_{2/3}$ value. This is essentially caused by the extremely high rate of mass-loss of the object ($\dot{M} \approx 8 \times 10^{-4} M_\odot$ yr^{-1}). Like IRS 7W, IRS 13E shows no stellar hydrogen features in its spectrum which confirms its highly evolved status, and hence it

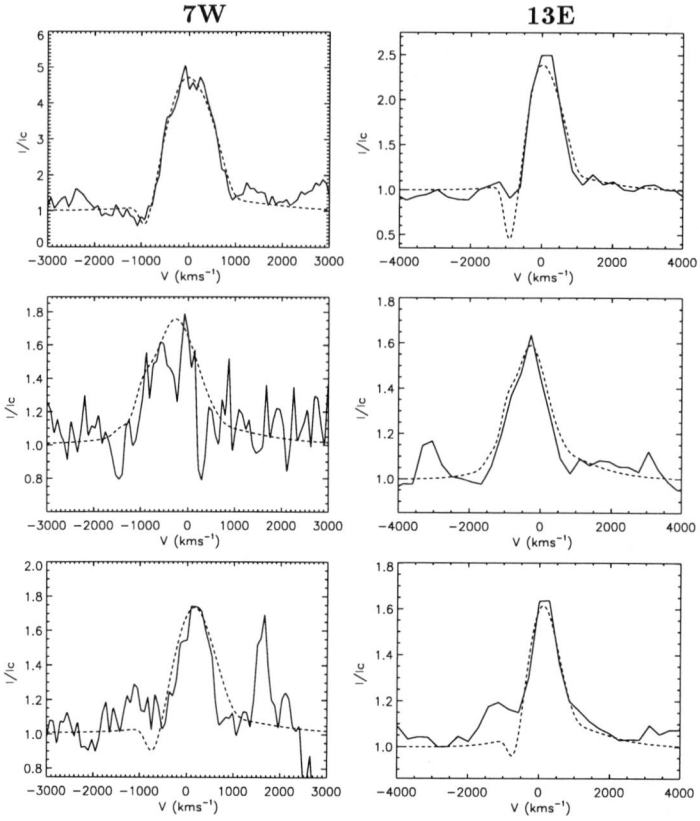

Figure 1. **left (from top to bottom):** IRS 7W observed **CGS4** (solid) and computed (dashed) He I 2.058μm, Bγ and He I 2.112μm profiles. **right (from top to bottom)** IRS 13E observed **3D** (solid) and computed (dashed) He I 2.058μm, Bγ and He I 2.112μm profiles. Diffuse emission has been removed for both objects.

may be classified as a WN9 or WN10 star. Interestingly, its performance number ($\eta \approx 17$) is slightly lower than that of IRS 7W, but still as high as those of WNE-w stars, and very similar to the values obtained for the AF star (Najarro et al. 1994). For the He II 2.189μm line we obtain a much weaker line flux (factor of 3) than observed. We attribute this result to the neglect of line blanketing (Hillier 1995). Nevertheless, it is important to stress that the presence in emission of both the He I 2.058μm and He II 2.189μm lines constitutes an important constraint for the effective temperature of the object as this occurs within an small effective temperature range ($\Delta T \leq 2000$ K).

Figure 1 shows the general good agreement of our computed line profiles for IRS 7W and 13E with the observed He I 2.058μm, Bγ and He I 2.112μm profiles. Both He I 2.058μm lines (including the strong electron scattering wings observed in IRS 13E) are well reproduced. For IRS 13E the computed absorption dip is clearly deeper than the observed one. This discrepancy may be due to stellar rotation. The fits to the observed He I (7−4) lines confirm the absence of H in the winds of IRS 7W and IRS 13E. Our models reproduce as well the central part ($\pm V_\infty$) of the 2.11μm observed feature corresponding to the He I 2.112μm line and reveal the important contribution of the N III/C III (8-7) lines (especially in IRS 13E).

IRS 16 sources The main and perhaps most interesting result from our analysis of the the IRS 16 sources, i.e., IRS 16 NE, C, SW and NW, is that they show similar He/H ratios (He/H$\sim$1, see Table 2). Since the values obtained for the stellar luminosities and effective temperatures of the objects are also very close (only IRS 16NW has a slightly lower L_* and T_{eff}), we may conclude that the brighter IRS 16 sources have similar ages. This result is also supported by the almost identical values obtained for the performance numbers (see Table 2), since η constitutes a powerful indicator of the evolutionary status of the star. The presence of hydrogen in the observed spectra is consistent with a WNL evolutionary phase, though the effective temperatures obtained for the IRS 16 objects is rather low when compared to the values derived for other later type (WN9-10) WNL stars (Crowther et al. 1995a). On the other hand, except for the enhanced He content, the stellar parameters of the IRS16 sources (R_*, L_*, $\dot{M}$ and V_∞) also resemble those of some known LBVs such as P Cygni (e.g., Najarro 1995, Langer et al. 1994). Recent work by Langer (private communication) shows that the inclusion of pulsational instabilities in the evolutionary calculations of massive stars leads to a more efficient mixing, and hence we may conclude that the IRS16 objects are just finishing their LBV phase. We retain, nevertheless, the Ofpe/WN9 classification for the IRS16 sources as their He content is close to that of the AF star and they show very similar stellar parameters to other known Ofpe/WN9 objects (Crowther et al. 1995a).

Figure 2 shows the excellent agreement between the observed **3D** and computed He I 2.058μm, Bγ and He I 2.112μm profiles for the IRS 16 sources. The obvious P Cygni shape of the observed He I 2.058μm profiles, confirms the wind nature of the lines. The presence of a blue absorption dip in the observed IRS 16SW Bγ line may be attributed to the problematic subtraction of the diffuse nebular emission. Once more, our results are able to reproduce the He I 2.112μm component quite well. Despite the relatively low S/N ratio, Figure 2 confirms the presence of the N III/C III (8-7) lines in three of the IRS 16 sources (NE, C and SW), while they seem to be absent in the spectrum of IRS 16NW. This result is consistent with the lower temperature ($\Delta T_{eff} \sim$1500 K) and wind density obtained for the latter, as we expect the strength of the N III/C III (8-7) lines to decrease with decreasing T_* and wind density. This would also explain the absence of these features in the spectrum of the AF star (lower T_{eff}) and their relatively large strength in the hotter, denser IRS 13E and IRS 7W.

IRS 15SW & IRS 15NE IRS15NE and IRS15SW may be considered as "transition" objects. These objects have effective temperatures only slightly higher ($T_{eff} \sim$25000 K) than those obtained for the IRS16 objects suggesting a Ofpe/WN9 classification, the main caveat for a WN9 to WN11 classification being the low effective temperature. However, they appear to be pure He sources (as do IRS7W and IRS13E-1) indicating a Wolf-Rayet phase. The latter could be due to the effective mixing through pulsation instabilities cited above. A further indication

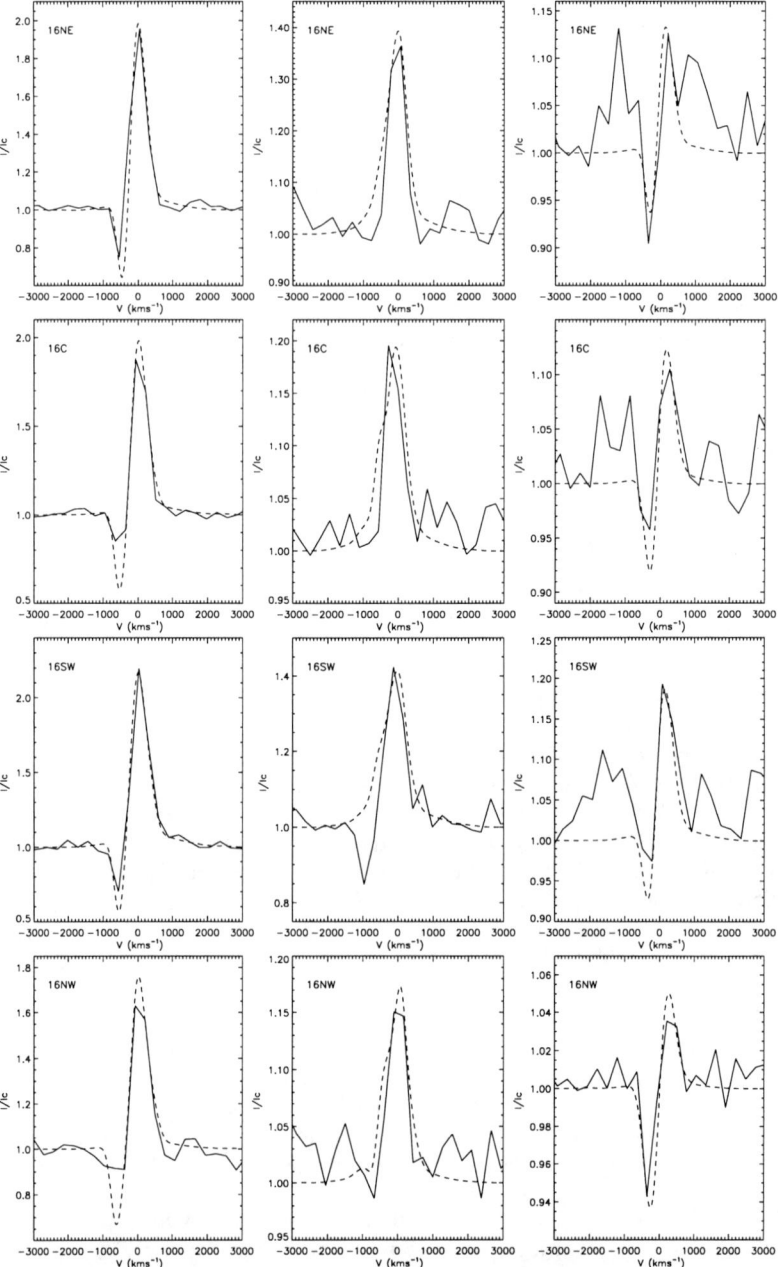

Figure 2. Observed **3D** profiles for the IRS 16 sources compared with the calculations (He I 2.058μm, Bγ and He I 2.112μm from left to right). Diffuse emission has been removed.

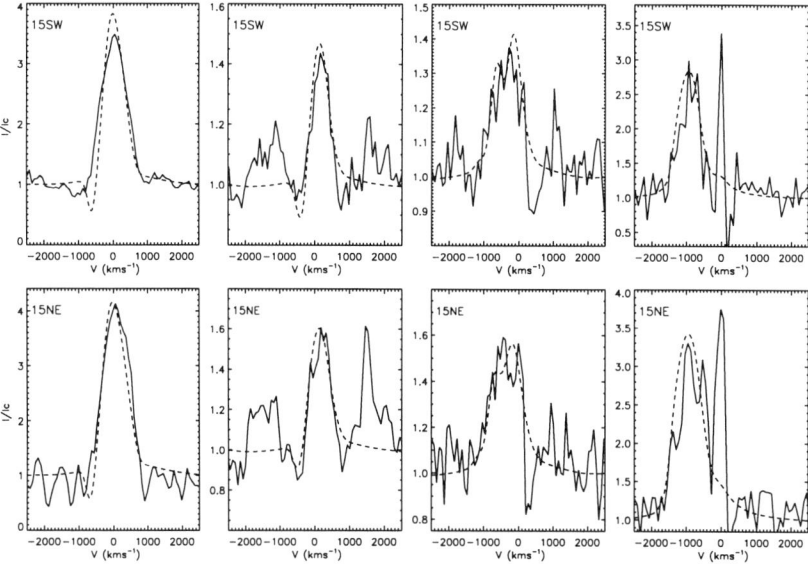

Figure 3. IRS 15SW and IRS 15NE observed **CGS4** (solid) and computed (dashed) He I 2.058μm, He I 2.112μm, Bγ and Pα profiles. Most of the diffuse emission has been removed.

for this "transition" status is given by the performance numbers and terminal velocities shown in Table 2.

Figure 3 shows the general good agreement of our model calculations with the observed IRS 15SW and IRS 15SW He I 2.058μm, He I 2.112μm, Bγ and Pα lines. The corresponding large slit width for the **CGS4** pixels (3.1$''$) hinders a clean subtraction of the relatively non-uniform background and diffuse emission, and therefore tends to blur the P Cygni dips of the He I lines (see Figure 3). This effect is even more extreme for the H lines (see the Bγ and Pα lines in Figure 3) where a "hole" or a narrow peaked feature at the wavelength of the hydrogen component may result from the presence of a non-uniform background. Despite these problems, our computed profiles are able to reproduce the shapes of the observed profiles satisfactorily (see Figure 3) and account for the broader blends due to the He I (4–3) and He I (7–4) lines. We note, once more, the clear presence of the N III/C III 2.103μm feature bluewards of the He I 2.112μm line.

Evolutionary status, ionizing flux and GC mass. Krabbe et al. (1995) have analyzed the nuclear star cluster using the star cluster models of Krabbe et al. (1994) and conclude that the observations fit a $7\pm1 \times 10^6$yrs old decaying burst in which $\sim$30000 stars formed very well. They obtain a model with a bolometric luminosity of L(Bol)$\sim 2.7 \times 10^7 L_\odot$ and a bolometric to Lyman luminosity ratio of ten. Such a model predicts 15 OB stars with L$\geq 3 \times 10^5 L_\odot$, 30 later type WR stars (WNL, WCL and the He I objects observed) and 2 to 4 KM supergiants with L(K)/L(Lyman)$\sim 3 \times 10^{-2}$. Further, the implied supernova rate (one in 4×10^4yrs) agrees with the interpretation of Sgr. A East (Genzel et al. 1994). Moreover, the presence of a fairly large number (~10) of moderately luminous late type stars in the central 8″ also indicates (see Haller et al. 1989) an earlier starburst about 10^8yrs ago. The above results clearly favor a model in which the central parsec of the Galaxy is powered by a cluster of hot stars, as earlier proposed by Rieke & Lebofsky (1982) and Allen & Sanders (1986). The eight objects studied here alone, already supply $\approx 3 \times 10^{50}$ Lyman continuum photons per second ($> 60\%$ of the total Lyman flux as estimated from thermal radio continuum measurements by Genzel et al. 1994). Therefore, the hot star cluster can fully account for the bolometric and Lyman ionizing luminosities of the central parsec. None of these objects contribute to the He I-continuum. Even IRS13E1 does not provide more than 1% of the flux required for the He-ionization of the SgrA West HII region, $\log Q(\text{He}^+) \approx 49$ (Krabbe et al. 1991). Hence, some hotter O and/or WR stars is required, the presence of which have been recently reported by Krabbe et al. (1995). Future, moderate S/N spectra of these objects are crucial to check whether they can or can not account for the He I-continuum in the Galactic center. Finally, we may use the derived radial velocities to obtain the velocity dispersion of the objects and therefore, by means of the virial theorem (and/or the Bahcall-Tremaine estimator, Bahcall et al. (1981), derive the enclosed mass in the central parsec. Krabbe et al. (1995) have obtained velocity measurements from Gaussian fits to the observed $P\alpha$, He I 2.058μm, He I 2.112μm and $B\gamma$ lines, and obtained a mass concentration of ~2 to $4 \times 10^6 M_\odot$, thus finding a high statistical significance for a dark mass concentration. The latter could be in the form of massive stellar remnants ($10 M_\odot$), i.e., the black hole cluster option suggested by Morris (1993), or more likely as a massive black hole (Genzel et al. 1994).

References

Allen, D.A., Sanders, R.H., 1986, NATURE, 319, 191
Bahcall, J.N., Tremaine, S., 1981, ApJ, 244, 805
Crowther, P.A., Hillier, D.J., Smith, L.J., 1995, A&A, 293, 172
Geballe, T.R., Genzel, R., Krabbe, A., Krenz, T., Lutz, D., 1994, in: Infrared Astronomy with Arrays: The Next Generation, ed. I. McLean, Kluwer, Dordrecht, p. 73
Genzel, R., Hollenbach, D.J., Townes, C.H., 1994, Rep.Progr.Phys., 57, 417
Haller, J.W., Rieke, M.J., 1989, in The Center of the Galaxy, ed. M.Morris, Kluwer, Dordrecht, 487
Hamann, W.R., Koesterke, L., Wessolowski, U., 1993, A&A, 274, 397
Hillier, D.J., 1985, AJ, 90, 1514
Hillier, D.J., 1987a, ApJS, 63, 947
Hillier, D.J., 1990, A&A, 231, 116
Hillier, D.J., 1995, in: IAU Symp. 163, Wolf-Rayet Stars: Binaries, Colliding Winds, Evolution, eds. K. A. van der Hucht and P. M. Williams, Kluwer, Dordrecht, p. 116
Krabbe, A., Genzel, R., Drapatz, S., Rotaciuc, V., 1991, ApJ, 382, L19
Krabbe, A., Stenberg, A., Genzel, R., 1994, ApJ, 424, 72

Krabbe, A., Genzel, R., Eckart, A., Najarro, F., Lutz, D., Cameron, M., Kroker, H., Tacconi-Garman, L.E., Thatte, N., Weitzel, L., Drapatz, S., Geballe, T., Sternberg, A., Kudritzki, R. , 1995, ApJ, 447, L95

Langer, N., Hamann, W.-R., Lennon, M., Najarro, F., Pauldrach, A.W.A., Puls, J., 1994, A&A, 290, 819

Morris, M., 1993, ApJ, 408, 496

Najarro, F., Hillier, D.J., Kudritzki, R.P., Krabbe, A., Genzel, R., Lutz, D., Drapatz, S., Geballe, T.R., 1994a, A&A, 285, 573

Najarro, F., 1995, PhD Thesis, University of Munich

Rieke, G.H., Lebofsky, M.J., 1982, in AIP Conference Proceedings 83: The Galactic Center, eds. G. Riegler, R. Blandford, New York, 194

Chemical Abundances of Stars at the Galactic Center

John S. Carr

Naval Research Laboratory, Washington, D.C.

K. Sellgren

Department of Astronomy, The Ohio State University, Columbus, OH

Suchitra C. Balachandran

Astronomy Department, University of Maryland, College Park, MD

Abstract. The first measurements of chemical abundances of stars in the Galactic Center cluster are presented. A detailed spectral analysis of two cool Galactic Center stars yields iron abundances that do not differ from similar supergiants and giants in the disk. In particular, we find that the young M supergiant IRS 7 has a solar iron abundance, [Fe/H] $= -0.06 \pm 0.14$.

1. Introduction

The Galactic Center is an unique environment in the Galaxy. The dense central star cluster is the site of a recent and intense burst of star formation that has produced an unusual collection of massive hot stars (Allen et al. 1990; Krabbe et al. 1991, 1995; Libonate et al. 1995; Blum et al. 1995a,b; Tamblyn et al. 1996). If this recent starburst is not a singular event but a recurring phenomenon, then the chemical evolution in the Galactic Center may have followed a course distinct from that in other parts of the Galaxy. Measurements of the present-day gas phase abundances in Sgr A West give conflicting results of $\sim 1 - 2$ times solar abundance (Shields & Ferland 1994). However, information on the chemical abundances in this unique stellar population of our Galaxy is severely lacking.

The elemental abundances in stars record the nucleosynthetic history of the gas from which they were formed. Many factors may have contributed significantly to the composition of star-forming gas in the Galactic Center. There are several potential sources for gas infall into the center, including molecular clouds from the inner disk migrating inward under the influence of a bar potential (Stark et al. 1991), and mass-loss from stars in the bulge (Blitz et al. 1993; Jenkins & Binney 1994). Observations of hot gas in the Galactic Center, probably fueled by supernova explosions (Yamauchi et al. 1991), suggest that gas may be driven out of the center by a "galactic wind". However, the effect of magnetic fields is unknown, and some or all of this gas may return to the center as a "galactic fountain" (Blitz et al. 1993). Elemental enrichment from supernovae and from extensive mass-loss from giants and supergiants in the Galactic Center can leave their own distinctive mark. An initial mass function weighted towards high mass stars, for example, would, over time, result in a proportionately larger enrichment in oxygen and α-elements via SN II. The extent of enrichment by processed and ejected material within the Galactic Center depends on how much enriched gas is retained for incorporation into the next starburst.

To understand the recent star formation history, it is important to determine whether the cool luminous stars at the Galactic Center are younger, high-mass supergiants or older, lower-mass stars on the asymptotic giant branch (AGB). Since CNO and s–process abundances are altered as a result of stellar evolution, these can provide clues to the evolutionary status of stars. The abundances influence our understanding of the Galactic Center in other ways. The overall metallicity will affect the mass-loss rates and evolutionary duration of the post-main-sequence phase of the hot massive stars. Unusually high or low metallicities could also affect comparisons of the luminosity functions of the Galactic Center and other stellar populations.

Given the deficiency in our knowledge about the chemical abundances of stars in the central stellar cluster, we decided to begin a program to obtain quantitative measurements of abundances for the brightest cool luminous stars in *the central few parsecs*. The advent of high-resolution infrared echelle spectrographs in the last few years has now made it possible to obtain spectra of bright M giants and supergiants in the Galactic Center in order to carry out a detailed abundance analysis. Our particular goals are to measure the mean value of [Fe/H] and its range and the relative abundances of elements, especially the α-elements (Mg, Si, Ca, Ti). In this contribution we present the first results on Galactic Center stellar abundances.

2. Data and Analysis

Spectra of the Galactic Center stars were obtained at the NASA Infrared Telescope Facility using CSHELL, the facility infrared echelle spectrograph. The nominal resolution of the spectra is 40,000 with a typical S/N ratio of 100. Telluric absorption lines were removed using spectra of hot stars. Because a single spectrum provides a limited wavelength coverage ($\sim$ 1000 km s^{-1}), exposures at different grating settings must be taken to obtain lines of interest for the abundance analysis. Sample CSHELL spectra are shown in Figure 1. In addition to the Galactic Center spectra, for comparative analysis we obtained CSHELL data for VV Cep (M2 I) and β And (M0 III) and archival KPNO 4-m FTS spectra of the following stars: α Boo (K2 III), β And (M0 III), α Ori (M1–2 Ia–ab), HR 6702 (M5+ II), and 30 Her (M6 III).

In order to determine stellar abundances, a knowledge of the star's effective temperature (T_{eff}), surface gravity ($\log g$), and microturbulent velocity (ξ) are required so that the appropriate model atmosphere can be used in the analysis. Gravities for Galactic Center stars can be estimated from their bolometric luminosity and a comparison of their position on the HR diagram to theoretical evolutionary tracks. Photometry cannot be used to determine T_{eff} since the observed colors are completely dominated by reddening. Instead, spectroscopic temperatures derived from high-resolution spectra or estimates based on the spectral type determined from low-resolution spectra must be used. The results

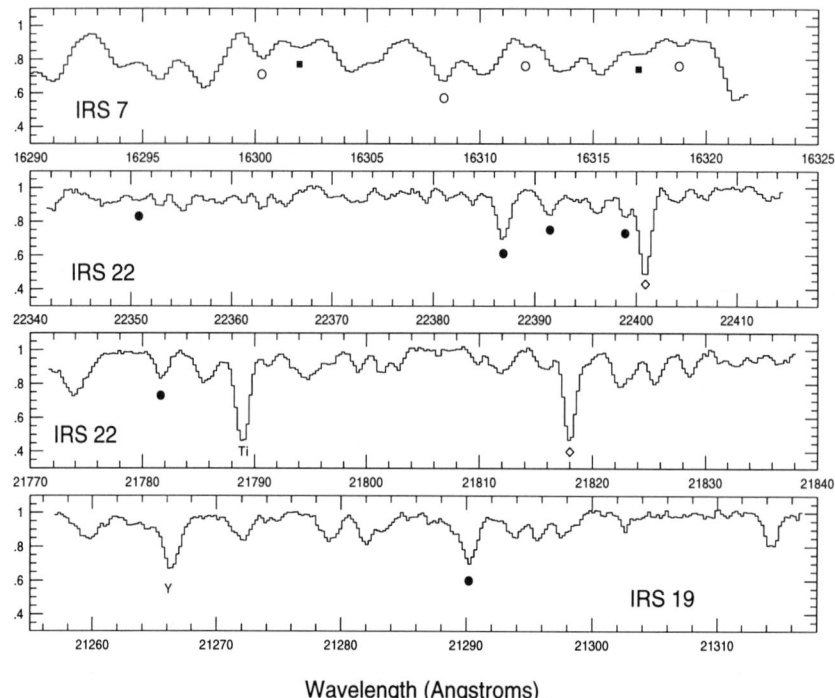

Figure 1. Sample CSHELL spectra of the Galactic Center stars IRS 7, 19 and 22. Various spectral lines are indicated by symbols: Fe (filled circles), Sc (open diamonds), CO (open circles) and OH (filled squares).

presented here were obtained using an updated version of the spectral synthesis program MOOG (Sneden 1973) and the spherical model atmospheres of Plez (1992).

The comparison field giants provide a sequence in $T_{\rm eff}$ and $\log g$ similar to the Galactic Center stars. The same analyses were carried out on the field stars to search for systematic effects and gauge the accuracy of the results. α Boo provides an important abundance reference point since it has previously been the subject of detailed optical analyses. Possible systematic errors in the abundances are a concern for these low gravity late-type stars. However, a comparison of abundances derived for similar stars (e.g., between IRS 7 and α Ori) gives a differential abundance free of systematic effects.

3. The M Supergiant IRS 7

The M supergiant IRS 7 is the most luminous of the cool Galactic Center stars. Its luminosity indicates an initial mass of 20–25 $M_\odot$ and an age of ∼8 Myr. As such, *the metallicity of this star should be representative of the stellar population in the recent starburst which took place in the last 10 Myr.*

For IRS 7, T_{eff} and ξ were derived from a spectroscopic analysis of CO, using weak to moderate strength second and first overtone lines the H and K bands. Unsaturated lines of CO provide a good estimate of the T_{eff} of late type stars: CO should be in LTE and the molecule provides a potentially large number of lines with a range in excitation potential and oscillator strength that can be used to fix both T_{eff} and ξ. The derivation of T_{eff} relies largely on the weaker second overtone lines because most first overtone lines are quite strong in M giants. Unfortunately, the second overtone occurs in the H band where line blending is more severe and where Galactic Center stars are a couple of magnitudes fainter than at K due to interstellar extinction. Synthetic spectra of the entire wavelength region of interest were calculated in order to identify largely unblended lines. The combination of line blending, large macroturbulent line broadening, and the limited CSHELL wavelength coverage greatly limit the number of usable CO lines.

The T_{eff} is determined by the requirement that the C abundance derived from individual weak CO lines be independent of the line excitation potential. Similarly, ξ is fixed by requiring the C abundance to be independent of line strength for weak to moderate strength CO lines of similar excitation potential. Using a Plez model atmosphere with $\log g = -0.5$ for IRS 7, gives $T_{\text{eff}} = 3470 \pm 250$ K and $\xi = 3.3 \pm 0.4$ km s^{-1}.

This analysis also provides the C abundance since CO is the principal species in which C is found in cool stars and its abundance is nearly independent of the O abundance in O-rich stars. We obtained $\log \epsilon(\text{C}) = 7.77 \pm 0.14$ (on a log scale where the abundance of H = 12.0). While T_{eff} and ξ for IRS 7 are similar to those found from the analysis of CO lines in the M supergiants α Ori and VV Cep, the carbon abundance is noticeably lower: [C/H] = -0.8 for IRS 7 versus -0.3 for α Ori and -0.4 for VV Cep (quantities in [ ] are a logarithmic abundance relative to the Sun). IRS 7 shows a CNO abundance pattern consistent with extreme mixing of material processed through the CNO cycle in the stellar interior: O (from OH lines) is highly depleted (~ 0.6 dex) and N (from CN lines) is enhanced about ~ 1.0 dex. A complete analysis of the CNO abundances and isotopic ratios of C and O, and their implications for massive star evolution, will be presented in a future paper.

With the stellar parameters defined, other abundances can be determined either from measurements of line equivalent widths or from spectrum synthesis. The iron abundances are based on equivalent widths of Fe I lines in the K band. We used solar gf-values based on the Kurucz (1991) solar model and equivalent width measurements from the Livingston & Wallace (1991) infrared solar atlas. Five usable lines of Fe I were measured in our spectra of IRS 7. From these, *we obtained an Fe abundance which is essentially solar*, [Fe/H] = -0.06 ± 0.14. The quoted uncertainty includes errors in all the stellar parameters (T_{eff}, ξ, $\log g$) and the error in the mean from the line-to-line scatter. The Fe abundance is relatively insensitive to T_{eff}; the total uncertainty is dominated by the uncertainty in ξ and the line-to-line scatter. An analysis of similar infrared Fe lines in α Ori also gives a solar abundance (see Table 1).

4. Results for Other Galactic Center Stars

High-resolution spectra of three other luminous M stars in the Galactic Center have been obtained: IRS 19, IRS 22 and IRS 24. With M_K between -9 and -10 these are either $9-12 M_\odot$ supergiants with an age of ~ 20 Myr or $5-7 M_\odot$ AGB giants with ages of $50-100$ Myr (see Blum et al. 1996), They have been assigned spectral types of M5 to M7 in the literature (Lebofsky et al. 1982; Sellgren et al. 1987). We currently have extensive spectral data for IRS 22 but a very limited set for IRS 19 and 24.

The high-resolution spectrum of IRS 24 is dominated by water absorption lines, making an abundance analysis infeasible. We do note that the water lines show a velocity redshift (~ 13 km s^{-1}) with respect to the other photospheric lines. This behavior is similar to that observed in the spectra of Mira variables (Hinkle & Barnes 1979; Wallace & Hinkle 1996). IRS 24 is observed to have strong water absorption in low-resolution spectra, and has recently been detected as both a H_2O and an OH maser (Lebofsky et al. 1982; Levine et al. 1995; Sjouwerman & van Langevelde 1996). *All of these suggest that IRS 24 is a long period variable on the AGB.*

Deriving T_{eff} from CO, as was done for IRS 7, is difficult for the other Galactic Center giants because they are too faint to easily obtain spectra of the H band where the crucial second overtone CO lines are found. However, the derived Fe abundance is not very sensitive to temperature over the expected range of T_{eff}. At $T_{eff} = 3200$ K, an uncertainty of $\pm$ 300 K leads to a 0.1 dex error in [Fe/H]. For IRS 22 and 19, we adopted $T_{eff} = 3200$ K, which is appropriate for a M6 giant, but considered a temperature range of 2900 to 3500 K. We derived log g = 0.0$\pm$ 0.2 and ξ = 2.9 $\pm$ 0.5 km s^{-1}. A total of 7 Fe I lines were measured for IRS 22, from which an abundance [Fe/H] = 0.13 $\pm$ 0.16 was obtained. Only one Fe I line was measured in IRS 19. Since this line has nearly the same equivalent width as in IRS 22, we surmise that IRS 19 would have the same Fe abundance if the stellar parameters are the same, but more lines are clearly needed to confirm this.

Table 1 compares the Fe abundances of IRS 7 and IRS 22 to those of the bright field M giants/supergiants obtained from the analysis of the archival FTS spectra. All of these stars have a Fe abundance close to solar. Possible systematic errors, such as non-LTE effects or uncertainties in the model atmospheres, could shift the abundance scale with respect to the Sun, but such systematic errors are largely eliminated in a *differential* comparison of similar stars. Hence, the importance of this comparison is that *the two analyzed Galactic Center stars have the same Fe abundance as disk supergiants and giants.*

5. Comparison to Low-Resolution Spectral Results

Sellgren et al. (1987) pointed out that the strengths of the Na and Ca features in low-resolution K band spectra were stronger in IRS 7 and a few other Galactic Center stars compared to similar field stars. This result has been confirmed and extended to additional Galactic Center stars by Blum et al. (1996). Given the near-solar Fe abundances measured for IRS 7 and IRS 22, do the larger equivalent widths of the Na and Ca features indicate the relative enhancement of

Table 1. Fe Abundances

Star	Type	# Lines	[Fe/H]
IRS 7	(M2 I)	5	-0.06 ± 0.14
IRS 22	(M5–7 II)	7	$+0.13 \pm 0.15$
α Ori	M2 Ia	9	-0.03 ± 0.11
30 Her	M6 III	7	-0.06 ± 0.11
HR 6702	M5+ II	7	$+0.12 \pm 0.11$
β And	M0 III	11	0.00 ± 0.06

some elements? A straightforward interpretation of these low-resolution features is not easy since these "lines" are a blend of Na, Ca, Sc, Ti, V, CN and other lines. Comparison of our high-resolution Galactic Center spectra with those of field giants does shed some light on this question.

The equivalent width of the "Na feature" measured in low-resolution spectra of IRS 7 is about 30 – 40% larger than in field stars of the same spectral type. In Figure 2 we compare our CSHELL spectrum of IRS 7 with an archival FTS spectrum of α Ori in the region of the Na doublet. The Na lines in IRS 7 and α Ori have very similar strengths, and a differential analysis yields the same Na abundance in both stars. The most obvious difference between the two spectra are the stronger Sc and V lines in IRS 7; CN lines account for most of the remaining differences. The stronger CN lines in IRS 7 are due to both a larger N abundance and a lower surface gravity. Spectrum synthesis shows that CN can account for about half of the difference in "Na feature" equivalent width between IRS 7 and field stars, with the remainder of the increase due to the Sc and V lines. Indeed, a careful comparison of the entire low-resolution FTS spectrum of IRS 7 (Sellgren et al. 1987) with standards from Kleinmann & Hall (1986) shows that the most prominent spectral differences (besides the CO bands) are stronger Sc lines in IRS 7. Our high-resolution spectra of IRS 19 and 22 also show stronger Sc lines than the similar spectral type field M giants HR 6702 and 30 Her.

Do the Galactic Center stars have an enhanced Sc abundance? We note that McWilliam & Rich (1994) find enhanced values of [Sc/Fe] in bulge K giants. Sc is considered to be an iron-peak element, so that the Sc abundance should follow that of Fe; but it has been suggested that substantial enhancements of Sc could be produced by mild s–processing (Smith & Lambert 1987). However, the Sc lines have large equivalent widths and are strongly effected by hyperfine structure (hfs). Because hfs constants for these infrared transitions have not been measured, a quantitative analysis cannot be carried out. The hfs produces a desaturating effect, which makes the lines particularly sensitive to atmospheric parameters like gravity. We note that the other element which shows a large equivalent width increase in IRS 7 is V, another odd-Z element with hfs. The low excitation potential of these lines also makes them potentially sensitive to conditions in the extended outer atmospheres of the stars. Hence, the interpretation of the strong Sc lines in the Galactic Center stars is not clear at this time. However, it is certainly clear that caution should be used in interpreting low-resolution spectra.

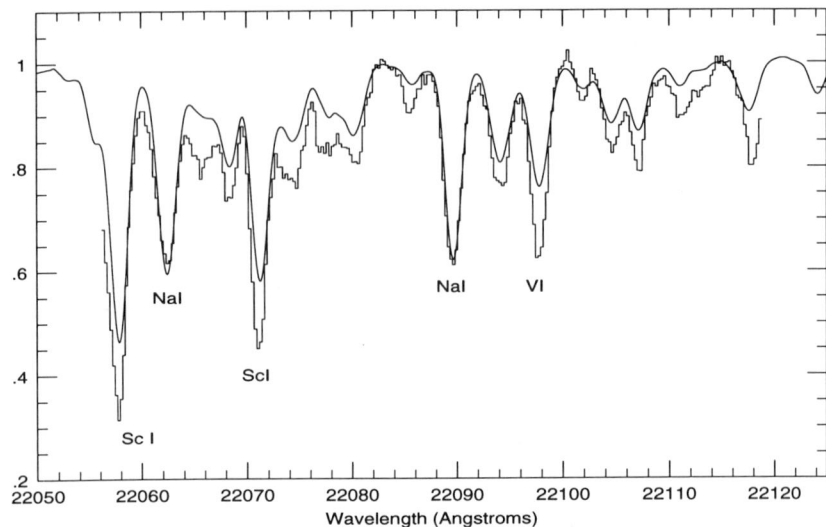

Figure 2. A spectrum of the Na doublet region in IRS 7 (*histogram*) compared to α Ori (*smooth line*). The spectrum of α Ori is smoothed to the same resolution as the IRS 7 data. The majority of unmarked lines are CN.

6. Conclusion

From these first measurements of chemical abundances in Galactic Center stars, we find that the Fe abundance in IRS 7 and IRS 22 do not differ significantly from similar field disk supergiants and giants we have analyzed. The solar Fe abundance of the M2 supergiant IRS 7, [Fe/H] = -0.06 ± 0.14, indicates that the most recent generation of star formation in the Galactic Center cluster formed out of gas which was not metal-rich.

However, a significant amount of work remains to be done to establish the abundance scale of stars in the Galactic Center. The foremost among them is to increase the sample of stars in order to determine the mean value of the metallicity and its range. With current instrumentation, high-resolution spectra of about the ten brightest cool Galactic Center stars can be obtained. Abundance measurements must also be extended to other elements, especially those (like the α-elements) that could show different abundance patterns.

Acknowledgments. This work was supported by NSF grant AST-9115236 and by a OSU Columbus Fellowship to JSC and AST 93-14851 to SCB.

References

Allen, D. A., Hyland, A. R., & Hillier, D. J., 1990, MNRAS, 244, 706
Blitz, L., Binney, J., Lo, K. Y., Bally, J., & P. T. P. Ho 1993, Nature, 361, 417

Blum, R. D., Sellgren, K., & DePoy, D. L. 1995a, ApJL, 440, L17
Blum, R. D., DePoy, D. L., & Sellgren, K., 1995b, ApJ, 441, 603
Blum, R. D., Sellgren, K., & DePoy, D. L. 1996, these proceedings
Hinkle, K. H. & Barnes, T. G. 1979, ApJ, 227, 934
Jenkins, A. & Binney, J. 1994, MNRAS, 270, 703
Kleinmann, S. G. & Hall, D. N. B. 1986, ApJS, 62, 501
Krabbe, A., Genzel, R., Drapatz, S., & Rotaciuc, V., 1991, ApJL, 382, L19
Krabbe, A., et al. 1995, ApJL, 447, L95
Kurucz, R. L. 1991, in Precision Photometry: Astrophysics of the Galaxy, ed. A. G. Davis Phillip, A. R. Upgren & K. A. Janes (Schenectady: Davis), 27.
Lebofsky, M. J., Rieke, G. H., & Tokunaga, A. T., 1982, ApJ, 263, 736
Levine, D. A., Figer, D. F., Morris, M., & McLean, I. S. 1995, ApJL, 447, L101
Libonate, S., Pipher, J. L., Forrest, W. J., & Ashby, M. L. N. 1995, ApJ, 439, 202
Livingstone, W. & Wallace, L. 1991, N.S.O. Technical Report #91-001, (Tucson: National Solar Observatory)
McWilliam, A. & Rich, R. M. 1994, ApJS, 91, 749
Plez, B. 1992, A&AS, 94, 527
Sellgren, K., Hall, D.N.B., Kleinmann, S.G., & Scoville, N.Z. 1987, ApJ, 317, 881
Shields, J. C. & Ferland, G. J. 1994, ApJ, 430, 236
Sjouwerman, L. O. & van Langevelde, H. J. 1996, ApJL, 461, L41
Smith, V. V., & Lambert, D. L. 1987, MNRAS, 227, 563
Sneden, C. 1973, Ph.D. thesis, Univ. of Texas
Stark, A. A., Gerhard, O. E., Binney, J., & Bally, J. 1991, MNRAS, 248, 14P
Tamblyn, P., Rieke, G. H., Hanson, M. M., Close, L. M., McCarthy, D. W., & Rieke, M. J. 1996, ApJ, 456, 206
Wallace, L. & Hinkle, K. 1996, ApJS, in press
Yamauchi, S., Kawada, M., Koyama, K., Kunieda, H. Tawara, Y., & Hatsukade, I. 1990, ApJ, 365, 532

Origin of the Peculiar He I Stars

Peter Tamblyn, Fulvio Melia, and G. H. Rieke

Steward Observatory, University of Arizona, Tucson, AZ 85721 USA

Abstract. The luminosities of the central emission stars are compared to Galactic and LMC samples. The absence of similar populations in M31 and M32 suggests they are a transient feature (Rieke & Rieke 1994). Models of extended star formation episodes can reproduce the gross features of the population (cf. Krabbe et al. 1995), but not so many very luminous emission stars. Various clues, including the stars' spatial distribution, suggest that unusual stellar development may alter the population. We have examined numerous steady-state and post-starburst models incorporating unique features of the region. Interactions of massive stars with one another, with Sgr A*, and with the mass dominating population are unlikely to influence the population substantially.

1. Introduction

With improved spatial resolution, the broad He I $2.058\,\mu$m emission line observed from the central stellar cluster (Hall et al. 1982) has been tied to individual blue, luminous stars (Krabbe et al. 1991; Krabbe et al. 1995; Tamblyn et al. 1996), including the bright components of IRS 16. Comparisons of the spectra of these sources with other luminous stars (Allen et al. 1990; Najarro et al. 1994; Libonate et al. 1995; Blum et al. 1995; Krabbe et al. 1995; Tamblyn et al. 1996) and detailed atmospheric models (Najarro 1995) have prompted comparisons to Wolf-Rayet, Ofpe, and Luminous Blue Variable stars. However, conflicts with the predictions of evolving starbursts and with a broad sample of luminous stars have led to doubts that this population represents a normally evolving, young starburst with a normal initial mass function (Tamblyn & Rieke 1993; Tamblyn et al. 1996). We consider various unique conditions in the central parsec and show that none of them are capable of modifying the evolution of a normal starburst to reproduce the observed population.

2. Are These Stars Expected?

Individually, the bright emission-line sources appear similar to some evolved massive stars, although the $2.058\,\mu$m feature is uncommon even among this class. The population seen in the central parsec appears more biased towards this emission feature than the populations in the neighboring star formation regions studied by Figer et al. (1995) and Cotera et al. (1996). The He I emission puts a lower limit on the stars' luminosities because stars cooler than 15,000 K do not exhibit this feature in strong emission. Without resorting to detailed models, the near-infrared brightness of IRS 16NE and this limit on T_{eff} imply that its luminosity must be at least $10^6 L_\odot$. Similarly, allowing the temperature to be a free parameter between 15,000 K and 35,000 K, the NIR brightnesses of the other emission sources define loci in the HR diagram. Comparisons to luminous stars in the Galaxy and the LMC (Figure 1) show that the GC stars are among the most extreme stars known. It is highly improbable that many of the Galaxy's most extreme stars would be in this sub-parsec region unless there is a mechanism in this region to prefer such stars. Our conclusions are strengthened if we make use of the stellar models of Najarro (1995; and this conference) which assign temperatures of 22,000 to 24,000 K to the most luminous three stars, requiring all three to have luminosities in excess of $10^6 L_\odot$. Note that Blum et al. (1995) conducted a similar analysis but came to different conclusions because they assumed the GC stars have strong IR excesses, which appears to be incompatible with their photometric colors.

The population can also be compared to the stars expected from a star formation episode (Tamblyn & Rieke 1993; Schaerer 1994; Krabbe et al. 1995; Tamblyn 1996), making use of the global constraints on the luminosity and ultraviolet radiation. Figure 2 shows the region of the HR diagram occupied by the brightest IRS 16 sources is rarely approached by evolutionary models and then only by rapidly evolving stars. The population of red supergiants (RSGs) is also readily seen to exist only in older bursts. No single burst age is expected to provide both types of luminous stars simultaneously, nor so many very luminous warm stars. Although the evolution of very massive stars is poorly understood, this conclusion seems robust to the remaining uncertainties and is consistent with the empirical comparison described above. These models also indicate uncomfortably tight constraints on the ages of the emission stars. It seems unlikely that we are fortunate enough to catch so many stars in a short evolutionary stage.

As presented by Rieke & Rieke (1994), we know enough about these stars to extrapolate to their appearance in visible wavelengths. In aggregate, we can compare with the populations in nearby galaxy nuclei where the individual stars are unresolvable. Assuming only that they are at least as hot as A stars, the cluster of emission stars produces at least 100 times as much blue light as any centrally confined populations in M31 or M32. This indicates that this stellar population is not common to galactic nuclei; it may result from an unknown feature of the GC which is different from the nuclei of M31 and M32 or it may be a transitory feature. Recent work by Massey (1996) indicates that the nucleus of M33 has characteristics reminiscent of LBV stars and may be similar to the GC.

3. Possible Origins

We have considered explanations for the central cluster beyond a normally evolving starburst to account for the large number of uncommon stars.

The extreme luminosities of the sources are easier to understand if they are compact sub-clusters rather than single stars. Indeed, IRS 13 and the IR source near Sgr A* have been shown to be such clusters (Eckart et al. 1993; Eckart

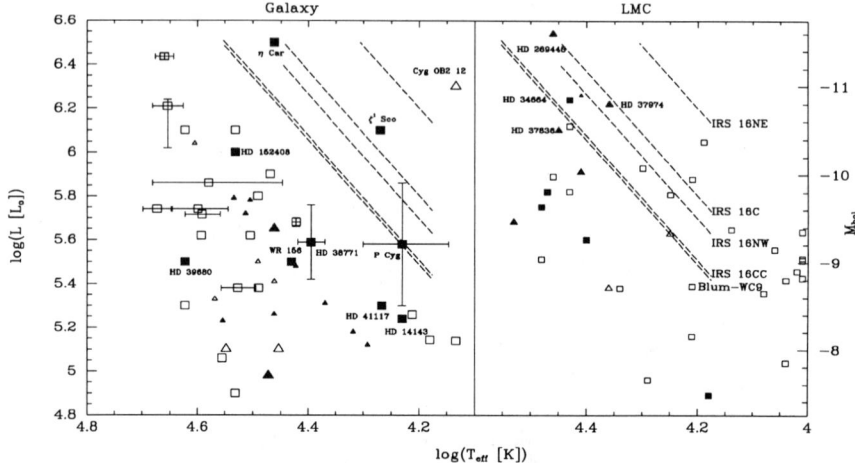

Figure 1. Luminous Galactic (left) and LMC (right) Emission-Line Stars Compared to the He I Emission Sources. The dashed lines are the loci near which several of the GC sources must lie. They have been labeled only in the right panel to reduce clutter. Filled symbols are comparison stars with the He I $2.058\,\mu$m emission line; open symbols are stars without the line or with only absorption in the line. Squares are used when both $T_{\rm eff}$ and L are available from the literature; small triangles are used when both had to be estimated from spectral type; large triangles are used when one had to be estimated. Error bars are indicated where available or where the literature has discordant values. The LMC stellar parameters are from McGregor et al. (1988). The IRS 16 luminosities assume blackbodies at a distance of 8 kpc with $A_K = 3.47$. Both vertical axes apply to both panels.

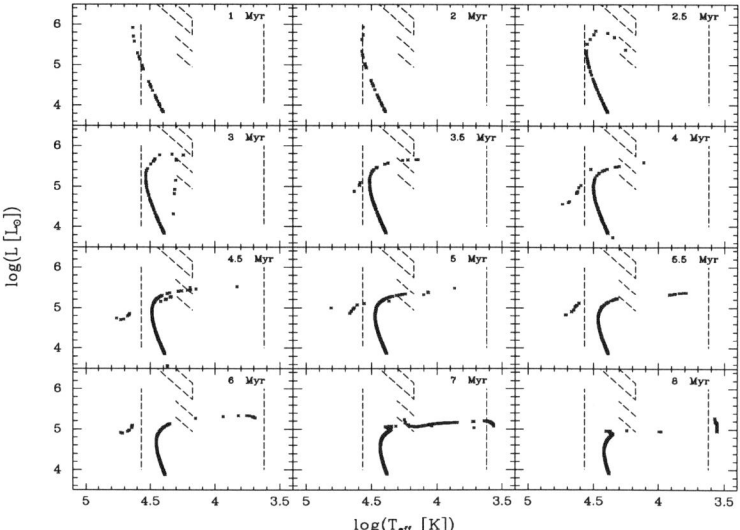

Figure 2. HR Diagrams of Synthetic Bursts. The source tracks for these randomly populated bursts are from Meynet et al. (1994) for Z=0.04 and enhanced mass loss. The dotted lines at $T_{eff} = 4,170$ and 37,000 K represent the regions of RSGs and the UV constraint. The rhombus is the location of the brighter He I sources with the cool edge at 15,000 K and top and bottom at $m_K = 8.8$ and 9.8 (d= 8 kpc, $A_K = 3.47$). Line segments at $m_K = 10.8$ and 11.8 (background) are also indicated for reference. Note that no stars are ever expected to match IRS 16NE well and that stars approach the region required to match IRS 16C,CC,NW, etc. only in bursts with ages $\sim$ 3 Myr but that RSGs do not develop for quite some time later. Also note the large number of O stars near the main sequence expected for young bursts and that they are fainter at K than the $1''$ seeing background.

this conference). However, the lunar occultation measurements (Simon et al. 1990; Simons et al. 1990) show that the dominant components of several of the brighter IRS 16 components are smaller than 200 AU. The detection of the 2.058 μm He I emission line from these sources is further evidence that they are dominated by a single star because stars with strong emission in this line are very luminous but shortlived, and there is no population of similarly bright hot stars without He I emission in the region.

Models in which the population of emission stars is maintained at present numbers have the appeal of not implying any fortuitous coincidence in observing these stars in a transitory state. One such model for the luminous sources presented by Morris (1993) is that the luminous sources are non-stellar accretion objects. However, this model requires instead some otherwise unknown feature to differentiate the nucleus of our galaxy from those of M31 and M32, as it is not obvious why such a steady-state population would not also be seen in those nuclei. A second steady-state model is that the luminous sources are the products of stellar mergers. However, several of the sources are now known to have luminosities of order $10^6 L_\odot$, and hence must have masses well over $30 M_\odot$. The stellar merger process in this environment is far too slow to build up appreciable numbers of such massive products (Lee 1993).

3.1. Persistent Star Formation

Some of the difficulties with single-age burst models are avoided by considering a star formation event which is extended slightly in time (cf. Krabbe et al. 1995). Specifically, one can imagine that the current RSGs are slightly older than the most luminous blue stars. However, some constraints remain. The recent level of star formation must be substantially reduced from that of a few Myr ago or the youngest stars would provide a hotter UV radiation field than that which dominates the region. Further, the comparison samples show that the emission stars are a minority of late-OI stars. Hence, they would be expected to be accompanied by bright blue sources without the He I emission feature, but these are not present in the central parsec. Hence, although introducing the free parameter of a time-variable star formation rate can significantly improve the fit of burst models with the gross properties of the observed cluster, the observed bias for He I emission would not be expected from any star formation history.

3.2. Unusual Evolution

Various features of the central parsec are unique in the Galaxy and might influence stellar evolution. Processes which can hasten the mass loss experienced by the most massive stars would be expected to expose the enriched inner regions earlier in the stellar life cycle, greatly increasing the fraction of a massive star's life in which it could exhibit helium emission features.

The central cluster has often been suspected of having super-solar metalicity. Increased atmospheric opacities would be expected to enhance the stellar mass-loss winds and super-metal-rich stellar evolution might be substantially different than that observed elsewhere in the Galaxy. However, the gas from which these stars formed has modest metalicity (Shields & Ferland 1994) and the best studied star, IRS 7, also has a metalicity near solar (Carr, this conference).

Interaction with a massive central object (Sgr A*) could strip material from an extended RSG, but the tidal radius for such an interaction is only 0.002 pc, and it is implausible that such a high fraction of the massive stars currently seen in the central 1/2 pc were this close to the central object during their late evolutionary phases.

Massive stars can influence one another if sufficiently close through direct mass exchange or stripping by winds or supernova ejecta (cf. Podsiadlowski et al. 1992, Livine et al. 1992). However, the separations at which these processes have been shown to be important are more typical of tight binaries than of the distances seen between the massive stars in the GC. Energy considerations alone are sufficient to show that the known massive stars have minimal impact on one another through stripping by stellar winds or supernovae.

The extreme space density of (unseen) low-mass stars or remnants implied by dynamical mass estimates differentiates the central parsec from even the neighboring star formation regions. Current limits permit a central density well over $10^7 \, M_\odot \, \text{pc}^{-3}$. In such an environment, an extended RSG will experience frequent collisions, some fraction of which can result in core ejection or tidal capture. Core ejection is a mechanism proposed by Tuchman (1985) in which the condensed core of an extended star is yanked from the dynamically slower envelope by the gravitational tug of a passing star or remnant if the imparted kinetic energy exceeds the core–envelope binding energy,

$$\Delta E_c = \frac{2G^2 M_c m^2}{v_r^2 b^2} \gtrsim 1.7 \times 10^{48} \, \text{erg}. \quad (1)$$

The exposed core of a RSG would be expected to have many observational features in common with a Wolf-Rayet star. However, plunging collisions with $b \lesssim 10 \, R_\odot$ are required, and we find that such collisions are far too uncommon to explain the observed population (Tamblyn 1996). Tidal capture is a process by which a passing star can be trapped in a very tight orbit (Fabian et al. 1975). Subsequent evolution of the primary can then drive the secondary into contact with the primary's envelope. Simulations (e.g., Taam & Bodenheimer 1991) show that substantial envelope material can be ejected from such a system by conversion of binary orbital energy. With Max Ruffert, we have conducted simulations of this capture mechanism appropriate for the conditions in the GC and find that this process is inefficient in this kinetically hot system. The timescale on which a $15 \, M_\odot$ RSG with radius $400 \, R_\odot$ is expected to capture companions is

$$\tau = 420 \left(\frac{n_*}{10^7 \, \text{pc}^{-3}}\right)^{-1} \left(\frac{m}{1 \, M_\odot}\right)^{-1.1} \left(\frac{\sigma_v}{140 \, \text{km/s}}\right)^3 \, \text{Myr}, \quad (2)$$

far too slow to have a substantial impact on this cluster (Tamblyn 1996).

3.3. Unusual Formation

Conditions in the central regions might also lead to unusual star formation as discussed by Morris (1993) and by Zinnecker (this conference). It may be that the inital mass function is severely skewed relative to what is seen in other regions, favoring the most massive stars. Alternatively, massive stars might preferentially form in tight binaries in which some of the processes described above can operate fast enough to influence the appearance of evolved massive stars. Further research in this direction is likely to benefit from consideration of other regions where star formation has occured in unusually dense regions. The dense cores of 30 Dor and NGC 3603 have recently been probed with *HST* and they have some characteristics in common with the central parsec of the Milky Way. The latter has a Wolf-Rayet fraction which increases in the densest regions and both have late-WN stars in tight binaries. Further, NGC 3603 also has 6 O3V-III stars (Drissen et al. 1996). These are generally thought to be precursors to the Wolf-Rayet stars seen in the coeval cluster. This suggests that age is not the distinguishing factor and that a second influence has driven some stars to appear as Wolf-Rayet stars at an early age.

4. CO-Depletion Region

Although the tidal capture mechanism has been shown to be too inefficient in this region to influence substantial numbers of massive stars, less massive stars have substantially longer lifespans. For a low-mass main-sequence star, the estimated timescale for capture of a companion into a tight orbit is 20 Gyr if the density of stars and remnants is $10^7 \, \text{pc}^{-3}$. The subsequent evolution of such a system is likely to inhibit the red giant phase of the primary. *HST* observations of globular clusters will allow quantitative tests of this hypothesis. In the GC, the expected consequence would be a region in which many of the low-mass stars have their envelopes ejected as they evolve to larger dimensions, especially in the densest central regions. This could explain the observed CO-depletion region (Sellgren et al. 1987) and provides additional detail on how stellar collisions may be responsible for this feature.

5. Conclusions

We have seen that the emission-line stars in the central parsec are problematic as a normal starburst population even though they have individual analogues among massive evolved stars elsewhere in the Galaxy. Comparison with the nuclei of M31 and M32 illustrate that the GC is either unique in its ability to sustain a population of luminous blue sources, or is in a time-dependent phase. Expected influences of the conditions in the central region are not sufficient to explain the unusual appearance of so many massive stars, but unusual star formation, such as a very top-heavy initial mass function or large fraction of tight binaries, may explain the development of such a cluster from a recent star formation episode.

References

Allen, D. A., Hyland, A. R., & Hillier, D. J. 1990, MNRAS 244, 706
Blum, R. D., DePoy, D. L., & Sellgren, K. 1995, ApJ 441, 603
Cotera, A. S., Erickson, E. F., Colgan, S. W. J., Simpson, J. P., Allen, D. A., & Burton, M. G. 1996, ApJ 461, 750

Drissen, L., Moffat, A. F. J., Walborn, N. R., & Shara, M. M. 1995, AJ 110, 2235

Eckart, A., Genzel, R., Hofmann, R., Sams, B. J., & Tacconi-Garman, L. E. 1993, ApJ 407, L77

Fabian, A. C., Pringle, J. E., & Rees, M. J. 1975, MNRAS 172, 15P

Figer, D. F., McLean, I. S., & Morris, M. 1995, ApJ 447, L29

Hall, D. N. B., Kleinmann, S. G., & Scoville, N. Z. 1982, ApJ 260, L53

Krabbe, A., Genzel, R., Drapatz, S., & Rotaciuc, V. 1991, ApJ 382, L19

Krabbe, A. et al. 1995, ApJ 447, L95

Lee, H. M. 1993, in The Nuclei of Normal Galaxies: Lessons from the Galactic Center, ed. R. Genzel & A. I. Harris (Dordrecht: Kluwer), 335

Libonate, S., Pipher, J. L., Forrest, W. J., & Ashby, M. L. N. 1995, ApJ 439, 202

Livine, E., Tuchman, Y., & Wheeler, J. C. 1992, ApJ 399, 665

Massey, P. 1996, ApJ submitted

McGregor, P. J., Hillier, D. J., & Hyland, A. R. 1988, ApJ 334, 639

Meynet, G., Maeder, A., Schaller, G. Schaerer, D., & Charbonnel, C. 1994 A&AS 103, 97

Morris, M. 1993, ApJ 408, 496

Najarro, F. 1995, Ph.D. thesis, Institut für Universität München

Najarro, F., Hillier, D. J., Kudritski, R. P., Krabbe, A., Genzel, R., Lutz, D., & Drapatz, S. 1994, A&A 285, 573

Podsiadlowski, Ph., Joss, P. C., & Hsu, J. J. L. 1992, ApJ 391, 246

Rieke, G. H. & Rieke, M. J. 1994, in The Nuclei of Normal Galaxies: Lessons from the Galactic Center, ed. R. Genzel and A. I. Harris (Dordrecht: Kluwer), 283

Schaerer, D. 1994, in Unsolved Problems in the Milky Way, ed. L. Blitz (Dordrecht: Kluwer), in press

Sellgren, K., Hall, D. N. B., Kleinmann, S. G., & Scoville, N. Z. 1987, ApJ 317, 881

Shields, J. C. & Ferland, G. J. 1994, ApJ 430, 236

Simon, M., Chen, W. P., Forrest, W. J., Garnett, J. D., Longmore, A. J., Grauer, T., & Dixon, R. I. 1990, ApJ 360, 95

Simons, D. A., Hodapp, K.-W., & Becklin, E. E. 1990, ApJ 360, 106

Taam, R. E. & Bodenheimer, P. 1991, ApJ 373, 246

Tamblyn, P. 1996, Ph.D. thesis, Univerity of Arizona

Tamblyn, P. & Rieke, G. H. 1993, ApJ 414, 573

Tamblyn, P., Rieke, G. H., Hanson, M. M., Close, L. M., McCarthy, D. W., & Rieke, M. J.

Tuchman, Y. 1985, ApJ 288, 248

2.2 μm Keck Images of the Galaxy's Central Stellar Cluster at 0."05 Resolution

B. L. Klein, A. M. Ghez, M. Morris, E. E. Becklin

UCLA Department of Physics and Astronomy,
8371 M.S. Bldg., Los Angeles, CA 90095

Abstract. We present preliminary results from a high angular resolution study of the inner $5''\times 5''$ of the central Galactic stellar cluster. This program was carried out at the W. M. Keck telescope using the facility near-infrared camera with a $K[2.2\mu m]$-band filter. A few thousand short exposure frames, combined using a shift-and-add algorithm, contribute to the final image which has the diffraction-limited resolution of 0."05. Even prior to deconvolution, sources as faint as $\sim 15^{th}$ Kmag are detected at the 5σ level and new isolated point sources are identified.

1. Introduction

Since the discovery of the near-infrared (NIR) emission from the Galactic Center (Becklin & Neugebauer 1968), the central stellar cluster has been the focus of many NIR studies. Recently, tremendous advances have been made in the angular resolution of NIR images of this region via methods which compensate for atmospheric distortions (Eckart et al. 1995, 1993; also Close et al. 1995, DePoy & Sharp 1991, Simons & Becklin 1996). Nonetheless, studies of the Sagittarius A* (IR) complex - the apparent concentration of infrared sources within $\sim 0.''5$ of SgrA* - remain confusion limited. The W.M. Keck telescope, which has a position-angle-dependent maximum baseline of 9.0-11.0 meters (Nelson 1989), offers a unique opportunity to probe the detailed structure of our Galaxy's central stellar cluster.

At Mauna Kea, the mean atmospheric seeing conditions constrain traditional images to angular resolutions of $\theta_{seeing} \sim 0.''5$ at 2.2 μm. Although this image size is excellent by conventional standards, it is still a factor of 10 worse than the diffraction limit of Keck, which at $\lambda = 2.2$ μm is $\theta_{diff} \sim \frac{\lambda}{D} \sim 0.''05$ (= 420 AU at 8.5 kpc). We exploit this high resolving power by employing high-resolution imaging techniques to recover the diffraction limit in post processing. The viability of diffraction limited imaging with the fully-phased Keck telescope has been demonstrated by Ghez (1996) and Matthews et al. (1996).

2. Observations

Our K-band observations incorporated three key elements for obtaining data suitable for the reconstruction of diffraction limited images.

- *Short Exposures:* 120 ms integrations were used in these observations. Under good seeing conditions at Keck this has proven to be sufficiently short to freeze the atmospheric distortions and preserve high-resolution information (see Figure 1).

- *Fine pixel scale:* A scale of 20 milliarcseconds per pixel is available with the facility camera, NIRC (256×256 InSb array). This is accomplished with the recently installed "image converter" (Matthews et al. 1996) that magnifies the field of view from the standard $38'' \times 38''$ to $5'' \times 5''$ which is Nyquist sampled at 2.2 μm.

- *Telemetry:* We collected $\sim$10,000 frames centered near the position of SgrA*. From this data, 2,741 of the best seeing quality frames were selected to be incorporated in the final image presented here (Figure 3).

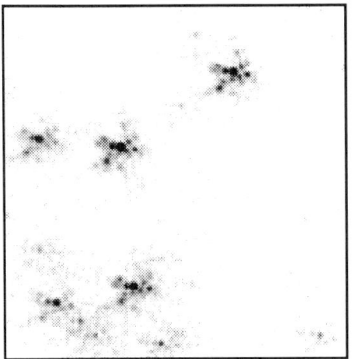

Figure 1: A single 120 ms short exposure (a specklegram) of the inner 3.5 arcseconds of the central stellar cluster. High resolution information is clearly contained in this image. It is worth noting that the speckle patterns for different sources are nearly identical across the field.

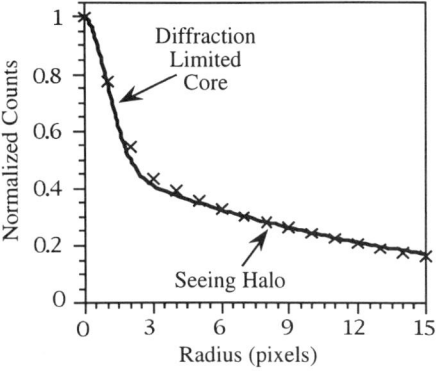

Figure 2: A model fit through measured data of the shift-and-add point spread function (PSF) radial profile. The diffraction limited core is built up from the brightest speckle in each contributed frame while the 'seeing halo' results from the fainter surrounding speckles. The core contains $\sim$10 percent of the flux and has the expected FWHM of $\theta = 0.''05$.

3. Shift-And-Add

One method for obtaining diffraction limited images from a series of short exposures (specklegrams) is "shift-and-add" (e.g., Christou 1991). In a short exposure such as the one shown in Figure 1, the image breaks up into a number of speckles, each of which can be thought of as a noisy diffraction limited image of the source. Shift-and-add builds up signal while preserving high resolution information by registering all the frames on the brightest speckle for each exposure. In this scheme the surrounding fainter speckles are treated as noise. The resulting image of a point source has a diffraction limited core on top of a "seeing halo" (see Figure 2).

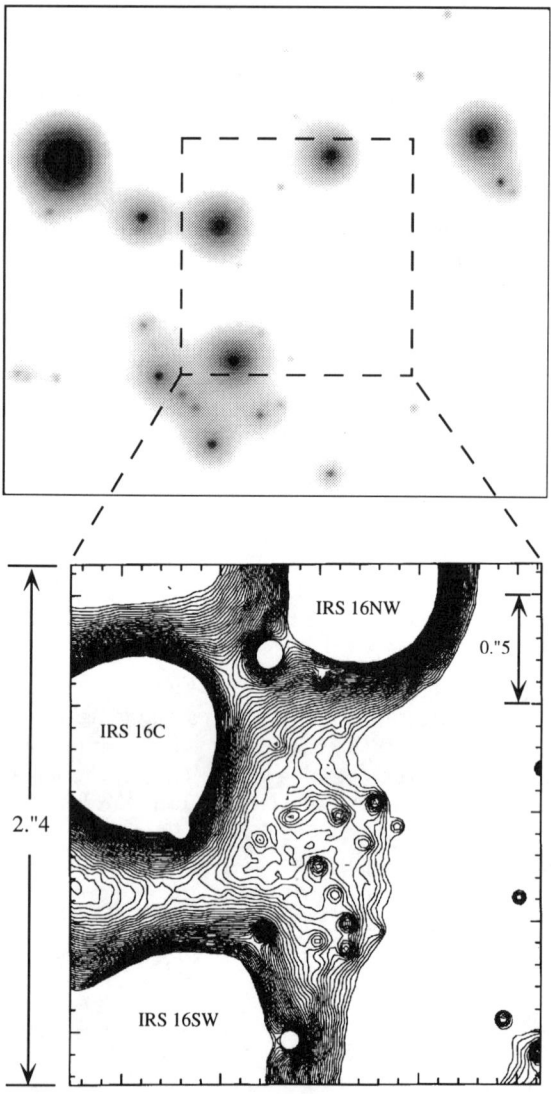

Figure 3: A shift-and-add image of the central $5'' \times 5''$ comprised of 2,741 seeing selected frames taken in June, 1995. The contours do not correspond directly to flux density levels since these images are not deconvolved. Nonetheless, comparisons with previously published maps at lower resolution indicate that sources as faint as 15th K mag are easily identified. At the resolution of $0.''05$ (=420 AU at 8.5 Kpc), new isolated point sources are detected.

4. Results

The results of shift-and-add are shown in Figure 3. The (linear) intensity display on top is scaled to show the PSF cores of the brighter members of IRS 16. The contour plot of the central 2.″4×2.″4 region is scaled to display the structure of the SgrA*(IR) complex. Sources as faint as $K \sim 15$ mag are detected at the 5σ level in these "raw" (i.e. not deconvolved) images. The images shown in Figure 3 are in general agreement with the earlier results of Eckart et al. (1995 and this volume: resolution 0.″15). Much of the extended emission seen in the previous maps is now resolved into point sources.

Acknowledgments. The authors are grateful to the excellent Keck staff whose hard work and diligence helped to ensure the success of these observations. In particular, many thanks to W. Harrison, A. Conrad, W. Wack, J. Aycock, T. Stickel and R. Campbel. We also thank our Caltech speckle-collaborators K. Matthews, G. Neugebauer, A. Weinberger for their support of this project. This project is funded through the NSF Young Investigator Award (AMG). BLK is also supported by a UCLA RA/Mentorship fellowship.

References

Becklin, E. E. & Neugebauer, G. 1968, ApJ, 151, 145
Christou, J. C. 1991, PASP, 103, 1040
Close, L. M., McCarthy, D. W., & Melia, F. 1995 ApJ, 439, 682
DePoy, D. L. & Sharp, N. A. 1991, AJ, 101, 1324
Eckart, A. et al. 1996, this proceedings
Eckart, A., Genzel, R., Hofmann, R., Sams, B. J. & Tacconi-Garman, L. E. 1995, ApJ, 445, L23
Eckart, A., Genzel, R., Hofmann, R., Sams, B. J. & Tacconi-Garman, L. E. 1993, ApJ, 407, L77
Ghez, A. M. 1996 in NATO Advanced Summer Institute, "Evolutionary Processes in Binary Stars", in press
Matthews, K., Ghez, A., Weinberger, A., & Neugebauer, G. 1996, PASP, in press
Nelson, J. 1989, AmSci, 77, 170
Simons, D. & Becklin, E. E. 1996, AJ, in press

Fourier Imaging Spectroscopy of the Galactic Center

D. A. Simons

Gemini 8 m Telescopes Project, 950 N. Cherry Ave., Tucson, AZ 85719

J. P. Maillard

Institut d' Astrophysique de Paris, CNRS, 98 bis Blvd. Arago, 75014, Paris, France

Abstract. We present results from a series of imaging spectroscopy observations of the Galactic center in the $\sim$2 μm region. The scans were made with the CFHT imaging FTS in June 1995 and, compared to the recent 3D results of Krabbe et al. (1995), have improved field of view, spatial, and spectral resolution. Here we only show results of our Brγ scan.

1. Observations

The CFHT imaging FTS (Maillard 1995) works by coupling the facility Fourier Transform Spectrometer (Maillard and Michel 1982) and "Redeye" infrared camera (Simons et al. 1993) to work as a single instrument. The spatial resolution of images is seeing limited with 0.33 arcsec/pixel sampling. The system is designed to reimage the CFHT f/35 focal plane onto an infrared array through the FTS optics by a special optical interface, creating a pair of complementary images that modulate in intensity as the interferometer is stepped through a scan. The infrared camera records an image at each interferometer step. From the recorded data, spectra at points in the field can be extracted through straightforward aperture photometry of complementary regions in the field of view. It is also possible to invert an entire raw data cube, making a four-dimensional processed cube (x, y, σ, intensity) from which monochromatic images can be extracted and manipulated. Some of the unique advantages of this instrument over other imaging spectrometers in use at observatories include:

1. Easily tuned spectral resolution from 1.0-2.5 μm up to R$\sim$10^4.
2. Both emission and absorption line observations are practical.
3. Wavenumber calibration is intrinsic to the data.

Scans were recorded of the Galactic center during the nights of 12-15 June 1995 at CFHT. A number of filters were used to isolate various spectral features of interest in and around IRS 16, including Brγ, He I, He II, H$_2$, and CO. Scan resolutions were typically $\sim$5000 with the prominent source IRS 7 located in the

~23" field of the instrument to help register the raw data cubes during post-processing. In this report preliminary results are given for the Brγ scan while the other scans through various bandpasses are further reduced and interpreted.

2. Brγ Results

Shown in Figure 1 is a coadded image made from 10 separate broadband exposures in the Brγ scan to illustrate the targets sensed in the IRS 16 region. Surrounding that image are spectra for various sources in the field calibrated in velocity around the Brγ line. They have a resolution of ~45 km/s. A number of objects are narrow-line emission sources while IRS 13, 16SW, and 16NE show broad-line emission, suggestive of stellar winds surrounding these sources. In order to boost spectral signal-to-noise all spectra were derived through ~2 arcsec apertures centered on sources, which is substantially broader than the ~0.8" resolution of the images. This helped reduce effects of time varying seeing and astigmatism from the telescope. In Figure 2 are narrow band images around the Brγ 0 km/s line which have had a mean continuum image subtracted off in order to show only Brγ emission. Residual flux around bright stars is due to seeing variations during the time the scan was made. A spot denoting the fixed position of IRS 7 is shown in all of the images. A streamer of gas with a continuously varying velocity is clearly seen. This emission tracks from a point near IRS 10, south and around IRS 16 in a comma shape. The basic results shown here agree well with past observations made by Herbst *et al.* (1993) who demonstrated good correlation between the Brγ and Ne II gas steamers and concluded this gas is orbiting around a point within ~1" of the radio position of Sgr A*. Additional scans were recorded in various bandpasses that are currently under analysis and not shown here.

References

Maillard, J.P. 1995, in Tridimensional Optical Spectroscopic Methods in Astrophysics, IAU Coll. 149, ed. G. Comte & M. Marcelin, ASP Conf. Series, 71, 316

Maillard, J. P. & Michel, G. 1982, in Instrumentation for Astronomy with Large Optical Telescopes, IAU Coll. No. 67, ed. C. M. Humphries, 92, Reidel, 213

Simons, D. A., Clark, C. C., Kerr, J., Massey, S., Smith, S. & Toomey, D. 1993, SPIE Proc. 1946, ed. A. L. Fowler, 502

Herbst, T. M., Beckwith, S. V. W., Forrest, W. J., & Pipher, J. L, 1993, AJ, 105, 956

Krabbe, A., Genzel, R., Eckart, A., Najarro, F., Lutz, D., Cameron, M., Kroker, H., Tacconi-Garman, L. E., Thatte, N., Weitzel, L., Drapatz, S., Geballe, T., Sternberg, A., & Kudritzki, R. 1995, ApJ, 447, L95

234   D. SIMONS & J. P. MAILLARD

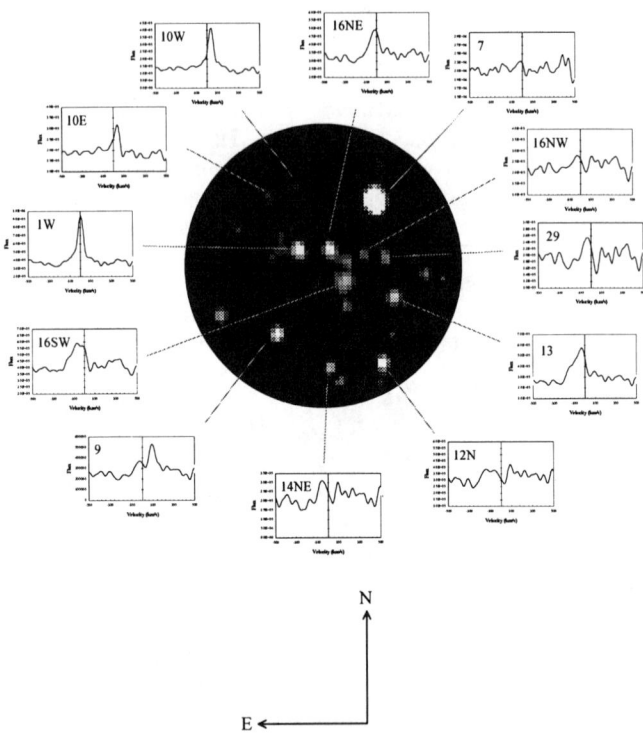

Figure 1. A montage of spectra for various sources in the IRS 16 region is shown.

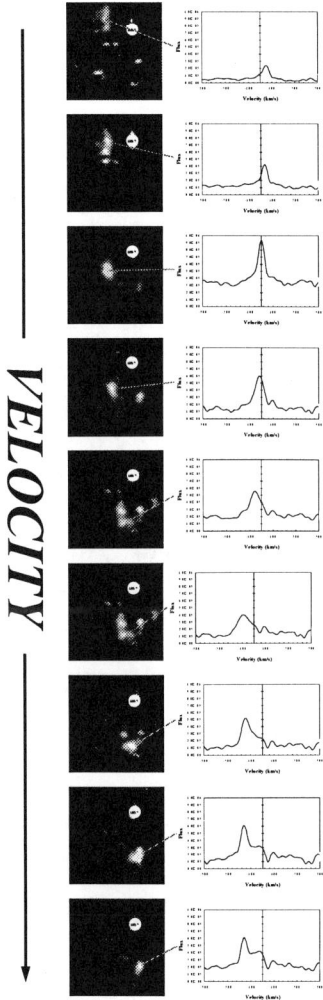

Figure 2. A series of images straddling the Brγ 0 km/s line is shown. A streamer of gas that loops around IRS 16 is detected.

Consequences of Extended Dark Mass in the Galactic Center Cluster

Peter J. McGregor, Geoffrey V. Bicknell, and Prasenjit Saha

Mount Stromlo and Siding Spring Observatories, Institute of Advanced Studies, The Australian National University, Private Bag, Weston Creek PO, ACT 2611, Australia

Abstract.
We review results from our mass model for the Galactic center region which indicate the existence of an extended distribution of dark mass in the central 0.8 pc of the Galaxy, coincidence with the cluster of young, massive He I emission-line stars. Specifically, the local M/L_K ratio decreases from ~ 0.3 at 2 pc radius to ~ 3 at 0.35 pc radius. This behavior cannot be due to the presence of a central massive black hole, but can result from an increasing concentration of dark stellar remnants towards the center. If so, the initial stellar mass function in the central region must be deficient in stars less massive than $\sim 1\ M_\odot$. Star formation is only likely to proceed within the core radius of the total mass distribution where the stellar density is approximately constant and tidal shear is correspondingly reduced. We suggest that starbursts, such as the one that occurred in the Galactic center $\sim 10^7$ yr ago, are a repetitive feature of the Galactic center as gas accumulates in the molecular ring, collapses into the central parsec, and forms stars in the region occupied by the cluster of He I emission-line stars and dark stellar remnants. Inward extrapolation of our mass model indicates that it should be possible to use proper motion measurements for stars within 0.02 pc (0.5″) of Sgr A* to discerning black hole masses down to $\sim 5 \times 10^5\ M_\odot$.

1. Introduction

Frequent reference has been made at this meeting and elsewhere to the existence of a 1–$3 \times 10^6\ M_\odot$ black hole in the Galactic center. This is an attractive proposition in terms of our expectations regarding the origin of nuclear activity in galaxies and the possible role of quasars in the formation of all galaxies. However, it is important to recognise that the basic evidence supporting the existence of a massive black hole in the center of our galaxy remains controversial. This evidence comes from dynamical estimates of the central mass profile, and from the existence of the Sgr A* radio source apparently coincident with the dynamical center of the Galaxy (Baker & Sramek 1987). Any central black hole is energetically insignificant, with the central stellar cluster being responsible for most of the ionizing flux, luminosity, and wind from the central region (Allen, Hyland, & Hillier 1990; Krabbe et al. 1991; Krabbe et al. 1995). The radio emission from Sgr A* can be explained in terms of low-level accretion onto a central black hole of mass $\sim 10^6\,M_\odot$ (Melia 1994; Narayan et al. 1995; papers in this volume). However, current low limits on the near- and mid-infrared flux from Sgr A* (Eckart et al. 1995; Stolovy, this volume) place tight constraints on these models. It is unclear whether accretion onto a significantly lower mass black hole can provide an alternative explanation for this unique object.

Dynamical estimates of the central mass profile are therefore the only means available for probing the presence of a central massive black hole. This has been done from studies of both gas (e.g., Serabyn et al. 1988; Goss, this volume) and stellar kinematics (McGinn et al. 1989; Sellgren et al. 1990; Krabbe et al. 1995; Haller et al. 1996). The latter studies have revealed the remarkable fact that the stellar velocity dispersion increases from ~ 50 km s^{-1} at 4 pc ($\sim 100''$) to ~ 120 km s^{-1} at 0.2 pc ($\sim 5''$) from the Galactic center. It is this feature of the stellar distribution which points most strongly to the presence of a massive compact object in the Galactic center.

The uncertainties due to non-gravitational forces in using gas kinematics to infer the central mass profile are well documented (e.g., Lacy 1989). Additionally, dynamical analyses of gas complexes to infer the enclosed mass profile require an inference about where along the line-of-sight the gas is located. In contrast, stellar dynamical studies are able to make the reasonable assumption of spherical symmetry, but suffer from uncertainties due to possible anisotropy in the stellar velocity dispersion. Any anisotropy is expected to be small since the stellar relaxation timescale at a radius of ~ 1 pc is of order 10^8 yr, which is much shorter than the ages of the $\sim 1\,M_\odot$ red giant stars that probably contribute the bulk of the K band light. However, this relaxation time may be comparable to the stellar age if the K band light is actually dominated by $5\,M_\odot$ red giants (see Rieke, this volume). Preliminary stellar proper motion measurements (Eckart, this volume) appear to confirm that the stellar velocity dispersion in the central region is not strongly anisotropic. While further proper motion measurements are awaited with interest, it seems safe to conclude that the only determinations of the central mass profile with sufficient precision to distinguish between a central massive black hole and a compact central star cluster will come from analyses of the stellar dynamics in the Galactic center.

With this in mind, Saha, Bicknell, & McGregor (1996) reanalysed published stellar kinematic data to determine how rigorously a central massive black hole is required by the available data. This was needed for three reasons. Firstly, Allen (1994) has shown that the cool stellar population displaying 2.3 μm CO absorption has a different spatial distribution from the total light. Secondly, the analyses by McGinn et al. (1989) and Sellgren et al. (1990) neglect the effect of projection on the velocity dispersion. This has a moderate effect on the inferred mass distribution. Thirdly, accurate mass and deprojected light distributions are necessary for the purpose of comparing enclosed mass and enclosed light, and so deriving the appropriate mass-to-light ratio. The correct interpretation of this mass-to-light ratio has an important bearing on whether one is justified in inferring the presence of a black hole.

In this paper, we review the results of the Saha et al. analysis and comment on the implications of these results for the mode of star formation occurring in the central parsec of the Galaxy, the star formation history in this region, and the potential for resolving the fundamental question of whether a massive black hole exists in the center of the Galaxy using the proper motion data which will soon be available for the stars within the central 1″ region. We adopt a distance to the Galactic center of 8.5 kpc.

2. A Galactic Center Mass Model

Saha et al. reanalysed the stellar hydrodynamics of the Galactic center cluster using observations of the radial distributions of the K band surface brightness Σ, projected rotational velocity v_p, and projected velocity dispersion σ_p. We adopted the standard assumptions that the stellar population is spherical, has isotropic velocity dispersion, and is in dynamical equilibrium. The plausibility of these assumption has been discussed by McGinn et al. (1989). The problem is under-constrained, so that different conclusions may be reached by adopting other sets of assumptions.

Under these assumptions, the gravitational potential is given by the spherically symmetric Jeans equation

$$\frac{1}{\rho}\frac{d}{dr}\left(\rho\sigma_r^2\right) - \frac{v_{\rm rot}^2}{r} = -\frac{GM(r)}{r^2} \tag{1}$$

from which $M(r)$, the total mass enclosed within radius r, can be inferred. Here ρ, σ_r and $v_{\rm rot}$ are the radial volume density, radial velocity dispersion, and rotational velocity functions, respectively, of a set of test particles within the potential.

The three dimensional variables were determined from the corresponding projected quantities by inverting the Abel integrals:

$$\begin{aligned}
\Sigma(r_p) &= 2\int_{r_p}^{\infty}\frac{\rho(r)}{\sqrt{r^2-r_p^2}}r\,dr, \\
\Sigma(r_p)\sigma_p^2(r_p) &= 2\int_{r_p}^{\infty}\frac{\rho(r)\sigma_r^2(r)}{\sqrt{r^2-r_p^2}}r\,dr, \\
\Sigma(r_p)v_p(r_p) &= 2\int_{r_p}^{\infty}\frac{\rho(r)v_{\rm rot}(r)}{\sqrt{r^2-r_p^2}}r\,dr,
\end{aligned} \tag{2}$$

r_p being the projected radius from the dynamical center. The deprojections were carried out by first transforming equations (2) to the variables $\ln r_p$ and $\ln r$ and then solving using a Fourier transform deconvolution method (Kalnajs, private communication).

The deprojection is unreliable when applied to noisy data so the surface brightness, projected rotational velocity, and projected velocity dispersion were modeled using smooth functions. The surface brightness distribution was modeled using a Reynolds-Hubble law such that

$$\Sigma(r_p) = \Sigma_0\left(1+\frac{r_p^2}{r_{p,0}^2}\right)^{-\alpha}. \tag{3}$$

The projected rotation velocity and velocity dispersion, v_p and σ_p, were fit with quartics in the variable $u = \ln(r_{p,0}+r_p)$.

Our understanding of the Galactic center region has improved recently with the realisation that the central $\sim$ 1 pc ($\sim 24''$) is populated by two distinct stellar populations (Allen, Hyland, & Hillier 1990; Krabbe et al. 1991). Burton & Allen (1992) used the presence or absence of the 2.3 μm CO absorption band in Galactic center cluster stars to distinguish between "cool" stars and "hot" stars. Allen (1994) found that the "hot" population is confined to within a radius of $\sim$ 1 pc ($\sim 24''$) of the Galactic center, with a core radius of ≤ 0.2 pc ($\sim 5''$), while the "cool" population is more extended with a core radius possibly as large as $\sim$ 0.6 pc ($\sim 15''$). Saha et al. used the "cool" star population as tracer particles for probing the gravitational potential, and used Allen's K band surface brightness distribution for this population in their analysis. The true core radius of this population is controversial, but does not affect their results outside a radius of ~ 0.2 pc ($\sim 5''$). Projected rotational velocity and velocity dispersion data for the "cool" star population were taken from McGinn et al. (1989) and Sellgren et al. (1990), and the Group I OH/IR star kinematic data from Lindqvist, Habing, & Winnberg (1992) were used to constrain the kinematic fits at large projected radii. The CO absorption band weakens within $8.5 \pm 1.5''$ of the Galactic center (Sellgren et al. 1990; Haller et al. 1996) so mass estimates based on the "cool" star population interior to ~ 0.35 pc must be treated with some caution due to the uncertain density distribution of the tracer stars.

Saha et al.'s numerical solution of the Jeans equation (1) leads to an enclosed mass of $\sim 3 \times 10^6 \, M_\odot$ within 0.35 pc, where the CO absorption band begins to weaken, and $\sim 10^7 \, M_\odot$ within 3 pc in reasonable agreement with the enclosed mass estimated from HCN measurements of the molecular ring (Güsten et al. 1987). Between the inner most velocity dispersion data point at ~ 0.2 pc ($\sim 5''$) and ~ 0.6 pc ($\sim 15''$), the enclosed mass profile increases approximately linearly with radius, and is consistent with an inward extrapolation to zero mass at the origin, i.e., no black hole. While this cannot be used to constrain the existence or otherwise of a central massive black hole, it at least demonstrates that a consistent model can be found for the available velocity dispersion data which does not contain a central massive black hole. The analysis led to a upper limit on any central black hole mass of $< 1.5 \times 10^6 \, M_\odot$.

Furthermore, Saha et al. showed that the plot of total enclosed mass versus total enclosed luminosity derived from their mass profile solution is not a straight line with non-zero intercept as expected for a central black hole immersed in a constant M/L stellar distribution, but instead is a curve that steepens towards decreasing radius and projects to the vicinity of the origin. Within the limitations of the assumptions of their dynamical model, they conclude that the local K band mass-to-light ratio, M/L_K, increases within a radius of ~ 0.8 pc from $\sim$ 0.3 at 2 pc radius to ~ 3 at 0.35 pc. The presence of a central massive black hole does not affect the local mass-to-light ratio away from the center. Consequently, the nature of the Galactic center stellar population must change within a radius of ~ 0.8 pc. This corresponds closely to the region of the compact cluster of He I emission-line stars (Allen, Hyland, & Hillier 1990; Krabbe et al. 1991), and suggests that the change in stellar population is associated with the existence of these massive young stars. A dark mass of $\sim 6 \times 10^6 \, M_\odot$ within 1 pc is implied.

A plausible explanation for this extended dark mass distribution, and its association with the cluster of young massive He I emission-line stars, is that it is comprised of stellar remnants that are the natural result of star formation in the central ~ 0.8 pc radius region of the Galaxy (c.f. Morris 1993). Saha et al. showed that if star formation in this central region has proceeded at an approximately constant average rate over the lifetime of the Galaxy, the observed dark-to-bright mass ratio would be obtained only if stars less massive than $\sim 1\ M_\odot$ were inhibited from forming in this central region. This is because the combined light of a large number of long-lived, low mass stars would exceed the limits required by the dark mass component. Shorter star formation durations in the central region, such as a single burst $\sim 10^7$ yr ago, require truncation of the initial mass function at even higher masses. Various data suggest that clouds in the molecular ring are hotter, denser, more massive, and have larger internal velocities than molecular clouds in the solar neighborhood (Blitz et al. 1993). While it is not known exactly what characteristics of molecular clouds define the mass spectrum of stars formed from them, it is commonly believed that such clouds generally produce high mass stars, with low mass star formation possibly being inhibited (Myers & Fuller 1993). There is, therefore, circumstantial evidence that supports the type of massive star formation in the central region of the Galaxy that would naturally account for the inferred extended distribution of dark stellar remnants within 0.8 pc of the center.

3. Star Formation History of the Central Region

It is well known that a massive molecular ring surrounds the central 4 pc diameter region of the Galaxy (e.g., Güsten et al. 1987), and that there is very little gas in the interior cavity (Lo & Claussen 1983; Jackson et al. 1993). The current low gas density in the central cavity is clearly insufficient to support star formation in this central region. However, the presence of many massive young He I emission-line stars, and the M supergiant IRS 7, in a compact cluster in the central ~ 1 pc region indicates that massive star formation did occur there $\sim 10^7$ yr ago, despite the large tidal shear which must be present near the Galactic center. The resolution of this apparent conflict lies in the realisation that massive stars are not present throughout the central 4 pc region interior to the present molecular ring, but instead occur only within the central ~ 1 pc region corresponding approximately to the core radius of the total mass distribution inferred by Saha et al. In this region, the inferred total enclosed mass distribution is dominated by dark mass and increases approximately linearly with radius, corresponding to a volume density which is approximately constant with radius. Sanders (1992) has noted that star formation is only likely to proceed inside the core radius of the total mass distribution where the stellar density is relatively constant and the tidal shear is correspondingly reduced. The existence of many young He I emission-line stars in this central region is clear evidence that this is the region where star formation has been able to occur. The association of the compact He I star cluster with the core region of the total mass distribution has the further implication that the He I stars are unlikely to have formed in the present molecular ring and migrated to the center. If they had, we might expect to see some He I emission-line stars at radii larger than ~ 1 pc.

The conclusion we reach from these considerations is that a burst of massive star formation occurred only inside ~ 0.8 pc from the Galactic center $\sim 10^7$ yr ago. The gas density in this central region must have been high at that time, and the most likely source of this gas is obviously infall from or the collapse of a structure similar to the present molecular ring. After the starburst was established, the winds from the newly born massive stars would have blown away much of the gas in the central region and a situation similar to that seen at the present time would have been re-established. In this scenario, the Galactic center region is not expected to be static; it is likely that the central core region has undergone repeated bursts of star formation as gas forms in a molecular ring which eventually collapses into the central core region. We speculate that this collapse may be rapid in comparison with the time needed to reform the molecular ring, and so we need not be viewing the Galactic center at a particularly special epoch. Cloud collisions and shock compression will be features of this collapse. Both are likely to favor the formation of massive stars in the central core region, and ultimately lead to a dark-to-bright mass ratio similar to that inferred by Saha et al.

4. Constraints On The Existence Of A Massive Black Hole

The current limit on any central black hole mass of $\sim 1.5 \times 10^6 \, M_\odot$ is constrained by the innermost stellar velocity dispersion data at ~ 0.2 pc. This can be simply seen by considering a dominant black hole of mass, $M_{\rm BH}$. Inside the core radius of the tracer population, the volume density is constant and the Jeans equation (1) reduces to

$$\sigma_r^2 = \frac{GM_{\rm BH}}{r}. \qquad (4)$$

In the Saha et al. model, the measured projected velocity dispersion of ~ 120 km s^{-1} at 0.2 pc deprojects to $\sigma_r \sim 180$ km s^{-1} which, according to equation (4), corresponds to a black hole mass of $1.5 \times 10^6 \, M_\odot$. Since it is not possible to constrain the distribution of mass interior to the innermost velocity dispersion data point in any dynamical model, it is apparent that the only way a more stringent limit can be placed on the existence or otherwise of a massive black hole in the Galactic center is to obtain velocity dispersion data at significantly smaller radii than the current ~ 0.2 pc ($\sim 5''$).

The high spatial resolution image of Eckart et al. (1995) clearly reveals the presence of an inner group of ~ 10 stars lying within a projected radius of ~ 0.02 pc ($\sim 0.5''$) of Sgr A*. These stars very likely lie within this radial distance of Sgr A*, and so considerable effort has already been expended in attempting to measure their proper motions for the purpose of constraining the mass distribution. We now consider to what extent one-dimensional velocity dispersion measurements based on these proper motion data may be able to distinguish between a massive central black hole and the extended dark mass distribution inferred from the Saha et al. mass model.

The Saha et al. mass model has been integrated inwards assuming different enclosed mass profiles in order to predict both the one-dimensional and projected velocity dispersion profiles that would result. In each case, the tracer stars are assumed to be distributed according to the "cool" star surface brightness profile of Allen (1994). The enclosed mass was assumed to vary linearly with radius inside 0.2 pc for enclosed masses above a lower mass limit corresponding to the adopted black hole mass. The predicted one-dimensional and projected velocity dispersion profiles at $r_p > 0.01$ pc ($> 0.24''$) for black hole masses of 0, 0.5, 1.0, and $1.5 \times 10^6~M_\odot$ are shown in Figure 1. From this figure, it is apparent that 1) a one-dimensional velocity dispersion of ~ 325 km s^{-1} is expected at 0.02 pc (0.5'') even if there is no central massive black hole, 2) a one-dimensional velocity dispersion of ~ 570 km s^{-1} at 0.02 pc is predicted if the maximum allowable $1.5 \times 10^6~M_\odot$ black hole is present in the Galactic center, and 3) it should be possible using proper motion data at a projected radius of 0.02 pc to confidently discern the presence of a central black hole of mass $> 5 \times 10^5~M_\odot$.

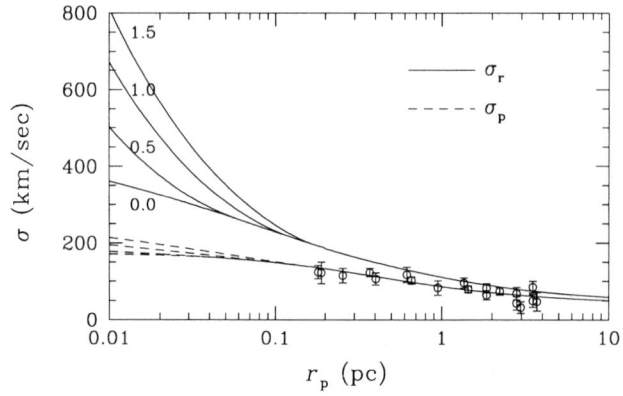

Figure 1. Predicted one-dimensional (solid lines) and projected (dashed lines) velocity dispersion profiles obtained by integrating the Jeans equation with enclosed mass profiles that combine the Saha et al. mass model with inward extrapolation with $M(r) \propto r$ and central black holes of masses of 0.0, 0.5, 1.0, and $1.5 \times 10^6~M_\odot$ as marked.

5. Conclusions

The Saha et al. Galactic center mass model demonstrates that the available kinematic data can be consistently modeled without recourse to a central massive black hole. Within the assumptions of spherical symmetry and isotropic velocity dispersion, evidence is found that the local M/L_K ratio increases inside ~ 0.8 pc in a way that cannot be explained by the existence of a central massive black hole. An extended distribution of dark stellar remnants coincident with the central cluster of young, massive, He I emission-line stars is an attractive explanation for this effect. This then requires that the stellar initial mass function in the central parsec of the Galaxy is deficient in stars less massive than $\sim 1~M_\odot$. Otherwise, the combined light of long-lived low mass stars would exceed limits set by the local M/L_K values derived from the mass model.

It is likely that the Galactic center region has undergone repeated starbursts as gas which has accumulated in the molecular ring collapses into the central parsec. Tidal shear outside the core radius of the total mass distribution is expected to disrupt individual gas clouds and inhibit star formation. Inside the core radius of the total mass distribution, the stellar density is approximately constant and the tidal shear is correspondingly reduced. It is here that star formation is most likely to occur, and the presence of many He I emission-line stars, and possibly also dark stellar remnants, in the core region support this assertion. After a short-lived starburst, winds from the massive stars formed are expected to clear away residual gas and re-establish the central cavity. A supply of molecular gas will then build up over time in a structure similar to the current molecular ring.

References

Allen, D. A. 1994, in The Nuclei of Normal Galaxies: Lessons from the Galactic Center, eds. R. Genzel & A. I. Harris, 293

Allen, D. A., Hyland, A. R., & Hillier, D. J. 1990, MNRAS, 244, 706

Backer, D. C., & Sramek, R. A. 1987, in The Galactic Center, AIP Conf. Proc. No. 155, ed. D. C. Backer (N.Y.: AIP), 163

Blitz, L., et al. 1993, Nature, 361, 417

Burton, M., & Allen, D. 1992, Proc. ASA, 10, 55

Eckart, A., Genzel, R., Hofmann, R., Sams, B. J., & Tacconi-Garman, L. E. 1995, ApJ, 445, L23

Güsten, R. et al. 1987, ApJ, 318, 124

Haller, J. W., et al. 1996, ApJ, 456, 194

Jackson, J. M., et al. 1993, ApJ, 402, 173

Krabbe, A., Genzel, R., Drapatz, S., & Rotaciuc, V. 1991, ApJ, 382, L19

Krabbe, A., et al. 1995, ApJ, 447, L95

Lacy, J. H. 1989, in The Center of the Galaxy, IAU Symposium 136, ed. M. Morris (Kluwer: Dordrecht), 493

Lindqvist, M., Habing, H. J., & Winnberg, A. 1992, A&A, 259, 118

Lo, K. Y., & Claussen, M. 1983, Nature, 306, 647

McGinn, M. T., Sellgren, K., Becklin, E. E., & Hall, D. N. B. 1989, ApJ, 338, 824

Melia, F. 1994, ApJ, 426, 577

Morris, M. 1993, ApJ, 408, 496

Myers, P. C., & Fuller, G. A. 1993, ApJ, 402, 635

Narayan, R., Yi, I., Mahadevan, R. 1995, Nature, 374, 623

Saha, P., Bicknell, G. V., & McGregor, P. J. 1996, ApJ, in press.

Sanders, R. H. 1992, Nature, 359, 131

Sellgren, K., McGinn, M. T., Becklin, E. E., & Hall, D. N. B. 1990, ApJ, 359, 112

Serabyn, E., Lacy, J. H., Townes, C. H., & Bharat, R. 1988, ApJ, 326, 171

Part 2. The Stellar Population
Section B. Star Formation and Hot Stars

Masers and Star Formation near the Galactic Centre

J.L.Caswell

Australia Telescope National Facility, CSIRO, PO Box 76, Epping, NSW 2121, Australia

Abstract. One clear signature of massive young stars is from molecular-line maser emission at radio frequencies. Surveys for masers, from hydroxyl, water and methanol, are thus a powerful means of probing the Galactic Centre for star-forming regions, and allow a census of such objects, unaffected by obscuration. Several species of maser are discussed here, with emphasis on a new survey for 6.668-GHz methanol emission, made with the Australia Telescope Compact Array. This survey reveals 23 masers in a 2-square degree region towards the Galactic Centre. Comparison of the space density of the maser sites with that in the spiral arms suggests that the current rate of massive star formation is neither enhanced nor suppressed in the Galactic Centre.

1. Introduction

In all parts of the Galaxy, the population of massive stars is of vital importance since it heavily influences the other constituents. In turn, the density and kinematics influence the rate of formation of massive stars. The centre of our Galaxy is a region of especially high density, rapid motions, and violent activity. Overall, it is not clear whether these extreme conditions might increase or decrease the rate of massive star formation relative to the spiral arms.

Unfortunately, young massive stars are not easily detected by direct observation: their optical emission is obscured by dust which surrounds the star and, for objects as distant as the Galactic Centre, additional obscuration arises from dust along the intervening path. Indirectly, the presence of a star is revealed by infrared emission from heated dust (which surrounds and obscures the star), and from continuum radio emission, which is present because the star ionises an enveloping compact HII region. However, the clearest signature of a young, massive star is often from molecular-line maser emission at radio frequencies. Indeed, the use of masers to study star formation near the Galactic Centre has a long pedigree: the first hint that masers might be used as tracers for recently formed stars was noted nearly 30 years ago, within a few years of the discovery of OH masers. Another decade later, Genzel & Downes (1979) suggested that masers (in this instance, water masers) might provide one of the best means to discover O and B stars very promptly after they had formed. They also remarked on an apparent deficiency of masers within 500 pc of the Galactic Centre, but this conclusion was based on very incomplete samples. By 1987, with the completion of much more extensive OH maser surveys, several more masers had been found near the Galactic Centre and the region no longer seemed deficient (Caswell & Haynes 1987). However, the statistics remained poor. Investigation of OH and water masers show them to be commonly pinpointing the same stars (Forster & Caswell 1989), but the water masers are generally stronger. In this respect water may be a better tracer of young stars than OH, but observations are generally more difficult at the higher frequency of the 22-GHz water transition, and extensive unbiased surveys for water have not been made.

Recently, the 6.668-GHz methanol transition has proved to be a maser typically 16 times stronger than OH (Caswell 1996a) and thus potentially our most valuable tracer of star formation sites. The survey of 6.6-GHz methanol masers by Caswell et al. (1995c) lists 245 masers, but was confined to the sites of previously suspected star-formation regions. More recently, a limited, but unbiased, exploratory survey was made (Caswell 1996a), using the Australia Telescope Compact Array (ATCA) between Galactic longitudes 331° and 340° where there is a high spatial density of star-forming regions in the Galactic spiral arms. It was found to be correspondingly rich in methanol maser sites, with a total of 57 detected in the narrow strip surveyed close to the Galactic plane. At least one thousand such sites are likely to be detected when larger scale methanol surveys are complete. Thus while not all star-formation sites harbour masers, a large number do.

In view of these facts, a survey for 6.668-GHz methanol masers has recently been conducted to probe the Galactic Centre for star-forming regions. The remainder of this report is largely concerned with its initial results. By directly comparing the Galactic Centre survey with the earlier methanol survey in the Galactic disk, we may assess whether regions of massive star formation show an enhanced concentration near the Galactic Centre, relative to the spiral arms.

2. Observations and results of the methanol survey

Here, I will be presenting preliminary results of the methanol survey, in advance of a full report (Caswell 1996b). The ATCA was used for one night, 1995 November 6, in a 6-km configuration yielding 15 baselines ranging from 627 to 5938 m. The correlator sampled two orthogonal linear polarizations, each processed to generate 2048 channnels across a 4-MHz bandwidth. The effective radial velocity coverage of the survey was from -70 to +100 km s^{-1}. The survey, at a grid of 85 positions symmetrical about the Galactic Centre, covered a roughly elliptical area, extending $\pm 0°.9$ in Galactic longitude, and $\pm 0°.5$ in Galactic latitude. The grid positions are shown in Figure 1.

Most positions were observed for four 1.3-minute observations spaced over 12 h and interspersed with position calibrations. Processing was completed within the AIPS reduction package. The synthesized beam has a width-to-half-power of 2 x 4 arcsec. The data were Hanning-smoothed to a frequency resolution of 4 kHz (0.18 km s^{-1}) and the resulting rms noise on an individual channel map was typically 160 mJy, allowing detection of sources to less than 1 Jy peak flux density. Where a maser site was detected, the maser spots with different velocities were found to be mostly in a single group (with size less than 1 arcsec), in accordance with previous work on methanol (Norris et al. 1993; Caswell, Vaile & Forster 1995a). Each group of maser spots is presumed to be

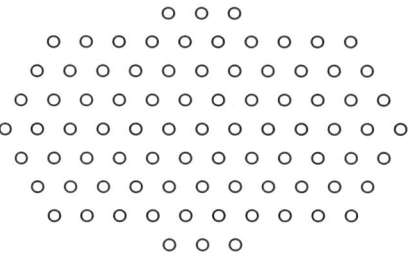

Figure 1. The field centres of the observing grid. The grid is symmetrical about the Galactic Centre, and extends from -0°.9 to +0°.9 in Galactic longitude and from -0°.5 to +0°.5 in Galactic latitude.

associated with a single ultracompact HII region. In two instances, two maser groups are present, with separations of 2.5 and 6 arcseconds respectively; in both instances, it is likely that each individual maser group delineates a separate massive star, formed at a similar epoch within the same cluster. A list of the 23 methanol maser groups present in the survey region is given in Table 1.

For each maser group, the Table summarizes the peak flux density of the strongest feature, its velocity, and its Galactic coordinates, used as a source name. Most of the masers have been previously detected in earlier less systematic or less sensitive surveys, and a major result from the present survey is the absence of any large previously unknown population. The methanol positions are not at the centre of the primary beam in the survey observations, but the methanol flux densities have been corrected for this. Nine of the methanol masers coincide with OH masers; preliminary estimates of the OH peak flux densities for these masers are given in column 5, and the OH observations and their implications are discussed in more detail in Section 5.

3. Remarks on individual sources

The strongest source detected in the survey is 359.615-0.243, known since 1991, but highly variable (Caswell, Vaile & Ellingsen 1995b) and now with peak flux density of 88.7 Jy, higher than in any previously reported spectrum. Other sources of special interest include:

359.138+0.031. OH and water masers coincide with the methanol maser (Forster & Caswell 1989).

359.436-0.104 and 359.436-0.102. Comparison with OH and water maser data from Forster & Caswell (1989) shows that the first of these (the stronger) coincides with an OH maser, and the second (offset 6 arcsec) coincides with a water maser. This an example that demonstrates the special value of methanol masers: conditions favouring OH and water masers overlap only partially, such that they are sometimes not present at the same site; however, the methanol masers seem to be present under a broader range of conditions, often encompassing both the OH- and the water-favoured sites.

Galactic Coordinates (l, b)	Flux density meth peak (Jy)	Velocity of peak (km.s-1)	References	Flux density OH peak (Jy)
359.138+0.031	15.6	-4	Caswell et al. 1995c	8
359.436−0.104	26.8	-52	Caswell et al. 1995c	3
359.436−0.102	4.4	-53.6		
359.615−0.243	88.7	22.5	Caswell et al 1995c	1.5
359.970−0.457	1.3	23	Caswell et al. 1995c	11
0.212−0.001	3.5	49.2	Walsh et al. 1995	
0.315−0.201	41.2	18	van der Walt et al. 1995	
0.316−0.201	1.3	21		
0.376+0.040	0.7	37	Caswell et al. 1995c	4.3
0.393−0.034	5.8	28.7		
0.496+0.188	10	0.8	Schutte et al. 1993	0.2?
0.645−0.042	65	49.1	#10,Houghton & Whiteoak 1995	
0.647−0.055	3.4	51	#11,Houghton & Whiteoak 1995	
0.651−0.049	31.7	48	#9,Houghton & Whiteoak 1995	
0.657−0.041	3	52	#8,Houghton & Whiteoak 1995	1.6
0.665−0.036	2.1	60.4	#7,Houghton & Whiteoak 1995	0.4?
0.666−0.029	33.7	72.2	#5,Houghton & Whiteoak 1995	
0.667−0.034	0.4	55.2	#6,Houghton & Whiteoak 1995	
0.672−0.031	4.5	58.2	#4,Houghton & Whiteoak 1995	1.8
0.673−0.029	0.4	66	#3,Houghton & Whiteoak 1995	
0.677−0.025	4.4	73.4	#2,Houghton & Whiteoak 1995	
0.695−0.038	26	68.5	#1,Houghton & Whiteoak 1995	
0.836+0.184	8.1	3.5	van der Walt et al. 1995	

Table 1. 6.668-GHz methanol masers near the Galactic Centre.

0.315-0.201 and 0.316-0.201. This pair of sources, separated by 2.5 arcsec, corresponds to the maser reported by van der Walt et al. (1995) and by Walsh et al. (1995).

Sgr B2 (the 11 sources 0.645-0.042 to 0.695-0.038). All of these sources were first detected in 1992 by Houghton & Whiteoak (1995). The present observations were made with 1/500 of the integration time used in their 1992 work and so the sensitivity is of course poorer. Table 1 shows that there have been no dramatic intensity changes between 1992 and 1995; our spectral resolution is 2.4 times sharper than that used by Houghton & Whiteoak and is chiefly responsible for the generally higher intensities (by typically 30 per cent) measured in the present spectra.

4. The distribution of the methanol masers

The distribution of the masers is shown in Figure 2. An important feature of the distribution is the absence of clustering towards the centre. Indeed no methanol maser lies closer than 30 pc from the centre. This is in contrast to less massive stars, detected in the IR, or as OH/IR masers, which show a concentration at the centre.

The survey region is symmetrical about the Galactic Centre, but the masers are chiefly located between longitude 0 and 0°.9, as is clearly seen in Figure 2: 18 lie between 0 and 0°.9, and only 5 between 359°.1 and 0°. The velocity distribution

Figure 2. Spatial distribution of the 23 methanol masers. The boundary of the surveyed region, which is symmetrical about the Galactic Centre, is also shown. The precise maser positions are given in Table 1 and the plotted positions are approximate, with separations of masers exaggerated if the symbols would otherwise be so close as to overlap. The eleven Sgr B2 masers are clearly seen as a compact cluster.

is shown in Figure 3 and reveals an even more pronounced asymmetry, with three at negative velocities and 20 at positive velocities. If Sgr B2 were disregarded, the asymmetries would be less pronounced: a ratio of 7 to 5 in longitude and a ratio of 9 to 3 in velocity.

5. Comparison with OH masers

The close relationship between methanol and OH masers was demonstrated by Caswell et al. (1995a): an investigation of 29 apparent OH/methanol associations showed 28 to be coincident to within an arcsecond. In the Galactic Centre region studied here, existing 1665-MHz OH searches include those of Caswell & Haynes (1983), supplemented by Forster & Caswell (1989) for accurate positions, the extensive study of Sgr B2 by Gaume & Mutel (1987), and some additional Parkes and ATCA observations (Caswell, unpublished). Thirteen 1665-MHz OH masers have now been detected, nine of them apparently coincident with a methanol maser. The remaining four are weak OH masers reported by Gaume & Mutel in the Sgr B2 region, with no detected methanol counterpart above 1 Jy. A further OH maser detected at the 6035-MHz transition has methanol but no detected 1665-MHz counterpart. Overall, the search for methanol masers has proved to be more productive than that for OH masers, as might be expected since they are typically more than an order of magnitude stronger.

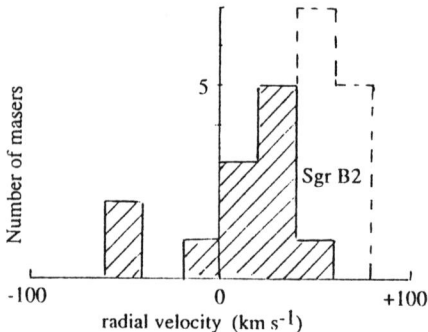

Figure 3. Histogram of velocities of the 23 methanol masers. The contribution of the 11 in Sgr B2 is shown with a broken line.

6. The abundance of star-formation regions near the Galactic Centre

One might expect the generally high density near the Galactic Centre to lead to an excess of star-formation regions, but this might be counteracted by a deficiency of gas resulting from earlier star-formation activity; the effects of violent activity near a galactic nucleus are similarly unpredictable, and could either promote star formation or inhibit it. An empirical determination could be a guide as to the relative influence of these effects.

Here we will assume that the methanol maser sites are indeed a measure of the present rate of massive star formation. The Galactic Centre may then be compared with a similar methanol survey of the inner spiral arms (Caswell 1996a). The longitude coverage of the Galactic Centre survey, $\pm 0°.9$, corresponds to a region of radius 150 pc at the distance of the Galactic Centre (8.5 to 10 kpc). The number of methanol masers (23) in this region may be slightly underestimated because of the limited velocity coverage of the survey. The sensitivity is similar to that of the disk survey from 331° to 340° longitude (57 sources total), where the average density is 16 masers in a $1°.8$ range of longitude. (Here the observed number of 11 has been increased by a factor of 1.5 to allow for the smaller latitude range covered; the correction factor is based on results from limited surveys of wider latitude range.) Thus, superficially, the space densities of 23 as opposed to 16 are similar, but with an insignificant excess in the Galactic Centre region. Note that for masers in the longitude region 331° to 340° the velocities indicate concentrations to two spiral arms, with very little contamination by foreground sources; any foreground contamination of the Galactic Centre region should be equally small and, in the search for an excess or deficit, will cancel out in the method of comparison. Within 30 pc of the Centre, it is striking that no masers are present, ruling out any dramatic excess, but not indicative of a significant 'hole' in the broader distribution.

As a consequence of the small total volume of the Galactic Centre region, and the relatively short lives of massive stars, one can expect that, at any epoch, the number of star-formation regions will be subject to large statistical fluctuations, and is not necessarily indicative of an intrinsic under- or over-abundance which might be caused by the unique conditions in this region. For example, Sgr

B2 accounts for nearly half the maser sites, but has only a 5-pc extent, compared to an extent of 300 pc for the whole survey. If Sgr B2 were absent (as viewed 100 million years past or future) we might regard the region as under-abundant; if two sites similar to Sgr B2 were present (as might occur at another epoch) we would regard it as over-abundant. This basic limitation in the statistics suggests that more sophisticated analyses are unwarranted in the absence of any marked over-abundance. In summary, the current formation rate of massive stars within a few hundred pc of the Galactic Centre is similar to that in the inner spiral arms. There is no evidence for an excess of star formation but, at present, Sgr B2 is an unusually active region, without parallel. No methanol masers have been detected within 30 pc of the Galactic Centre.

7. Comparison with OH/IR maser surveys near the Galactic Centre

The OH/IR stars are most readily detected as prominent masers on the 1612-MHz transition of OH. They correspond to stars of only a few solar masses and the maser emission occurs only near the end of their lifetime so, mostly, they trace stars that are at least a billion years old. A search for such masers towards the Galactic Centre, in a region of similar extent to the present survey, has been reported by Lindqvist et al. (1992), with additional data reported by L. Sjouwerman and A. Winnberg at the present meeting. More than 100 such stars are present in this region, a consequence of the high density of low-mass stars and the lifetime of the maser emission. (At the time of the formation of these low-mass stars it is likely that small numbers of massive stars were also formed, but these would have short lifetimes and would long since have disappeared in a supernova phase, with their cores perhaps now present as pulsars.) In contrast to this population of low-mass stars that formed more than 1 billion years ago, the survey of methanol masers is believed to be tracing massive stars with ages of less than a million years, and thus relates to a very different and complementary aspect of the Galactic Centre.

8. Conclusions and future work

The methanol survey described here has yielded 23 methanol masers. These results suggest that the current formation rate of massive stars near the Galactic nucleus (within a radius of $1°$, i.e. 150 pc) is similar to that in the spiral arms several kpc from the Centre. No methanol masers have been detected within 30 pc of the Centre. Future work will include a better understanding of the mass range and evolutionary stage of the stars being traced by methanol masers. Studies of counterparts in the infra-red, and the radio continuum ultra-compact HII regions, are needed for this. Such studies will also permit a critical assessment of whether the methanol masers are uniquely associated with young massive stars (unlike the OH and water maser populations which are contaminated by some objects of other types). It would also be of interest to extend the methanol survey to lower flux densities, perhaps corresponding to somewhat lower mass stars.

References

Caswell, J.L. 1996a, MNRAS, 279, 79
Caswell, J.L. 1996b, MNRAS, in press
Caswell, J.L., Haynes, R.F. 1983, Aust.J.Phys., 36, 361
Caswell, J.L., Haynes, R.F. 1987, Aust.J.Phys., 40, 215
Caswell, J.L., Vaile, R.A., Forster, J.R. 1995a, MNRAS, 277, 210
Caswell, J.L., Vaile, R.A., Ellingsen, S.P. 1995b, Publ.Astron.Soc.Aust., 12, 37
Caswell, J.L., Vaile, R.A., Ellingsen, S.P., Whiteoak, J.B., Norris, R.P. 1995c, MNRAS, 272, 96
Forster, J.R., Caswell, J.L. 1989, A&A, 213, 339
Gaume, R.A., Mutel, R.L. 1987, ApJS, 65, 193
Genzel, R., Downes, D. 1979, A&A, 72, 234
Houghton, S., Whiteoak, J.B. 1995, MNRAS 273, 1033
Lindqvist, M., Winnberg, A., Habing, H.J., Matthews, H.E. 1992, A&AS, 92, 43
Norris, R.P., Whiteoak, J.B., Caswell, J.L., Wieringa, M.H., Gough, R.G. 1993, ApJ, 412, 222
Schutte, A.J., van der Walt, D.J., Gaylard, M.J., MacLeod, G.C. 1993, MNRAS, 261, 783
van der Walt, D.J., Gaylard, M.J., MacLeod, G.C. 1995, A&AS, 110, 81
Walsh, A.J., Hyland, A.R., Robinson, G., Bourke, T.L., James, S.D. 1995, Publ.Astron.Soc.Aust., 12, 186

Object #17 and Quintuplet Clusters near the Galactic Center

Tetsuya Nagata

Department of Physics, Nagoya University Nagoya 464-01, Japan

Abstract. Various observations of two star clusters, Object #17 and Quintuplet, which are located within 0.2° of the Galactic center are compared. Both of them show the 3.4 μm absorption feature which have been detected only in sources reddened heavily by diffuse interstellar dust. Although the position angle of infrared polarization of the #17 cluster (central part) is slightly different from the other Galactic center objects, the two star clusters seem to be actually close to the Galactic center, not mere projection of nearby objects. Object #17 shows He II (3.09 μm) and other emission lines, and several stars in the Quintuplet cluster also have emission lines though the brightest five members have no detectable emission lines. According to recent CS J=1-0 line observations of a field of 3' centered at Object #17 with the Nobeyama millimeter interferometer, #17 is located on the southwestern edge of a negative-velocity cloud. Thus it is possible that this cluster is responsible for the ionization of the thermal Arched Filaments to the west of it. However, the lack of warm dust around Object #17 in recent KAO observations is inconsistent with this picture.

1. Introduction

In a polarimetric survey in the region within 20' of the Galactic center, Kobayashi et al. (1983) found objects GCS 3 and GCS 4 at ($l = 0.16°$, $b = -0.06°$), which have very red $H - K$ colors. These objects were actually the two brightest members of the Infrared Quintuplet cluster (Nagata et al. 1990, Okuda et al. 1990). Recently, Wolf-Rayet stars and luminous blue variable stars were discovered in this cluster (Figer et al. 1995, 1996). Similarly, Object #17 found at (0.12°, 0.02°) in a near-infrared survey by Nagata et al. (1993) has turned out to be a cluster of emission line stars (Nagata et al. 1995, see also Cotera et al. 1996).

These two star clusters, along with the central parsec cluster, seem to provide evidence for recent occurrence of massive star formation near the Galactic center. The Quintuplet cluster is relatively well studied, but the location of the #17 cluster along the line of sight and its possible interaction with the thermal Arched Filaments are still uncertain. Here, I shall present new data on #17 and discuss its properties in comparison with the Quintuplet.

2. Their Location along the Lines of Sight

There are several lines of evidence that the Quintuplet cluster is located near the Galactic center. Absorption lines of the P and R branches of the CO $v = 0 - 1$ band are clearly seen in each member of the Quintuplet (Okuda et al. 1990). The similarity of the band in the five objects and the low temperature of the absorption are indicative of their interstellar, not circumstellar, origin. In addition, Okuda et al. (1990) inferred that the Quintuplet is inside 250-pc and 3-kpc ring structures of the Galaxy on the basis of their velocity-resolved CO absorption spectra.

Their polarization, 3.4 μm absorption feature (see next subsections), 9.7 μm absorption feature, and continuum reddening further support the idea that the Quintuplet is near the Galactic center. The Quintuplet members have similar optical depths of the 9.7 μm silicate absorption feature; Okuda et al.(1990) derived a value $\tau_{9.7} \sim 2.2$ assuming that the source spectra contains no intrinsic emission or absorption. This assumption seems reasonable especially if they are WCL-type Wolf-Rayet stars (see Figer, Morris, & McLean 1996). Although this $\tau_{9.7}$ is slightly smaller than the optical depth of the central parsec cluster (Roche & Aitken 1985), it is consistent with an extinction $A_V > 20$. The five brightest plus a few members have very red colors in the near infrared ($H - K > 3.5$), but other members have $H - K$ colors consistent with the Galactic center objects (Moneti, Glass, & Moorwood 1994).

On the other hand, Nagata et al. (1995) estimated the extinction toward Object #17 to be $A_K = 2.7$, and it is consistent with the idea that #17 is near the Galactic center. However, this extinction could be due to local to the #17 cluster, and it might be located relatively close to us. After all, they seem to be very young stars, and it is possible for them to be still in their placental cloud. Let us examine its location with its polarization and 3.4 μm absorption. (We will be able to use the ISO data for the silicate absorption of Object #17 in the near future.)

2.1. Infrared Polarization

Kobayashi et al. (1980, 1983) showed that both the polarization degree and its position angle are fairly uniform in the K band within 20' of the Galactic center. In addition, Kobayashi et al. (1983) found that there is a significant change in the posision angle as a function of the observed $H - K$ color. The position angles of less heavily obscured objects lying at distances < 5 kpc from us are $170° \pm 10°$, whereas those of objects closer to the Galactic center are $\sim 17°$. K-band polarization of these objects are attributed to dichroic absorption by grains aligned to the Galactic magnetic field. Thus, the direction of the magnetic field seems to change at a distance of ~ 5 kpc from us.

Polarization properties of the Quintuplet members have been studied in some detail. Okuda et al. (1990) showed that the five members have very similar polarization in the near-infrared. Polarization degrees of these objects are 8 % at H, 5 % at K, and 2 % at L', and the position angles are $10° - 20°$. Spectropolarimetric observations of GCS 3-II and GCS 4 showed that polarization degrees up to 2.5 μm can be represented very well by a power law $p(\lambda) \propto \lambda^{-2}$ (Nagata, Kobayashi, & Sato 1994; #24 & #26 in Figure 1), and that

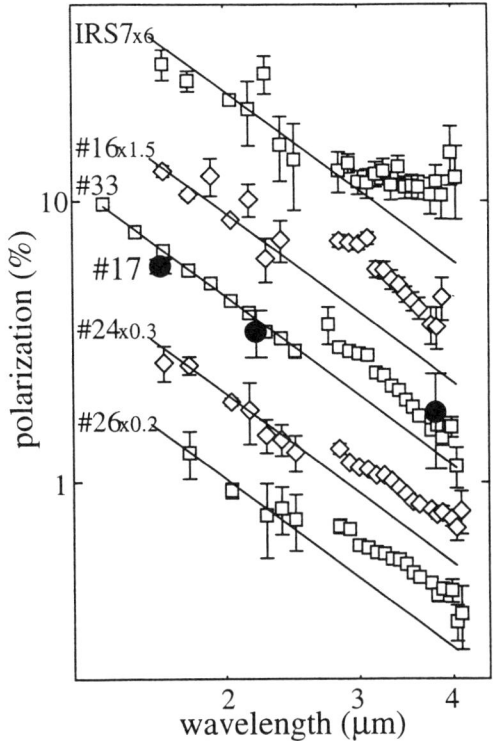

Figure 1. Polarization degree of Object #17 (filled circles) with other Galactic center objects in Nagata et al. (1994).

there is large polarization excess in the silicate absorption band at 10 μm (Roche 1995). These polarization specta show no significant rotation of the position angle from the near-infrared to the mid-infrared. The observed polarization is consistent with its interstellar origin, and provides further evidence that the Quintuplet cluster is actually near the Galactic center.

We made polarimetry of Object #17 with a 7.8″ beam. We used the Kyoto Polarimeter on the 3.8-m UKIRT, and centered the aperture on the K-band signal peak. Probably the stars 7, 8, and 10 (see Figure 4) dominate the measured flux. The polarization degree and position angle are 5.9 % (33°), 3.4 % (29°), and 1.8 % (35°) in the H, K, and L' band, respectively. The polarization degrees are plotted on the spectropolarimetry of other objects near the Galactic center in Figure 1. The decrease of the polarization with the wavelength and no significant rotation of the position angle are consistent with the idea that the polarization of #17 is mainly interstellar in origin. However, its position angle is slightly different from other objects in the central degree of the Galaxy. Further polarization measurements of Object #17 are needed.

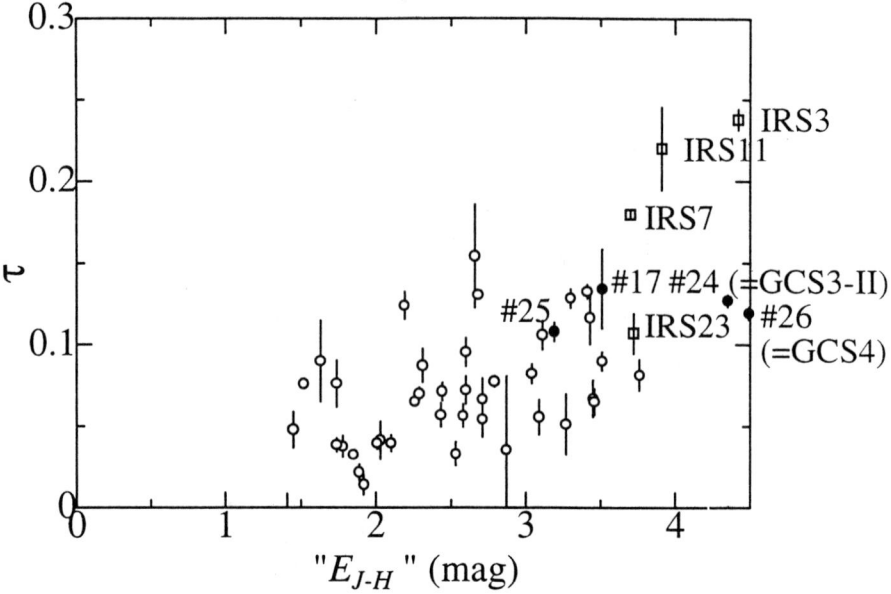

Figure 2. 3.4 μm absorption vs. color excess of objects within 1° of the Galactic center.

2.2. 3.4 μm Absorption

An absorption feature at 3.4 μm seen in the Galactic center object IRS 7 has been studied in other lines of sight also (Sandford, Pendleton, & Allamandola 1995; Imanishi et al. 1996). The absorption is attributed to C-H stretching vibrations in interstellar grains in diffuse interstellar medium. This absorption feature has never been detected in molecular cloud sources although a long-wavelength wing of the 3.1 μm ice absorption feature (e. g., Smith, Sellgren, & Tokunaga 1989) and a shallow 3.47 μm feature (Brooke, Sellgren, & Smith 1996) exist in them. Furthermore, toward the central persec cluster of the Galaxy, the optical depth of the 3.4 μm absorption per unit continuum extinction in the optical and near-infrared $\tau_{3.4}/A_V$ seems to be about twice as large as that in the solar neighborhood (Pendleton et al. 1994). We examine the depth of this feature in the spectra of the Quintuplet and Object #17 clusters.

Figure 2 shows the result of photometry and spectroscopy of 50 bright near-infrared objects within a degree of the Galactic center (Nagata et al. 1993, 1996). The color excess E_{J-H} has been calculated on the assumption that the object is either an early-type star (for which $(J-H)_0=0$) or a late-type giant (for which $(J-H)_0=0.88$); this assumption might not be valid for some objects because many long-period variables are probably in the list, and this might have increased the scatter of the data points.

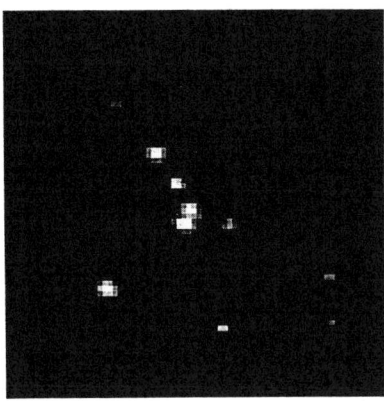

Figure 3.
Object #17 image. Brγ (2.17μm) line minus continuum.

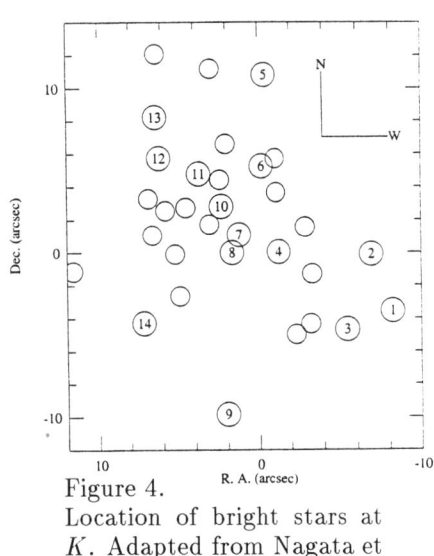

Figure 4.
Location of bright stars at K. Adapted from Nagata et al. (1995)

Several objects in the central parsec of the Galaxy (IRS 3, 7, & 11; see also Sandford et al. 1995) have large $\tau_{3.4}$, but IRS 23 seem to have slightly smaller $\tau_{3.4}$. Note that IRS 19 has very deep ice absorption with no detectable 3.4 μm absorption (McFadzean et al. 1989). Some variations of $\tau_{3.4}$ might exist even in the central cluster, or these objects with smaller $\tau_{3.4}$ could be in front of the central cluster. The 3.4 μm absorption to the central cluster might have a small local component.

The two brightest objects of the Quintuplet (#24=GCS 3-II & #26=GCS 4) have $\tau_{3.4}$ comparable to that of IRS 23. They have large color excesses presumably due to circumstellar dust. If we follow the reasoning of Pendleton et al. (1994) and assume they are WC stars (Figer et al. 1996), this circumstellar dust do not contribute much to their $\tau_{3.4}$. In fact, Object #25, which is probably a luminous blue variable (Figer et al. 1995) in the Quintuplet cluster, has a similar $\tau_{3.4}$ though it has smaller E_{J-H}.

Object #17, observed with a 2.7″ beam and probably containing stars 7 & 8, shows similar $\tau_{3.4}$, and it is among the largest $\tau_{3.4}$ outside the central cluster. This strongly suggests that #17 is actually located near the Galactic center.

3. Emission Lines

Figure 3 shows a continuum-subtracted Brγ image of Object #17 (Nagata et al. 1995). Cotera et al. (1996) detected Brγ emission in 13 stars; 12 of them also have He I (2.11 μm) emission, and two show fainter He II (2.19 μm) emission (see also Cotera 1995). An estimate of mass loss rate from the brightest star was made by Nagata et al. (1995) on the basis of a simple model. The resultant mass loss rate $\dot{M}$ is $\sim 2 \times 10^{-5} M_\odot$ yr^{-1}, which is similar to that found for

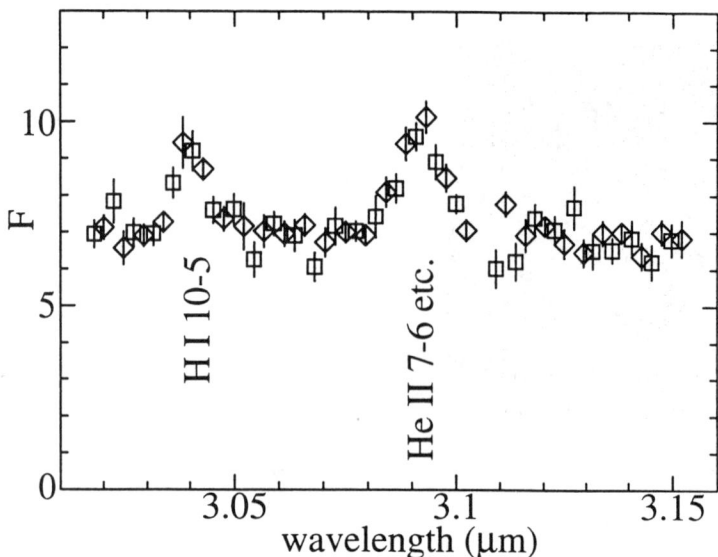

Figure 5. He II/C IV (3.09μm) line spectrum of Object #17.

the brightest He I emission-line star in the central parsec (Krabbe et al. 1991; Najarro et al. 1994).

The spectrum in Figure 5 is again the central 2.7″ of Object #17, and with a spectral resolution $\lambda/\Delta\lambda \sim 750$. It shows moderately strong He II/C IV (3.09 μm) emission. This emission line complex is very strong in some Wolf-Rayet stars (Werner, Stauffer, & Becklin 1990), but its strength varies from star to star, and some late-type Wolf-Rayet stars seem to have no detectable emission (Figer 1995). Nagata et al.(1995) mentioned that its strength in the center of Object #17 might be a little too weak for them to be Wolf-Rayet stars, but in fact it is quite comparable to those in some Wolf-Rayet stars. The linewidth ($\sim 10^3 \mathrm{kms}^{-1}$) is also consistent with their Wolf-Rayet star identification.

4. CS Cloud

Kobayashi et al. (1996) made CS J=1-0 line observations of a field of 3′ centered at the #17 cluster with the Nobeyama millimeter interferometer. A negative-velocity cloud has been detected which is elongated in the NNW-SSE direction and highly clumped. The dense clumps have large velocity widths of 15 − 20 kms^{-1}. This might be the result of disturbance by Object #17, which is located on the southwestern edge of this cloud. Therefore, it is possible that #17 is responsible for the ionization of the thermal Arched Filaments to the west of it.

However, the non-detection of warm dust near Object #17 in KAO observations by Latvakoski et al. (1996) is inconsistent with this picture. If the #17 cluster is located in the southwest of the molecular cloud, it should be shrouded by a warm-dust region, but only the Arched Filaments are bright in their 31.5 and 37.7 μm maps. Thus, the negative velocity cloud and the #17 cluster might be a chance projection of unrelated objects. The relationship between the molecular cloud and #17 remains to be elucidated.

References

Brooke, T. Y., Sellgren, K., & Smith, R. G. 1996, ApJ, 459, 209
Cotera, A. S. 1995, Ph. D. Thesis, Stanford University
Cotera, A. S., Erickson, E. F., Colgan, W. J., Simpson, J. P., Allen, D. A., & Burton, M. G. 1996, ApJ, in press
Figer, D. F. 1995, Ph. D. Thesis, University of California at Los Angeles
Figer, D. F., McLean, I. S., & Morris, M. 1995, ApJ, 447, L29
Figer, D. F., Morris, M., & McLean, I. S. 1996, in this volume
Imanishi, M., Sasaki, Y., Goto, M., Kobayashi, N., Nagata, T., & Jones, T. J. 1996, AJ, in press
Kobayashi, N., Nagata, T., Kawabe, R., & Tsuboi, M. 1996, in preparation
Kobayashi, Y., Kawara, K., Kozasa, T., Sato, S., & Okuda, H. 1980, PASJ, 32, 291
Kobayashi, Y., Okuda, H., Sato, S., Jugaku, J., & Dyck, H.M. 1983, PASJ, 35, 101
Krabbe, A., Genzel, R., Drapatz, S., & Rotaciuc, V. 1991, ApJ, 382, L19
Latvakoski, H., Stacey, G., Hayward, T., & Gull, G. 1996, in this volume
McFadzean, A. D., Whittet, D. C. B., Longmore, A. J., Bode, M. F., & Adamson, A. J. 1989, MNRAS, 241, 873
Moneti, A., Glass, I., & Moorwood, A. F. M. 1994, MNRAS, 268, 194
Nagata, T., Woodward, C. E., Shure, M., Pipher, J. L., & Okuda, H. 1990, ApJ, 351, 83
Nagata, T., Hyland, A. R., Straw, S. M., Sato, S., & Kawara, K. 1993, ApJ, 406, 501
Nagata, T., Kobayashi, N., & Sato, S. 1994, ApJ, 423, L113
Nagata, T., Woodward, C. E., Shure, & Kobayashi, N. 1995, AJ, 109, 1676
Nagata, T., Sato, S., & Hyland, A. R. 1996, in preparation
Najarro, F., Hillier, D. J., Kudritzki, R. P., Krabbe, A., Genzel, R., Lutz, D., Drapatz, S., & Geballe, T. R. 1994, A&A, 285, 573
Okuda, H., Shibai, H., Nakagawa, T., Matsuhara, H., Kobayashi, Y., Kaifu, N., Nagata, T., Gatley, I., & Geballe, T. R. 1990, ApJ, 351, 89
Pendleton, Y. J., Sandford, S. A., Allamandola, L. J., Tielens,. A. G. G. M., & Sellgren, K. 1994, ApJ, 437, 683
Roche, P. F., & Aitken, D. K. 1985, MNRAS, 215, 425
Roche, P. F. 1995, in Polarimetry of the Interstellar Medium, W. G. Roberge & D. C. B. Whittet, in press
Sandford, S. A., Pendleton, Y. J., & Allamandola, L. J. 1995, ApJ, 440, 697

Smith, R. G., Sellgren, & Tokunaga, A. T. 1989, ApJ, 344, 413
Werner, M. W., Stauffer, J., & Becklin, E. E. 1990, in Evolution of the Universe of Galaxies, PASP conf. ser. Vol 10, p. 167

Hot Stars in the Quintuplet

Donald F. Figer, Mark Morris, and Ian S. McLean

Division of Astronomy, Department of Physics & Astronomy, University of California, Los Angeles, CA 90095

Abstract.
We present K-band spectra of newly identified hot stars in the Quintuplet cluster, as well as template spectra for 34 Galactic Wolf-Rayet stars. Five of the new stars are WR types (3 WC and 2 WN), while 14 others are OB supergiants; three of the WR stars are probably the hottest identified stars within 50 pc of the Galactic Center. The newly identified stars increase the estimated ionizing flux from this cluster by about an order of magnitude with respect to earlier estimates, to $8.2(10^{49})$ photons s^{-1}, or about one third of what is required to ionize the "Sickle" (G0.18-0.04). In addition, we propose that the 5 original enigmatic members of the Quintuplet-proper are dusty WCL stars, similar to the dozen or so known examples in the Galaxy.

1. Introduction

The Quintuplet is a cluster of young stars located approximately 35 pc, in projection, from the Galactic Center. In addition to the 5 bright stars for which the Quintuplet was named (Nagata et al. 1990; Okuda et al. 1990), there is a clustering of many more stars in the vicinity. Our image clearly shows a diffuse background of light from unresolved stars which spans a diameter of $\approx 50''$. Beyond this, cluster members are indistinguishable, in continuum images, from the already crowded field of stars in this part of the sky.

While the nature of the 5 Quintuplet members has been unclear, several of the cluster stars are now known to be evolved massive stars only a few million years old (Figer, McLean & Morris 1995; hereafter FMM95; Moneti, Glass & Moorwood 1994; Harris et al. 1994; Cotera et al. 1996). This cluster is proving to be key to understanding many important questions, i.e. 1) Is the cluster in the inner parsec of the Galaxy really "unique", i.e., are extraordinary physical processes required to explain the exsistence of the central cluster? 2) Do hot stars in the Quintuplet ionize G0.18-0.04 (the "Sickle") and G0.15-0.05 (the "Pistol")? 3) Does the initial mass function (IMF) in the Galactic Center favor high-mass stars? 4) Is the Quintuplet population consistent with stellar evolution models which predict that WR/(WR+O) and WC/(WR+O) should be elevated in higher metallicity regions?

2. Observations and Data Reduction

All data were taken with the UCLA double-beam near-infrared camera (McLean et al. 1993; McLean et al. 1994) at the University of California Observatories' 3-m Shane telescope. H and K grisms were used to produce spectra with R ≈ 525 (Figer 1995). The data were reduced according to the procedure in Figer (1995). Table 1 gives the identifications, coordinates, photometry, spectral classifications, and estimated ionizing fluxes for the target stars. Reference codes are shown at the bottom of the table. K-band magnitudes were converted from K' using the relation in Wainscoat & Cowie (1992).

3. K-band Spectra

K-band spectra for Galactic WR stars are shown in Figure 1. The emission lines tend to follow the expected trend of greater equivalent width for higher ionization species with earlier subtype. K-band classification is more effective in distinguishing different subtypes amongst the WN sequence than the WC sequence. Both WN9 stars lack HeII emission at 2.189 μm while WN8 stars show a hint of it. This is the strongest feature in earlier WN stars. The WC stars tend to all have similar spectra except for the latest types (WC8 and WC9). WR112 and WR118 have featureless spectra, presumably due to dilution by dust emission (Williams, van der Hucht & The 1987). Spectra for the target stars are shown in Figure 2. They have been dereddened as in FMM95.

4. Spectral Classifications

The WR stars were classified by comparing their spectra to the spectra in Figure 1 and by comparing their flux excesses at 3.09 μm to those measured in Galactic WR stars (Figer 1995). The new WN9 stars, qF256 and qF274, have line widths similar to qF320 (FMM95-1), and they lack HeII emission (2.189 μm). The new WN6 star, qF353e, has prominent emission at 2.189 μm which is comparable to the emission line strength near 2.166 μm; it also has a considerable excess at 3.09 μm. The new WC stars all have similar spectra, lacking the prominent emission at 2.058 μm which is usually seen in WC9 stars (FMM95). Two of the stars, qF309 and qF235, are classified as earlier than WC8 for their excesses at 3.09 μm, while qF151 has very little excess there. Together, the WN6 and the two "<WC8" stars are probably the hottest identified stars within 50 pc of the Galactic Center, although their ionizing fluxes are quite meager owing to their small radii (Schmutz 1996).

The "OBI" stars were classified using the new atlases from Hanson & Conti (1996) and Tamblyn et al. (1996). The "<BOI" stars have a featureless continuum. Their K-band magnitudes put them in the supergiant class, and their featureless spectra can only be fit by stars earlier than BOI. The 'OBI" stars have Brγ and HeI (2.058 μm) in emission with HeI (2.112/2.113 μm) in absorption. The spectra are similar to those of HD207329 (B1.5IB:e; Tamblyn et al. 1996) and BD+36 4063 (ON9.7Ia; Tamblyn et al. 1996 and Hanson & Conti 1996). It should be noted, though, that ON9.7Iab stars in Hanson & Conti (1996) have all three diagnostic features in absorption. The early BI stars were classified

by measuring the equivalent widths in these three spectral lines and comparing these values to those in the atlases.

5. Mass and Age of the Quintuplet

We can estimate the total cluster mass by integrating the IMF from the upper mass cutoff down to the lowest mass star identified in Table 1. Assuming, conservatively, that 32 stars in the table have masses between $m_u = 100$ M$_\odot$ and $m_o = 20$ M$_\odot$, we calculate $M_{cluster} \approx 5500$ M$_\odot$ (3700 M$_\odot$) for $m_l = 0.1$ M$_\odot$ (1 M$_\odot$) and an IMF slope of -2. We can make an independent estimate by assuming that the cluster is bound against tidal disruption, the orbital velocity is equal to the line-of-sight velocity (130 km s^{-1}; Figer 1995), and the projected distance from the GC (35 pc) is equal to the orbital radius. For these parameters, the orbital time for the cluster is $\approx 1.7(10^6)$ yrs assuming a circular orbit. The enclosed mass at this radius is $\approx 2(10^8)$ M$_\odot$ (McGinn et al. 1989). Using the tidal equation, we find $M_{total} = (2M_{r<35pc}) \times (r_{Quin}/35 \text{ pc})^3 = 9{,}300$ M$_\odot$, where r_{Quin} is the average distance of the stars in the table from the center of the cluster and is ≈ 1 pc. If O-stars are still present, then the cluster age is between 2.5 to 4.7(10^6) yrs depending on the mass-loss rates and the metallicity, assuming an IMF slope of -2 (Meynet 1995). Otherwise, the age may be up to 8(10^6) yrs.

6. Ionizing Flux

Harris et al. (1994) estimate that the Sickle requires a Lyman continuum flux of $Q_o \approx 3(10^{50})$ s^{-1}, and, according to Yusef-Zadeh, Morris & van Gorkom (1989), the Pistol requires $Q_o \approx 3.9(10^{48})$ s^{-1}. Timmermann et al. (1996) use the radio flux at 32 GHz to estimate that the Quintuplet produces $Q_o \approx 1.6(10^{50})/\eta$ s^{-1}, where η is less than 1 and accounts for dust absorption and deviations from an ionization-bounded region. We have estimated contributions to Q_o from each classified star in the Quintuplet. The results are given in Table 1 and the total is $Q_{o,tot} \approx 8.2(10^{49})$ s^{-1}. The OB stars presented here represent some of the previsouly predicted population (Harris et al. 1994; FMM95; Timmermann et al. 1996), and it is still possible that main sequence O-stars might be contributing to Q_o. FMM95 estimate an ionizing flux of 3.4(10^{48}) s^{-1} from FMM95-3 (the LBVc). This is adequate to ionize the Pistol, although contributions from other hot stars in the Quintuplet may be important, i.e. qF151 (WC8).

7. The Quintuplet-proper Members

The Quintuplet-proper members (QPMs) have remained a mystery since their discovery. Some have suggested that they are protostars, or at least not giants or supergiants (Okuda et al. 1990; Nagata et al. 1990; Glass, Moneti & Moorwood 1990). We now suggest that these objects are dusty WCLs (c.f. Abbott & Conti 1987; Williams, van der Hucht & The 1987; Cohen 1995). DWCL stars represent relatively short-lived phases of evolution when the coolest WC types (WC8 and WC9) tend to form dust shells. Williams, van der Hucht & The (1987) find that 19/27 of the WC8 and WC9 they studied have circumstellar dust emission. The near-infrared emission from these shells is well-fit by a black-body of 780-1650 K. Spectra for WR112 and WR118 in are representative of the class.

We have calculated apparent K-band magnitudes that various Galactic WC9 stars would have if they were in the Quintuplet. The most striking result is that m_K spans a very large range, ≈ 3 (WR112) to 12 (WR92). This is, presumably, due to the different amounts of dust emission from each star. The QPMs have $m_K \approx 6$ to 9, similar to the expected m_K for WR118, and featureless K-band spectra, also similar to WR118. As a test of our hypothesis, we will obtain J-band spectra of the sources so that the classical emission-line spectra might be seen (Figer 1996).

8. The Quintuplet and the Central Cluster

The Galactic Center emission-line stars have been regarded as "exotic" for their spectral characteristics in the K-band (Allen, Hyland & Hillier 1990; Krabbe et al. 1991; Libonate et al. 1995; Krabbe et al. 1995; Blum, DePoy & Sellgren 1995; Tamblyn et al. 1996). While other stars around the Galaxy and in the Large Magellanic cloud are similar, the ensemble of stars in the center, as a collection, is peculiar, but not unique. The Quintuplet cluster contains many similar stars, as can be seen in Figure 2 and FMM95.

The WC9 stars recently found in the center (Blum, Sellgren & DePoy 1995; Krabbe et al. 1995) have counterparts in the Quintuplet (here and FMM95-2). IRS16NE, which has LBV-like spectral characteristics (Tamblyn et al. 1996), is similar to the LBVc in the Quintuplet (FMM95-3). Some of the Ofpe/WN9 types in the center (IRS16 components) are similar to q8, q10 (Geballe et al. 1994; Figer 1995), and FMM95-1. IRS33E has a spectrum similar to the "OBI" stars in the Quintuplet (Figer 1995; Najarro 1995; Genzel et al. 1996). IRS7, the supergiant in the central cluster, is similar to q7 in the Quintuplet (Moneti, Glass & Moorwood 1994). Finally, the QPMs share similar spectral energy distributions (Okuda et al. 1990; Becklin et al. 1978) and K-band spectra (Figer 1995) with the very red sources in the Galactic Center, c.f. IRS8; both groups of stars may be DWCLs. This cursory comparison will be expanded in Figer, McLean & Morris (1996).

References

Abbott, D. C. & Conti, P. S. 1987, ARA&A, 25, 113

Allen, D. A., Hyland, A. R. & Hillier, D. J. 1990, MNRAS, 244, 706

Becklin, E. E., Matthews, K., Neugebauer, G & Willner, S. P. 1978, ApJ, 219, 121

Blum, R. D., Sellgren, K. & DePoy, D. L. 1995, ApJ, 440, L17

Blum, R. D., DePoy, D. L. & Sellgren, K. 1995, ApJ, 441, 603

Cohen, M. 1995, ApJS, 100, 413

Cotera, A. S., Erickson, E. F., Colgan, S. W. J., Simpson, J. P., Allen, D. A., Burton, M. G. 1996, preprint

Figer, D. F., McLean, I. S. & Morris, M. 1995, ApJ, 447, L29

Figer, D. F. 1995, PhD Thesis, University of California, Los Angeles
Figer, D. F., McLean, I. S. & Morris, M. 1996, in preparation
Figer, D. F. 1996, in preparation
Geballe, T. R., Genzel, R., Krabbe, A., Krenz, T. & Lutz, D. 1994, in Infrared Astronomy with Arrays, I. S. McLean, Dordrecht: Kluwer, 73
Genzel, R., Thatte, N., Krabbe, A., Eckart, A., Kroker, H. & Tacconi-Garman 1996, ApJ, submitted
Glass, I. S., Moneti, A. & Moorwood, A. F. M. 1990, MNRAS, 242, 55P
Hanson, M. M., Conti, P. S. & Rieke 1996, preprint
Harris, A. I., Krenz, T., Genzel, R., Krabbe, A., Lutz, D., Poglitsch, A., Townes, C. H. & Geballe, T. R. 1994, in The Nuclei of Normal Galaxies: Lessons from the Galactic Center, R. Genzel & A. I. Harris, Dordrecht: Kluwer, 223
Krabbe, A., Genzel, R., Drapatz, S. & Rotaciuc, V. 1991, ApJ, 382, L19
Krabbe, A., et al. 1995, ApJ, 447, L95
McGinn, M. T., Sellgren, K., Becklin, E. E. & Hall, D. N. B. 1989, in IAU Symp. 136, The Center of the Galaxy, M. Morris, Dordrecht: Kluwer, 501
McLean, I. S., et al. 1993, in Infrared Detectors and Instrumentation, A. Fowler, Bellingham: SPIE , 513
McLean, I. S., et al. 1994, in Instrumentation in Astronomy VIII, D. Crawford, Bellingham: SPIE, 457
Meynet, G. 1996, preprint
Moneti, A., Glass, I. S. & Moorwood, A. F. M. 1994, MNRAS, 268, 194
Nagata, T., Woodward, C. E., Shure, M., Pipher, J. L. & Okuda, H. 1990, ApJ, 351, 83
Najarro, F. 1995, PhD Thesis, Ludwig-Maximilian University
Najarro, F., Hillier, D. J., Kudritzki, R. P., Krabbe, A., Genzel, R., Lutz, D., Drapatz, S. & Geballe, T. R. 1994, A&A, 285, 573
Okuda, H, Shibai, H., Nakagawa, T., Matsuhara, H., Kobayashi, Y., Kaifu, N., Nagata, T., Gatley, I. & Geballe, T. R. 1990, ApJ, 351, 89
Panagia, N. 1973, AJ, 78, 929
Schmutz, W. 1996, private communication
Schmutz, W., Leitherer, C. & Gruenwald, R. 1992, PASP, 104, 1164
Tamblyn, P. & Rieke, G. H. 1993, ApJ, 414, 573
Tamblyn, P., Rieke, G. H., Hanson, M. M., Close, L. M., McCarthy, D. W. & Rieke, M. J. 1996, preprint
Timmermann, R., Genzel, R., Poglitsch, A., Lutz, D., Madden, S. C., Nikola, T., Geis, N. & Townes, C. H. 1996, preprint
Vacca, W. D., Garmany, C. D. & Shull, J. M. 1996, preprint
Wainscoat, R. J. & Cowie, L. L. 1992, AJ, 103, 332
Williams, P. M., van der Hucht, K. A. & The, P. S. 1987, A&A, 182, 91
Yusef-Zadeh, F., Morris, M. & van Gorkom, J. H. 1989, in IAU Symp. 136, The Center of the Galaxy, M. Morris, Dordrecht: Kluwer, 275

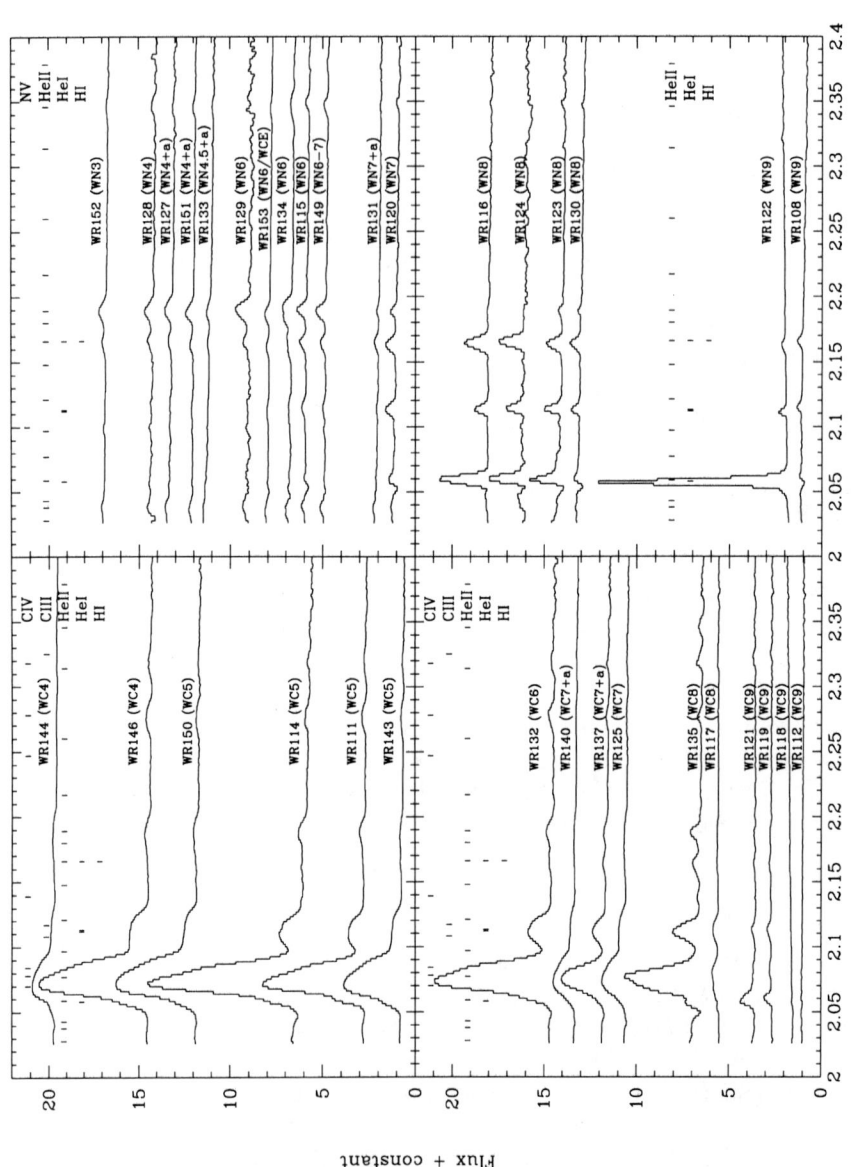

Figure 1. K-band spectra of WR stars in the Galaxy.

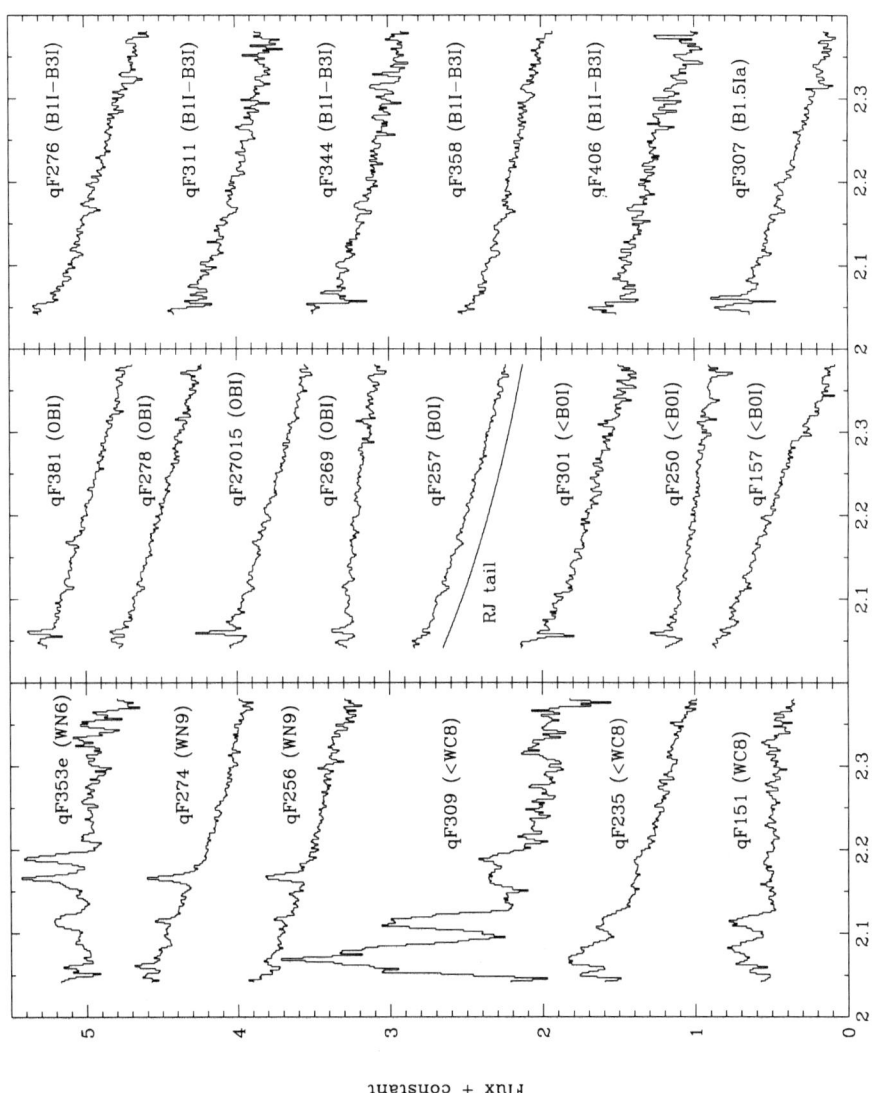

Figure 2. K-band spectra of target stars in the Quintuplet.

Table 1: Massive Stars in the Quintuplet

Nag90	MGM94	qF ID#	RA (1950) h	m	s	DEC (1950) deg	amin	asec	K'	H	K	(H-K)	Sp. Type	where classified	N(Ly c)	notes
		76	17	43	5.1	-28	49	11.6	11.8	13.5	11.4	2.1	WC9	FMM95-2	5.0E+48	BSD95
	seren	134	17	43	4.8	-28	48	56.9	7.5	9.1	7.1	1.9	LBVC	FMM95-3	3.4E+48	P73
		151	17	43	4.4	-28	48	53.8	11.0	13.3	10.4	2.9	WC8	this work	5.0E+48	BSD95
		157	17	43	3.5	-28	48	52.0	10.7	11.8	10.4	1.4	<B0I	this work	3.4E+48	P73
		178	17	43	9.3	-28	48	46.2	12.2	13.7	11.9	1.8	nebular?	this work		
	q7	192	17	43	6.2	-28	48	43.6	conf.	conf.			MIa	MGM94		
		197	17	43	4.9	-28	48	42.9	13.8	14.2	13.7	0.6	nebular?	this work		
GCS4	q3	211	17	43	5.5	-28	48	39.0	8.3	10.7	7.7	3.0	DWCL?	this work	5.0E+48	BSD95
GCS3-2	q2	231	17	43	4.3	-28	48	34.0	sat	9.8			DWCL?	this work	5.0E+48	BSD95
qG	q11a	235	17	43	4.8	-28	48	33.5	9.8	11.2	9.4	1.8	<WC8	this work	7.5E+48	S96
	q8	240	17	43	5.5	-28	48	31.2	9.7	10.9	9.4	1.4	WN9/Ofpe	Har94	1.0E+48	Naj94
qF	q10	241	17	43	4.7	-28	48	30.2	9.1	10.6	8.7	1.9	WN9/Ofpe	Har94	1.0E+48	Naj94
		242	17	43	4.1	-28	48	30.0	10.9	12.6	10.4	2.2	low S/N	this work		
GCS3-4	q1	243	17	43	3.7	-28	48	29.9	sat	11.1			DWCL?	this work	5.0E+48	BSD95
qB	q6	250	17	43	5.0	-28	48	27.9	9.8	11.7	9.4	2.3	<B0I	this work	3.4E+48	P73
GCS3-1	q4	251	17	43	4.4	-28	48	27.6	sat	10.5			DWCL?	this work		
		252	17	43	4.1	-28	48	27.1	10.4	13.1	9.7	3.3	low S/N	this work		
		256	17	43	6.1	-28	48	25.2	10.7	11.9	10.4	1.5	WN9	this work	1.0E+48	Naj94
qD	q13	257	17	43	4.8	-28	48	25.5	conf.	conf.			B0I	this work	3.4E+48	P73
GCS3-3	q9	258	17	43	3.9	-28	48	24.7	9.8	13.0	9.0	4.0	DWCL?	this work	5.0E+48	BSD95
qA		269	17	43	5.1	-28	48	23.1	11.1	12.8	10.7	2.1	OBI	this work	3.4E+48	VGS95
qC	q15	270	17	43	4.7	-28	48	21.9	conf.	conf.			OBI	this work	3.4E+48	VGS95
qC	q5	270	17	43	4.7	-28	48	21.3	conf.	conf.			late	this work		
		274	17	43	7.1	-28	48	22.4	10.7	11.9	10.4	1.5	WN9	this work	1.0E+48	Naj94
		276	17	43	3.0	-28	48	22.3	11.1	12.5	10.8	1.7	B1I-B3I	this work	1.5E+46	P73
qE	q12	278	17	43	4.7	-28	48	27.5	9.6	10.6	9.3	1.3	OBI	this work	3.4E+48	VGS95
		301	17	43	5.6	-28	48	15.0	11.0	12.3	10.6	1.7	<B0I	this work	3.4E+48	P73
		307	17	43	5.1	-28	48	13.5	9.7	10.8	9.4	1.4	B1.5Ia	this work	2.8E+46	P73
		309	17	43	7.1	-28	48	12.1	12.0	13.5	11.6	1.9	<WC8	this work	7.5E+48	S96
		311	17	43	3.3	-28	48	12.5	11.5	12.8	11.1	1.7	B1I-B3I	this work	1.5E+46	P73
		320	17	43	3.7	-28	48	9.7	10.9	12.4	10.5	1.9	WN9	FMM95-1	1.0E+48	Naj94
		344	17	43	6.3	-28	48	2.6	11.7	13.2	11.3	1.8	B1I-B3I	this work	1.5E+46	P73
		353E	17	43	0.8	-28	47	58.3	12	13.1	11.5	1.6	WN6	this work	1.0E+48	Naj94
		358	17	43	6.2	-28	47	58.2	10.7	12.3	10.3	2.0	B1I-B3I	this work	1.5E+46	P73
		381	17	43	3.1	-28	47	52.2	10.3	11.5	9.9	1.6	OBI	this work	3.4E+48	VGS95
		406	17	43	3.5	-28	47	43.5	11.4	12.9	11.0	1.9	B1I-B3I	this work	1.5E+46	P73
													total ionizing flux		8.2E+49	

Notes:
- BSD95 Blum, Sellgren & DePoy (1995)
- Har94 Harris et al. (1994)
- MGM94 Moneti, Glass & Moorwood (1994)
- Nag90 Nagata et al. (1990)
- Naj94 Najarro et al. (1994)
- P73 Panagia (1973)
- S96 Schmutz (1996), also see Schmutz et al. (1992)
- VGS95 Vacca, Garmany & Shull (1996)
- sat. = saturated, see MGM94 or Nag90 for photometry
- conf. = confused, see MGM94 for photometry
- nebular = emission lines might be due to nearby ionized gas

qF134 and qF274 are referred to as "Pistol A" and "Pistol B" in Cotera et al. (1996)
qF134, qF211, and qF231 are referred to as Objects 25, 26, and 24 in Nagata et al. (1993)

The Chemical Composition near the Galactic Centre – A Study of Four Blue Supergiants

S. J. Smartt and P. L. Dufton

Department of Pure and Applied Physics, The Queen's University of Belfast, BT9 1NN, N. Ireland

D. J. Lennon

Max-Planck-Institute for Astrophysics, Karl-Schwarzschild-Str. 1, D-85740 Garching, Germany

Abstract. High resolution optical spectra of four blue supergiants which lie within 4.8 kpc of the Galactic centre are presented and model atmosphere analyses are used to quantify their photospheric metal abundances. Three of these stars have metal rich atmospheres and we conclude that they show a present day composition typical of the inner regions of the Galaxy, allowing a lower limit to be set on abundances within 1 kpc of the centre.

1. Introduction

Early OB-type stars have been extensively used as probes of the present day composition of the interstellar medium in the Galactic anti-centre direction, however due to extinction little work has been done on hot Population I objects towards the Galactic centre. Simpson et al. (1995), from far infra-red studies of H II regions, derived abundance gradients which suggest that the metallicity at the Galactic centre is between 0.6 – 0.8 dex greater than solar (from nitrogen, sulphur, and neon results). Shaver et al. (1983) analysed an extensive sample of well observed emission line nebulae and estimated abundance gradients in the region $6 < R_g < 14$ kpc and an extrapolation of their oxygen gradient (for example) predicts an enhancement of 0.6 dex in the central regions. We have begun a programme to observe A & B type supergiants and B-type main-sequence stars in regions of low extinction with subsequent model atmosphere analyses yielding reliable differential abundances.

2. Selection of targets and spectroscopic observations.

We selected 4 targets classified as early B-type supergiants from Sembach et al. (1983). Spectroscopic observations were taken on 4 different observing runs; three at the Anglo-Australian Observatory (3.9m AAT) and one at the South African Astronomical Observatory (1.9m) to provide high signal-to-noise spectra ($S/N > 150$) in the range 3900 – 4900Å at high resolution.

3. Method of analysis & classification of the spectra.

The accurate and realistic modelling of the physical processes occurring in the atmospheres of blue supergiants is notoriously difficult. We do not attempt to derive truly realistic physical models from which accurate *absolute* physical parameters and metallicities can be derived. Rather we shall quantify the differences in metal line strengths between our Galactic centre stars and relatively nearby MK spectral standards. Using the atlas of Lennon et al. (1992 & 1993) which presents spectra of Galactic blue supergiants from O9.5 – B8, we have classified our Galactic centre stars (GCS).

4. LTE model atmosphere techniques and differential abundances.

Details of the line-blanketed LTE model atmosphere techniques can be found in Smartt et al. (1996a). The effective temperatures and helium abundances were estimated using the He I/He II ionization equilibria, and the hydrogen Balmer lines allow the determination of logarithmic surface gravity. Microturbulence was included as an empirical fitting parameter to ensure that no systematic trends occurred in the relation abundance vs. line strength for a given element (the O II lines provide the best means to estimate this). Metal line strengths allowed a line by line differential abundance analysis to be carried out for each GCS with respect to the corresponding Galactic standard and the results are listed in Table 2 (n is the number of lines used and results are on a logarithmic scale).

Table 1. Atmospheric parameters and spectral types.

Star	Spectral Type	T_{eff}	$\log g$	[He]	$\xi(\mathrm{kms}^{-1})$
GALACTIC CENTRE STARS					
HD148422	B0.5 Ib	25 000	3.10	10.74	30
HD178487	B0.5 Ib	27 000	3.30	10.77	30
HD163522	B1 Ib	22 500	2.80	10.68	30
HD179407	B1 Ib	24 500	3.15	10.70	30
STANDARD GALACTIC COMPARISON STARS					
HD213087	B0.5 Ib	26 500	3.10	10.73	30
HD192422	B0.5 Ib	26 500	3.10	10.64	30
HD24398	B1 Ib	22 000	2.90	10.70	30

Table 2. LTE Differential metal abundances.

Species	n	HD148422	n	HD178487	n	HD179407	n	HD163522
O II	29	0.00 ±0.16	29	+0.06 ±0.17	12	+0.16 ±0.10	34	+0.27 ±0.21
Mg II	1	−0.10	1	+0.24	1	+0.51	1	+0.28
Al III	0	—	0	—	0	—	2	+0.07 ±0.23
Si III	6	−0.02 ±0.11	6	+0.05 ±0.09	6	+0.38 ±0.11	6	+0.34 ±0.14
Si IV	1	+0.25	1	+0.19	1	−0.15	1	+0.62
S III	3	0.00 ±0.10	3	+0.10 ±0.15	2	+0.27 ±0.03	2	+0.31 ±0.34
Fe III	1	+0.06	2	+0.21 ±0.22	0	—	1	+0.38

5. Non-LTE differential analysis of HD163522 and comparison with LTE results

As a further check on the validity of the above results we have undertaken a non-LTE differential analysis of the star HD163522, to show that our results are not seriously compromised by the neglect of these effects (model details can be found in Lennon et al. 1991). We have estimated the effective temperature, logarithmic surface gravity and helium abundance simultaneously as their derivation is interdependent, following similar procedures to Kudritzki (1980), with microturbulence estimated as in the LTE case. Detailed non-LTE multilevel line formation calculations were preformed for O II and N II (again see Lennon et al. 1991 for details) and the differential abundances are in Table 3. A comparison with the LTE results indicates that the differential abundances derived in section 4 are not seriously in error if we assume LTE in this analysis.

Table 3. NLTE Differential metal abundances for HD163522

Method	O II	N II
Non-LTE	+0.30 ±0.17	−0.02 ±0.16
LTE	+0.29 ±0.21	+0.06 ±0.07

6. Spatial positions and origins.

The absolute visual magnitudes for the GCS are taken from the spectral classification – absolute magnitude calibration of Walborn (1972) and along with the visual magnitudes and extinctions quoted in Sembach et al. (1993) allow a calculation of their distances ($R_\odot$). Their corresponding Galactic positions are listed in Table 4 (where R_g and z are Galactocentric distance and perpendicular distance below the Galactic plane respectively). The stars lie significantly below the plane of the Galaxy and it is likely that they were formed in, and subsequently ejected from the disk. Given a maximum lifetime of ~ 10 Myrs for stars of this mass (i.e 20–25 $M_\odot$ – see for example Schaller et al. 1992) and an estimate of each stars peculiar radial velocity with respect to a "local standard of rest" (i.e. a point which is rotating with a standard Galactic rotation law e.g. Kerr & Lynden-Bell 1986) we can estimate their expected regions of origin. The V_{pec} of HD148422 is negligible i.e. it appears to be moving with a relatively normal galactic rotational velocity and hence it is likely that this star was ejected from the plane relatively near the point of projection. If we assume that the other components of the peculiar velocities of HD178487 and HD163522 are of similar magnitude to the measured line of sight peculiar velocity, then a space velocity of $\sim 110\,{\rm kms^{-1}}$ is implied. We can thus define a circular locus in the disc from which these stars could feasibly have originated (calculated including the effect of the Galactic gravitational potential e.g. from House & Kilkenny 1980), and the radii of these loci are listed as r in Table 4. The V_{pec} of HD179407 is larger than the others, and we suggest an ejection velocity of $\geq 220\,{\rm kms^{-1}}$ leading to a value of 1.7 kpc for r.

Table 4. Distances and Galactic positions.

Star	m_v	$R_\odot$(kpc)	l°	b°	z(kpc)	R_g(kpc)	V_{pec}(kms^{-1})	r(kpc)
HD148422	8.60	5.2	329.9	−5.6	−0.5	4.8	+2	∼0
HD178487	8.66	4.8	25.8	−8.6	−0.7	4.6	−62	1.2
HD179407	9.41	7.6	24.0	−10.4	−1.4	3.5	−128	1.7
HD163522	8.46	5.5	349.6	−9.1	−0.9	3.2	+46	1.2

7. Conclusions: Abundances near the Galactic centre.

The metallicity of HD148422 appears very similar to B-type stars in the solar neighbourhood. HD178487 has a marginally higher metal content than normal, showing that abundances slightly higher than solar exist in the intermediate region between the Sun and the Galactic centre. The other two stars in the study show definitive evidence of higher than normal metallicities. HD163522 has metal abundances higher than normal by 0.27 – 0.38 dex with HD179407's abundances up by between 0.16 – 0.5 dex.

We can conclude that the differential abundances derived (particularly for HD163522) provide an accurate lower limit to those at the Galactic centre, however a larger sample of early-type stars is required to tie down the relation between Galactocentric distance and metallicity in the inner few kiloparsecs. Our results show that the metallicity at the centre of the Galaxy is greater than solar by at least approximately 0.3 dex (a factor of 2).

References

House, F., Kilkenny, D., 1980, A&A 81, 251
Kerr, F.J., Lynden-Bell, D., 1986, MNRAS 221, 1023
Kudritzki, R.P., 1980, A&A 85, 174
Lennon, D.J., Kudritzki, R.P., Becker, S.T., Butler K., Eber, F., Groth, H.G., Kunze, D., 1991, A&A 252, 498
Lennon, D.J., Dufton, P.L., Fitzsimmons A., 1992, A&AS 94, 569
Lennon, D.J., Dufton, P.L., Fitzsimmons A., 1993, A&AS 97, 559
Schaller G., Schaerer, D., Meynet, G., Maeder, A., 1992, A&AS 96, 269
Sembach, K.R., Danks, A.C., Savage, B.D., 1993, A&AS 100, 107
Shaver, P.A., McGee, R.X., Newton, M.P., Danks A.C., Pottasch S.R., 1983, MNRAS 204, 53
Simpson, J.P., Colgan, S.W.J., Rubin, R.H., Erickson, E.F., Hass, M.R., 1995, ApJ 444, 721
Smartt, S.J., Dufton P.L., Rolleston W.R.J., 1996a, A&A *in press*
Walborn, N.R., 1972, AJ 77, 312

Part 2. The Stellar Population
Section C. Evolved Stars and Stellar Remnants

Really Cool Stars at the Galactic Center

R. D. Blum[1]

JILA, University of Colorado, Campus Box 440, Boulder CO, 80309, USA

K. Sellgren & D. L. DePoy

Astronomy Dept., The Ohio State University, 174 W. 18th Ave, Columbus, OH, 43210

Abstract. Absorption strengths for strong atomic and molecular features have been measured from K–band spectra of Galactic center (GC) M stars. We classify the GC stars as supergiants, long period variables (LPVs), and asymptotic giant branch (AGB) stars. Estimates of initial masses and ages for the GC stars suggest multiple epochs of star formation have occurred in the GC over the last 7–100 Myr.

1. Introduction

Blum et al. (1996) and Sellgren et al. (1996, these proceedings) presented near infrared photometry for the Galactic center (GC) stellar population (within the central $\sim$ 4–5 pc). Analysis of the $JHKL$ photometry showed an excess component of bright stars in the stellar population compared to the population in the nearby bulge field known as Baade's window (BW, $l, b = 1°, -4°$).

This excess of bright stars has been known for some time (e.g. Haller & Rieke 1989) and is generally attributed to younger, more massive stars than exist in BW, resulting from more recent star formation. Substantial evidence for very recent star formation ($\lesssim$8 Myr) has come from studies of hot emission–line stars in the GC (Forrest et al. 1987; Allen et al. 1990; Krabbe et al. 1991; Libonate et al. 1995; Blum et al. 1995a,b; Krabbe et al. 1995; Figer 1995; Tamblyn et al. 1996).

But, as pointed out by Blum et al. (1996), based on published spectra and the spectra we present here, the emission–line stars are not the most conspicuous stars in the GC bright component. The brightest GC stars are largely stars identified as being cool M stars. An important distinction between the cool and hot stars is that the luminous cool stars can trace the most recent epochs of star formation (red supergiants) as well as older ones (M giants and intermediate age asymptotic giant branch stars), while the hot stars trace only the former.

In this paper, we extend the work begun by Lebofsky et al. (1982, hereafter LRT) and Sellgren et al. (1987, hereafter S87) which sought to identify the

---
[1] Hubble Fellow

most luminous M stars in the GC as massive red supergiants or less massive giants and so trace star formation there. Our new spectroscopy (this paper) and photometry (Blum et al. 1996) are of higher angular resolution than these earlier studies. The spectra presented here represent GC stars in the brightest five magnitudes of the observed K−band luminosity function (Blum et al. 1996).

2. Observations and Data Reduction

The GC observations were obtained on the nights of 11 − 13 July 1993 on the 4−m telescope at the Cerro Tololo Inter−American Observatory (CTIO) using the Ohio State Infrared Imager and Spectrometer (OSIRIS). Observations and data reduction of the OSIRIS GC K−band spectra taken in July 1993 have been discussed in detail by Blum et al. (1995b). Briefly, these R (= $\lambda/\Delta\lambda$) ∼ 570 (19.4 Å pix^{-1}) spectra were extracted from 0.4″ pix^{-1} long−slit images (slit oriented E–W) of the central 102″ × 10″ of the Galaxy (slit width ∼ 1.2″). We also obtained long−slit images centered on several additional stars up to ∼ 30″ from the center. Similar OSIRIS spectra of a number of disk M giants were obtained in 1994 June.

In addition to the OSIRIS spectra, we make use of GC spectra from S87 and Levine et al. (1995); disk M giants and M supergiants from Kleinmann & Hall (1986, hereafter KH) and Terndrup et al. (1991); bulge M giants from Terndrup et al. (1991); and disk Miras (long period variables, or LPVs) from Johnson & Méndez (1970) and Wallace & Hinkle (1996). Where appropriate, these additional spectra have been re−binned to the OSIRIS spectral sampling.

3. Results

Figure 1 shows a comparison of measured CO vs. H_2O for the GC stars which have A_K determined from two or more infrared colors (Blum et al. 1996) and the comparison stars. The bulge giants, disk giants, and LPVs show a correlation between CO and H_2O. A small sample of CO and H_2O data for disk supergiants is also shown in Figure 1. These data show a different relation between CO and H_2O than the giants and LPVs. CO absorption strength increases with decreasing T_{eff}, decreasing gravity, increasing [C/H], and increasing microturbulence (Baldwin et al. 1973; McWilliam & Lambert 1984). This last parameter increases with increasing luminosity (McWilliam & Lambert 1984; McWilliam & Rich 1994). A large increase in luminosity (accompanied by a decrease in gravity and increase in microturbulent velocity) leads to significantly higher CO absorption. H_2O absorption strength also increases with decreasing T_{eff}, but decreases with increasing luminosity (KH; Wallace & Hinkle 1996). It is this contrasting luminosity dependence in CO and H_2O that leads to the separation of giants and supergiants in Figure 1. Note that the H_2O measure is sensitive to the derived A_K. If the derived extinction for a GC star is too high (GC star photometry, A_K, taken from Blum et al. 1996), the H_2O strength in this diagram is underestimated.

The GC stars and the BW stars both appear to have stronger Na and Ca absorption strengths than the field M giants and M supergiants (Figure 2). KH identified Na and Ca as two strong atomic features in the K−band for a large

COOL STARS

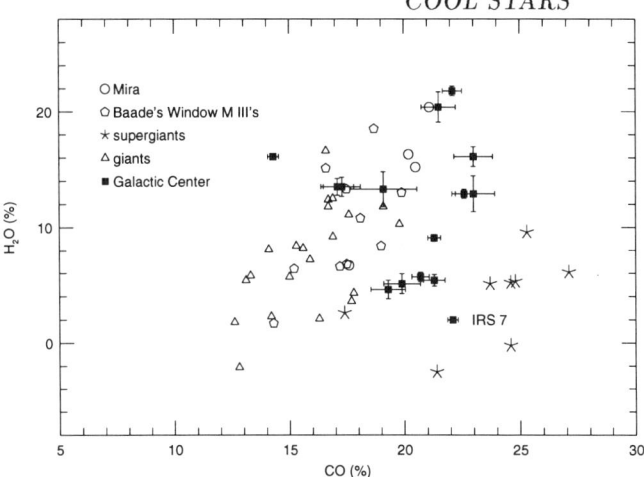

Figure 1. Measured CO vs. H_2O strength for Galactic center stars (*filled squares*), disk giants (*open triangles*), disk supergiants (*asterisks*), Baade's window M giants (*open pentagons*), and disk Miras (*open circles*). The error bars reflect measurement uncertainty and uncertainty in A_K. The large value of H_2O for some of the stars suggests they are long period variables (LPVs); see text.

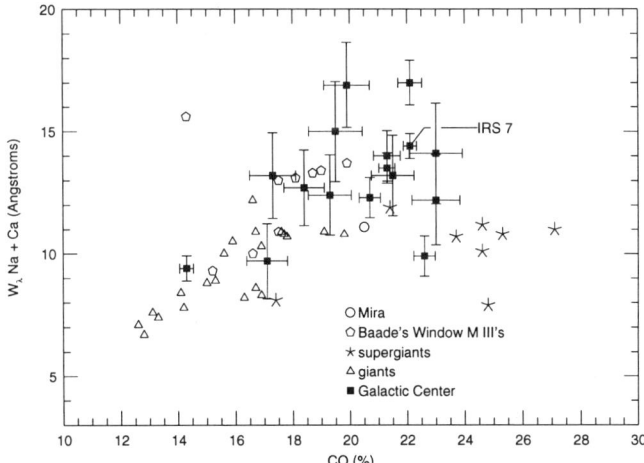

Figure 2. Sum of "Na" and "Ca" equivalent widths vs. measured CO strength. Symbols are the same as for Figure 1. GC stars identified as asymptotic giant branch (AGB) stars or long period variables (LPVs) have similar "Na" plus "Ca" absorption as bulge stars and are enhanced relative to disk stars. GC stars identified as supergiants (e.g. IRS 7) have significantly higher "Na" plus "Ca" than disk supergiants, and the difference is not easily understood in terms of luminosity and T_{eff}; see text.

range of dwarves, giants, and supergiants. Recent higher spectral resolution data indicates that the situation is more complex for M supergiants and late M giants. The high resolution spectra ($R \geq 45,000$) of M giants and M supergiants presented by Wallace & Hinkle (1996) show that Sc I contributes half or more to the total equivalent width of both our "Na" and "Ca" measures. Other significant contributors are Ti I, Si I, and V I, to "Na", and Ti I and Fe I to "Ca" (Wallace & Hinkle 1996). A high resolution ($R = 40,000$) spectrum suggests Sc is likely the largest single atomic contributor to "Na" in IRS 7 (J. Carr, private communication). In addition, both "atomic" absorption features may also be affected by CN absorption.

4. Discussion

4.1. The CMD, CO, and H_2O

Figure 3 presents the de–reddened color–magnitude diagram (CMD) for the GC cool stars and other well studied populations of supergiants, giants, and LPVs for comparison (Humphreys 1978; Elias et al. 1986; Hughes & Wood 1990; Wood et al. 1992; Glass et al. 1995; Zijlstra et al. 1996). The magnitudes and colors for the Milky Way M4–M7 IIIs were derived from M_V and MK spectral type (Thé et al. 1990) and $V - K$ and $H - K$ colors from Lee (1970) and Frogel et al. (1978).

Only one M star in our spectroscopic sample of the central ~ 5 pc must be a massive supergiant based on its luminosity. This is IRS 7. IRS 7 has M_{bol} = -9.0 mag ($BC_K = 2.6$ mag for M2 I, Elias et al. 1986). The remainder of the GC stars are within the theoretical upper limit on bolometric luminosity ($M_{bol} = -7.0$; Paczyński 1970) for lower mass asymptotic giant branch (AGB) stars if we apply the BC_K (2.9–3.2 mag) derived from known LPVs. However, inspection of Figures 1 and 3 suggests that three other GC stars (IRS 19, 22, and 28) may be early–type M supergiants based on the lack of strong H_2O absorption, presence of strong CO absorption, and M_K (-9.8, -9.0, and -8.9 mag, respectively).

LPVs are characterized by strong CO and H_2O absorption and photometric variability. Ten of the cool GC stars presented here have large or extreme H_2O absorption (Figure 1). In addition to having strong H_2O, like disk LPVs, IRS 9, 12N, and 24 have recently been identified as photometric variables at K (Haller 1992; Tamura et al. 1996) and J (IRS 9 and 12N, Blum et al. 1996). IRS 23 is just below Haller's (1992) three sigma cut–off for variability at K. The spectral characteristics of IRS 9 and 12N, coupled with the photometric variability, strongly support the suggestion by Tamura et al. (1996) that these two stars are LPVs like the Miras. A similar case now exists for IRS 24 and most likely IRS 23. IRS 24 has also been identified with an H_2O maser (Levine et al. 1995) and an OH maser (Sjouwerman & van Langevelde 1996). We believe these maser identifications strengthen the case for IRS 24 being an LPV, but see Levine et al. (1995) for a different interpretation. The remainder of the GC stars in Figure 1 have CO, H_2O, and M_K consistent with AGB stars and/or red giant branch stars.

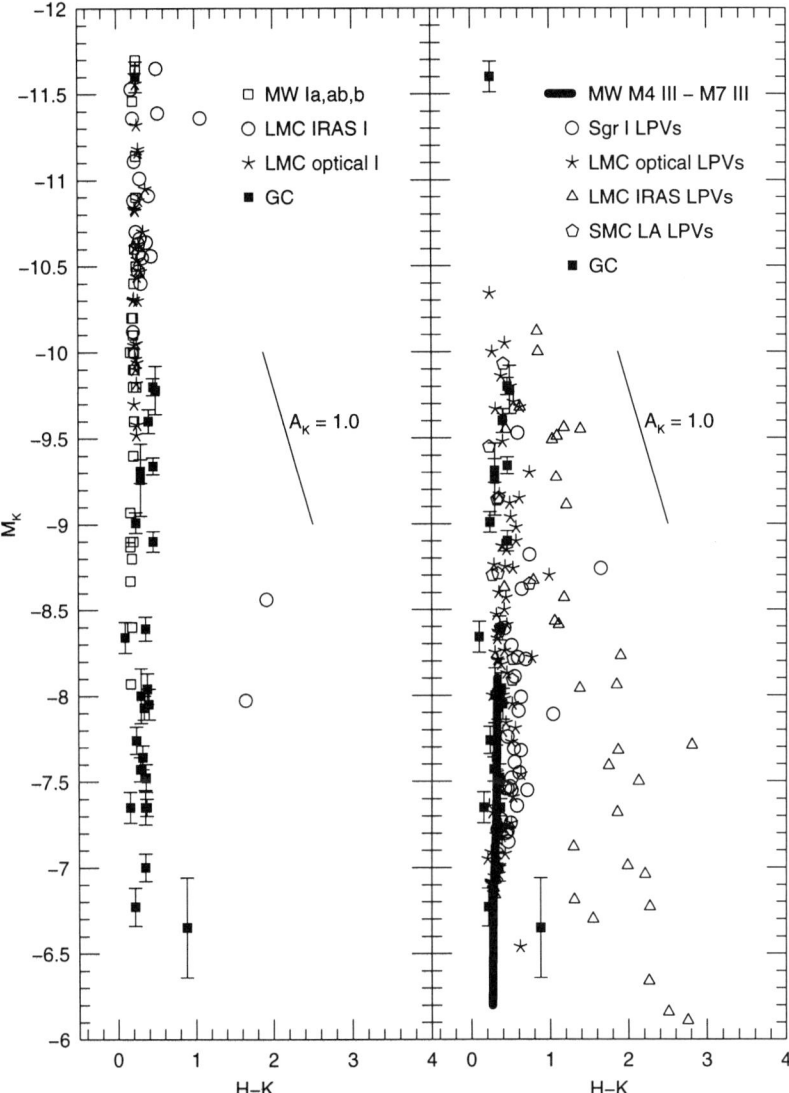

Figure 3. Color–magnitude diagram for the Galactic center cool stars and comparison stars. The Galactic center stars are plotted in each panel along with known supergiants (*left panel*) and LPV/giant stars (*right panel*). IRS 7 is the most luminous Galactic center star. For clarity, not all the fainter stars in the LPV data sets are plotted. A sun-to-Galactic center distance of 8 kpc (Reid 1993) was adopted for the Sgr I and Galactic center stars. The Large Magellanic Cloud (LMC) distance was taken as 46.8 kpc (Reid & Strugnell 1986). The red colors of some of the LPVs and the IRAS sources are due to circumstellar dust shells.

4.2. "Na" and "Ca"

Figure 2 suggests that the GC and BW stars have higher "Na" and "Ca" than the disk stars of similar CO strength. Figure 2 is strongly affected by T_{eff} and by luminosity. However, the luminosity effect is almost entirely in the CO strength which separates the disk and bulge stars from the disk supergiants. The absorption strength of the "Ca" and "Na" lines is observed to increase with decreasing T_{eff} (KH; Terndrup et al. 1991; Ramírez et al. 1996; all three analyses are affected by the other atomic contributors discussed above). The high resolution data of Wallace & Hinkle (1996) also show this trend for the various contributors to "Na" and "Ca" which were identified above. The correlation of "Na" and "Ca" with T_{eff} is clearly seen in Figure 2 for the disk giants and supergiants where CO, for a fixed luminosity class, now represents a measure of T_{eff}.

For the bulge stars, Terndrup et al. (1991) argued for enhanced *abundances* of Na and Ca by considering a range of temperatures for both disk and bulge stars and showing that, on average, the Na and Ca strengths for a given temperature were higher for bulge stars. This effect can be seen in Figure 2 for the bulge M giants. Consider stars with CO greater than 15 %, which corresponds to the T_{eff} range (actually $J - K \geq 1.00$ mag) for the Terndrup et al. (1991) bulge and disk stars analyzed here. The bulge stars have mean "Na" plus "Ca" which is higher than the disk stars, 12.48 ± 0.67 Å compared to 10.05 ± 0.31 Å.

If we consider the GC AGB/LPV stars in the same range of CO strength as above, we find that "Na" plus "Ca" (12.29 ± 0.82 Å) is also stronger than for the disk stars, and similar to the value for the bulge stars ("Na" and "Ca" are stronger individually too). We caution, however, that we are primarily comparing disk giants to GC AGB stars, as our disk sample only includes one LPV.

The GC supergiants show a clear enhancement of "Na" and "Ca" absorption over the disk supergiants, as previously noted for IRS 7 by S87. The mean value of "Na" plus "Ca" for the four GC stars is 14.4 ± 1.33 Å compared to a maximum of 11.9 Å for the disk supergiants (SAO 11969). A high-resolution spectrum of the "Na" feature in IRS 7, however, shows that this enhancement is dominated by strong Sc I absorption (Carr, private communication). Sc is an iron-peak element, and so should follow the iron abundance, but Carr et al. (1996) find [Fe/H] near solar in IRS 7. The strong "Na" and "Ca" absorption in the GC supergiants, therefore, is currently a puzzle.

4.3. Masses and Ages

The M1–2 I classification for IRS 7 (LRT; S87; and consistent with the CO strength derived here) implies $T_{\text{eff}} = 3550$ K (Johnson 1966). The detailed analysis of weaker CO lines from high resolution spectra (Carr et al. 1996) is consistent with this temperature. Adopting 3500 K and using $M_{\text{bol}} = -9.0$ mag ($BC_K = 2.6$ mag, Elias et al. 1986) suggests an initial mass of ~ 20–25 $M_\odot$ and age of 7–9 Myr for the $Z = 0.02$ evolutionary tracks of Schaller et al. (1992). The solar metallicity tracks were chosen based on the Carr et al. (1996) finding that [Fe/H] in IRS 7 is nearly solar. Figure 1 suggests IRS 19, 22, and 28 may have similar T_{eff}. Taking the values $M_{\text{bol}} = -7.2, -6.4$, and -6.4, respectively ($BC_K = 2.6$ mag) and $T_{\text{eff}} \sim 3500$ K gives masses of 12–15 $M_\odot$ for IRS 19 and

9–12 $M_\odot$ for IRS 22 and 28, again for Schaller et al. (1992) $Z = 0.02$ tracks. Corresponding ages are $\sim$ 12–18 Myr and 18–29 Myr.

The situation for the candidate LPVs (IRS 9, 12N, 23 and 24) is more speculative. Few models exist for evolution near the top of the AGB; such models depend on empirically determined mass–loss rates (Wood 1990). Comparison to the $Z = 0.016$ (largest Z for which models were computed) models of Vassiliadis & Wood (1993) will allow for a rough estimate. These models suggest that a star with initial mass of 5 $M_\odot$ will evolve on the AGB to $M_{bol} \sim -6.5$ mag while still being an optically visible Mira. Using the BC_K derived from the Sgr I LPVs (3.2 mag, Glass et al. 1995), the GC LPV candidates have $M_{bol} \gtrsim -6.5$ mag. These stars thus probably had progenitor masses $\lesssim 5$ $M_\odot$. The corresponding age from the Vassiliadis & Wood model is 120 Myr, similar to the 107 Myr age of a 5 $M_\odot$ star from the Schaller et al. (1992) models. The difference in age for these luminous M stars suggests that there have been multiple, recent epochs of star formation in the GC.

Acknowledgments. We thank D. Terndrup, D. Levine, D. Figer, and L. Wallace for supplying us with spectra of cool stars.

References

Allen, D. A., Hyland, A. R., & Hillier, D. J., 1990, MNRAS, 244, 706
Baldwin, J. R., Frogel, J. A., & Persson, S. E. 1973, ApJ, 184, 427
Blum, R. D., Sellgren, K., & DePoy, D. L. 1995a, ApJ, 440, L17
Blum, R. D., DePoy, D. L., & Sellgren, K., 1995b, ApJ, 441, 603
Blum, R. D., Sellgren, K., & DePoy, D. L. 1996, ApJ, in press
Carr, J. S., Sellgren, K., & Balachandran, S. 1996, in *9th Cambridge Workshop on Cool Stars, Stellar Systems, and the Sun*, ed. R. Pallavicini, in press
Elias, J. H., Frogel, J. A., & Humphreys, R. M. 1985, ApJS, 57, 91
Figer, D. F. 1995, Ph.D. Dissertation, University of California at Los Angeles
Forrest, W. J., Shure, M. A., Pipher, J. L., & Woodward, C. E., 1987, in *The Galactic Center*, ed. Backer, D. C., AIP, New York, p. 153
Frogel, J. A., Persson, S. E, Aaronson, M., & Matthews, K. 1978 ApJ, 220, 75
Glass, I. S., Whitelock, P. A., Catchpole, R. M., & Feast, M. W. 1995, MNRAS, 273, 383
Johnson, H. 1966, ARA&A, 4, 193
Johnson, H. L. & Méndez, M. E. 1970, AJ, 75, 785
Haller, J. W. & Rieke, M. J. 1989, In *The Center of the Galaxy*, IAU Symp. 136, ed. M. Morris (Dordrecht: Kluwer), 487
Haller, J. 1992, Ph.D. Dissertation, University of Arizona, Tucson
Hughes, S. M. G. & Wood, P. R. 1990, AJ, 99, 784
Humphreys, R. M. 1978, ApJS, 38, 309
Jones, T. J., Hyland, A. R., Wood, P. R., & Gatley, I. 1983, ApJ, 273, 669
Kleinmann, S.G. & Hall, D.N.B. 1986, ApJS, 62, 501 (KH)
Krabbe, A., Genzel, R., Drapatz, S., & Rotaciuc, V., 1991, ApJ, 382, L19

Krabbe, A., et al. 1995, ApJ, 447, L95
Lebofsky, M. J., Rieke, G. H., & Tokunaga, A. T., 1982, ApJ, 263, 736 (LRT)
Lee, T. A. 1970, ApJ, 162, 217
Levine, D., Figer, D. F., Morris, M., & McLean, I. S. 1995, ApJ, 447, 101
Libonate, S., Pipher, J. L., Forrest, W. J., & Ashby, M. L. N. 1995, ApJ, 439, 202
McWilliam, A. & Lambert, D. 1984, PASP, 96, 882
McWilliam, A. & Rich, R. M. 1994, ApJS, 91, 749
Paczyński, B., 1970, Acta Astron., 20, 47
Ramírez, S. V., DePoy, D. L., Frogel, J. A., Sellgren, K., & Blum, R. D. 1996, in preparation
Reid, M. J. 1993, ARA&A, 31, 345
Reid, I. N. & Strugnell, P. R. 1986, MNRAS, 221, 887
Schaller, G., Schaerer, D., Meynet, G., & Maeder, A. 1992, A&AS, 96, 269
Sellgren, K., Hall, D.N.B., Kleinmann, S.G., & Scoville, N.Z. 1987, ApJ, 317, 881 (S87)
Sellgren, K., Blum, R. D., DePoy, D. L. 1996, these proceedings
Sjouwerman, L. O. & van Langevelde, H. J. 1996, ApJ, 461, L41
Tamblyn, P., Rieke, G. H., Hanson, M. M., Close, L. M., McCarthy, D. W., Jr., & Rieke, M. J. 1996, ApJ, 456, 206
Tamura, M., Werner, M. W., Becklin, E. E., & Phinney, E. S. 1996, ApJ, in press
Terndrup, D.M., Frogel, J.A., & Whitford, A.E. 1991, ApJ, 378, 742
Vassiliadis, E. & Wood, P. R. 1993, ApJ, 413, 641
Wallace, L. & Hinkle, K. 1996, ApJS, in press
Wood, P. R., 1990, in *From Miras to Planetary Nebulae: Which Path for Stellar Evolution?*, eds. M. O. Mennessier & A. Omont, Yvette Cedex: Editions Frontières, p. 67
Wood, P. R., Whiteoak, J. B., Hughes, S. M. G., Bessell, M. S., Gardner, F. F., & Hyland, A. R. 1992, ApJ, 397, 552
Zijlstra, A. A., Loup, C., Waters, L. B. F. M., Whitelock, P. A., van Loon, J. T., & Guglielmo, F., 1996, MNRAS, 279, 32

Interstellar Extinction and the Luminosity Function of Galactic Center Stars

K. Sellgren, R. D. Blum[1,2], and D. L. DePoy

Astronomy Dept., Ohio State University, Columbus, OH 43210 USA

Abstract. We present J, H, K, and L photometry for stars within $\sim 1'$ of the Galactic Center (GC). We use the observed $J - H$, $H - K$, and $K - L$ colors, and assumed intrinsic colors, to derive the interstellar extinction for ~ 1100 stars. Our mean value of the K extinction, A_K, is 3.3 mag. This is similar to previous results, but we find that A_K is highly variable, with some stars likely to be seen through $A_K > 6$ mag. The de-reddened K ($K_\circ$) luminosity function contains GC stars which are significantly brighter (> 1.5 mag at K) than the most luminous bulge stars in Baade's Window, confirming previous work done at lower spatial resolution. The *observed* flux of all GC stars with $K_\circ < 7.0$ mag is $\sim 25\%$ of the total K flux in the $2' \times 2'$ field. Our observations confirm the recent finding (Tamura et al. 1996) that several bright M stars in the GC are variable. Our photometry also establishes the M supergiant IRS 7 as variable.

1. Observations and Data Reduction

We obtained our primary set of Galactic Center (GC) images at J, H, and K in 1993 July, using the Ohio State Infrared Imager (OSIRIS) on the CTIO 4-m. Our images are $\sim 2' \times 2'$, with $0.39''$ pixels and $1.0''$ seeing. We also analyzed the GC images of DePoy & Sharp (1991), obtained in 1989 September at H and K and in 1990 April at J and L. We obtained additional GC images with OSIRIS on the CTIO 4-m in 1993 July, using two narrowband filters at 2.19 and 2.27 μm, and with OSIRIS on the Perkins 1.8-m, in 1993 May at H and K, and in 1995 April at J. All data were analyzed with DAOPHOT (Stetson 1987). Our observations and results are described in more detail in Blum, Sellgren, & DePoy (1996).

2. Extinction

We derived the reddening for individual GC stars from our measured $J - K$, $H - K$, and $K - L$ colors, by assuming intrinsic colors corresponding to late-type M giants (Frogel & Whitford 1987) and the Mathis (1990) interstellar extinction

---

[1] Hubble Fellow

[2] Current address: JILA, University of Colorado, Campus Box 440, Boulder, CO 80309, USA

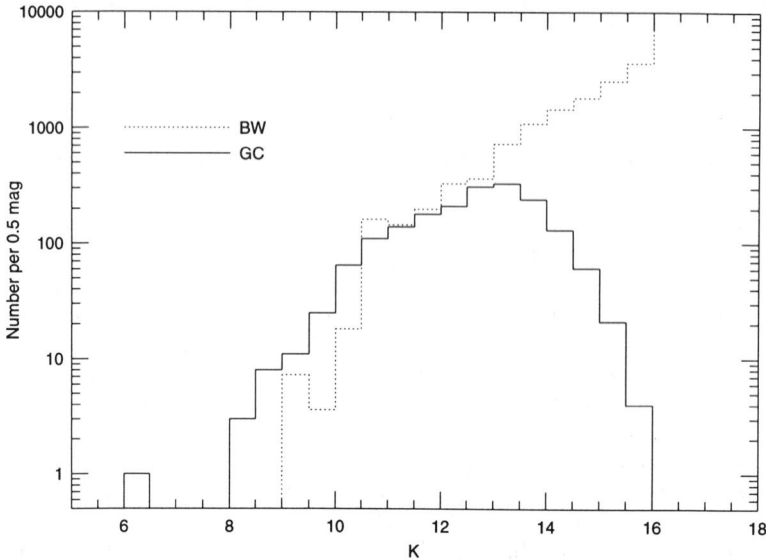

Figure 1. The observed K luminosity function (*solid histogram*) of the Galactic Center (GC) and the bulge K luminosity function (*dotted histogram*) in Baade's Window (BW). The BW relation has been reddened by $A_K = 3.5$, and normalized to the GC near $K = 11.5$.

law. The mean A_K for stars detected at J, H, and K is 2.8 ± 0.7 mag, and is 3.6 ± 0.8 mag for stars detected only at H and K. The mean value of A_K for all stars with one or more observed near-infrared colors is 3.3 ± 0.9 mag.

3. Variability

Our photometry shows that IRS 7 has varied in brightness at J, H, and K by 0.8, 0.5, and 0.3 mag, respectively. We also confirm the variability of IRS 9 and 12N found at K by Tamura et al. (1996) by finding that IRS 9 and 12N have varied at J by 1.0 and 1.7 mag, respectively, compared to previous photometry.

4. Artificial Star Experiments

To determine the completion limits at H and K, we combined the observed GC luminosity function for $K \leq 11.5$ (Figure 1) with a renormalized bulge luminosity function for $K > 11.5$ (Tiede et al. 1995) from Baade's Window (BW), reddened by $A_K = 3.5$ mag. We then constructed artificial images in which we

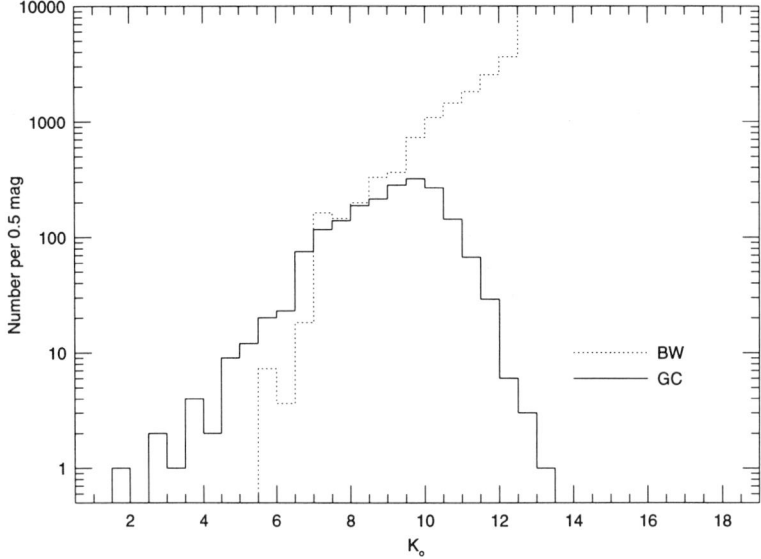

Figure 2. Dereddened K luminosity function (*solid histogram*) within $\sim 1'$ of the Galactic Center (GC), compared to the renormalized bulge luminosity function (*dashed line*) in Baade's Window (BW).

randomly sampled this luminosity function, with a radial surface density distribution proportional to $R^{-0.8}$ and a core radius of $4''$ (Becklin & Neugebauer 1968; Eckart et al. 1993, 1995), to distribute stars on the artificial frame. We find that our GC frames are complete to $K \leq 12.0$ and $H \leq 14.25$.

To assess the effect of image crowding on the bright end of the observed luminosity function, we also constructed artificial frames using only the reddened BW luminosity function, normalized to the observed flux of the GC frames. Photometry of these artificial frames predict $\sim 1\%$ of the brightest stars in our real GC images to have K too bright by 0.5–0.75 mag due to chance alignments of bright stars. DePoy et al. (1993) demonstrated that observations of a BW-like stellar population at high stellar surface densities led to blending of groups of stars which would then be falsely identified as single, more luminous stars. Our artificial star experiments show that this is not a significant problem in our GC images.

4.1. The K–band Luminosity Function

In Figure 2 we show the de–reddened K (K_o) luminosity function (KLF) for the GC, complete to $K_o = 8.5$ mag. We compare this to the KLF for BW (Frogel & Whitford 1987; Tiede et al. 1995), normalized by the observed K flux in the

GC. This normalization is consistent with assuming all the dynamically inferred mass within $\sim 1'$ of the GC (Genzel, Hollenbach, & Townes 1994) is due to stars with the mass-to-K-light ratio of BW, 1.2 $M_\odot/L_{\odot K}$.

The GC KLF shows an excess of bright stars at $K_o < 7$ mag, compared to the renormalized BW KLF (Figure 2). The BW KLF has stars as bright as $K_o = 5.5$ mag while the GC KLF extends to $K_o \approx 2.0$ mag. The renormalized BW KLF has 30 stars with $K_o \leq 7.0$; the GC KLF has 149 stars in this range. An excess of luminous stars in the GC, relative to the luminosity function of BW, has been observed previously (Lebofsky & Rieke 1987; Rieke 1987, 1993; Haller & Rieke 1989; Haller 1992). We suggest that this excess of bright GC stars is the result of recent star formation, as their brightness suggests they are more massive, and hence younger than the older BW stars. The bright GC stars apparently trace at least two star formation episodes, as they include not only emission-line stars and M supergiants with an age of 4-8 Myr, but also asymptotic giant branch stars with an age of 100 Myr (see Blum et al. 1996 for references).

References

Becklin, E. E. & Neugebauer, G. 1968, ApJ, 151, 145

Blum, R. D., Sellgren, K., & DePoy, D. L. 1996, ApJ, in press

DePoy, D. L. & Sharp, N. A., 1991, AJ, 101, 1324

DePoy, D. L., Terndrup, D. M., Frogel, J. A., Atwood, B., & Blum, R. 1993, AJ, 105, 2121

Eckart, A., Genzel, R., Hofmann, B., Sams, B. J., & Tacconi-Garman, L. E., 1993, ApJL, 407, L77

Eckart, A., Genzel, R., Hofmann, B., Sams, B. J., & Tacconi-Garman, L. E., 1995, ApJ, 445, L23

Frogel, J. A. & Whitford, A.E. 1987, ApJ, 320, 199

Genzel, R., Hollenbach, D. & Townes, C. H. 1994, *Reports on Progress in Physics*, 57, 417

Haller, J. 1992, Ph.D. Dissertation, University of Arizona, Tucson

Haller, J. W. & Rieke, M. J. 1989, In *The Center of the Galaxy*, IAU Symp. 136, ed. M. Morris (Dordrecht: Kluwer), 487

Lebofsky, M. J., & Rieke, G. H. 1987, in *The Galactic Center*, ed. D. C. Backer (AIP: New York), p. 79

Mathis, J.S. 1990, ARA&A, 28, 37

Rieke, M. J. 1987, in *Nearly Normal Galaxies from the Planck Time to the Present*, ed. S. M. Faber (Springer-Verlag: New York), p. 90

Rieke, M. J. 1993, in *Back to the Galaxy*, eds. S. Holt & F. Verter (AIP Press: New York), p. 37

Stetson, P. B. 1987, PASP, 99, 191

Tamura, M., Werner, M. W., Becklin, E. E., & Phinney, E. S. 1996, ApJ, in press

Tiede, G. P., Frogel, J. A., & Terndrup, D. M. 1995, AJ, 110, 2788

The Nature of OH/IR Stars in the Galactic Centre Region

Joris Blommaert,

Sterrewacht Leiden, P.O. Box 9513, 2300 RA Leiden, The Netherlands and ISO Science Operations Centre, Astrophysics Division of ESA, Villafranca, Spain

Wil van der Veen,

Columbia University, Dept. Of Astronomy, 538 West 120th Str., New York, NY 10027, USA

Harm Habing,

Sterrewacht Leiden, P.O. Box 9513, 2300 RA Leiden, The Netherlands

Huib-Jan van Langevelde,

JIVE, P.O. Box 2, 7990 AA Dwingeloo, the Netherlands

and Loránt Sjouwerman

Sterrewacht Leiden, P.O. Box 9513, 2300 RA Leiden, The Netherlands and Onsala Space Observatory, 439 92 Onsala, Sweden

Abstract. We report on infrared observations of stars in a field of 30' near the galactic centre. All these objects were previously detected as OH (1612 MHz) maser sources. The resulting infrared colours and luminosities are compared with results for other samples of OH/IR stars in the Galaxy. We find a large range of luminosities and thus ages. The question of the existence of two distinct populations of OH/IR stars near the galactic centre is addressed. The dust-to-gas mass loss ratio, which depends on metallicity, is on average twice as high for the group with high circumstellar shell expansion velocities than for the group of stars with lower expansion velocities.

1. Introduction

As low and intermediate mass stars approach the end of their lives they go through a relatively short phase lasting about $10^5 - 10^6$ yr which is known as the Asymptotic Giant Branch (AGB) phase. Miras, carbon stars and OH/IR stars are all examples of AGB stars. This phase is characterized by long-period pulsations (P= 200 - 2000 days) with bolometric amplitudes up to 1 mag and high mass loss rates ($> 10^{-7} M_\odot/yr$). The mass loss also dominates the appearance of the star. The circumstellar shell totally obscures the underlying red giant star, and the shell itself is detectable as a strong infrared source at wavelengths around 10 µm, where the stellar luminosity is reemitted by the cool dust shell. In the shell several types of masers may be present depending on the physical conditions in the shell. Most often the OH maser at 1612 MHz is the strongest of these masers. The fact that these giants are very bright ($L_* \gtrsim 5000 L_\odot$) and are observable at infrared and radio wavelengths makes them interesting objects for studying the galactic centre which is very obscured at the visible wavelengths.

OH/IR stars span a broad range of ages; the age difference between low mass (1 $M_\odot$, 10^{10} yrs) and higher mass (6 $M_\odot$, a few 10^8 yrs) OH/IR stars is more than an order of magnitude. It was shown (Baud et al. 1981) that the expansion velocity of the circumstellar shell is usually larger for OH/IR stars with smaller velocity dispersions. Because the final outflow velocity is determined by radiation pressure on dust in the circumstellar shell(Schutte and Tielens, 1989 and Habing et al. 1994), it is expected that both the total luminosity and the metallicity of the star play a role in this.

In a survey for OH 1612 MHz masers with the VLA, 134 OH/IR stars were detected (Lindqvist et al. 1992a). The first goal of this survey was to study stellar velocities in the galactic centre region (Lindqvist et al. 1992b). The kinematical behaviour of this sample gave an indication that also there different populations of OH/IR stars may be involved: stars with high expansion velocities seem to rotate in a flatter system than those with low expansion velocities. The aim of the present work[1] is to study the luminosities, masses and ages of the OH/IR stars in the galactic centre.

2. Observations

We selected 32 OH maser sources, all roughly within a 28' × 28' area, centered on SgrA. Twenty-four of these OH masers were monitored for OH variability (Van Langevelde et al. 1993) allowing us to determine mean bolometric magnitudes corrected for variability.

The sample was observed in 1990, 1991 and 1993 with the ESO 2.2 m and 3.6 m at La Silla (Chile) and the 3.8 m United Kingdom Infrared Telescope (UKIRT) at Mauna Kea (Hawaii) in the wavelength range of 1.2 to 20µm.

For all sources we had accurate VLA positions ($\sim 0.2''$) which facilitated their identification. Observations were done using IR cameras at the shortest wavelengths (up to 4µm) and traditional single channel photometers at longer wavelengths.

To convert the observed magnitudes at various wavelengths to mean bolometric magnitudes corrections were made for interstellar extinction and source variability.

---

[1] Based on observations collected at the European Southern Observatory, La Silla, Chile and UKIRT, Mauna Kea, Hawaii.

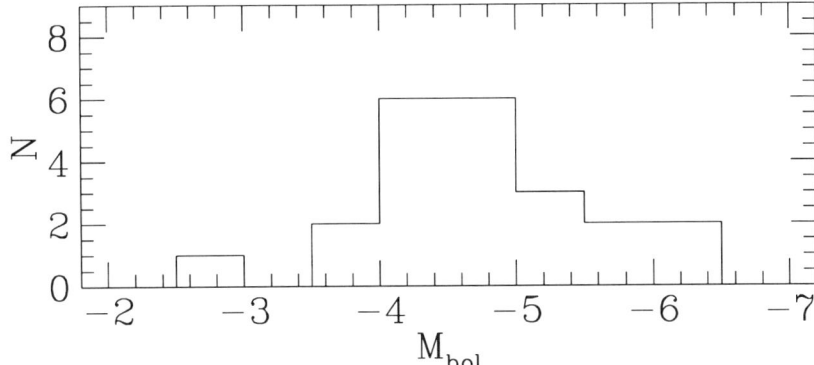

Figure 1. Distribution of bolometric magnitudes

3. Kinematic populations

Lindqvist et al. (1992b) showed that within the original sample of 134 stars statistically significant kinematic differences are seen when the OH/IR stars are divided into two groups based on the expansion velocities of their shells (V_{exp} respectively smaller and larger than 17 km/s). The high expansion velocities generally trace more luminous, and hence one assumes, more massive and thus younger OH/IR stars.

The distribution of the observed mean M_{bol} as shown in Fig. 1 indeed shows a quite large range ($-6 \lesssim M_{bol,mean} \lesssim -4$), indicating a range in masses ($1 \lesssim M_i \lesssim 4 M_\odot$) and thus ages.

In Fig. 2 two quantities are plotted against the expansion velocity. The comparison between circumstellar expansion velocity and radial velocity dispersion is taken from Lindqvist et al. (1992b). Each of the black dots represent an average of about 10 stars. Although the scatter is considerable there is a clear trend towards smaller velocity dispersions for OH/IR stars with larger expansion velocities.

The group of OH/IR stars with smaller expansion velocities has a narrow range of mean M_{bol} with an average value of -4.5. This is about the same value as was found by van der Veen and Habing (1988) for a sample of OH/IR stars in the bulge, suggesting that this subgroup may be related to that OH/IR star population. The dispersion that was found by Lindqvist et al ($\sigma = 82$ km/s) may be smaller than what is found for the bulge but their determination excluded the extreme velocity sources. By including them we find ($\sigma = 110$ km/s) in agreement with the value for the bulge. Also the expansion velocities here have the same range as what is found in the bulge.

The group of OH/IR stars with high expansion velocities, with a flatter kinematical distribution, however, have a broader distribution showing the same range in luminosities as was found for the total sample. It thus contains objects with the same luminosities as the smaller expansion group but almost half of it's group has luminosities higher than 10,000 $L_\odot$ ($M_{bol} < -5$). This "disk-like" population thus contains a large spread in ages.

It is not clear that this disk-like population is connected with the galactic exponential disk. On basis of modeling of the distribution of OH/IR stars de-

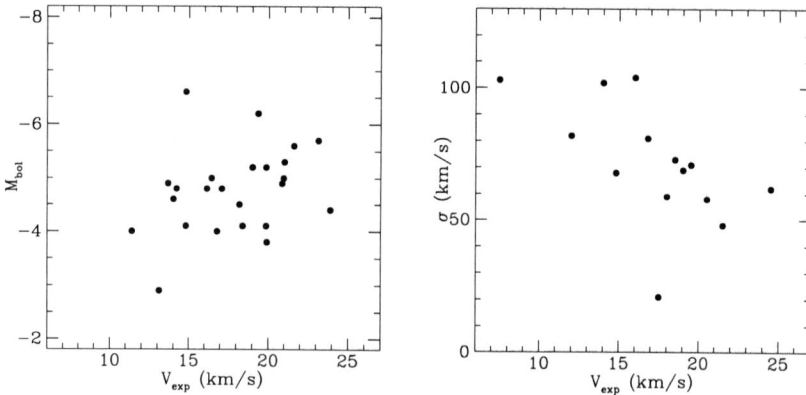

Figure 2. Distribution versus the circumstellar shell expansion velocity of a) the bolometric magnitudes (this study) and b) radial velocity dispersions (Lindqvist et al. (1992b))

tected in an IRAS based survey (te Hekkert et al. 1991) by Dejonghe (1992) one expects to find about 100 objects at the centre of the Galaxy. Blommaert et al. (1994) however, show in a VLA survey along the galactic plane that a simple continuation of the exponential disk as observed at longer longitude and higher latitude is not likely as the number of OH/IR stars drops rapidly for $l \leq 10°$.

3.1. Metallicities

The expansion velocity of the circumstellar shell is dependent on both luminosity and metallicity (Schutte and Tielens, 1989 and Habing et al. 1994). The group of OH/IR stars with high expansion velocities is really extreme in this sense. We find that their luminosities are certainly not extreme in comparison with what is found for OH/IR stars in the disk, nevertheless no other OH/IR stars in our Galaxy are known with expansion velocities as high as these in the galactic centre. It indicates that the metallicities of this group must be very high.

For both groups the luminosities are different, but not enough to explain the difference in expansion velocity. It would then follow that the metallicity is the main reason for the difference in the expansion velocities. This would be reflected in the ratio of the gas- ($\dot{M}_g$) and dust-mass loss ($\dot{M}_d$),

$$\mu = \dot{M}_g/\dot{M}_d \propto L_*^{0.5} v_{exp}^{-2}$$

which is believed to depend on metallicity (van der Veen 1989). The $\langle L_*^{0.5} v_{exp}^{-2} \rangle$ of the high expansion velocity group is only half as high as that of the low expansion velocity group (0.21 (±0.05) against 0.38 (±0.10)) indicating that the kinematically more "flattened" group is more metal-rich than the low expansion velocity group. This is also found by Spaenhauer et al. (1992) who claims that for K and M giants in Baade's window the dispersions in latitude are significantly smaller than in longitude but that this is strongest for the metal-rich ones.

References

Blommaert, J.A.D.L., Van Langevelde, H.J., Michiels, W. 1994, AA, 287, 479
Baud, B., Habing, H.J., Matthews, H.E., Winnberg, A. 1981, AA, 95, 156
Dejonghe, H. 1992, In: Winberger and Acker (eds) Proc. IAU Symp. 155, Planetary Nebulae, Kluwer Academic Publishers, p. 541
Habing, H.J., Tignon, J., Tielens, A.G.G.M., 1994, AA, 286, 523
Lindqvist, M., Winnberg A., Habing H.J., Matthews H.E. 1992a, AAS, 92, 43
Lindqvist, M., Habing H.J., Winnberg A.1992b, AA, 259, 118
te Lintel Hekkert, P., Casswell, J.L., Habing, H.J., Haynes, R.F., Norris, R.P. 1991, AAS, 90, 327
Schutte, W.A. Tielens, A.G.G.M., 1989, AJ, 288, 618
Spaenhauer, A., Jones, B.F., Whitford, A.E. 1992, AJ, 103, 297
van der Veen, W.E.C.J., 1989, AA, 210, 127
van der Veen, W.E.C.J., Habing, H.J. 1990, AA, 231, 404
Van Langevelde, H.J., Janssens, M., Goss, M., Habing, H.J. 1993, AAS, 101, 109

A new Sample of OH/IR Stars in the Galactic Center

Loránt Sjouwerman[1] and Anders Winnberg

Onsala Space Observatory, 439 92 Onsala, Sweden

Huib Jan van Langevelde

Joint Institute for VLBI in Europe, P.O. Box 2, 7990 AA Dwingeloo, the Netherlands

Harm Habing

Sterrewacht Leiden, P.O. Box 9513, 2300 RA Leiden, the Netherlands

Michael Lindqvist[1]

Onsala Space Observatory, 439 92 Onsala, Sweden

Abstract. We report on the preliminary results of an extensive search for OH/IR stars in the Galactic center region. The main goal is to use the larger sample of OH/IR stars to probe the gravitational potential in the Galactic center (see Lindqvist, Habing & Winnberg 1992b). The search in the 1612 MHz line of OH was performed by observations taken with the Australia Telescope Compact Array and by means of concatenating Very Large Array data sets used for a monitoring program (Van Langevelde et al. 1993). The volume surveyed is over 80 pc in diameter centered on Sgr A* and has a velocity coverage of -550 to $+600$ km s^{-1}. The 1σ noise is about 5 mJy, at least four times as deep as earlier surveys. So far we have found about 50 new OH/IR stars, almost as many as we expected. The newly found OH/IR stars seem to be similar in their observable properties to the OH/IR stars already known in the Galactic center. In our data we have found OH counterparts for two H$_2$O masers detected by Levine et al. (1995) and by Yusef-Zadeh & Mehringer (1995). These H$_2$O masers are not the argued clues for recent star formation: the objects are old OH/IR stars (Sjouwerman & Van Langevelde 1996).

1. Introduction

An OH/IR star is a far-evolved variable star, with a main-sequence mass of 1-8 M$_\odot$. It has reached the asymptotic giant branch (AGB) phase of its evolution. Due to heavy mass-loss, the star is obscured in the optical by a circumstellar envelope. These envelopes are environments in which masers can form, in par-

---

[1]Sterrewacht Leiden, P.O. Box 9513, 2300 RA Leiden, the Netherlands

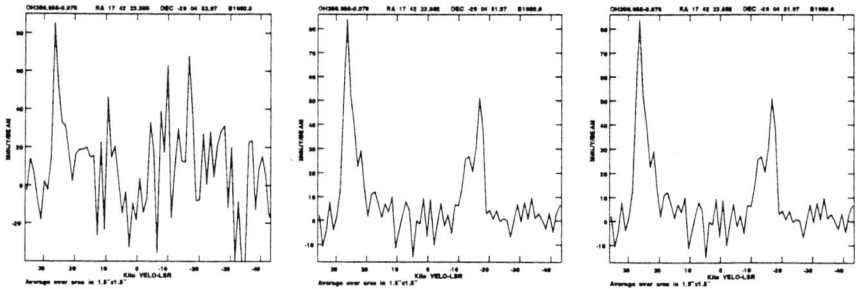

Figure 1. VLA concatenation for OH359.855-0.078 at different stages.

ticular the 1612 MHz satellite line of the hydroxyl (OH) molecule. The OH spectrum has a typical double peaked line profile, reflecting the red and blue shifted OH maser emission. The stellar radial (line-of-sight, LSR) velocity is taken as the mean of the red and blue shifted peaks. See Habing (1996) for a review.

Assuming that all normal stars will pass through this evolved evolution stage, OH/IR stars make very good objects to probe our Galaxy. The reason for this is that OH/IR stars are easy to detect. The Galaxy is mostly transparent at 1.6 GHz and the masers are relatively strong. Moreover, the stellar radial velocity can easily be determined from the spectrum. The population of OH/IR stars in the Galactic center (GC) is of special interest as it can provide clues to the gravitational potential in the very center of our Galaxy. Furthermore, the stars are located at the same distance, making studies of intrinsic physical properties of this stellar evolution stage possible.

From previous surveys already about 150 OH/IR stars are known within ~100 pc of Sgr A*, the radio source that is believed to be (close to) the dynamical center of our Galaxy (Baud et al. 1981, Habing et al. 1983, Lindqvist et al. 1992a, Van Langevelde et al. 1992). Unfortunately, this sample suffers from small number statistics when used as test particles in the Galactic potential. With an extended sample of OH/IR stars we hope a better analysis can be performed with smaller statistical errors, both on dynamics and physical properties (Lindqvist, Habing & Winnberg 1992b, Lindqvist et al. 1996). An increased number of OH/IR stars within 50 pc of Sgr A*, and in particular the role of high velocity stars, will further constrain future modeling of the GC: with or without a massive black hole.

2. Observations

In order to determine the phase-lag for the most luminous OH/IR stars close to the GC, Van Langevelde et al. (1993) monitored OH/IR stars in one Very Large Array (VLA) primary beam (which contains Sgr A*). In the period from February 1988 till January 1991, this field was observed at 19 different epochs for two hours each. Due to interference, the noise in a single map varies from 13 to 60 mJy, with an average of 20 mJy. By adding these data sets, that only differ in date, VLA configuration (resolution) and noise level, theoretically one would obtain a data set with an rms noise per channel of about 5 mJy. Although the VLA data sets are not centered at the GC (in sky coordinates and radial velocity), one should be able to use these data sets to find an additional number of 'weak' OH/IR stars close to the GC. We have concatenated 17 usable data sets and analyzed the maps. We have searched for emission stronger than 40 mJy, made spectra for these 5σ detections and listed them in a table (Sjouwerman et al. 1996).

To overcome the asymmetry problems, and additionally to look for high velocity OH/IR stars (over ± 200 km s^{-1}, which is the VLA velocity coverage) we used the Australia Telescope Compact Array (ATCA) in July 1994. The maps were searched for double peaks down to about 25 mJy (Sjouwerman et al. 1996). The total area searched is out to 40 pc from Sgr A* with LSR velocities between -550 and $+600$ km s^{-1}, completely overlapping the VLA observations. Typical noise levels (1σ) reach 5 mJy, four times less than the Lindqvist et al. (1992a) survey. Following Lindqvist et al. (1996), a reasonable estimate of the increase in the number of OH/IR stars, adopting a derived luminosity function, would yield about 60 new OH/IR stars within 30 pc from Sgr A*.

3. Preliminary results

Figure 1 concerns the OH source 359.855-0.078, a typical weak (ie. low flux density) OH masering OH/IR star in the direction of the GC. We use it to demonstrate and justify the concatenation process in order to find weak and yet unknown OH/IR stars in the VLA data sets.

On the left, a spectrum of an ordinary VLA monitor data set is given. This source is too weak to be monitored by Van Langevelde et al. (1993) as the signal hardly can be distinguished from the noise. This data set is as sensitive as the original OH/IR star survey of Lindqvist et al. (1992a) and one might only suspect that there is an OH/IR star to be found here. Second, an intermediate result of the concatenation project is shown. This spectrum is taken from the concatenation of six VLA B-array monitor data sets. From this spectrum it becomes clear that this star has a second peak at a velocity of -18 km s^{-1}. The right panel shows a spectrum of this source in the final data set. Apart from an increase in signal to noise, the spectrum looks more like a genuine OH/IR star spectrum than the former ones. This is also caused by averaging the variable flux density of the source over the 17 epochs. The rms noise is 8 mJy.

The ATCA data were reduced in a standard way. Searching the cube was done by creating a map containing the maximum intensity of all velocity channels at each pixel representing the sky. This map was searched for pixels with more

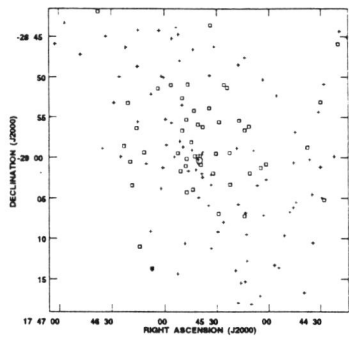
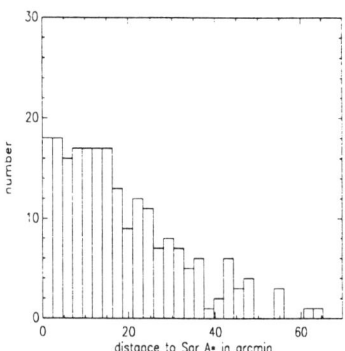

Figure 2. OH/IR stars in the Galactic center. Left: squares are new detections, crosses are previously known OH/IR stars and the circle in the middle shows the position of Sgr A*. Right: number of OH/IR stars versus distance to Sgr A* in arcminutes.

flux than a certain level and the pixel and its flux was stored. Later, we grouped adjacent pixels in 'islands' and made spectra at the maximum intensity of each island. Plausible double peaked spectra were regarded as detections of OH/IR stars. We listed about 50 detections, the squares in the left hand panel of figure 2. Crosses are previously known OII/IR stars; the circle in the middle shows the position of Sgr A*. Most of the newly found double peaks can be found both in the concatenated VLA data as well as in the ATCA observations. A number of new detections are corresponding to K-band variables found by Glass et al. (1996). Two previously unknown OH/IR stars have radial velocities over ±200 km s^{-1}.

As the space and velocity distributions and the expansion velocity distribution for the newly found double peaked OH masers compare very well with the known OH/IR stars, we expect to have found similar objects, though with a less luminous OH maser (in our direction). Adopting the same metallicity as for the previously known, strong OH masering OH/IR stars, we derive a similar bolometric luminosity distribution and hence main sequence masses for these stars. We found almost as many OH/IR stars as expected: about 50 instead of 60. This probably means that the cut-off at the low luminosity side of the luminosity distribution is caused by the sensitivity limit; we have not reached the low luminosity cut-off of the real OH maser luminosity distribution. The right hand panel of figure 2 shows the number of stars versus the distance to Sgr A*. The numbers have not been corrected for completeness yet.

Finally, we found OH/IR stars where Levine et al. (1995) as well as Yusef-Zadeh & Mehringer (1995) claim to have found H$_2$O masers associated with young stars or star forming regions (Sjouwerman & Van Langevelde 1996). The table focuses on detections in H$_2$O of Levine et al. (1995) and Yusef-Zadeh &

TABLE 1
H_2O AND OH DETECTIONS

survey	Right Ascension & Declination (1950)		(″)	Radial Velocity (km s^{-1} wrt. LSR)			(km s^{-1}) resolution	Line flux (mJy km s^{-1})
				low	mean	high		
the LFMM source: OH 359.956−0.050								
H_2O LFMM	17 42 32.00	−28 58 47.8	(±0.2)	...	45.3	55.8	(5.0)	...
H_2O Y-ZM	17 42 31.87	−28 58 46.8	...	...	44.7	...	(2.6)	603
OH VLA	17 42 31.95	−28 58 47.9	(±1.9)	34.3	48.5	62.7	(1.2)	610
OH ATCA	17 42 31.96	−28 58 50.0	(±3.3)	34.2	50.2	66.2	(1.5)	190

Mehringer (1995) combined with our OH counterparts for this source. Looking at the data in the table, we claim the maser emission is from one source only; from the double peaked OH maser spectra we conclude the source is an evolved intermediate mass AGB star with an age of about one Gyr.

References

Baud, B., Habing, H.J., Matthews, H.E., & Winnberg, A. 1981, A&A 95, 156
Glass, I.S., Matsumoto, S., Ono, T., & Sekiguchi, K. 1996, these proceedings
Habing, H.J. 1996, submitted to A&A
Habing, H.J., Olnon, F.M., Winnberg, A., Matthews, H.E., & Baud, B. 1983, A&A 128, 230
Levine, D.A., Figer, D.F., Morris, M., & McLean, I.S. 1995, ApJ 447, L101
Lindqvist, M., Winnberg, A., Habing, H.J., & Matthews, H.E. 1992a, A&AS 92, 43
Lindqvist, M., Habing, H.J., & Winnberg, A. 1992b, A&A 259, 118
Lindqvist, M., et al. 1996, A&A in prep.
Sjouwerman, L.O., et al. 1996, A&A in prep.
Sjouwerman, L.O., & Van Langevelde, H.J. 1996, ApJ 461, L41
Van Langevelde, H.J., Brown, A.G.A., Lindqvist, M., Habing, H.J., & De Zeeuw, P.T. 1992, A&A 261, L17
Van Langevelde, H.J., Janssens, A.M., Goss, W.M., Habing, H.J., & Winnberg, A. 1993, A&AS 101, 109
Yusef-Zadeh, F., & Mehringer, D. 1995, ApJ 452, L37

Spectroscopy of New Planetary Nebulae close to the Galactic Center

G.C. Van de Steene

European Southern Observatory, Casilla 19001, Santiago 19, Chile

G.H. Jacoby

National Optical Astronomy Observatories, P.O. Box 26732, Tucson, AZ 85726, U.S.A.

Abstract. We report on the spectroscopic observation and confirmation of 16 newly discovered planetary nebula found in the galactic bulge using on-band/off-band imaging at [S III]λ9532 of which 3 were also recently identified by Kohoutek (1994). An additional 80 new PN candidates were identified but not yet observed spectroscopically.

1. Introduction

Planetary Nebulae (PN) are bright emission line objects, observable at large distances throughout the Galaxy. As such they are useful probes of abundance gradients, the chemical enrichment history of the ISM, the effects of metallicity on stellar evolution, and kinematics. However, it has long been noticed that there is a lack of PN within 2 degrees of the galactic plane, especially in the direction of the bulge. Very large extinction towards the bulge accounts for this. This is unfortunate, because bulge PN are particulary important for the study of PN evolution. PN are concentrated towards the galactic center and it has been shown that about 90% of small PN within 10 degrees of the galactic center are actually close to it (Pottasch & Acker, 1989). As bulge PN can be assumed to be at the same well known distance, it is possible to also determine their distance dependent physical parameters, crucial for studying PN properties and evolution.

2. Imaging: Observations and Results

We imaged a 4 x 4 degree field centered on the galactic center in [S III]λ9532 and a continuum band (a Schott RG850 filter) during June - July 1994 and 1995 at KPNO with the 60 cm Schmidt telescope and a 2048 x 2048 pixels thick STIS CCD. The field of view is 65' x 65' (2"/pix) and the typical integration times were 40 min per field. The [S III]λ9532 filter is about 30 Å wide.

This survey has uncovered 96 new PN candidates in addition to the 34 previously known in this central bulge region (Acker et al. 1992, Kohoutek 1994). In the surveyed region we have retrieved all but one (Te 2022) PN listed in the Catalog of Galactic PN (Acker 1992). Te 2022 appears to be a strong

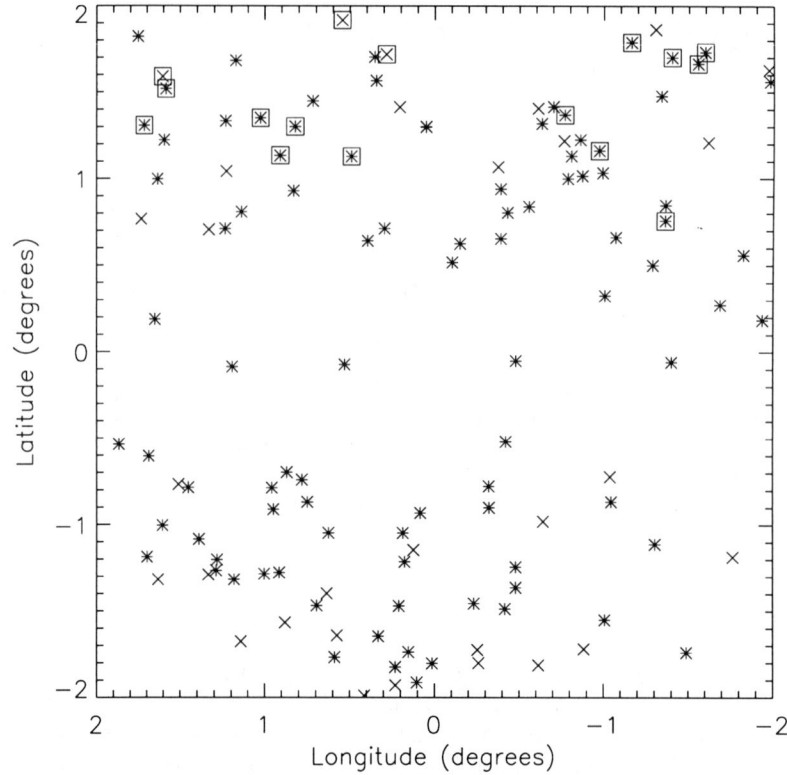

Figure 1. Galactic distribution of the 34 known PN (crosses) and 96 new PN candidates (stars). The stars framed in a box are the PN candidates confirmed spectroscopically.

continuum source and it doesn't show up as an emission line object. We plan to obtain a spectrum of this object in June to confirm its nature. In addition we found that 4 PN candidates had already been identified by Kohoutek (1994).

At the distance of the galactic center, PN would be faint or undetected by IRAS. Only 20% of the new PN candidates have a counterpart in the IRAS PSC catalog within $\sim 30''$. Of these only 50% have two or more flux values well determined. For 3 PN candidates the flux value at 60 micron is 5 times or more the flux value at 25 micron and could be H II regions. Only one of the PN observed spectroscopically has one IRAS flux of 4.8 mJy at 25 μm. One object having an IR flux of 1179 10^{-13} Wm^{-2} is definitely too bright to be a PN at the distance of the galactic center. It is located in the galactic plane. Its IRAS colors are between the PN and H II regions. It appears that the contamination of our sample by H II regions and foreground objects is probably very small.

The strong decrease in detection rate within 1 degree of the plane is likely to be due to the increased extinction, since this is the location of the CO molecular cloud. This covers ~ 25 % of the surveyed region. Assuming that there would be 160 PN within 2 degrees of the galactic center, that this region has ~ 1.5 10^9 M_o, that the average PN observable lifetime is 8,000 years, a constant PN formation rate per mass throughout the Galaxy and a galactic mass of 1.4 10^{11} M_o, there would be roughly 15,000 PN in the Galaxy Pottasch (1992). Other values for the number of PN in the Galaxy range from 4,000 to 140,000 PN, with a recent determination 24,000 to 35,000 (Pottasch 1996).

3. Spectroscopy: observations and results

We obtained spectra betwen 4000 and 9600 Å with a resolution of 2.8 Å per pixel. Observations were done in June 1995 at the 1.52m ESO telescope with the Boller & Chivens spectrograph and a Ford Aerospace 2048L CCD. The spatial scale is 0.″82 per pixel, and the slit width was 2″. The typical integration time was 1h.

These new bulge PN show very large extinction. Hβ was detected for 9 of the 16 PN. For 3 PN we could estimate the extinction from Paschen lines. A_V minimum = 5.9 mag ; A_V maximum = 11.8 mag (Paschen line); A_V median = 7.8 mag The [O III]λ5007, usually the brightest optical line and often used in PN searches, is weak and even undetected in 2 PN. These PN are by far brightest at [S III]λ9532 (which is in part also a selection effect.)

The excitation class (EC) was calculated for the 9 PN for which Hβ and [O III]λ5007 were detected, according to the formulation by Dopita & Meatheringham (1991). The PN seem to be of rather low excitation: four are of low excitation having EC < 1.2 (T_{eff} < 45,000 K), two have 2 < EC < 3 (50,000 K < T_{eff} < 65,000 K), and three have 4 < EC < 5 (80,000 K < T_{eff} < 100,000 K). The latter group would be expected to show He IIλ4686 at lower extinctions. He II was not detected for any of the PN.

The radial velocities of the PN are presented in Figure 2. This statistically large sample will allow us to investigate bulge kinematics and abundances.

4. Future work

The prime goal is to calculate full photoionization models (van Hoof & Van de Steene, 1995) for the determination of abundances and distance independent parameters. The high extinction and faintness of the nebulae and consequently the small number of emission lines detected in the spectra are a serious drawback and so we will only be able use the brightest PN to reach this goal. Determining chemical abundances of these PN will allow us to investigate whether bulge PN derive only from the low end of the metallicity range in metal-rich populations (Ferguson & Davidsen, 1993) under the assumption that PN cannot form from metal-rich stars. With this PN sample, it also will be possible to look for a smooth, vertical abundance gradient in the bulge, which is expected if the bulge consists of a population having a large range in age and metallicity. This statistically large sample will allow us to investigate the bulge kinematics. The significant increase in the number of bulge PN will allow a much better determination of the total number of PN in the galaxy and the amount of mass returned to the interstellar medium

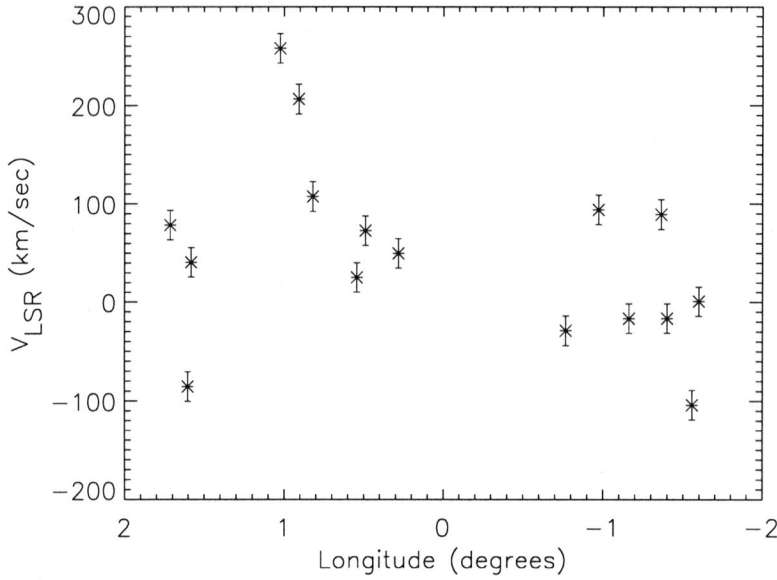

Figure 2. The radial velocities corrected to LSR versus longitude. The velocities were determined from the [O III]λ5007 (if strong enough), Hα, [S III]λ9069 and λ9532. The typical one sigma error-bar is 15 km/sec. The velocity dispersion is 92 km/sec.

Acknowledgments. This survey was motivated by a suggestion from Stuart Pottasch. We thank P. van Hoof for the use of his program to determine the velocities. This research has made use of the Simbad database, operated at CDS Strasbourg (France). This research is based on observations obtained at the European Southern Observatory.

References

Acker, A., Marcout, J., Ochsenbein, F., Stenholm, Tylenda, R., 1992, Strasbourg-ESO catalogue of galactic planetary nebulae, ESO
Dopita, M.A., Meatheringham, S.J., 1991, ApJ 367, 115
Ferguson, H.C., Davidsen, A.F, 1993, ApJ 408, 92
Kohoutek, L., 1994, Astron. Nachr. 315, 235
Pottasch, S.R., 1992, A&AR 4, 256
Pottasch, S.R., 1996, A&A 307, 561
Pottasch, S.R., Acker, A., 1989, A&A 221, 123
van Hoof, P.A.M., Van de Steene, G.C., 1996, A&A submitted

Part 2. The Stellar Population
Section D. Surveys

The 15μm ISOGAL Survey of the Inner Galactic Disk

Alain Omont[1]
Institut d'Astrophysique de Paris, C.N.R.S., 98bis Bouleavrd Arago, 75014 Paris, France

Abstract. We have undertaken a 15μm ISOCAM survey of about 12 deg^2 distributed along the inner Galactic disk ($|b| < 0.5$, $|l| < 30$). The sensitivity and the pixel area ($6''\times 6''$) are 2 orders of magnitude better than IRAS. The data will be combined with those of the IJK DENIS survey also in progress. The revolutionary performances of ISOCAM should allow to detect M giants up to the Galactic Center. They will be good tracers of inner obscured Galactic structures: central regions close to the Galactic Center, bar, old disk, molecular ring, young disk and spiral arms, etc. ISOGAL will also detect several thousands of rarer infrared bright stars, will map the spectacular 15μm diffuse emission and, combined with DENIS, will provide information on the interstellar extinction at small angular scales. The data reduction is particularly difficult because of the complex time behaviour of the detectors, persistence effects after observation of strong sources, patchy background emission and source confusion. A preliminary analysis of the first data seems to meet the expectations.

1. Introduction

With its much better sensitivity than previous infrared surveys, IRAS had a major impact on the knowledge of infrared stellar properties and on the census of infrared stars. In particular, it achieved a practically complete census of infrared AGB stars in the solar neighborhood up to a few kiloparsecs, detected the brightest of them at distances comparable to the Galactic Center, and made tremendous progresses in the study of young stellar objects (YSO) still enshrouded in molecular clouds or in YSO circumstellar disks. It has provided a mass of detailed and completely new information about circumstellar matter in these various classes of objects, initiated a large number of follow up studies on rich sample of infrared sources and allowed the identification of numerous peculiar objects. In addition, the brightest sources, mostly AGB stars, well observable up to the Galactic bulge, can be used as tracers of stellar populations and of Galactic structures (disk and bulge (Habing et al. 1985), bar (Weinberg 1992), etc.). Similarly, young IRAS sources are one of the best tracers of regions of star formation, even in the heavily obscured inner disk.

---

[1]on behalf of the ISOGAL team which includes: P.I.: A. Omont. Co-Is: J. Blommaert, C. Césarsky, N. Epchtein, M. Felli, R. Genzel, G. Gilmore, F. Guglielmo, H. Habing, M. Perault, S. Price, A. Robin, G. Simon

IRAS has also provided an extremely rich information, partially unexpected, on the infrared diffuse emission of the interstellar medium in the four IRAS bands. This has revolutionarised our knowledge of the strucrure of the interstellar medium and of its dust content, including the smallest "dust" particles, probably large aromatic molecules ("PAHs") (see e.g. Désert et al. 1990).

Thirteen years after IRAS, the ISO satellite has capacities much upgraded with respect to the latter. The gain is particularly impressive with ISOCAM in the wavelength range 3–20μm, with $\sim$1000 detectors. It achieves a sensitivity of a few mJy in a few secondes of integration, i.e. $\sim$100 times better than IRAS. Unfortunately such revolutionary performances have some drawbacks. in particular a very complex time behavior of the detectors which need a long time to reach a stabilised signal and which display strong and non linear persistence effects after the observation of a strong source.

Indeed, contrary to IRAS, ISO and ISOCAM were not devised for a survey mode. In particular, a minimum duration of a few seconds is mandatory for the integration time on each sky position. A comparable time is needed to move the telescope on distances of a few arcmin comparable to the angular size of the detector array. It is thus impossible to observe one square degree in much less than five hours, which of course strongly limits the total area observable in any survey.

Nevertheless, there was such a strong interest to take opportunity of the ISO-CAM capacities to address the problems of Galactic structures and stellar populations in the most obscured regionsof the inner Galactic disk, that we have successfully proposed a relatively large survey of $\sim$10-14 deg^2 at 15μm.

2. Scientific Goals

Probably more than 80-90% of the sources detected by ISOGAL will be M giants. They are good tracers of the stellar densities. The ISOGAL data at 15μm, combined with the DENIS and 2MASS surveys, will provide for them reliable identifications and actual magnitudes and colors corrected for extinction (within the limits of the ISOGAL photometry accuracy). The main product of ISOGAL will thus be detailed maps of (projected) stellar densities, which will trace the main structures of the inner Galaxy. One can expect in addition a rich (statistical) information on these stellar populations at various locations, as well as the detection of many rarer and infrared luminous stellar sources, important tracers of stellar evolution there. The small pixel size of ISOGAL will allow quite detailed studies of the 15μm diffuse emission, especially in regions of star formation, as well as mapping the interstellar extinction at scales comparable to that of CO surveys.

2.1. Galactic Structures

In spite of this importance, the central parts of the Galactic disk and bulge remain poorly studied. This is due to a mix of extinction at short wavelengths and poor spatial resolution in available studies at longer wavelengths. No effectively reliable information is available concerning the large scale spatial structure (inner spiral arms, ring, bar, central massive cluster, bulge,...) and the true distribution of reddening, in the inner Galaxy. There is no extant description of the distribution of stellar populations in the central Galactic disk and bulge. This is particularly true for the short-lived high luminosity stages of stellar evolution, which are just those which can dominate the infrared emissivity of galaxies. It is in the central regions of the Galaxy that ISO data combined with the coming data of near infrared surveys (DENIS and 2MASS) will provide the most new information, where a vast improvement over available data in both spatial resolution and sensitivity is possible and essential. The near IR surveys by themselves will be strongly impeded by reddening in the central regions of the Galactic disk. The ISOGAL survey of a large area of the central Galactic Plane, complemented by the shorter wavelength ground-based survey data which we are currently obtaining (DENIS), will greatly improve our understanding of the basic structural components of our Galaxy, and of the earliest and latest stages of stellar evolution. Only ISOCAM has the resolution and the sensitivity to do that at wavelengths free from extinction.

Central Regions. Although the central stellar cluster itself, with its quite extreme conditions of stellar density, will be hardly observable with ISOGAL because of too many saturating sources, ISOGAL is essential to address the conditions in the transition regions between the immediate vicinity of the Galactic Center and the Galactic disk and bulge. The dominant (by luminosity) M giants are ideal to trace the density distribution in these regions. However, these stars must be distinguished reliably from other sources and mapped over sufficient extent to overlap with useful low resolution available data, especially from DIRBE. ISOGAL has the resolution and sensitivity to do that at wavelength free from extinction. Indeed, the extinction in the central region is so extreme that two ISO passbands are required to complement available and future near infrared data.

The Galactic Bar. It has long been suspected by those modelling the inner gas dynamics of the Galaxy that a bar is required (e.g. Liszt et al. 1991, Blitz et al. 1993). The important point for the present is that the best available models are in only moderate agreement with the data. Improved understanding of the central regions of the Galaxy awaits much improved determination of the underlying mass distribution from surface density data for old stars. This must be achieved without contamination from reddening, irregular foreground disk structure (spiral arms, patchy dust, star formation regions,...), in the plane of the disk. For this purpose, the ISO observations will have a unique value: Weinberg (1992) derived the existence of the bar, not entirely convincingly, based on IRAS observations out of the plane. ISO, due to its superior resolution and sensitivity, will much increase the number of sources in Weinberg's flux and color limited sample by obtaining observations in the plane and correlating them with the reprocessing of the IRAS survey data of Price et al. along the plane (see e.g. Price 1994).

The Inner Old Disk. Determination of the inner disk mass profile, which contains most of the Galactic stellar mass, is fundamental for a global understanding of the Galaxy. However, its structure is very poorly known. Determinations of the radial scale length of an assumed exponential distribution differ by a factor of nearly two, as do determinations of the inner scale height, while the possibility of a central disk hole continues to be considered seriously. Reliable determination of large scale disk structure requires detection of a tracer of old disk populations which is detectable and identifiable through the obscuration, over an angular scale wide enough to determine the large scale underlying old disk structure. Early MIII's are ideal; they will be the dominant stellar population detected by ISOGAL.

The Molecular Ring. CO observations show the Galaxy has a molecular ring, some 3–4kpc from the centre. In addition to its intrinsic scientific interest in determining the recent star formation history and dynamical significance of the ring, careful mapping and subtraction of sources and extinction associated with the molecular ring is critically important for a reliable determination of background galactic structure. The true nature of the stellar populations in the ring is extremely difficult to discern without ISO mid-IR data, as considerable irregular extinction is expected to be associated with the molecular material, while IRAS has seen only the most luminous sources.

The Central Young Disk and Spiral Structure. The line of sight of the ISO-GAL fields contains asymmetric structures – spiral arms – on just the scale of relevance. For example, the OGLE studies of the stellar populations in Baade's Window suggest a very large perturbation in the stellar density and dust extinction associated with the foreground Sagittarius spiral arm, some 2kpc from the Sun (Paczynski et al. 1994). Similar results for both young and old disk stars have been suspected for many years, and have recently been supported by near IR data on M51 (Rix and Rieke 1993), which show a very substantial perturbation of the old disk distribution by spiral arms. The Sagittarius arm appears to be a major deviation away from symmetry in the underlying disk and appears interesting for study of young and high mass stellar populations. Until this foreground structure is mapped through the extinction, and distinguished reliably from the central galaxy, any analysis of large scale galactic structure or of the stellar populations associated with arm and inter
-arm regions must remain provisional. It is clear that there is a very strong interest to add a wavelenth free of absorption, such as 15μm, to a multiwavelength study in order to overcome the ambiguities due to absorption.

2.2. Stellar Populations in the Inner Galaxy

Our ISO survey data, to be complemented by DENIS, will allow for the first time the following classes of sources to be identifiable individually (see the table below for order of magnitude estimates of the various classes): ISOGAL will see predominantly MIII giants up to the centre, probing the dominant mass distribution for the many studies noted above; AGB stars including the warmer AGB stars not seen by IRAS, through much of the Galaxy, thus complementing the IRAS sample, extending it closer to the Plane, and in the central regions where their number and properties are still quite uncertain, and providing a wealth of sources for later study of variability, OH maser emission, etc; Planetary Nebulae throughout the Galaxy, constraining chemical evolution, mass loss histories, calibrations of the PN distance scale, etc; supergiants through the Galaxy, being especially valuable for identifying regions of recent star formation, and perhaps allowing identification of many very dusty supergiants which are likely supernova progenitors; foreground K giants, to determine the old disk mass associated with foreground spiral structure, and so help determine the mass perturbation associated with spiral arms; foreground T Tauri stars, allowing comparison of their distribution with that of known GMC and regions of formation of higher mass stars; non dusty OB stars as far as the molecular ring, mapping the high mass star formation in the ring, and in closer structures in the line of sight; and HII regions, complementing the IRAS survey (see Section 2.3).

Finally we note serendipity: the regions of the Galaxy covered in this survey contain most of the baryonic mass of the Galaxy, most of the stars, and the most extreme objects of every known type. It is probable that new discoveries of rare and short-lived objects await, given the substantial gain of ISO over previous satellites.

Table 1. Rough estimates of expected source counts. Number of stars brighter than $m_{12}=9$ ($S_{15\mu m} > 4$ mJy) in 1 deg^2 centered in l and with $0 < b < 0.5$, without taking into account spiral arms (computed by F. Guglielmo and A. Robin)

l(deg)	25	20	15	10	05	03
Total number	5100	6300	7700	9300	12600	21400
classes IV/V	27	28	28	29	29	29
T Tauri	22	23	23	24	24	24
Supergiants	27	29	31	31	30	30
KIII	240	250	270	280	290	300
MIII	4400	5700	7000	8500	11800	20500
M AGB	155	195	253	338	450	~ 600

2.3. Diffuse 15μm Emission and Interstellar Extinction

IRAS has shown the ubiquity in the Galaxy of the far infrared diffuse emission even in its 12μm band. The latter is particularly strong in theregions with relatively strong, but not too strong, UV radiation at the boundary of molecular clouds. However, it displays a complex behaviour close to massive stars, believed to be related to the destruction of one of its main carrier, posibly PAHs (Boulanger et al. 1990).

As shown by the first data, ISOGAL is bringing a very rich information about the spatial distribution of the 15μm enission, especially close to young stellar objects (YSOs) traced as strong point sources by IRAS. It will determine which part is unresolved, and what is the spatial distribution of the remaining part, tracing the distribution of circumstellar material on scales of fraction of a parsec around these sources, and the way it is excited by the central source. Désert et al. (1990) have proposed that the "12μm" diffuse emission detected by IRAS is dominated by PAHs. It is one of the important goals of ISO to confirm that and to check the detailed spectral distribution among the PAHs bands. Other possible contributors to the 15μm diffuse emission are very small grains or warm dust close to YSOs. Mapping the 15μm diffuse emission at 6" scale will help disentangling these different possible contributions in the various cases of the many YSOs observed with ISOGAL (see e.g. modelling such emission by Siebenmor-

gen 1993).
Although the main PAH feature at 11.3μm is outside the ISOGAL band (12–18μm), the latter includes features in the range 12–15μm believed to be associated with PAHs and hydrogenated amorphous carbon. Combined with the ISO dedicated programmes aimed to precisely identify the carriers of the diffuse emission, ISOGAL will offer a unique global picture of its small scale distribution with a lot of information on the small scale structure of the warm dense interstellar medium and dust/PAH chemistry.

On slightly larger angular scales, ISOGAL combined with DENIS will also trace the total amount of (cold) dust through the extinction measured on the line of sight of indivudual stars. Given the number of sources expected from ISOGAL (typically one per arcmin^2), one can hope to map the extinction on arcminute scales at least (possibly even on smaller scales from DENIS data alone well calibrated with ISOGAL). This is quite comparable to the scale of some CO surveys. Comparison with the CO or ^{13}CO intensities will allow some 3D determination of the extinction and give information on the CO/dust ratio in various galactic environments.

3. Technical Details and Difficulties, and Preliminary Results

Most of the difficulties are related to the time behaviour of the detectors and to their high sensitivity itself. Additional difficulties are coming from the strong and patchy Galactic background and the possible source confusion in the most crowded areas, which will limit the effective sensitivity.

The mapping mode of ISOGAL is devised so that most sky positions are observed twice, yielding a total effective integration time per sky position of $\sim$16 sec. With a smooth background, one could thus hope a 5σ sensitivity of $\sim$4 mJy, i.e. a magnitude limit at 15μm of $\sim$ 9. However, it is feared that the flat field uncertainty generated by the patchy background and the persistence effects in the detectors could degrade the sensitivity by possibly typically $\sim$ 1 magnitude. Anyway, the actual sensitivity cannot exceed the confusion limit estimated from source count models to be also $\sim$9 mag in the central regions.

It is known that, after any modification of the illumination, the ISOCAM detectors need a long time to reach their new stabilised signal. As mentioned, the ISOGAL integration time is not long enough to allow such a complete stabilisation. The actual source fluxes will thus be derived from an extrapolation inferred from models of the time behaviour of the detectors, implying some uncertainty. For strong sources, the persistence effects are immediately visible on the raw data as "ghost" sources generated in the subsequent images by the pixels which have seen them. Even the effects on pixels which have seen a strong source only in a slew motion can be quite visible. Softwares have been elaborated to efficiently correct such persistence effects. However, it is impossible to avoid that they degrade the ultimate quality of the data, especially in crowded regions.

Anyway, it is mandatory to avoid the strongest sources, above $\sim$10 Jy per pixel, which would seriously degrade the detector performances for a long time. Therefore, we have decided to avoid every IRAS source with a 12μm flux larger than 6 Jy (to take into account a possible variability). This has imposed to devise rasters avoiding such strong sources in each of our observed fields. Our observations are thus biased against regions where the density of strong sources is larger, especially regions of star formation.

The observed regions are regularly distributed along the galactic plane, with $|b| < 0.5°$, mostly with $|l| < 30°$. More precisely, they include a central field with $|l| < 1.5°$ and 24 fields with an area ~ 0.5 deg^2, symmetric in longitude, centered at $|l|$ = 2.75, 4.25, 5.75, 7.25, 9.75, 12.25, 15.75, 18.75, 22.25, 25.75, 30.0 and 45.0.

The first ISOGAL observations were performed early in the ISO mission in January-February 1996, in fields with large negative longitudes visible at the time, for a total area of almost 1 deg^2. The data processing is in progress, mostly by M. Perault, P. Seguin and G. Simon. The data seem to meet the expectations with the problems mentioned above. The rms is below 1 mJy in non crowded and quiet regions. Diffuse emission is spectacular, as well as persistence effects ! There is a good matching of strong sources with DENIS K sources, except maybe for a few YSOs. Further observations of ~ 5 deg^2, with positive longitudes or near the Galactic Center, have been scheduled in early April 1996.

References

Blitz, L., Binney, J., Lo, K.Y., Bally, J., Ho, P.T.P. 1993, Nature 361, 417

Boulanger, F., Falgarone, E., Puget, J.L., Helou, G. 1990, ApJ 364, 136

Désert, X., Boulanger, F., Puget, J.L. 1990, A&A 237, 215.

Habing, H.J., Olnon, F.M., Chester, T., Gillett, F., Rowan-Robinson, M., Neugebauer, G. 1985, A&A 152, L1.

Liszt, H.S., MNRAS 252, 210

Paczynski, B. et al. 1994, AJ 107, 2060

Price, S. 1994, in "Science with Astronomical Near-Infrared Surveys", Epchtein, N., Omont, A., Burton, B., Persi, P. eds, Astrophys. Space Sci. 217

Rix, H-W. and Rieke, M.J. 1993, ApJ 418,123

Siebenmorgen, R. 1993, ApJ 408, 218

Weinberg, M.D. 1992, ApJ 384, 81.

Variable Stars close to the Galactic Centre

I.S. Glass
South African Astr. Observatory, PO Box 9, Observatory, South Africa

S. Matsumoto
Institute of Astr., University of Tokyo, Bunkyo-ku, Tokyo 113, Japan

T. Ono
Nishi-Harima Astr. Observatory, Sayo-cho, Hyogo 679-53, Japan

K. Sekiguchi
National Astr. Observatory, Mitaka, Tokyo 181, Japan

Abstract.
A large-format infrared camera has been used for repeated observations of a 24 × 24 arcmin^2 region around the Galactic Centre. The survey is carried out at K (2.2μm), with single-epoch coverage at H (1.65μm) and J (1.25μm). So far, the observations have extended over two seasons and it is intended that they will continue for one more.

The brightest ~200 objects within each 5 × 5 arcmin^2 sub-field have been examined for variability. Preliminary processing of the data has revealed about 280 likely variables in the overall field. Their positions have been determined relative to stars in the STScI Guide Star Catalog. They include several previously known OH/IR stars.

1. Introduction

The general neighbourhood of the Galactic Centre is obscured in visible light and its stellar content can only be examined at infrared wavelengths beyond 1.6μm. A survey of a 1° × 2° region including the Centre and with sufficient resolution (6″ × 12″, RA × Dec) to resolve individual stars was reported by Glass, Catchpole & Whitelock (1987), and other non-quantitative surveys have been reported from time to time. Even at the resolution mentioned, however, crowding towards the Centre presents a serious problem. Large corrections had to be applied when analyzing the luminosity distribution of stars towards the Centre (Catchpole, Whitelock & Glass, 1990).

The IRAS survey did not have sufficient resolution to avoid severe crowding effects near the Centre, and its effective sensitivity at 12μm to stars in this region dropped by about a factor of 10. Analysis of the types of objects it did detect (Glass, 1988; Moneti, Glass & Moorwood, 1992) shows that they were predominantly of nebular character, such as compact H2 regions.

The knowledge of individual objects near the Centre is very limited, apart from the cluster in the Centre which has been examined in great detail using techniques of high spatial resolution (Eckart et al, 1993) and particularly interesting small fields have been looked at with infrared arrays; for example the AFGL 2004 (Quintuplet) cluster (Glass, Moneti & Moorwood, 1990; Okuda et al, 1989; Moneti, Glass & Moorwood, 1992, 1994) and the cluster of hot stars near G0.121+0.017 (Nagata et al, 1995; Cotera et al,). Photometry of a limited sample of bright stars has been published by Nagata et al (1993).

The variable star population of the Galactic Bulge is of great interest in kinematical and other studies. The easiest component to identify and study has been the dust-rich population of AGB variables found by the IRAS satellite. An investigation of the overall population of long-period variables near the Centre, in the Sgr I window at $l = 1.46, b = -2.68$, has been made by Glass et al (1995, GWCF). There is some suggestion from this study and similar ones in more distant parts of the Bulge that the proportion of the longest period sources rises as the Centre is approached, although whether the ratio depends primarily on latitude or distance has not been established. Closer to the Centre, in a 5×5 arcmin^2 field, Haller & Rieke (1989) have detected some Mira-like variables, but few details have been published.

The OH/IR population near the Centre has been surveyed by Lindqvist et al (1992). Follow-up infrared work has been done by Jones et al (1994), van Langevelde (1992) and Blommaert (1992). The concentration of these sources towards the centre appears to be co-extensive with an apparent population of luminous late-type stars found by Catchpole et al (1990). They have been found to be kinematically separate from the previously known sample of Bulge/Disc sources.

The study of the variable star content of the Centre with reasonable efficiency has been made possible by the availability of the large-area PtSi infrared array camera PANIC (PtSi Astronomical Near-Infrared Camera) with 1040×520 pixels on the 0.75m reflector at Sutherland. This was constructed as a joint project of the Institute of Astronomy, University of Tokyo, the National Astronomical Observatory of Japan and the South African Astronomical Observatory. The instrument has been described in detail by Glass, Sekiguchi & Nakada (1995).

In this paper, we present some results from a "first pass" reduction of the data. Some of the conclusions remain tentative and await completion of the observational programme and a more comprehensive analysis.

2. Observations

The programme consists of repeated observations in the K band of twenty-five sub-fields of 5×5 arcmin^2 having a slight overlap. Supplementary observations at J and H were also obtained, but generally only once for each field.

Observations are made every month or two during the Galactic Centre season. It is intended that they should continue for at least three years. Should the stellar images be larger than about 1.8 arcsec FWHM, they are repeated within the same observing run under better conditions, if possible, because the sensitivity limit and the ability to resolve stars in a crowded field is degraded by spread of the images.

Although the observing frequency was mainly set by the desire to study long-period Mira-like variables, the survey will also be able to detect long-period cepheids if they exist in this region.

The observations were made with two different detectors, the changeover being at the start of June 1995. The standard exposure time was 5 mins with the first detector and 10 minutes with the second which had lower Q.E. at K but better cosmetics.

3. Data Reduction

This can be conceived as in three parts, a preliminary one which takes care of the detector-related problems associated with this programme, standard processing by DoPHOT and an analysis by means of the STAR set of programmes written by Balona (1995).

3.1. Pre-DoPHOT Processing

The background to be removed from the exposures includes telescope and sky radiation, amplifier radiation and dark current. The exposures are reduced at present in groups of five. First the members of the group are averaged and the differences between the individuals and the average are found. Each frame is then corrected by adding a constant to each pixel so that all frames have the same modal pixel value. The median average frame is then found. This should contain no stars and is subtracted from the individual frames. Cosmetic corrections are then made (necessary often for the earliest frames).

3.2. DoPHOT

The coordinates and intensities of the stars are extracted from each exposure using DoPHOT (Schechter, Mateo & Saha, 1993). Saturated images, unresolved doubles etc are identified in the output files.

3.3. Post-DoPHOT processing

All tabulated stellar images are first placed on a uniform coordinate system based on a particular fiducial image and a new catalog file is produced for each image. Next, all the photometry is placed on a uniform instrumental system by iterated comparison of bright stars. The bright non-variable star residuals are usually around 0.03 mag. Finally, a table of photometry is constructed for each individual star.

The light curves of the brightest stars are plotted and examined by eye to select the variables.

The photometry is then placed on a standard system, currently by comparison with objects common to the Glass, Catchpole & Whitelock (1987) survey, as reduced by Ono (1994).

The positions of the variables are found with reference to the Guide Star Catalog of the HST via the Digitized Sky Survey in a two-stage least squares fitting process.

4. Discussion

The present reduction of the data is somewhat experimental and has been made not only to obtain preliminary results but also in order to identify and correct problems. The methods used have therefore varied somewhat from sub-field to sub-field, so that the uniformity of, for example, the surveys for variable stars is not perfect.

The single most severe problem encountered is related to flat-fielding and photometric calibration. Although the arrays show little or no variation of sensitivity on spatial scales of a few tens of pixels, there is some evidence from comparisons of results obtained with different chips that they have variations of up to 15% from one side to the opposite. This is partly confirmed from a field where photometer observations of a number of stars exist, although the accuracy of the discrete-object photometry may be compromised by crowding. It is expected, however, that this problem will be solvable.

In addition to the variable star results discussed below, it is likely that the following information will be forthcoming:

1. Population studies, such as the variation of number densities of AGB and giant stars with distance from the Centre of the Galaxy.

2. Study of very luminous stars towards the Galactic Centre. For example, there is evidence for MIa supergiants both in the Centre Cluster and in the AFGL 2004 cluster. Such stars are not, however, common.

3. Mapping of interstellar extinction: It is well-known that the grosser festures of the extinction are well-correlated with the column density of low-velocity ^{13}CO (Glass, Catchpole & Whitelock, 1987). However, the resolution of the CO maps is much lower than the infrared ones.

4. Study of objects with colours similar to those of the long-period variables but which do not vary.

5. Variable Stars

Examination of 200 of the brightest stars from each sub-field has led to the identification of 280 probable long-period variables. Their average K mags and H-K colours place them in the part of the colour-magnitude diagrams occupied by known Miras and OH/IR stars. About 90% of them can be regarded as certainly variable, having amplitudes in excess of 0.6 mag. The next season's data as well as improved processing techniques should remove almost all incorrectly classified stars as well as improving coverage at the edges and in the corners of the fields, missed in some of the earlier exposures.

The periods of the variables and their full amplitudes have not yet been determined since the coverage has long seasonal gaps and many of them evidently have periods of over 400 days. A further season's observing and the improved data-processing procedure will enable reasonably reliable periods etc to be found, although the time coverage could usefully be extended even further.

6. Extinction

The colour-magnitude (H-K, K) diagram of stars from the sub-field GC 14, whose centre is about 10 arcmin E of the Centre, is shown in Figure 1. This particular field has been chosen because the interstellar absorption in its direction is relatively uniform.

The densest part of the C-M diagram is clearly the top end of the AGB, with extinction of about $A_V = 25$ mag. There are considerable numbers of bluer objects, presumably mostly in the foreground. The other fields show wider AGBs because of their more variable extinction.

The distribution in the line of sight of the absorbing material is poorly known. Low-velocity ^{13}CO is fairly ubiquitous and often correlates spatially with the dark clouds seen in front of the Galactic Centre on the large-scale 2.2 μm maps, which suggests that much of it is associated with the molecular ring extending from the solar circle half way to the Centre. However, Neckel & Klare (1980) from studies of O and B stars suggest that the extinction in the direction of the GC rises rapidly up to $A_V = 3$ at about 1 kpc but remains fairly constant thereafter out to at least 5 kpc. The polarizations of GC sources also show a sharp change at a distance corresponding to $E_{H-K} \sim 0.75$ or $A_V \sim 12$, suggesting a discontinuity in the interstellar medium in the inner galaxy (Kobayashi et al, 1983).

The infrared colour-magnitude diagrams of all the sub-fields seem to show a minimum extinction towards the red giants of about 25 mags, with excursions in many of them to much higher reddenings. Of course, they do not show the very highest reddenings because the stars become too faint to be observed at H and sometimes even K. The uniformity of the minimum extinction suggests that it originates in the foreground, perhaps in the molecular ring. Although the AGB stars must have some distribution in depth, they are predominantly well inside the molecular ring.

7. Discussion of the Variable Star Results

The position of the variable stars in the 25 sub-field colour-magnitude diagrams is towards the high luminosity end and somewhat to the red of the AGB.

Figure 1 also shows the distribution of the long-period variables in the Sgr I field at about 3° from the Centre, taken from GWCF (1995), with a correction applied for its relatively low interstellar reddening of $A_V = 1.78$. The non-variable part of the AGB extends about half way up the parallelogram occupied by the long-period variables. There is a considerable component of bright stars blueward of the AGB, presumably in the foreground. The non-variable stars are from a field of about 200 arcmin^2, while the variables are from the whole Sgr I field of 1200 arcmin^2. It is suprising that more foreground bright blue stars are not seen in GC 14 (25 arcmin^2); their absence may be indicative of nearby absorption. Some of the long-period variables lie redward of the parallelogram and owe their extreme colours to circumstellar extinction.

The estimated position of the parallelogram after 25 mags of extinction is also shown. The variables of GC 14 are typical, in that they lie from the middle to the top of the parallelogram and redward of it. The lack of shorter-period, less luminous, variables is very noticeable. A visual examination of the light curves, incomplete though they are, indeed suggests that the periods are 300d or longer, with at least half in excess of 400d. The existence of a few shorter-period stars cannot be excluded completely because they are fainter and generally have lower amplitudes. However, a search of GC 14 stars down to $K_{av} = 10.0$ did not reveal any short-period candidates.

The LPVs of the Galactic Centre fields are thus skewed in their period distributions towards the longer periods and tend also to be redder, when compared with Sgr I. They are thus likely to be undergoing greater mass loss than their Sgr I counterparts.

Of the 46 OH/IR stars from Lindqvist et al (1992) in the field, only 17 have been detected so far. These objects are often heavily reddened by their own circumstellar shells and frequently have large amplitudes and long periods, so that the remainder may have been too faint to find. Some were searched for individually (using their known positions) amongst the light curves of stars fainter than the brightest 200 of each field. It is possible that more will be found later, when they are at more favourable phases. At least 15 of the new sample reported by Sjouwerman et al (this conference) have also shown up in this study.

The space density of Mira variables in the solar neighbourhood is about 600 kpc^{-3} according to Jura, Kleinmann and co-workers (see GWCF, 1995). If we take the dimensions of the volume of the Bulge covered in the present survey as about 56 pc in each direction, the space density of Miras becomes about $2 \ 10^6$ kpc^{-3}, or about 3000 times greater than locally.

It is interesting to examine the total mass lost from these variables to the interstellar medium. Although relations between periods or colour indices and the mass-loss rates of Miras have been found, these parameters are not yet available for the present sample. Instead, we can make an estimate of the mass loss from the K amplitudes of the variables, based on a very approximate relation derived from Figure 20 of Whitelock et al (1994). For the ~280 variables we obtain ~$3 \ 10^{-4}$ M$_\odot$ yr^{-1}. The largest contributors to the rate are the longest-period variables, in particular the OH/IR stars. Considering that only 37% of these were detected, the total mass-loss rate could be more than twice the figure stated. In addition, some 10% to 20% percent of the Miras in the area may have escaped detection because of heavy foreground reddening in some directions. These will not, however, affect the total mass-loss rate significantly.

There are, of course, other mass-losing stars such as the emission-line objects found in the three clusters and the field objects with protostellar colours found by Moneti et al (1992, 1994). The very red GC cluster sources, i.e. IRS 1, 3 and 10, may also be losing mass rapidly. However, the HeI-emission stars in the GC cluster alone may contribute more than 10^{-3} M$_\odot$ per year (Genzel et al, 1994). The total input to the interstellar medium in the survey volume may therefore be several times 10^{-3} M$_\odot$ per year when all sources are taken into account.

Acknowledgments. We wish to thank Drs Y. Nakada and T. Tanabe for their help in setting up this programme. Dr L.A. Balona's STAR suite of programmes has proved invaluable. We also thank B.S. Carter for some of the

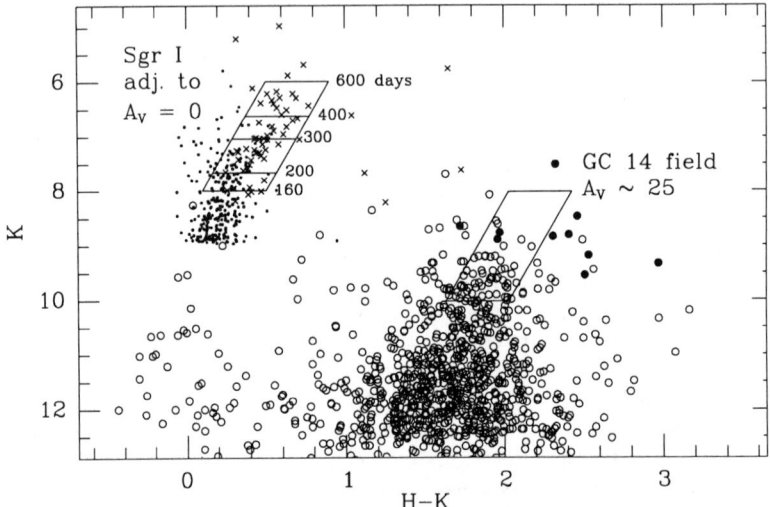

Figure 1. Composite plot of the brighter stars (points) in part of the Sgr I field (adjusted for zero interstellar reddening) and those in the GC 14 sub-field of the present study. All the measured long-period variables from the Sgr I field are plotted as crosses and the superimposed parallelogram encloses most of them. The location of the parallelogram corresponding to $A_V = 25$ is also given. The long-period variables of GC 14 are shown as solid circles and the other stars as open ones.

observations and Dr J. Menzies for help with programming problems. Dr T. Lloyd Evans kindly commented on the draft paper.

References

Balona, L. 1995, in A.G. Davis Philip, et al (eds) New Developments in Array Technology and Applications, Kluwer, Dordrecht, p187
Blommaert, J.A.D.L. 1992, PhD Thesis, Leiden
Catchpole, R.M. Whitelock, P.A. Glass, I.S. 1990, MNRAS, 247, 479
Cotera, A.S. Erickson, E.F., Colgan, S.W.J. Simpson, J.P. Allen, D.A. Burton, M.G. in press
Eckart, A. Genzel, R. Hofmann, R. Sams, B.J. Tacconi-Garman, L.E. 1993, ApJ, 407, L77
Genzel, R., Hollenbach, D.J. Townes, C.H. Eckart, A. Krabbe, A. Lutz, D. Najarro, F. 1994, in Genzel, R. Harris, A.J. eds. The Nuclei of Normal Galaxies, Kluwer, Dordrecht
Glass, I.S. 1988, MNRAS, 234, 115
Glass, I.S. Moneti, A. Moorwood, A.F.M., 1990, MNRAS, 242, 55p; erratum 244, 767
Glass, I.S. Catchpole, R.M. Whitelock, P.A. 1987, MNRAS, 227, 373
Glass, I.S. Sekiguchi, K. Nakada, Y. 1995, in A.G. Davis Philip et al (eds) New Developments in Array Technology and Applications, Kluwer, Dordrecht, p109
Glass, I.S. Whitelock, P.A. Catchpole, R.M. Feast, M.W. 1995, MNRAS, 273, 383 (GWFC)
Haller, J.W. Rieke, M.J. 1989, in Morris, M. ed. The Center of the Galaxy, Kluwer, Dordrecht, p487
Jones, T.J. McGregor, P.J. Gehrz, R.D. Lawrence, G.F. 1994, AJ, 107, 1111
Kobayashi, Y. Okuda, H. Sato, S. Jugaku, J. Dyck, H.M. 1983, PASJ, 35, 101
Lindqvist, M. Winnberg, A. Habing, H.J. Matthews, H.E. 1992, A&AS, 92, 43
Moneti, A. Glass, I.S. Moorwood, A.F.M. 1992, MNRAS, 258, 705
Moneti, A. Glass, I.S. Moorwood, A.F.M. 1994, MNRAS, 268, 194
Nagata, T. Hyland, A.R. Straw, S.N. Sato, S. Kawara, K. 1993, ApJ, 406, 501
Nagata, T. Woodward, C.E. Shure, M. Kobayashi, N. 1995 AJ, 109, 1676.
Neckel, Th. Klare, G. A&AS, 1980, 42, 251
Okuda, H. Shibai, H. Nakagawa, T. Matsuhara, H. Kobayashi, Y. Hayashi, M. Kaifu, N. Nagata, T. Gatley, I. Geballe, T. 1989, in The Center of the Galaxy, ed. Morris, M. Kluwer, Dordrecht, p281
Ono T. 1994, M Sc Thesis, Dept of Earth and Astr. Sci., Tokyo Gakugei Univ.
Schechter, P.L. Mateo, M. Saha, A. 1993, PASP, 105, 1342
van Langevelde, H.J. 1992, PhD Thesis, Leiden
Whitelock, P.A. Menzies, J. Feast, M. Marang, F. Carter, B. Roberts, G. Catchpole, R. Chapman, J. 1994, MNRAS, 267, 711

The Stellar Content of the Bulge: NTT and HST Photometry

A. Vallenari

Astronomical Observatory, Padua, Italy

C. Chiosi and G. Bertelli

Department of Astronomy, Padua, Italy

Y. K. Ng

Institute d'Astrophysique, Paris, France

Abstract.
We derive photometric data for the stellar content in a field 9×8 arcmin of the Baade Window (BW8) in V- and I-passbands. About 36000 stars are measured down to V $\sim$ 22 mag. A small part of this field has been also observed by Holtzman et al. (1993) with the WFC of the Hubble Space Telescope (HST) down to V $\sim$ 22. We make use of the data to construct a composite CMD and we analyze the luminosity function of this sample of stars arguing that a young population is present, whose age and metallicity are in the range 9-10 Gyr and $0.005 \leq Z \leq 0.03$, respectively. In addition to this an old component is also present, age and metallicity in the range 13-16 Gyr and $0.0005 \leq Z \leq 0.06$, respectively, but its relative proportion with respect to the young one cannot exceed 30% of the total. Most likely it is much less than this upper limit.

1. Introduction

A systematic analysis of the stellar content towards the Centre of the Galaxy has been undertaken (Bertelli et al. 1995a,b; Ng et al. 1995, 1996) to study the galactic structure; to obtain information about the age and the metallicity of the stars in various galactic components (halo, bulge and disc); to obtain information about the past history of star formation and chemical enrichment.

In brief, Bertelli et al. (1995) analyzed three regions of the Bulge located near the clusters NGC6603 (l=12.9 b=-2.8), Lynga 7 (l=328.8 b=-2.8) and Terzan 1 (l=357.6 b=1.0). First they trace the position of a molecular ring between 3.5 and 4 kpc in the direction of NGC6603, of a stellar ring between 3.0 and 4.0 kpc, of the Sagittarius arm in the direction of NGC6603 (5.0 to 7.0 kpc) and Lynga 7 (4.2 to 7.0 kpc). Second, they recognize the presence near the Galactic Centre of an old and metal rich population (see also the detailed study of the CMD of Terzan 1 by Bertelli et al. 1996 in which evidences for the existence of old, high metallicity stars in the so-called hot horizontal branch stage are found), and finally give a hint that the Galactic Bar points its nearest side toward positive galactic longitude.

There is no general agreement on the age and the spread in age of stars of the Galactic Bulge (see Rich 1992 for a review of the subject). While all determinations of ages agree that the Bulge ought to be older than 5-8 Gyr, they diverge on whether the presence of a component substantially younger than 15 Gyr is required (cf. Renzini 1993; Terndrup 1988; Holtzman et al. 1993; Paczynski et al. 1994; Ortolani et al. 1995; Bertelli et al. 1995, 1996; Ng et al. 1995,1996).

It goes without saying that determining the age and the age range of the stellar populations in the Bulge is of paramount importance as it would provide clues on the mechanism of galaxy formation, in particular it would answer the question whether the Bulge formed before the Halo or viceversa.

In order to cast light on this important issue, we tried to derive good quality CMDs for the stars in the Baade Window, from which direct estimates of the age might eventually be possible. Ground-based photometric observations provide rich samples of stars in the brightest regions of the CMD. However, they are hampered by crowding effects at the fainter magnitudes. In contrast, HST is particularly useful to study faint main sequence stars, because its high resolution reduces the crowding effects, and its small field of view limits the contamination by disk stars.

In this paper, first we present ground based observations of the Baade Window region named BW8 (cf. Udalski et al. 1993), and second make use of the central area of the field, for which HST data are available, to derive a composite CMD.

2. Observations and Data Reduction

V and I photometric data for the region BW8 centered at $\alpha = 18^h 03^m 24^s$ and $\delta = -29°52'03''$(2000) have been obtained with NTT equipped with EMMI at La Silla in August 1995. The field of view is about 9×8 arcmin. The night was photometric with seeing of about $1.1''$. The data have been reduced with Daophot and Allstar in the Midas package. About 36000 stars are found down to $V \sim 22$. The data are reasonably complete down to $V \sim 19.5$ - 20 mag, where completeness is about 50%.

The central part of our frame has also been observed by Holtzman et al. (1993) with the HST-WFC1 in the F555W and F785LP pass-bands. These data have been retrieved from the HST archive and reduced using INVENTORY in MIDAS environment for the aperture photometry (an aperture diameter of 3 pixels). About 12000 stars are measured down to $V \sim 24.5$. The HST data are fairly complete down to $V \sim 21$. At fainter magnitudes, the data lose completeness, which amounts to 50 % at $V \sim 22$ mag.

The NTT and HST CMDs have been used to derive a composite CMD for the small area in common. Stars brighter than V=18.5 are from the NTT data whereas those fainter than this limit are from the HST sample. The composite CMD for two of the four chips is shown in Figure 1a.

We adopt the reddening E(B-V)=0.6 and the distance to the Galactic Centre of 8 kpc or the distance modulus $(m-M)_0$=14.52 (Reid 1989, Rodger 1986).

3. Discussion of the CMDs and LFs

At this stage of the analysis, the age cannot be easily determined owing to the well known age-metallicity degeneracy and the lack of information on the termination magnitude of the bulge main sequence stars. To address this point we look at the LFs and derive the mean magnitude of the stars in the red clump and the termination magnitude of the main sequence stars. The ground based CMD presents well developed red giant branch (RGB) and asymptotic giant branch (AGB) sequence. Performing star counts along this vertical strip, we notice a peak at V∼16.7-16.9, which is provisionally adopted as the magnitude characterizing the red clump. As far as the main sequence termination is concerned, the star counts (see the solid lines in the Figure 1b referring to the data of two WFC1 chips) hint that the transition magnitude spans the range V=18.5 to V=19.5 at most. The difference between the red clump and main sequence magnitudes ΔV_{TM}^{HB} goes from 1.8 to 2.8 mag, respectively. Considering residual contamination from disk stars, a spread in metallicity, and effects induced by binary stars or blue stragglers blurring the main sequence termination by say 0.7 mag, we are only marginally consistent with $\Delta V_{TM}^{HB} \simeq 3.5$ that would be required for the age of the population mix in the Bulge being in the range 13-15 Gyr. However such a possibility cannot be excluded (Ortolani et al 1995).

In order to check further the above reasoning we have used the synthetic HRD technique developed by Ng et al. (1995) and performed simulations varying the age, the law of star formation $\Psi(t)$, the metal enrichment $Z(t)$, and using the Padua library (Bertelli et al. 1994) of isochrones. The contribution of the disk population is calculated as in Ng et al (1995). Two cases are considered. **a) Old stars only.** In the first simulation, the bulge stars span the age and metallicity ranges 13-16 Gyr and $0.0005 \leq Z \leq 0.06$, respectively. **b) Old and Intermediate age stars.** In the second simulation, in addition to the old component, we include also a population made of intermediate age stars like those found in the Bar, for which Ng et al. (1996) get an age of 9-10 Gyr and a metallicity in the range $0.005 \leq Z \leq 0.03$. The resulting LFs for the main sequence stars (thin lines) are shown in Figure 1b and compared with the observational one (thick line). The comparison must be limited to stars brighter than V≃20.0 because the observational LF is not corrected for completeness. Many simulations at varying the ratio between young and old stars indicate that the old component cannot exceed 30% of the total, possibly being significantly lower than this limit.

4. Conclusions

The preliminary analysis of the composite CMD of the field BW8 (NTT + HST data) finds the presence of a significant young component (identified with the Bar) together with an old component whose relative contribution, however, cannot exceed 30% of the total. This agrees with the stellar population probabilities for the source of each OGLE and MACHO microlensing event reported in Table 2 of Ng et al. (1996), where the mean Bar/Bulge/Halo probabilities are in the ratio 2/1/0.25 (see Ng et al. 1996 for more details).

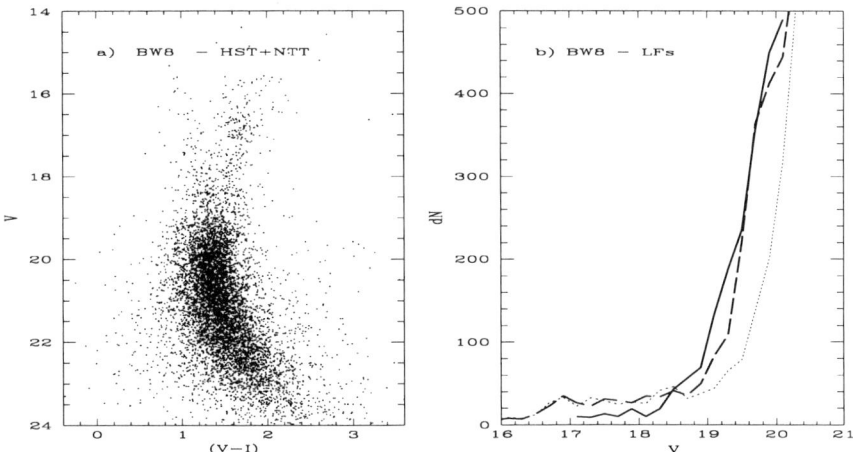

Figure 1. Figure 1a: CMD of BW8, from NTT and HST data. Figure 1b: The observational luminosity function corrected for completeness (solid line) with two simulations referring to the case a) (dotted line), case b)(dashed line) as described in the text.

References

Bertelli G., Bressan A., Chiosi C., Fagotto F., Nasi E., 1994, A&AS 106, 275
Bertelli G., Bressan A., Chiosi C., Ng Y.K., Ortolani S. 1995, A&A, 301, 381
Bertelli G., Bressan A., Chiosi C., Ng Y.K., 1996, A&A in press
Harris H.C., Baum W.A., Hunter D.A., Kreidl T.J., 1991, AJ 101, 677
Holtzman J.A., Light R.M., Baum W.A., et al., 1993, AJ 106, 1826
Ng K.Y., Bertelli G., Bressan A., Chiosi C., Lub J. 1995, A&A 295, 655
Ortolani A., Renzini A., Gilmozzi R., et al., 1995, Nature, 377,701
Paczynski B., Stanek K.Z., Udalski A., et al., 1994, ApJ 435, L113
Reid M.J., 1989, In The Center of the Galaxy, IAU Symp. 136, M. Morris (ed.), Kluwer, 37
Renzini A., 1993, Proceedings IAU Symp. 153, "Galactic Bulges", 17-21 August 1992, Ghent (Belgium), H. Dejonghe and H. Habing (eds.), 235
Rich R.M., 1992, In the center, bulge and disc of the Milky Way, L. Blitz (ed.), Kluwer, p. 47
Rodger A.W., Harding P., Ryan S., 1986 AJ 92, 600
Terndrup D.M, 1988 AJ 96, 884
Udalski A., Szymanski M., Kaluzny J., et al., 1993, Acta Astron. 43, 289

Part 2. The Stellar Population
Section E. Dynamical Studies

Dynamics of the Core Region

R. H. Miller

University of Chicago, Astronomy Center, 5640 Ellis, Chicago 60637

Abstract. Core regions of galaxies are not likely to be in a dynamically steady state. The nucleus orbits around a galaxy's mass centroid, and the orbits show complicated motions (nutation, precession, change of orbital amplitude, etc.). A massive object (such as a black hole) near the center should orbit around the mass centroid as well. Additional studies show that objects, such as a globular cluster, that may be captured into the core region, can survive for very long times, possibly creating long-lived doubles like M31's double nucleus. Additionally, the galaxy can ring like a bell in a global normal mode. These possibilities must be included in interpreting observations of the Galactic core region.

The dynamics of the core region of galaxies allows considerably more freedom than is normally used in interpreting observations. Books on galaxy dynamics tell you only about dynamical situations simple enough to treatment by admit analytic theory. Real galaxies have no such constraints. Three examples of dynamical phenomena we have noticed in galaxy core regions are described in this report. They suggest some of this dynamical freedom, but they certainly do not exhaust the possibilities.

1. Orbiting Cores

The nucleus of a galaxy orbits around the galaxy's mass centroid (Miller & Smith 1988, 1992). These orbital motions appear overstable in numerical experiments started with a galaxy's nucleus at rest atop its mass centroid. The amplitude doubles in 6 − 10 orbital periods, and the orbital periods are given by the local (nuclear) $G\rho$. Orbits precess, nutate, and change their amplitudes, but the periods remain fairly constant. Amplitudes reach a core radius. The nucleus is essentially an orbiting density wave in this picture.

The entire nuclear region partakes of this orbital motion. A massive "thing," (such as a supermassive black hole) located near a galaxy's mass centroid, will also orbit (§1.1.).

In external galaxies, these center motions show up observationally as a shift of the nucleus away from the center defined by nearby isophotes (as in M33, M101, NGC 3379, NGC 3384, and a host of other galaxies) or as a velocity shift of the nucleus relative to the galaxy as a whole (as in M87; Jarvis & Peletier 1991).

One manifestation of this effect is that a particle initially at rest at the center of a galaxy starts to oscillate. Its amplitude continues to grow thereafter. An example is shown in the upper track of Figure 1, which shows one component of

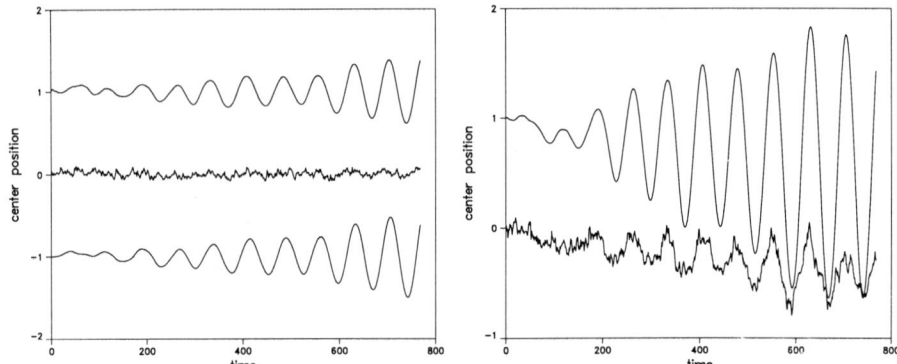

Figure 1. (Left Panel) x–projection of the position of a particle initially at rest at the center vs. time. See text for details.

Figure 2. (Right Panel) x–projection of the position of a particle initially at rest at the center of the galaxy in a different experiment with fewer particles.

the particle's position plotted as a function of the time. The amplitude doubles in about 6 – 10 cycles of the basic oscillation. In three dimensions, the orbital amplitude grows and the center follows a rather complicated trajectory.

A model reproduces the motion of that centermost particle quite well. It is based on the guess that the extremum of the galaxy's gravitational potential moves. We refer to the location of this extremum as the galaxy's "potential center." The middle track in Figure 1 shows the same component of the (vector) displacement of the potential center as a function of time. A bit of structure can just be seen above the noise at late times in the middle track.

Suppose that particle, initially at rest at the center, is pushed around by the time-varying force implied by the wandering potential center. It gets tugged first one way, then another. Within a core radius the potential is approximately harmonic. (There is no massive "thing" such as a black hole to destroy its harmonic character.) The bottom track of Figure 1 shows the motion of a harmonically bound hypothetical test particle subject to these time-varying forces. The three tracks of Figure 1 have been shifted vertically for clarity. They are all plotted to the same vertical scale.

The agreement is good, and it suggests that this model describes the physical situation. The frequencies (spring constants in the model) must be tuned rather carefully to produce agreement this good. A change of one or two timesteps per period (out of 70) changes the response amplitude noticeably. That gives a sensitive means to determine the frequency, even though the actual frequency drifts somewhat during an experiment.

Figure 2 shows the same phenomenon in data from a different experiment, one that has a larger initial amplitude. Motions of the potential center are much larger. A systematic drift shows up as well. The centermost particle and potential center drift together, verifying that the centermost particle follows the crest of the density wave that defines the galaxy's nucleus. Tracks in Figure

2 are arranged as they were in Figure 1. The growth rate is unchanged in this experiment with larger initial amplitude, even though the excursion of the potential center is noticeably greater. The drift of the center results from the initial condition.

Overstability (instability) implies that a galaxy cannot exist in Nature with its nucleus at rest atop its mass centroid. And it can't be at rest if the nucleus is anywhere else, either. Real galaxies probably do not form with the nucleus at rest atop the mass centroid. There is no way to get it there afterward, much less to keep it there. The central regions of any galaxy, including our own, are not likely to be at rest. The nucleus orbits the Galaxy's mass centroid. This is the first of dynamical effects we want to mention that could (and should) be considered in interpreting observations of the Galactic center region.

1.1. Amplitudes and Velocities

A nagging question concerning center motions is how far off center might a massive black hole be, and how fast might it move? We argued in our 1992 paper (Miller and Smith, 1992) that a black hole might wander off center until a sphere, centered on the mass centroid and whose radius is equal to the black hole's distance from the mass centroid, would contain a mass about equal to that of the black hole. For the numbers at the Galactic center (few times $10^6 M_\odot$ for the black hole and a few times $10^6 M_\odot/\mathrm{pc}^3$ mass density), this is about 0.6 pc. At that density, the black hole could move as fast as 70 km/sec, but it might also move much more slowly. It is likely to move slowly if near maximum excursion, more rapidly if near the mass centroid.

2. Lessons from A Second Core

The double peak in the brightness near the center of M31, discovered by Lauer et al. (1993), once more brings attention to multiple-core systems. The two peaks are separated by about 0.49 arcseconds on the sky. This is 1.8 pc at the distance of M31. They were found photometrically on the HST. Lauer et al. point out that the fainter of these two peaks (P2 in their notation) lies very near a center defined by nearby isophotes, and they infer that it is a true nucleus. The brighter (P1) peak lies off-center. Near infrared photometry by Rich et al. (1993) implies that doubling is not caused by an absorbing dust lane cutting across a single peak. Velocity maps of M31's nuclear region show that the brighter peak partakes of the general rotation (Bacon et al. 1994). King et al. (1994) point out that P2 (the object near the center) is bright in the far ultraviolet, while P1 can scarcely be found in that band. They infer that P1 is a pile of ordinary stars, like the bulge population of M31, while something strange may be going on in P2, as befits the nucleus of a galaxy.

A system with two or more peaks is likely not to be in a steady state from a dynamical point of view, and this raises the question how long such a system might survive. There is perceived to be a problem with the life expectancy of a system with two peaks because of energy transfer from orbital motion to internal degrees of freedom within the cores. This transfer is mediated by tidal stresses.

Tidal damage need not be as severe as generally thought. Results from numerical experiments demonstrate that a second core, once it has gotten down

Figure 3. Trajectories of one of the cores projected onto its orbital plane. The core orbits in a background potential.

into the central regions of a galaxy, can orbit for quite a long time (possibly thousands of orbits) under physical conditions that exist there. The potential generated by a nearly uniform mass density, like that present near the center of a galaxy, produces tidal stresses which are compressive in all directions. These tidal forces help a second core withstand the buffeting it suffers as it orbits through these regions. No massive "thing" (such as a black hole) is included in this picture. More detail is given below (§2.1.).

The nuclear regions of a galaxy with two cores can be considered to consist of two piles of stars, not necessarily identical. One sits at the nucleus, and the other moves through a nearly uniform common background. The principal effect of the background is that it provides a potential within which the cores orbit. We are led to investigate encounters of two piles of stars in a background. Both cores are self consistent, self-gravitating, dynamical systems, able to respond to tidal forces. From a dynamical point of view, the situation is akin to a collision of two galaxies in a background potential. The background includes any dark matter present in the nuclear regions of the galaxy. We refer to the two piles as "cores."

The two cores spiralled in to merge very quickly in a control experiment with no background potential. They merged in about $1\frac{1}{2}$ orbits. This happens because tidal distortions feed energy from the orbit into the internal structure of the cores.

By contrast, the orbits of the two cores show essentially no decay in a second experiment with a background potential, such as would be generated by a uniform mass density extending over the region in which the cores orbit. The trajectory of one of the cores through 28 complete orbits is shown in Figure 3. The distance between the centroids of the two cores is shown decreases so slowly that an extrapolation leads us to expect at least 1000 orbits before the two would get close enough to merge. There may eventually be a capture, but not for a very long time.

These experiments show that the normal background potential in the central region of a galaxy allows the cores to orbit for a long time without appreciable damage. The dynamical response of the cores to this potential makes them much more robust. That normal background potential is produced by the uniform mass distribution in the central region of the galaxy.

The experiments reported here are a kind of "proof of principle," illustrating the physical principles involved. We want to call attention to this aspect of the physical conditions in the core region, an aspect that is often overlooked but which may be quite important from a dynamical point of view.

2.1. Tidal Stresses

A galaxy without a "thing" at its center (such as a supermassive black hole) has nearly uniform mass density inside a core radius, and produces a gravitational potential that varies quadratically with distance from the center. This follows from the Poisson equation.

Tidal stresses produced by this nonzero mass density are compressive in all directions. The tidal stresses are given by the second partial derivatives of the potential. They form a tensor, and in spherical symmetry the tensor is reduced to diagonal form with one principal axis in the radial direction and the other two perpendicular to a radius. That quadratic part of the potential is oftentimes ignored (Ed Spiegel called this the "Jeans Swindle," since it is built into the usual derivations of the Jeans instability; the term has been adopted by others without attribution). The contribution of local mass density to tidal stresses near the Galactic Center has been ignored a couple of times at this meeting.

A black hole, combined with uniformly distributed matter, continues to produce stretching tidal stresses until a distance r is reached such that a sphere of radius r contains *twice* the mass of the black hole. With the numbers used earlier for the Galactic center region (§1.1.), this is about $r = 0.8$ pc. Inside r, tidal stresses are stretching in the radial direction, while they are compressive in orthogonal directions. Beyond r, tidal stresses are compressive in all directions.

In more general cases, tidal stresses can be estimated from the rotation curve, but even that requires some assumptions about the form of the potential outside the plane in which the material orbits.

3. Global Normal-Mode Oscillations

These oscillations, being global, do not strictly relate to the Galactic center region, but they suggest some interesting dynamical possibilities. We were motivated to search for global normal modes by local disturbances that were evident from the center motions work described in §1. Local disturbances are difficult to picture in a self-gravitating system with its long-range forces. We lucked out in stumbling onto a model that supported oscillations with surprisingly large amplitude. This provided good signal-to-noise and consequently permitted a fairly thorough experimental study.

You have all heard of "virialization," the notion that a galaxy will fairly quickly settle down into a quiescent state in which the "virial theorem" applies.

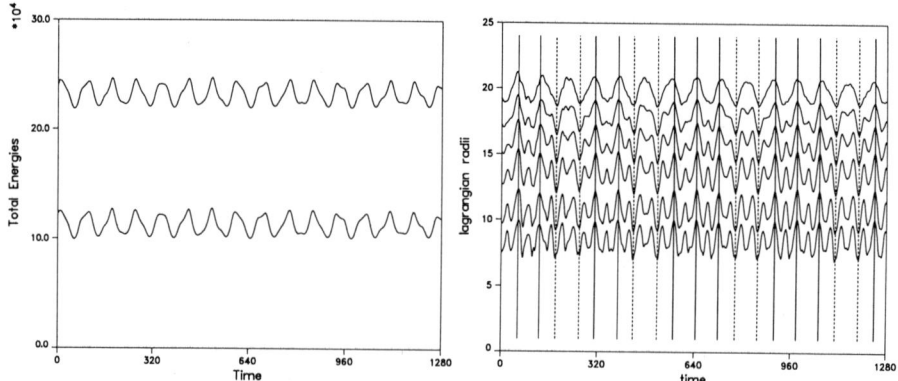

Figure 4. (Left Panel) Total kinetic energy (lower track) and potential energy (upper track; sign reversed) vs. time for an experimental galaxy mode. Peak-to-peak kinetic energy variations are about 22% of the mean value. Again, $T_{cr} = 25$, and energies are plotted in arbitrary units.

Figure 5. (Right Panel) Lagrangian radii vs. time for the same experiment as shown in the left hand panel. See text for details.

It seems less widely appreciated that the virial *equations* permit oscillatory solutions. No dynamical principle forbids oscillations.

The experimental situation is clear from Figure 4, which shows the total kinetic energy (lower track) and the total potential energy (with its sign reversed; upper track) as functions of time for this experimental galaxy. Kinetic and potential energies are not separately conserved; only their sum is conserved. The peak-to-peak variation of the total kinetic energy is 22% of the mean total kinetic energy. The oscillation continues with no perceptible damping for the duration of the experiment, about half a Hubble time in this case. We could have seen a 5% decrease, but none could be detected. A damping time of 5 − 10 Hubble times is implied. This oscillation amplitude is several hundred times that expected for $1/\sqrt{N}$ noise (an equipartition argument). It is much too large (and much too regular) to be noise caused by the limited number of particles (400 000 in the experiment shown).

Plots of the Lagrangian radii (Figure 5) provide interesting additional information about these oscillations. Lagrangian radii are familiar to astronomers, since the usual stellar structure equations are written in terms of a mass variable, R(M), the radius of a sphere, centered on the configuration, that contains a given fraction of the total mass. That's a Lagrangian radius, even if not described by that term in the stellar structure case. The bottom track in Figure 5 shows the radius of a sphere that contains 1/16 of the total mass. The second track shows 1/8, the third 1/4, the fourth 3/8, the fifth 1/2, and the top 5/8.

Several important properties of the oscillations can be seen in Figure 5. First, the bottom track (innermost of the tabulated radii) shows a rapid oscillation (34 full cycles), while the top track (outermost) shows an oscillation at about half the frequency (15 full cycles). Smooth trends leading from the one oscillation to the other can be followed through the intermediate tracks. The lack of damping is also evident in this figure.

Solid vertical lines in Figure 5 show moments when all tabulated Lagrangian radii show maxima; dashed vertical lines show simultaneous minima. These maxima and/or minima are nearly simultaneous in spite of incommensurate frequencies. Simultaneous maxima and minima imply normal modes. Waves (or wave packets) travelling inward or outward would give sloping lines connecting maxima or minima. Vertical lines imply infinite phase velocity or, more plausibly, normal modes. If so, two different normal modes are present—one with period 38 that dominates near the center and another with period 83 that dominates the outer regions.

We have been able to work out the structure of these two modes. The graver (fundamental) mode is an homologous expansion/contraction of the entire galaxy (a "breathing mode"), while the higher mode is spherically symmetrical with a radial node. Some other modes show up as well, but with lower amplitude and with geometrical shapes that make them more difficult to study.

This same pattern shows up in a wide range of experiments, although the second radial mode is not always this strong. The top track of Figure 5 follows the energy plots of Figure 4 rather accurately (but upside down), showing that the fundamental mode couples directly into the total kinetic or total potential energy. This coupling is much weaker for the second mode.

We conclude that a galaxy, once a low-order normal mode is excited, rings like a bell, supporting surprisingly large amplitudes and essentially remaining undamped over a Hubble time. This yet another dynamical freedom galaxies are free to exploit, one that must be borne in mind as you interpret your observations.

3.1. Significance for the Galactic Center

Radial flows ban be driven by galactic oscillations. They are not permitted in the usual picture, in which a galaxy is considered stationary, because material would systematically flow toward or away from some regions (such as the center), either piling up or leaving a hole. This is one of the important new features introduced by oscillations. Piling up is avoided in an oscillating case because the flow reverses after a while.

Explosions are often invoked to explain radial flows. They can drive outflows, but inward flows are not readily explained by explosions.

Depending on the modal structure, various kinds of flow patterns are possible. While we have not yet figured out how to construct a good picture of the velocity-longitude diagram for CO near the Galactic Center by means of a combination of oscillations and of center motions, that does not mean that it is difficult to do so.

References

Bacon R., Emsellem E., Monnet G., & Nieto J.-L. 1994, A&A, 281, 691
Jarvis B. J. & Peletier, R. F. 1991, A&A, 247, 313

King, I. R., Stanford, S. A., & Crane, P. 1995, AJ, 109, 164

Lauer, Tod R., Faber, S. M., Groth, E. J., Shaya, E. J. Campbell, B. , Code, A. D., Currie, D. G., Baum, W. A., Ewald, S. P. Hester, J. J., Holtzman, Jon A., Kristian, J., Light, R. M., Lynds, C. R., O'Neil, E. J., & Westphal, J. A. 1993, AJ, 106, 1436

Miller, R. H. & Smith, B. F. 1988, In Applied Mathematics Fluid Mechanics, Astrophysics, a Symposium to Honor C. C. Lin, D. J. Benney, F. H. Shu, and Chi Yuan, editors, Singapore: World Scientific 366

Miller, R. H. & Smith, B. F. 1992, ApJ, 393, 508, (Fiche 122–B7)

Rich, R. M., Mould J. R., & Graham, J. R. 1993, AJ, 106, 2252

The Dynamics of the Galactic Center: Origin of the Mini–Spiral

A.M. Fridman[1], O.V. Khoruzhii, V.V. Lyakhovich

Institute of Astronomy, Russian Academy of Sciences, 48 Pyatnitskay St., Moscow, 109017, Russia

L. Ozernoy[2]

Institute for Computational Sciences & Informatics and Department of Physics & Astronomy, George Mason University, Fairfax, VA 22030-4444, USA

O.K. Sil'chenko

Sternberg Astronomical Institute, Moscow State University, Moscow 119899, Russia

and L. Blitz

Department of Astronomy, University of Maryland, College Park, MD 20742

Abstract. It is shown that the shape of the rotation curve of the central gaseous ring (a Keplerian law inside $\simeq 0.5$ pc and a constant velocity outside $\simeq 1$ pc) can lead to the generation of an one-arm spiral wave by the mechanism of over-reflection instability. The shape of this spiral coincides with that for the Northern Arm and the Western Arc as well as does not contradict to the recent data by Roberts & Goss (1993) and Latvakoski et al.(1996).

1. Introduction

The subject of this paper is dynamics of the so-called 'mini-spiral' – the gaseous filamentary structure in the central one-parsec cavity that represents the innermost part of the circumnuclear disk at the Galactic center. The 'mini-spiral' consists of several ionized filaments (the Western Arc, the Bar, the Northern Arm, the Eastern Arm), which most likely are ionization fronts at the interfaces between low- and high-density regions (Telesco et al.1996). There have been several different approaches to find out what might be the origin of the 'mini-spiral',

---

[1] also Sternberg Astronomical Institute, Moscow State University, Moscow 119899, Russia

[2] also Laboratory for Astronomy and Solar Physics, NASA, Goddard Space Flight Center, Greenbelt, MD 20782, USA

which include (i) an inflow, (ii) an outflow, and (iii) a disk instability; those approaches are reviewed in our recent paper (Fridman et al.1996a) and below. Here, we would like to explore in more detail whether an instability mechanism is able to describe quantitatively the main features of the 'mini-spiral'.

There are two basic mechanisms for generation a spiral structure in a rotating gaseous disk – gravitational (Lin & Shu 1964, Lin, Yuan, & Shu 1969, Bertin 1994) and hydrodynamical ones (Fridman 1979, 1986, 1990, Fridman et al.1994). Since self-gravitation in the gas inside the central cavity is negligent, a hydrodynamical approach can only be of interest. As we show here, a specific instability mechanism – the so-called *over-reflection instability* leads to generation of an one-arm spiral density wave which is consistent with the recent observational data on the 'mini-spiral'.

2. Evidence for the presence of a single central disk

Observations of various components (dust, molecular gas, atomic gas, and ionized gas) all favour the flat subsystems in the central region to make up a single, differentially rotating disk. As Fig.1a shows, the innermost part of the dust ring overlaps with the atomic/ionized gas whereas its outer part overlaps with the molecular ring.

Table 1 shows that, for each flat subsystem (including the 'mini-spiral'), the position angles as well as the inclinations are close to each other. Both this and Fig.1a indicate that all the flat subsystems make up a single, although highly inhomogeneous, disk (ring). Fig.1b shows that the thickness of the circumnuclear disk increases with radius like that, on a much larger scale, for the gaseous disks in the Milky Way and other spiral galaxies.

The rotation velocity of the circumnuclear disk and its interior is nearly constant as a function of radius everywhere exept perhaps the innermost (≤ 1 pc) central part (Genzel et al.1985). This can be easily understood from a simple estimation using the mass distribution. Indeed, the density of the central stellar cluster is given asymptotically by $\rho_* \propto r^{-2}$ (Eckart et al.1993, see Fig.1c), whence it follows from the equilibrium condition, $V_\varphi^2/r = GM(r)/r^2$, that, at those outer radii, the rotation velocity must be constant: $V_\varphi^2 \propto M(r)/r \propto \rho_* r^2 =$ const. The presence of a central massive black hole or a dense stellar core would produce the Keplerian rotation curve near the center (Fig.1d). As a result, one could expect the rotation curve of the disk close to the center of the Galaxy to have a kink.

Although the overall dynamics in the central region of the Galaxy may be as complicated as its morphology, the corner-stone of many dynamical problems is apparently the origin of the 'mini-spiral'.

3. Hypotheses on the nature of the 'mini-spiral'

Available hypotheses on the nature of the 'mini-spiral' can be divided into the three categories:

1) *Inflow* (Lo & Claussen 1983; Ekers et al.1983; Quinn & Sussman 1985; Serabyn & Lacy 1985; Serabyn et al.1988; Jackson et al.1993). It is assumed

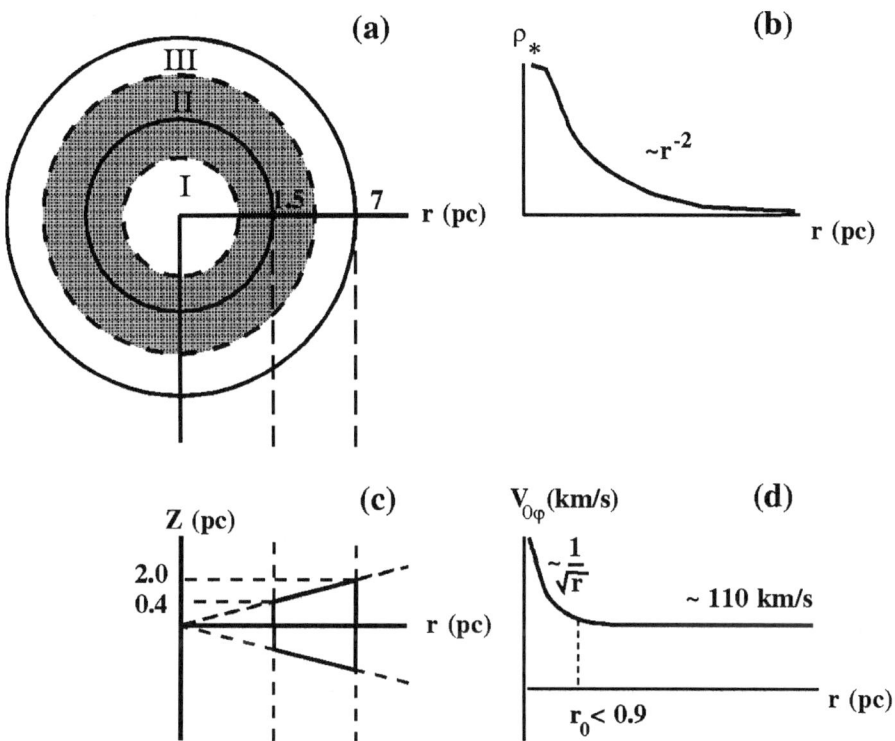

Figure 1. A sketch of the flat subsystems in the innermost central region of the Galaxy. (a) Arrangement of different components. I — the circumnuclear (molecular) ring: inner radius $r_{in} \simeq 1.5$ pc (consensus), outer radius $r_{out} \simeq 7-12$ pc (Genzel 1989, Roberts & Goss 1993, Zylka et al.1995); II — the dust ring: $r_{in} \simeq 0.9 \pm 0.2$ pc, $r_{out} \simeq 6$ pc (Davidson et al.1992); III — atomic gas: total mass $M \simeq 300$ $M_\odot$ (Jackson et al.1993), density $\rho \simeq 10^4$ cm^{-3} (Zylka et al.1995); ionized gas: $\rho_{average} \simeq 2.5 \cdot 10^3$ cm^{-3}, $\rho_{mini-spiral} \simeq 10^4$ cm^{-3} (Zylka et al.1995). (b) The thickness of the molecular ring according to Güsten et al.(1987). (c) The density run of the stellar cluster. (d) The azimuthal (rotation) velocity of the disk if a massive black hole sits in the center of the stellar cluster.

that the gaseous clouds in the central region lose, due to collisions, their angular momenta and fall into the central parsec forming a 'mini-spiral'. As soon as they arrive there, a strong UV radiation from the center dissociates and ionizes the molecular gas. Accounting for the atomic gas alone, its mass $\approx$ 300 $M_\odot$ inside the cenral 1.5 pc (Zylka et al.1995, Jackson et al.1993) and the inflow time $\sim 10^4$ yr yield the inflow rate $\approx (2-3) \cdot 10^{-2}$ $M_\odot$/yr (Jackson et al. 1993).

Table 1. Position angles and inclinations of the flat subsystems (including the 'mini-spiral') in the innermost central region of the Galaxy.

Object	Position angle	Inclination	References
Circumnuclear			Serabyn et al.1986
(molecular)	$15° \pm 10°$	$65° \pm 5°$	Güsten et al.1987
ring			Sutton et al.1990
Dust ring	$\approx 22°$	$60° \pm 5°$	Davidson et al.1992
Atomic ring	$\approx 25°$	$70° \pm 5°$	Jackson et al.1993
Ionized ring	$\approx 19°$	$61° \pm 2°$	Schwarz et al.1989
Western Arc	$\approx 22°$	$\approx 56°$	Roberts & Goss 1993
Northern Arm	$\approx 20°$	$65° \pm 5°$	Serabyn & Lacy 1985

This inflow cannot reach an immediate vicinity of the central black hole (BH) and feed the latter owing to the ram pressure of the wind from the hot stars in the central parsec (Ozernoy & Genzel 1996). However, the wind itself can serve as a source of feeding the BH due to Bondi–Hoyle accretion, especially if the BH mass, M_{BH}, is high enough (Ozernoy 1989). If M_{BH} were as high as $\sim 10^6$ $M_\odot$, the X-ray emission of Sgr A* would significantly exceed what is observed from the direction toward the Galactic center (Mastichiadis & Ozernoy 1994, Goldwurm et al.1994)[1]. Although Narayan et al. (1995) argue that the advection of radiation could appreciably reduce the accretion luminosity, a low efficiency alone seems to be unable to resolve the above contradiction (Ozernoy & Genzel 1996).

2) *Outflow from the center* (Brown 1982, Ekers et al.1983; Heyvaerts, Norman & Putritz 1988; Schwarz, Bregman & van Gorkom 1989). An assumption that the 'mini-spiral' is a double-side precessing jet (Brown 1982) is in contradiction with the absence of symmetry (e.g. Lo 1986): Sgr A* as a natural source of the outflow is displaced from the center of the 'mini-spiral'.

3) *The 'mini-spiral' as the wave*: The Northern Arm and Western Arc are a single structure rotating in the circumnuclear disk plane + some flows (Lacy, Achtermann, & Serabyn 1991). Analysis of the line-of-sight velocity field of the 'mini-spiral' has lead Lacy et al.(1991) to the conclusion that the Northern Arm and Western Arc are a coherent structure in the form of an Archimedes spiral rotating in the plane of the circumnuclear disk. At the same time, the presence of a steep velocity gradient in the region where the Eastern Arm and Bar join with the Northern Arm (Fig.2) indicates, according to Lacy et al.1991, the absence of any dynamical connection between those structures.

---

[1] It is worth noting that Lynden-Bell & Rees (1971) suspected the existence of a black hole at the Galactic center *before* the actual candidate for it (Sgr A*) was revealed.

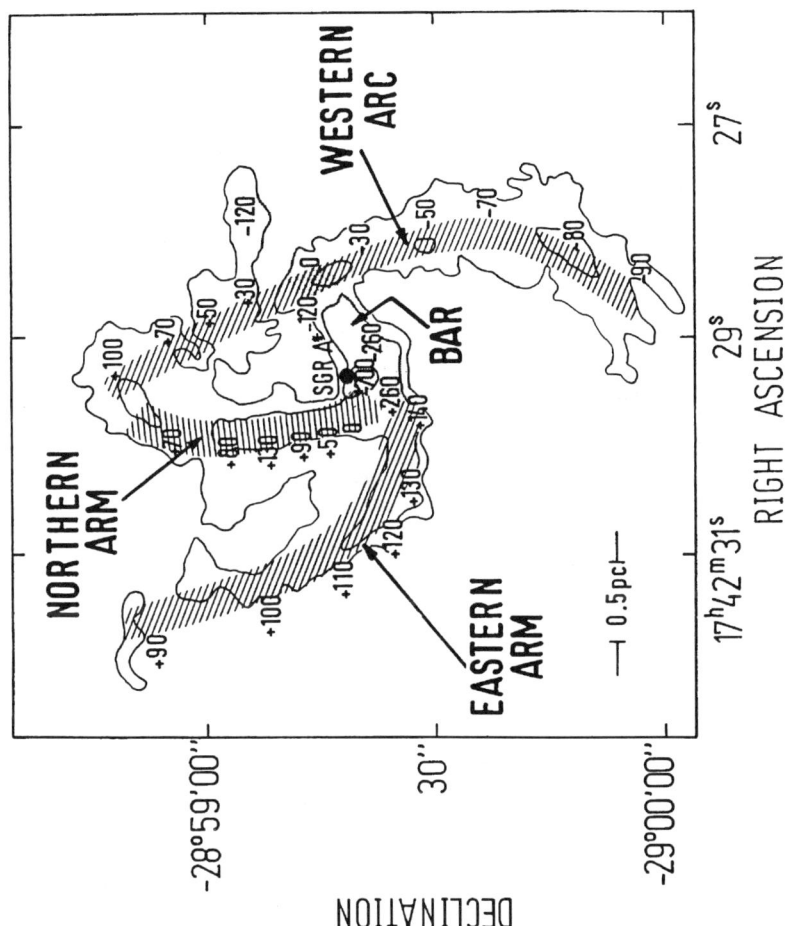

Figure 2. Line-of-sight velocity field of the 'mini-spiral' (Genzel & Townes 1987).

Roberts & Goss (1993) have reported the presence of double-peaked emission lines in the region where the Northern Arm crosses the Western Arc; they interpret this fact as an evidence for independent origins of these two features, i.e., an absence of a dynamical connection between them. We have investigated in detail their excellent data on the ionized gas velocities for the central 100×100 arc sec and have revealed multiple sites of the double-peaked emission unrelated to the crossing of the Northern Arm with the Western Arc. The brightest emission line profiles along the two features look quite regularly and allow to trace a *single* spiral arm that includes both the Northern Arm and the Western Arc. We speculate that a starburst in the center of the Galaxy (arguments for that are summarized by Lipunov et al.1996) might produce gas flows and filaments which, being superimposed on the regular spiral arm, could result in double-peaked emission lines. An example of such filaments may be a straight 'ridge' of emission which intersects (in projection) the northern half of the 'mini-spiral'; the most pronounced double-peaked profiles are located there. We note in passing that asymmetric and/or double-peaked profiles of emission lines are observed in the grand-design spiral arms of the Sc galaxy NGC 6181 as well (Fridman et al.1996b); perhaps, they are related to shock waves in the sites of an intensive star formation.

Recently, new images of the Galactic center were obtained by Latvakoski et al.(1996) with the Kuiper Widefield Infrared Camera at 31.5 and 37.7 microns. Although it is not easy to interpret these data unambiguously, a continuous transition from the Northern Arm to the Western Arc is clearly seen in these images. One can argue that this corroborates the presence of an one arm spiral at the Galactic center (distinction between images of a circle and a tightly wound spiral is generally rather difficult, though).

4. Possible mechanism of the spiral generation

Here we explore the possibility of the one-arm spiral generation in the central gaseous disk whose rotation velocity profile has a kink.

Suppose, a kink is located at the point $r = R_0$ on the rotation curve. It means that, at this point, there is a jump of the derivative of the rotation velocity: $V'_{0\varphi}(R_0) \propto \Theta(r - R_0)$ and, respectively, $V''_{0\varphi}(R_0) \propto \delta(r - R_0)$, where Θ is the Heaviside step function, δ is the Dirac delta-function. In this case, the hydrodynamical equations that describe the rotating disk dynamics can be reduced to the Schrödinger equation with the potential in the form of a δ-well. The latter has always at least one level, i.e., in the region of $r = R_0$ there is a spiral wave. The generation mechanism for this spiral wave is a corotation resonance. Indeed, the energy of a sound wave in the flow is given by (Landau & Lifshitz 1984) $E = E_0\omega/(\omega - \vec{k}\vec{v}(r)) = E_0\omega/(\omega - m\Omega_0(r))$, $E_0 > 0$, where ω is the eigenfrequency, $\vec{k}$ is the wave vector, and we use $\vec{k}\vec{v} = k_\varphi v_{o\varphi} = (m/r) \cdot (r\Omega_0)$, m being the azimuthal wave number. Evidently, energy E is positive if $\omega/m > \Omega_0(r)$.

In Fig.3 (Fridman et al.1994) the form of the potential well corresponds to the rotation velocity profile shown in Fig.1d. The wave function has an exponential cutoff at the both edges of the potential well. Therefore, if the corotation resonance were absent, the reflection coefficients at both sides of the potential

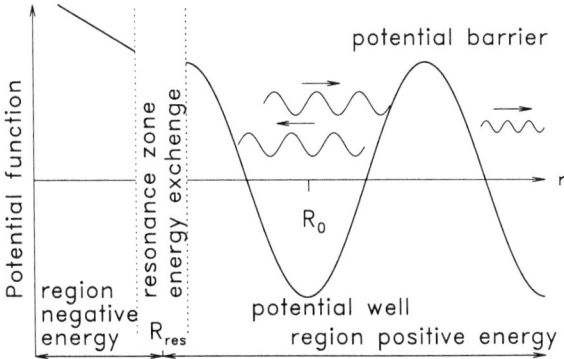

Figure 3. A sketch of the over-reflection phenomenon (Fridman et al. 1994).

barrier would be equal, with an exponential accuracy, to one. In reality, the wave reflected from the corotation resonance gets from it an additional positive energy. As a result, the reflection coefficient at the left potential barrier exceeds one[2], i.e. an *over-reflection instability* happens. Orr (1907) was the fist who discovered an over-reflection phenomenon in hydrodynamics, and in physics of galaxies an over-reflection was investigated, for the first time, by Goldreich & Lynden-Bell (1965) and by Julian & Toomre (1966).

The kink on the rotation curve shown in Fig.4a causes the generation of the one-arm spiral wave that coincides with the Northern Arm and Western Arc (Fig.4b).

It is possible that the spiral density wave is not bounded by the central cavity, but continues into the circumnuclear (molecular) ring. The presence of a density wave in the circumnuclear ring could clarify the fact that such ring parameters as the position angle and inclination both depend on the radius (Güsten et al.1987). A similar situation takes place in other galaxies whenever the gravitational potential has an axial asymmetry (Zasov & Silchenko 1996).

Acknowledgments. We are grateful to Miller Goss, Doug Roberts, and Harri Latvakoski who kindly provided their data to us; A.F., O.Kh., and V.L. acknowledge the financial support by the RFBR grant 96-02-17792.

References

Bertin, G. 1994, in Physics of the Gaseous and Stellar Disks of the Galaxy, ed. I.R. King, ASP Conference Series, v. 66, p. 35

Brown, R.L. 1982, ApJ, 262, 110

---

[2]The potential well lies on the right-hand side of the corotation circle, where the wave energy is positive since $\omega/m > \Omega_0(r)$.

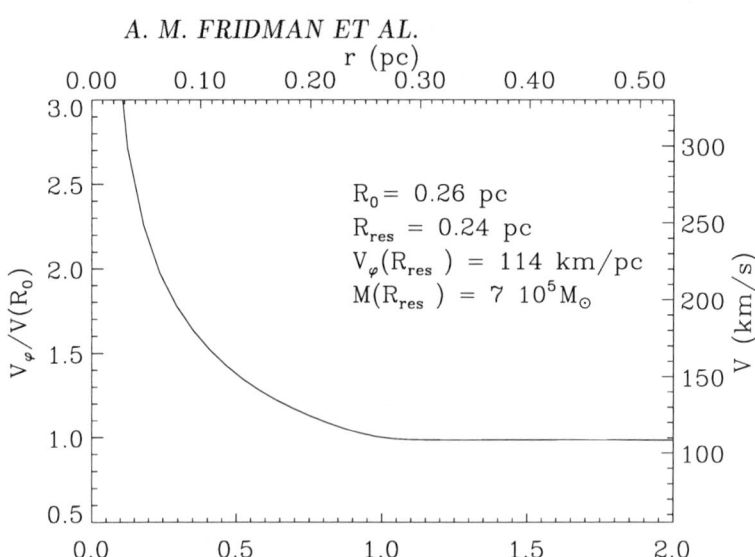

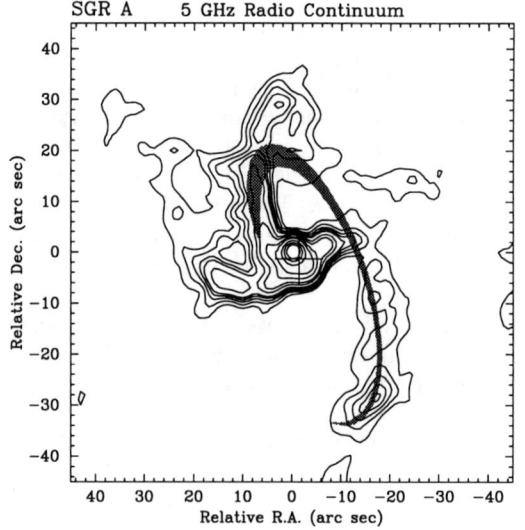

Figure 4. (a) The rotation curve used for modelling of the spiral wave. (b) The mini-spiral: the contour map represents the observational data (Blitz et al. 1993), the filled region is the computer simulation (Fridman et al.1996a), the cross is the center of rotation. The offset for the rotational center with respect to Sgr A* is 0."6W and 1."3S [the rotational center coincides with the center of the stellar cluster, 0."6W ± 0."7, 1."3S ± 1."0 (3σ errors), as given by Eckart et al. 1993]. The sound speed $c_s = 50$ km/s. The inclination of the disk is 60°, the position angle of the disk is 20°.

Davidson, J.A., Werner, W.M., Wu, X., Lester, D.F., Harvey, P.M., Joy, M., & Morris, M. 1992, ApJ, 387, 189
Eckart, A., Genzel, R., Hoffmann, R., Sams, B.J., & Tacconi-Garman, L.E. 1993, ApJ, 407, L77
Ekers, R.D., van Gorkom, J.H., Schwartz, U.J., & Goss, W.M. 1983, A&A, 122, 143
Fridman, A.M. 1979, Sov. Phys. Usp. 21, 536
Fridman, A.M. 1986, Sov. Astron. 30, 525
Fridman, A.M. 1990, in Dynamics of Astrophysical Discs, ed. J. Sellwood, Cambridge: Cambridge Univ. Press, p. 185
Fridman, A.M., Khoruzhii, O.V., Lyakhovich, V.V., Ozernoy, L., & Blitz, L. 1994, in Physics of the Gaseous and Stellar Disks of the Galaxy, ed. I.R. King, ASP Conference Series, v. 66, p. 285
Fridman, A.M., Khoruzhii, O.V., Lyakhovich, V.V., Ozernoy, L., & Blitz, L. 1996, In Proc. Nobel Simposium 98, eds. P.O. Lindblad, l. Sjouwerman, Springer–Verlag
Fridman, A.M., Khoruzhii, O.V., Lyakhovich, V.V., Silchenko, O.K., & Zasov, A.V. 1996, submitted
Genzel, R. 1989, in Proc. IAU Symp. 136 "The Center of the Galaxy", ed. M. Morris, (Kluwer: Dordrecht), p. 393
Genzel, R., & Townes, C.H. 1987, ARA&A, 25, 377
Genzel, R., Watson, D. M., Crawford, M. K., & Townes, C.H. 1985, ApJ, 297, 766
Goldreich, P., & Lynden-Bell, D. 1965, MNRAS, 130, 125
Goldwurm, A., Cordier, B., Paul, J., Ballet, J., Bouchet, L., Roques, J.-P., Vedrenne, G., Mandrou, P., Sunyaev, R., Churasov, E., Gilfanov, M., Finogenov, A., Vikhlinin, A., Dyachkov, A., Khavenson, N., & Kovtunenko, V. 1994, Nature, 371, 586
Güsten, R., Genzel, R., Wright, M. C. H., Jaffe, D.T., Stutzki, J., & Harris, A. I. 1987, ApJ, 318, 124
Heyvaerts, J., Norman, C., & Pudritz, R. 1988, ApJ, 330, 718
Jackson, J.M., Geis, N., Genzel, R., Harris, A.I., Madden, S., Poglitsch, A., Stacey, G., & Townes, C.H. 1993, ApJ, 402, 173
Julian, W.H., & Toomre, A. 1966, ApJ, 146, 810
Lacy, J.H., Achtermann, J.M., & Serabyn, E. 1991, ApJ, 380, L71
Landau, L.D. & Lifshitz, E.M. 1984, Fluid Mechanics, Oxford: Pergamon Press
Latvakoski, H., Stacey, H., Hayward, T., & Gull, G. 1996, these proceedings
Lin, C.C., & Shu, F.H. 1964, ApJ, 140, 646
Lin, C.C., Yuan, C., & Shu, F.H. 1969, ApJ, 155, 721
Lipunov, V.M., Ozernoy, L.M., Popov, S.B., Postnov, K.A., & Prokhorov, M.E. 1996, ApJ, 466, No.3
Lo, K.Y. 1986, Science, 233, 1394
Lo, K.Y., & Claussen, M.J. 1983, Nature, 306, 647
Lynden-Bell, D., & Rees, M.J. 1971, MNRAS, 152, 461

Mastichiadis, A., & Ozernoy, L.M. 1994, ApJ, 426, 599
Narayan, R., Yi, I., & Mahadevan, R. 1995, Nature, 374, 623
Orr, W.McF. 1907, Proc. Roy. Irish Acad., Sect A., 27, 9 and 69
Ozernoy, L.M. 1989, in Proc. IAU Symp. 136 "The Center of the Galaxy", ed. M. Morris, (Kluwer: Dordrecht), p. 555
Ozernoy, L. & Genzel, R. 1996, in Proc. IAU Symp. 169 "Unsolved Problems of the Milky Way", ed. L. Blitz (in press)
Quinn, P.J. & Sussman, G.J. 1985, ApJ, 288, 377
Roberts, D.A. & Goss, W.M. 1993, ApJS, 86, 133
Schwarz U. L., Bregman J.D., & van Gorkom J.H. 1989, A&A, 215, 33
Serabyn, E., Güsten, R., Walmsley, C.M., Wink, J.E., & Zylka, R. 1986, A&A, 169, 85
Serabyn, E., & Lacy, J.H. 1985, ApJ, 293, 445
Serabyn, E., Lacy, J.H., Townes, C.H., Bharat, R. 1988, ApJ, 326, 171
Sutton, E.C., Danchi, W.C., Jaminet, P.A., & Masson, C.R. 1990, ApJ, 384, 503
Telesco, C.M., Davidson, J.A., & Werner, M.W. 1996 ApJ, 456, 541
Zylka, R., Mezger, P.G., Ward-Thompson, D., Duschl, W.J., & Lesch, H. 1995, A&A, 297, 83

A Large Proper-Motion Survey in Plaut's Low-Extinction Window

R. A. Méndez

European Southern Observatory, Karl-Schwarzschild Straβe 2, D-85748, Garching b. München, GERMANY

R. M. Rich[1]

Columbia University, Astronomy Department, 538 W. 120th St. Box 42 Pupin, New York, NY 10027, USA

W. F. van Altena and T. M. Girard

Yale University, Astronomy Department, P. O. Box 208101, New Haven, CT 06520-8101, USA

S. van den Bergh

Dominion Astrophysical Observatory, 5071 W. Saanich Rd., Victoria, BC, V8X 4M6, CANADA

S. R. Majewski

University of Virginia, Department of Astronomy, P. O. Box 3818, Charlottesville, VA 22903-0818, USA, and Carnegie Observatories, 813 Santa Barbara Street, Pasadena, CA 91101, USA

Abstract. We present preliminary results from the deepest and largest photographic proper-motion survey ever undertaken of the Galactic bulge. Our first-epoch plate material (from 1972-3) goes deep enough ($V_{lim} \sim 22$) to reach below the bulge main-sequence turnoff. These plates cover an area of approximately $25' \times 25'$ of the bulge in the low-extinction ($A_v \sim 0.8$ mag) Plaut field at $l = 0°, b = -8°$, approximately 1 kpc south of the nucleus. This is the point at which the transition between bulge and halo populations likely occurs and is, therefore, an excellent location to study the interface between the dense metal-rich bulge and the metal-poor halo.

In this conference we report results based on three first-epoch and three second-epoch plates spanning 21 years. It is found that it is possible to obtain proper-motions with errors less than 0.5 mas/yr for a substantial number of stars down to V= 20, without color restriction. For the subsample with errors less than 1 mas/yr we derive proper-motion dispersions in the direction of Galactic longitude and latitude of 3.378 ± 0.033

---

[1] Visiting Astronomer, Cerro Tololo Inter-American Observatory which is operated by AURA, Inc. under cooperative agreement with the National Science Foundation

mas/yr and 2.778 ± 0.028 mas/yr respectively. These dispersions agree with those derived by Spaenhauer et al. (1992) in Baade's window.

1. Introduction

Historically, tremendous observational effort has been invested in understanding the formation of the halo and chemical evolution of the disk. Ideally, one wants to measure proper-motions, radial velocities, and abundances for members of a stellar population, inspired by the seminal effort of Eggen, Lynden-Bell & Sandage (1962). While the halo does not appear to exhibit clear correlations of abundances with kinematics (cf. Carney et al. 1990), there is some indication that metal-rich bulge stars have a smaller velocity dispersion (Rich 1990). In much larger samples of bulge giants in fields away from the minor-axis, Minniti (1993, 1996) finds abundance/kinematics trends that suggest that more metal-rich bulge giants have greater rotational support. Unfortunately, fields distant from the minor axis may be contaminated with giants at the tangent point arising from the disk population (Tiede and Terndrup 1996). The next step in confirming Minniti's findings is the careful study of a minor-axis field, including proper-motions.

These issues can be settled by correlating abundances with proper-motion and radial velocity dispersions. The geometry for viewing the bulge is favorable: we are 8.5 kpc distant from a system which has most of the mass contained within 1 kpc. We are therefore privileged to study the bulge from an almost extragalactic perspective.

Given its extreme crowding, high extinction, and southerly declination, the bulge has received substantially less observational attention than the globular cluster system. Only one proper-motion study has been undertaken (Spaenhauer et al. 1992). While a landmark achievement, their work addresses the extremely crowded Baade's Window field ($l = 0°, b = -4°$). There were three first-epoch plates, all obtained on the Palomar 200-inch telescope. Because the plates were B plates, very few late M giants were measured; these stars dominate the bulge asymptotic giant branch, and it is of great interest to compare their kinematics with bulge K giants.

Our study addresses the tangential kinematics of the Galactic bulge in Plaut's (Plaut 1970, 1971) low-extinction window ($E_{B-V} = 0.25$ mag, van den Bergh & Herbst 1974). This window, centered at $l = 0°, b = -8°$, provides a unique place to look at the Galactic bulge and its transition into the the halo of the Galaxy, as the line-of-sight crosses the Galactic minor-axis some 1.3 kpc to the South of the nucleus. Furthermore, this region has smaller reddening and is less crowded than Baade's window ($E_{B-V} = 0.42$ mag, Blanco and Blanco 1985).

The field in this study is of critical importance as it lies at the edge of the bulge (as defined, for example, by the *COBE-DIRBE* map). It has extinction $A_v = 0.8$ mag, lower than most bulge fields, and it will contain a substantial number of halo giants, allowing one to probe the bulge/halo transition. We know little about the field population of the inner halo; this study will also fill that gap in our knowledge.

We expect to measure CCD photometry in the B, V, and I passbands, and proper-motions for an unbiased sample of approximately 30,000 stars in our minor-axis field. We hope to further obtain radial velocities and low resolution abundances for about 5,000 stars. A large, unbiased sample is important because much of the outcome depends on dividing the data into subsamples as a function of abundance or kinematics.

2. Plate material, measurements, and the proper-motions

Our project is based on a unique sample of twenty photographic plates of a Galactic bulge field on the minor-axis at $b = -8°$ obtained by Sidney van den Bergh in 1972-3 using the Kitt Peak 84-inch and the 200-inch telescopes (van den Bergh & Herbst 1974). The 100-inch telescope at Las Campanas has been used to obtain thirteen second epoch plates in 1993; deep intermediate epoch plates of this field (1979) were obtained by Jeremy Mould also at Las Campanas (these intermediate epoch plates have not been used in this preliminary study).

The plates were digitized using the the Yale PDS 2020G laser interferometer/microdensitometer measuring machine in raster-scan mode. The aperture size was 33 μm, while the step size was 30 μm. At this step size, the faintest images will be properly sampled for the astrometric centroiding (our plates have scales on the order of 10 arc-sec/mm). All scans were performed under the best possible thermal conditions to avoid instrumental drifts during the scan. Each scan took between 8 and 10 hours. Depending on the plate scale, the digitized frames had between $4,200 \times 4,200\, pixels^2$ and $5,600 \times 5,600\, pixels^2$. The PDS system outputs an integer fits-file with 16 bits/pixel that is easily converted into other formats for later analysis. At this conference we report on preliminary results from reductions of only three first-epoch and three second-epoch plates.

Analysis of the Yale-PDS microdensitometer data routinely yields star centroids to 1/20 of a pixel (20 mas). A star measured on five plates in each color will have its position known at least 2–3 times better than this, and over the 21 yr baseline we expect errors no larger than 0.5 mas/yr in each color. This corresponds to approximately 20 km/s at the distance of the Galactic Center. This is comparable to Spaenhauer et al., and matches the accuracy with which radial velocities in our spectroscopic follow-up will be measured.

The first step in the process of obtaining proper-motions was to create an input catalogue of approximate stellar positions. For this purpose we selected a relatively deep second-epoch master plate with the best image and fog characteristics, requiring also that the full FOV would be available in this master plate. Then, we created a master list of candidates to perform the astrometric solution. This was done by running DAOFIND from the DAOPHOT package within IRAF. Approximate "image" parameters (FWHM) and "frame" parameters (the rms variation on the plate fog) were computed from the digitized frame of the master plate. The master list contains a little less than 100,000 detections at an 8σ level. Visual inspection of the detections against the digitized plate confirmed that no obvious stars will be left-out of the master list, even in cases of relatively high crowding.

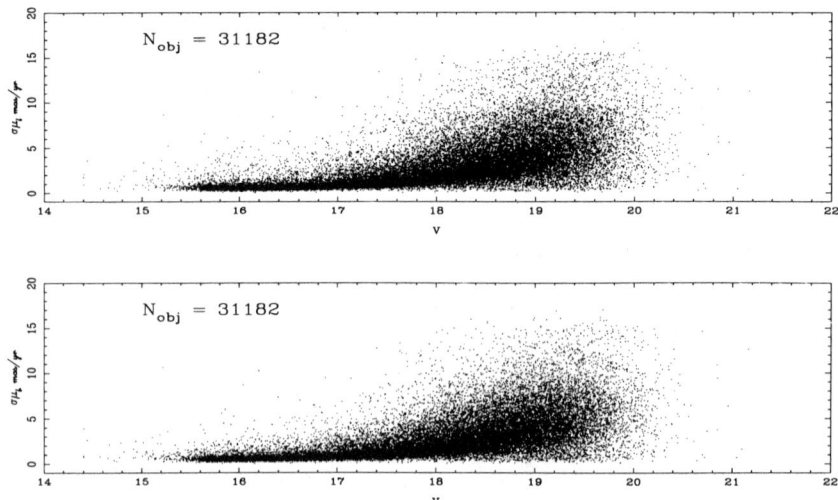

Figure 1. Proper-motion error *vs.* V magnitude. Upper panel is for Galactic longitude, lower panel is for Galactic latitude. Note the rapid increase in the errors for $V > 18$.

This master list was then used to "feed" the Yale Image Centering routine described by Lee & van Altena (1983), which provides image centers, instrumental magnitudes and centering error estimates.

The next step was to reduce all of the plates to the master plate. No preliminary corrections for atmospheric refraction or distortion were performed in this solution. Refraction is likely to be an important effect, particularly for the first-epoch plates. However, this effect would be absorbed in the plate constants when converting to the master plate. The centered list on each plate was converted to the master plate using interactive software designed to recognize and isolate outliers, and to include distortion terms also in an interactive way until the residuals show no systematic trends.

The plate transformations described above indicate that we are indeed able to reach the expected level of accuracy. For example, plate-to-plate transformations between pairs of same-epoch plates have an rms scatter of 35 mas (i.e., 25 mas on each plate). On the other hand, transformations between first and second epoch plates indicate an rms dispersion of 80 to 100 mas. This higher dispersion is due to the proper-motion dispersion of the stars: For a population of velocity dispersion of 120 km/sec at 8 kpc, we expect an rms dispersion of 75 mas (21 yr baseline). Convolving the 75 mas with our measurement error gives our measured 80 mas dispersion. Evidently, the full analysis (including a pre-correction for optical distortion and atmospheric refraction) will give more precise results, but these preliminary reductions show that we can achieve the required accuracy from the available plate material. Once all the plates had been transformed to the master plate, and distortions had been removed, we computed the proper-motions by performing a *weighted* least-square linear fit to position *vs.* time, taking into account the individual plate solutions. This solution gave not only proper-motions but also error estimates for the derived motions. The individual proper-motion errors were computed from the scatter of the residuals about the best-fit line as well as from the formal error of the slope in the linear fit. Figure 1 shows our proper-motion errors in Galactic longitude (μ_l) and latitude (μ_b) as a function of apparent V magnitude. For $V > 18$ the errors increase sharply. Also, we have very few objects with $V > 20$, mainly because of the magnitude limit for the 1st-epoch 84-inch plates. The use of the 200-inch plates will allow us to go to $V \approx 22$ for a selected sample of stars. Table 1 indicates the mean proper-motion errors in μ_l and μ_b as a function of V magnitude, as well as the number of stars in each magnitude interval. For $V > 18$ the errors become of the same order as the expected proper-motion dispersion ($\Sigma_\mu \approx 3\ mas/yr$). At this magnitude (or fainter), the sample with proper-motion errors less than about 1 mas/yr will only contain objects far from the mean error, and their errors are probably not very well determined. This translates into a spuriously large proper-motion dispersion for the fainter objects (see Table 2).

Table 1. Proper-motion error vs. V magnitude.

Magnitude range	Number of stars	σ_{μ_l} mas/yr	σ_{μ_b} mas/yr
$14.0 \lesssim V < 16.0$	1530	0.75	0.69
$16.0 \lesssim V < 17.0$	3944	0.91	0.88
$17.0 \lesssim V < 17.5$	3023	1.47	1.44
$17.5 \lesssim V < 18.0$	4102	2.18	2.13
$18.0 \lesssim V < 18.5$	5904	3.12	3.05
$18.5 \lesssim V < 19.0$	6392	4.10	4.00
$19.0 \lesssim V < 22.0$	6287	5.32	5.23

3. Analysis of the proper-motions

The vector-point diagram for proper-motion error cuts at 1 mas/yr is shown in Figure 2. This figure compares well with Figure 1 in Spaenhauer et al. (1992). This suggests that the proper-motion dispersion for these two fields is approximately the same, as it is indeed the case (Table 2). The combination of a large sample and small measurement errors means that the proper-motion dispersions are known with great accuracy. Following Spaenhauer et al. (1992) the true (error-corrected) proper-motion dispersion, Σ_μ, is given by:

$$\Sigma_\mu^2 = \frac{1}{(n-1)} \sum_{i=1}^{n} (\mu_i - \bar{\mu})^2 - \frac{1}{n} \sum_{i=1}^{n} \sigma_{\mu_i}^2 \qquad (1)$$

where n is the sample size, μ is one component of the proper-motion, and σ_{μ_i} is the error of a single proper-motion measurement. The error in Σ_μ in any subsample of n stars is given by:

$$\xi_\Sigma = \left(\frac{1}{2n} \Sigma_\mu^2 + \frac{1}{2n^2 \Sigma_\mu^2} \sum_{i=1}^{n} \frac{\sigma_{\mu_i}^4}{n_i} \right)^{1/2} \qquad (2)$$

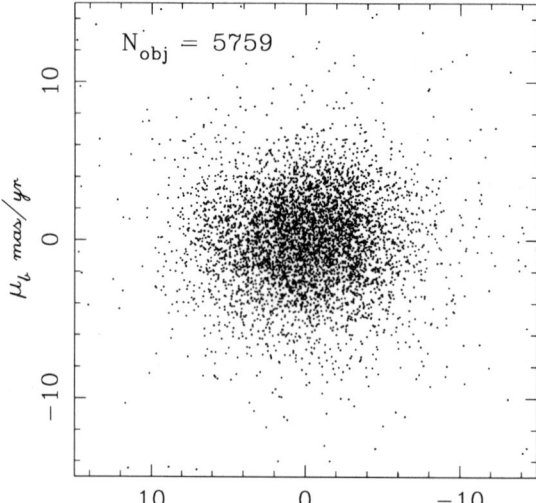

Figure 2. Vector-point diagram for objects with errors less than 1 mas/hr in each coordinate. This Figure compares very well with Figure 1 in Spaenhauer et al. (1992).

where n_i is the number of plates on which star i is measured. Spaenhauer et al. (1992) measured proper-motion dispersions of approximately 3 mas/yr. The above equations show that for a typical error of 0.5 mas/yr for a single measurement (Spaenhauer et al.'s level) our results would be unaffected by measurement errors. For subsamples of 200 stars, we estimate that the error in each subpopulation dispersion would be approximately 0.1 mas/yr (4 km/s at the distance of the Galactic bulge).

We have computed (intrinsic) proper-motion dispersions along Galactic longitude (Σ_{μ_l}) and latitude (Σ_{μ_b}) using Equations (1) and (2). In order to properly handle outliers, dispersions were determined in an iterative way following a procedure similar to the technique of outlier elimination using probability plots (Méndez & van Altena 1996). The results for the dispersions are shown in Table 2.

Table 2 shows how important it is to have a large sample of small-error proper-motions; the dispersions are determined with very high accuracy, typically the errors are less than 2%. Also, it can be seen that the proper-motion dispersion *does not* seem to change with apparent magnitude within the uncertainties (except for the large-error bin at $18 \leq V < 22$). There may be an indication that Σ_{μ_l} is slightly increasing with apparent magnitude, but the change is only at the 2σ level. Since we *do* expect to see a mixture of populations along the line-of-sight, each with different dispersion and mean rotational motion, this result implies that the contamination by these populations is rather minimal. The stellar ratio of disk/bulge, thick-disk/bulge, and halo/bulge is expected to change as a function of V magnitude, and so, therefore, is the proper-motion dispersion. We do not expect the disk or thick-disk to make a very large contribution (e.g., at Baade's window their contribution is less than 20%, and our field is at twice the Galactic latitude). However, we *do* expect to have a rather significant contribution from stars in the inner halo, which we do not seem to detect. Table 2 also shows that the proper-motion dispersions Σ_{μ_l} and Σ_{μ_b} are

Table 2. Intrinsic proper-motion dispersion vs. V magnitude (only stars with errors less than 1 mas/yr in each coordinate included in solutions).

Magnitude range	No. stars in l	Σ_{μ_l} mas/yr	No. stars in b	Σ_{μ_b} mas/yr
$14.0 \leq V < 16.0$	1136	3.279 ± 0.069	1136	2.811 ± 0.059
$16.0 \leq V < 16.5$	1442	3.272 ± 0.061	1438	2.872 ± 0.053
$16.5 \leq V < 17.0$	1211	3.432 ± 0.070	1206	2.674 ± 0.055
$17.0 \leq V < 18.0$	1299	3.464 ± 0.068	1293	2.748 ± 0.054
$18.0 \leq V < 22.0$	236	5.293 ± 0.244	238	5.270 ± 0.240
$14.0 \leq V < 18.0$	5088	3.378 ± 0.033	5077	2.778 ± 0.028
Spaenhauer et al.	429	3.2 ± 0.1	429	2.8 ± 0.1

different in Galactic latitude and longitude at a 14σ level, which is a much more definitive result than that of Spaenhauer et al.'s (shown on the last line of Table 2), who suggested differences at the 3σ level. On the other hand, our results do agree with Spaenhauer et al.'s results within their rather large uncertainties in both Σ_{μ_l} and Σ_{μ_b}.

The interpretation of the values listed in Table 2 in terms of velocity dispersions for the bulge stars is complicated due to the expected contamination from halo stars. A simple two-component model predicts that the velocity dispersions for bulge stars in Galactic longitude and latitude (Σ_{B_l} and Σ_{B_b} respectively) are given by:

$$\Sigma_{B_l}^2 = (1+x)\Sigma_l^2 - x\Sigma_{H_l}^2 - \frac{x}{1+x} <V_B>^2 \qquad (3)$$

and

$$\Sigma_{B_b}^2 = (1+x)\Sigma_b^2 - x\Sigma_{H_b}^2 \qquad (4)$$

where Σ_l and Σ_b are the total velocity dispersions in Galactic longitude and latitude respectively, Σ_{H_l} and Σ_{H_b} are the halo velocity dispersions in Galactic longitude and latitude, $<V_B>$ is the mean rotation for the bulge, and x is the ratio of the number of halo to bulge stars in our sample.

Assuming 8.5 kpc as the distance to the Galactic center, and the mean values derived from Table 2, Equations (3) and (4) imply that $\Sigma_{B_l} \approx 140\ km/s$ and $\Sigma_{B_b} \approx 120\ km/s$. The larger Σ_{B_l} is consistent with rotation broadening and anisotropy of the Galactic bar (Zhao 1996, Zhao et al. 1996), but it does not necessarily argue for triaxiality. Zhao et al. (1994) have shown that only a strong *vertex deviation* of the bulge velocity ellipsoid (i.e., a non-zero cross-term $<V_r V_l>$) will be a definitive indication of triaxiality.

4. Conclusions and the future

We can obtain proper-motions for a large sample of bulge stars with errors small enough to allow a meaningful kinematical study of the bulge. It is found that the proper-motion dispersion in our field is comparable with that found by Spaenhauer et al. (1992) in Baade's window. Our dispersions are consistent with broadening by rotation and anisotropy of the Galactic bar as predicted by the dynamical models of Zhao et al. (1996).

Radial velocities from our spectroscopic survey will be extremely important as a complement to the proper-motions to confirm the presence of a bar. Triaxiality will take the form of a vertex deviation in the Σ_r vs. Σ_l plane, as suggested from the Spaenhauer et al. data analysed by Zhao et al. (1994), and from a larger spectroscopic follow-up of Spaenhauer et al.'s sample by Rich et al. (1996).

If the bulge collapsed and spun up as metallicity increased, we should see Σ_r and Σ_b decrease with higher metallicity (Minniti 1993, 1996). In a rapidly rotating population, integration through the line of sight will reveal that Σ_l will be artificially broadened (Zhao et al. 1994, 1996). Applied to our large minor-axis sample this analysis will help constrain the formation/enrichment history of the bulge.

Acknowledgments. We are grateful to Jeremy Mould for lending us the deep intermediate-epoch plates taken by him in 1979. We are also grateful to Kyle Cudworth, Michael Irwin, Dante Minniti, and HongSheng Zhao for useful discussions. TMG and WFvA acknowledge partial support from the National Science Fundation and NASA. SRM was supported by Hubble Fellowship Grant Number HF-1036.01-92A awarded to the Space Telescope Science Institute which is operated by the Association of Universities for Research in Astronomy, Inc. for NASA under Contract No. NAS5-26555.

References

Blanco, V. M. and Blanco, B. M., 1985, Mem. Soc. Astron. Ital., 56, 15
Carney, B. W., Latham, D. W., Laird, J. B., 1990, AJ, 99, 572
Eggen, O. J., Lynden-Bell, D., and Sandage, A. R. ApJ, 136, 748
Lee, J. -F., and van Altena, W. F., 1983, AJ, 88, 1683
Méndez, R. A., and van Altena, W. F., 1996, AJ, submitted
Minniti, D., 1993, Ph.D. Thesis, University of Arizona
Minniti, D., 1996, ApJ, 459, 175
Plaut, L., 1970, A&A, 8, 341
Plaut, L., 1971, A&AS, 4, 75
Rich, R. M., 1990, ApJ, 326, 604
Rich, R. M., Terndrup, D. M., and Sadler, E. M., 1996, in preparation
Spaenhauer, A., Jones, B. F., Whitford, E., 1992, AJ, 103, 297
Tiede, G. P., and Terndrup, D. M., 1996, in preparation
van den Bergh, S., and Herbst, E., 1974, AJ, 79, 603
Zhao, H. S., Spergel, D. N., and Rich, R. M. 1994, AJ, 108, 2154
Zhao, H. S., 1996, M.N.R.A.S., submitted
Zhao, H. S., Rich, R. M., and Biello, J., 1996, ApJ, in press

Galactic Cores as Separate Stellar Subsystems

Olga K. Sil'chenko

Sternberg Astronomical Institute, Moscow 119899, Russia

Abstract. Cores distinguished both dynamically and chemically are found in some early-type spiral galaxies. There are some evidences that formation of decoupled cores may be a result of galactic interaction or merging, followed by a secondary star formation burst in the nuclei.

1. Introduction

Kinematically decoupled galactic cores became known in 1988, when Kormendy (1988a, b), Dressler & Richstone (1988), and Jarvis & Dubath (1988) have found that very central parts of rotation curves in M 31 and NGC 4594 are distinguished by a fast rotation. The same thing was found in a dozen of ellipticals by Jedrzejewski & Schechter (1988) and Bender (1988).

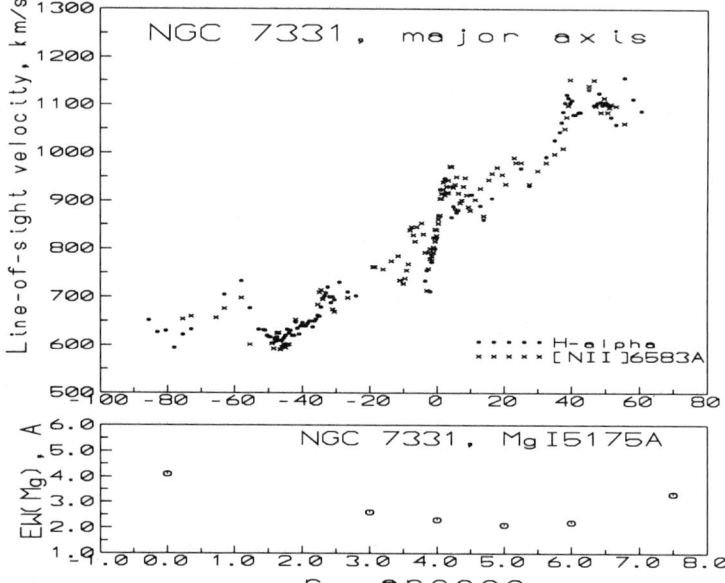

Figure 1. *a* – Line-of-sight velocity profile along the major axis for NGC 7331; *b* – Magnesium-line strength radial profile for NGC 7331

Firstly kinematically decoupled nuclei were interpreted as supermassive black holes – so called "dead quasars". But later Bender & Surma (1992) have found a sharp drop of Mg-line strength outside of decoupled cores in 4 elliptical galaxies; so kinematically decoupled cores in some cases appear to be also chemically decoupled ones. We have faced kinematically decoupled cores in spiral galaxies during our investigation of galactic rotation curves at the 6m telescope in 1985-1988. We have found six galaxies in which nuclear regions with radii of 100-300 pc were distinguished by fast rotation (Afanasiev et al. 1989). In some of them possessing extended bulges the dynamically decoupled cores have to be compact nuclear disks – such is the case of NGC 7331 (Fig. 1). A special observational task to find stellar population differences between decoupled cores and their outskirts was formulated for the Multi-Pupil Field Spectrograph of the 6m telescope. First observations in 1989 gave positive results: three early-type spiral galaxies with dynamically decoupled nuclei showed Mg-line strength drops beyond the unresolved nuclei (Sil'chenko et al. 1992). To make the search of chemically decoupled nuclei more effective, I have used catalogues of multi-aperture photoelectric data (Longo & Vaucouleurs 1983, 1985). where galaxies with distinct red nuclei were separated; among 234 early-type galaxies about 25% of ellipticals and lenticulars and more than 50% Sa-Sb galaxies appear to have such nuclei. A list of galaxies with photometrically distinct nuclei seen on the northern sky includes 34 objects (Sil'chenko 1994). Now I present first results of 3D spectroscopy for some of them.

2. Observations

Observations were made with the Multi-Pupil Field Spectrograph (MPFS) of the 6m telescope of the Special Astrophysical Observatory (Russia, Nizhnij Arkhyz). The array of 8×12 square microlenses allowed to obtain 96 spectra from a rectangular area, say, in the center of a galaxy. One spatial element is $1.3'' \times 1.3''$. The dispersions were 1.4 - 2.0 Å/px (spectral resolution 5 - 10 Å). We exposed two spectral ranges: 4800–5400 Å to obtain radial dependencies of absorption-line equivalent widths (mainly of H_β, $MgI\lambda5175$, and $FeI\lambda5270$) and 6250–6800 Å to obtain two-dimensional gas velocity fields by measuring emission lines H_α and $[NII]\lambda6583$. The detectors were IPCS of 512×512 up to May of 1994, and later CCD of 530×580. The spectra registered with CCD in the green were also used to derive stellar velocity fields by cross-correlation. So, the results of one exposure with MPFS include surface brightness maps, both in continuum and in emission lines, two-dimensional velocity fields, both for gas and stars, and radial profiles of absorption-line strengths. The accuracy of velocities is about 20 km/s, that of equivalent widths – 0.15 Å (CCD) and 0.30 Å (IPCS).

3. Results

NGC 2685 is a well-known peculiar S0 galaxy with an outer polar ring well-detected in neutral hydrogen and broad-band filters. The wide-spread hypothesis of polar-ring genesis is a capture of gas-rich satellite, destroyed after capturing by tidal forces. So, galaxies with polar rings are *a priori* past-interacting ones. We have found a chemically decoupled unresolved nucleus in NGC 2685 (Fig. 2a). The galaxy was observed twice: with IPCS and with CCD; the results agree well (some discrepancy at the radius of $2.6''$ results from different seeing quality). Observations in the red have revealed that in the center ionized gas rotates perpendicularly to stellar component rotation. So the polar ring of NGC 2685 penetrates into very center of the galaxy being in reality rather a polar disk. Obviously, the polar-ring gas may supply a material for a secondary star formation burst which has produced the chemically decoupled core of NGC 2685.

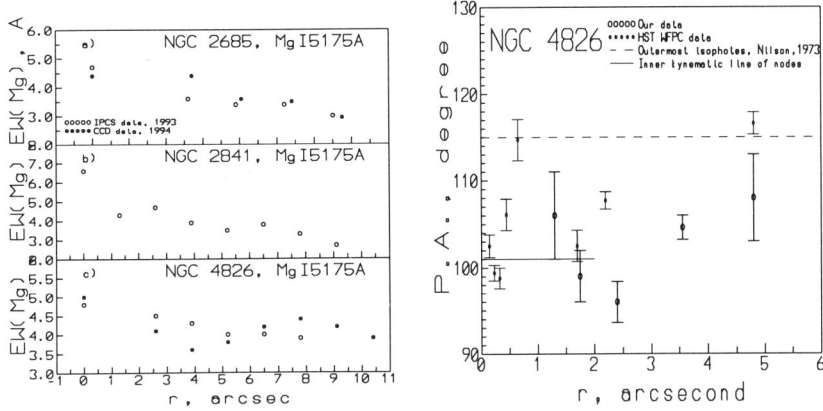

Figure 2. Left – Mg-line profiles for 3 disk galaxies; right – measurements of the isophote major-axis P.A. for the center of NGC 4826

If NGC 2685 is known to be a peculiar galaxy, a giant Sb spiral NGC 2841, on the contrary, looks like a regular, quite isolated galaxy. Rotation of stars in the center of NGC 2841 was studied more than once; a central part of the galaxy with the radius of $\sim 7''$ (the core) is kinematically decoupled. We have found that the galaxy possesses also a chemically decoupled unresolved nucleus (Fig. 2b). The bulge demonstrates its own Mg-line gradient, but being extrapolated to the center, it gives an equivalent-width value lower by more than 3σ than the real Mg-line equivalent width in the nucleus. And again a chemically decoupled nucleus is accompanied by a nuclear gaseous "polar disk": its central kinematical line of nodes is perpendicular to a photometric line of nodes. A radius of the nuclear polar disk is estimated as $4''-5''$, so it cannot be identified with the unresolved chemically distinct nucleus. But it is a sign of past interaction and of possible merging accompanied by strong dissipative effects.

Sab galaxy NGC 4826 became famous two years ago when a switch of gas rotation sense at the radius of about 1.2 kpc was found in it (Braun et al. 1994, Rix et al. 1995). The inner gas rotates together with stars, and outer gas counterrotates. A suggestion was made that outer gas was accreted by NGC 4826 during some interaction (or gas-rich satellite merger). In NGC 4826 we have also found a nucleus distinguished by stellar population properties (Fig. 2c); as a stellar rotation velocity has a distinct maximum at the radius of $4''$ (Rix et al. 1995), we can state that the central core is also decoupled by increased density; so a secondary star formation burst in the nucleus is very probable. In my paper on NGC 4826 (Sil'chenko 1996) a hint on rotation plane warp by $15°$ in the very center was not persisted. Now I state a reality of this warp: in Fig. 2(right) one can see that HST WFPC data on NGC 4826, together with ours, confirm an orientation of the very inner isophotes in the position angle $\sim 100°$. As the kinematical line of nodes inside $r = 2''$ is also at $P.A. = 101°$ ($281°$), we conclude that nuclear ionized gas rotates circularly in the plane inclined by $15°$ to the global plane of the galaxy.

4. Conclusions

We have found several chemically and dynamically decoupled cores in early-type disk galaxies. Evidently, they result from secondary star formation bursts in the galactic nuclei, because a dissipationless merging of a stellar system is improbable for a galaxy with a thin cold disk. In all three cases we see obvious signs of past interaction. In addition, in NGC 7331 (Fig. 1) a counterrotating bulge was detected (Prada et al. 1996). There must be a link between galaxy interaction and decoupled core occurence. Perhaps, interaction stimulates strong gas inflow in galactic nuclei, which in turn produces a secondary nuclear star formation burst.

Acknowledgments. I am grateful to observers of the 6m telescope V. L. Afanasiev, A. N. Burenkov, S. N. Dodonov, S. V. Drabek, and V. V. Vlasyuk for the assistance during observations. The WFPC images of NGC 2841 and NGC 4826 were used from the HST Archive.

References

Afanasiev V. L., Sil'chenko O. K., Zasov, A. V. 1989. A&A, 213, L9

Bender R. 1988. A&A, 202, L5

Bender R., Surma P. 1992. A&A, 258, 250

Braun R., Walterbos R., Kennicutt R. C., Tacconi L. J. 1994. ApJ, 420, 558

Dressler A., Richstone D. O. 1988. ApJ, 324, 701

Jarvis B. J., Dubath P. 1988. A&A, 201, L33

Jedrzejewski R., Schechter P. L. 1988. ApJ, 330, L87

Kormendy J. 1988a. ApJ, 325, 128

Kormendy J. 1988b. ApJ, 335, 40

Longo G., de Vaucouleurs A. 1983. A General Catalogue of Photoelectric Magnitudes and Colors in the U, B, V System. Austin: Univ. Texas Press

Longo G., de Vaucouleurs A. 1985. Supplement to the General Catalogue of Photoelectric Magnitudes and Colors of Galaxies in the U, B, V System. Austin: Univ. Texas Press

Prada F., Gutierrez M., Peletier R. F., McKeith C. D. 1996. ApJ, in press

Rix H.-W. R., Kennicutt R. C., Braun R., Walterbos R. 1995. ApJ, 438, 155

Sil'chenko O. K., Afanasiev V. L., Vlasiuk V. V. 1992. AZh, 69, 1121

Sil'chenko O. K. 1994. AZh, 71, 706

Sil'chenko O. K. 1996. Pis'ma v Astron. Zh., 22, 124

On Masses of Equilibrium Configurations

L.V.Verozub

*Department of Physics and Astronomy, Kharkov State University
Kharkov 310077 Ukraine (verozub@gravit.kharkov.ua).*

Abstract. Proceeding from the our gravitation equations (see Verozub 1991) we argue that the theory in principle allows equilibrium stable configurations of a degenerate fermion gas with very large masses. This are objects with low luminosities . It is an alternative to the hyposesis about a massive black hole in the Galaxy center.

Proceeding from the newtonian gravity law and Einstein's equations it is considered that masses of equilibrium configurations cannot go over several Sun masses. In the paper (Verozub 1991) new vacuum gravitational bimetric equations was proposed, which in flat space-time have no physical singularity for the spherically symmetric field. If the distance r from an attractive mass M is much larger than the Shwarzshild radius α, then their physical consequences coinside with the ones in Einstein's theory. However, they are quite different at r of the order α or less than that. There is no events horizon at $r = \alpha$. The gravitational force affecting the a test particle of the mass m in rest is given by

$$F = -\frac{GmM}{r^2}(1 - \alpha/f), \qquad (1)$$

where G is the gravitational constant, $\alpha = 2GM/c^2$, c is the speed of light, $f = (\alpha^3 + r^3)^{1/3}$. Fig. 1 shows the plot of the function $F_1 = -(1/2\,\bar{r}^2)(1-\alpha/f)$ against the distance $\bar{r} = r/\alpha$.

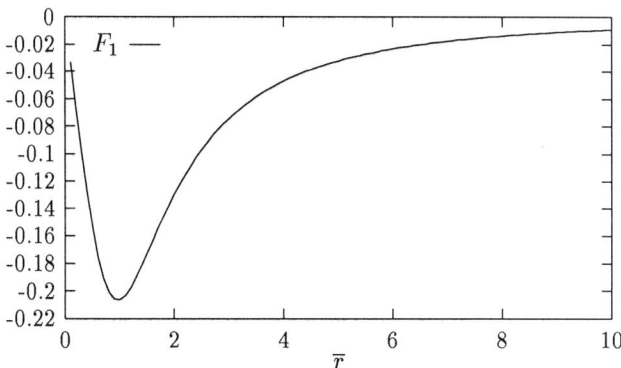

Fig.1 The plot of the function F_1 against the $\bar{r} = r/\alpha$

It follows from Fig.1 that the $|F|$ reaches its maximum at r of the order of α and tends to zero at $r \to 0$. It would therefore be interesting to know what the limiting masses of the equilibrium configurations the gravitational force $F(r)$ (1) can admit. To answer this question we start from the equation

$$\frac{dp}{dr} = -\frac{G\rho M}{r^2}(1 - \alpha/f) \qquad (2)$$

In this equation p is the pressure, $M = M(r)$ is the matter mass inside of a sphere of the radius r, $\rho = \rho(r)$ is the matter density at the distance r from the center, α and f is the function of $M(r)$.

Suppose the equation of state is $p = K\rho^\Gamma$, where K and Γ are the constants. For numerical estimates we shall use their values (Shapiro, Teukolski 1983):

For a degenerated electron gas:
$\Gamma = 5/3$ $K = 1 \cdot 10^{13}$ SGS units at $\rho \ll \rho_0$, where $\rho_0 = 10^6\ gm/cm^3$,
$\Gamma = 4/3$ $K = 1 \cdot 10^{15}$ SGS units at $\rho \gg \rho_0$.

For degenerated neutron gas:
$\Gamma = 5/3$ $K = 5 \cdot 10^9$ SGS units at $\rho \ll \rho_0$, where $\rho_0 = 5 \cdot 10^{15} gm/cm^3$,
$\Gamma = 4/3$ $K = 1 \cdot 10^{15}$ SGS units at $\rho \gg \rho_0$.

for rough estimates we replace dp/dr by $-p/r$, where p is the average matter pressure and R is its radius. Under the circumstances we obtain from eq.(2)

$$\frac{p}{\rho c^2} = \frac{\alpha}{2R}(1 - \alpha/f). \qquad (3)$$

If $R \gg \alpha$, then the term α/f is negligible. Setting $M(R) \approx \rho R^3$ we find the mass of equilibrium states as a function of ρ :

$$M = (K/G)^{3/2} \rho^{(\Gamma - 4/3)(3/2)}. \qquad (4)$$

It follows from eq.(4) that there is the maximal mass (Chandrasekhar 1935) $M = (K/G)^{3/2}$ at $\rho \gg \rho_0$.

However, according to eq.(3), there are also equilibrium configurations at $R < \alpha$. In particular, at $R \ll \alpha$ we find from eq.(3) that the masses of the equilibrium configurations are given by

$$M = c^{9/2} 10^{-1} K^{-3/4} G^{-3/2} \rho^{-(\Gamma - 1/3)(3/4)}. \qquad (5)$$

These are the configurations with very large masses. For example, the following equilibrium configurations can be found:

the nonrelativistic electrons: $\rho = 10^5 gm/cm^3$, $M = 1.3 \cdot 10^{42} gm$, $R = 2,3 \cdot 10^{12} cm$,

the relativistic electrons: $\rho = 10^7 gm/cm^3$, $M = 2.3 \cdot 10^{40} gm$, $R = 1.3 \cdot 10^{11} cm$,

the nonrelativistic neutrons: $\rho = 10^{14} gm/cm^3$ $M = 3.9 \cdot 10^{35} gm$, $R = 1.6 \cdot 10^7 cm$.

The reason of the two types of configurations existence can be seen from Fig. 2, where for $\rho = 10^{15}\ gm/cm^3$ the plots of right-hand and left-hand sides of Eq.(3) against the mass M are given.

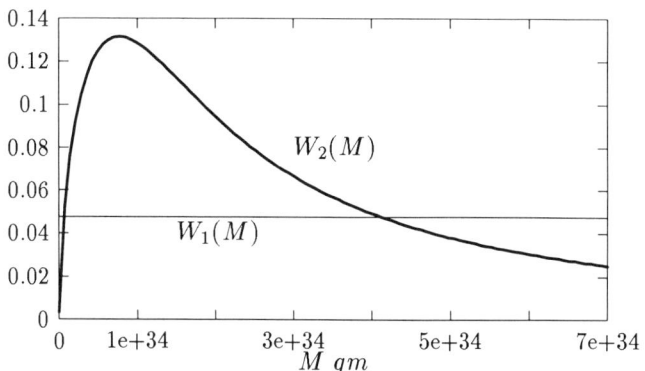

Fig 2. The plot of right-hand ($W_2(M)$) and left-hand ($W_1(M)$) sides of Eq.(3) against M.

The following conclusions can be made after considering the plots of the above kind:

1. There are no equilibrium configurations whose the density is larger than a certain value $\rho_{max} \sim 10^{16} gm/cm^3$.

2. For each value of $\rho < \rho_{max}$ there are two equilibrium states (with $R > \alpha$ and $R < \alpha$).

Are the configurations with largre masses stable?

The total energy of the degenerate gase is $E = E_{int} + E_{gr}$, where E_{int} is the intrinsic energy and E_{gr} is the gravitational energy. The gravitational energy of a sphere is

$$E_{gr} = \int_\infty^R dM(r)\, \chi(r)\, M(r), \qquad (6)$$

where

$$\chi(r) = \int_\infty^r dr'\, (r')^{-2}(1 - 1/f),$$

$\alpha = 2GM(r)/c^2$, $f = (\alpha(r)^3 + (r')^3)^{1/3}$,

$$M(r) = 4\pi \int_0^r dr'\, \rho\, (r')^2.$$

The function $\chi(r)$ is approximately

$$\chi(r) = (1/r)(1 - \exp(-r/\alpha)). \qquad (7)$$

Therefore, at $p = const$ up to a constant of the order one

$$E_{gr} = -\frac{GM^2}{R}(1 - \exp(-R/\alpha)). \qquad (8)$$

The intrinsic energy $E_{int} = \int u \, dM$, where u is the energy per the mass unit. For the used equation of state $u = K(\Gamma - 1)^{-1}\rho^{\Gamma-1}$. Thus, up to a constants of the order of one

$$E = KM\rho^{\Gamma-1} - GM^{5/3}\rho^{1/3}[1 - \exp(-QM^{-2/3}\rho^{-1/3})], \qquad (9)$$

where $Q = c^2/2G$. As an example, Fig.3 and Fig.4 show the plot of the function $E = E(\rho)$ for the nonrelativistic neutron gas of the mass $M = 10^{36}$ gm and $M = 10^{33}$ gm (neutron stares) correspondingly.

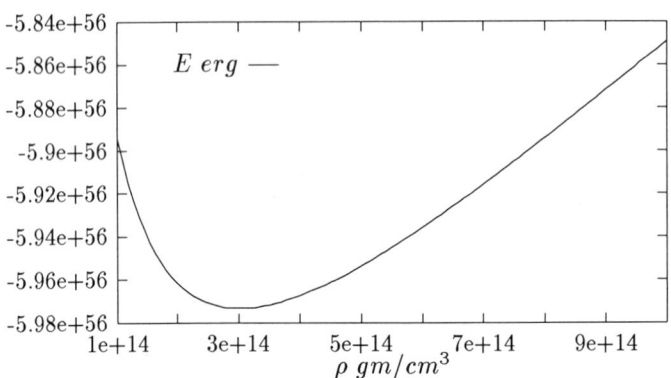

Fig. 3 The plot of the function $E = E(\rho)$ for the neutron configuration of the mass $M = 10^{36}$ gm.

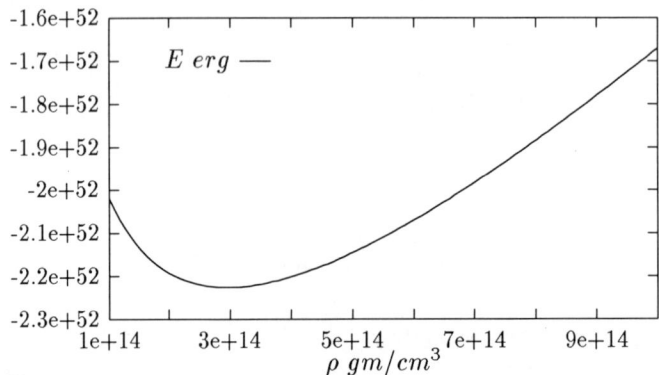

Fig.4 The plot of the function $E = E(\rho)$ of the neutron star of the mass $M = 10^{33}$ gm.

The analysis of such plots show that the function $E = E(\rho)$ has the minimum. Thus, the above equilibrium states of large masses are stable.

The gravitational potential on the surface of a stable massive configuration of the degenerate fermion gas of the order of

$$V = (GM/R)[1 - \exp(-R/\alpha)] \tag{10}$$

It follows from the virial theorem that the above objects of large masses (their $R \ll \alpha$) are the ones with low temperatures . They, probably, refer to "dark" matter of the Universe. If their luminosity are caused by an accretion, then the Eddington limit of luminosity is approximately

$$\mathcal{L} = \mathcal{L}^0_{Edd}[1 - \exp(-R/\alpha)], \tag{11}$$

where $\mathcal{L}^0_{Edd} = 1 \cdot 10^{39} M\ erg/s$. Hence, if $R/\alpha \ll 1$, their luminosity $\mathcal{L} \ll \mathcal{L}^0_{Edd}$.

The motion of test particle not far from the above objects is considered in the paper (Verozub 1996)

References

Verozub L.V. 1991 , Phys.Lett. A ,156 , 404 .
Shapiro S. and Teukolsky S. 1983, Black Holes, White Drafts and Neutron Stars.
Chandrasekhar S. 1935 , Mon. Not. Roy. Astron. Soc. ,95 , 226 .
Verozub L.V. 1996 Astr. Nachr. (in press) (E-preprint astro-ph/9510044).

MOND and the Feeding of the Monster: The Conveyor Belt

Fernando J. Selman

Las Malvas 228, Depto 901, Santiago, Chile

Abstract. In a recent work Milgrom's MOdified Newtonian Dynamic (MOND) theory of gravitation was used by the author to derive a formula for dynamical friction using Kalnajs' Fourier transform formalism. This formula has many similarities with the classical Chandrasekhar expression but has one very important difference: it leads to dynamical friction with no mass segregation. This may have important consequences for our understanding of the fueling of the central engines of galaxies in general, and of our galaxy in particular.

1. Introduction

The possibility has been explored in the past that a way to feed the bright nuclei of galaxies and their putative black holes is through the cannibalizing of the members of their globular clusters systems (GCSs) that sink to the center of the galaxy through the action of dynamical friction, see for example Lauer and Kormendy (1986). This is one of the *conveyor belt* processes in charge of bringing matter to the neighborhood of the center. This process was shown, by the same authors, to be unable to explain the extent of the core radius of the GCS in M87. More recently it has been shown that there is not enough mass in the *missing* globular clusters to account for the large observed mass of the central massive dark object (MDO) (McLaughlin 1995). An even stronger argument against this hypothesis is provided by the observed lack of luminosity segregation in all the GCSs observed so far (e.g.,Harris 1991). Because luminosity segregation is the clear signature of dynamical friction this process has all but been discarded.

It has been recently shown by the author that in the context of Milgrom's MOND theory one obtains a dynamical friction deceleration that is independent of the mass of the object being decelerated, that is, we have dynamical friction without mass segregation (Selman 1996). In this poster the consequences of such an effect on the above conveyor belt process are explored.

2. MOND: A Brief Summary

For a review of MOND see Milgrom (1989c). MOND elevates the fact that the rotation curves of disk galaxies are flat to a manifestation of a physical law. In one possible interpretation it proposes that the inertial mass of a particle depends on the strength of the field it is immersed in. Thus, the rotation curves of disk galaxies are flat because the most distant stars feel a weaker field and thus have a smaller inertial mass than they would otherwise have and can consequently move faster and still be bounded to the galaxy. MOND introduces a new fundamental constant of physics, a_0, with dimensions of acceleration, and with a value of $2 \times 10^{-8} cm s^{-2}$. The remarkable fact of MOND is that it can fit the rotation curve data for all the galaxies with this single value for a_0 (but see Lake (1989)).

Among the succeses of MOND we can mention: It provides the best statistical fit of the rotation curves data of disk galaxies with fewer parameters than Newtonian physics plus dark matter (Kent 1987). It naturally explain the Tully-Fisher, Faber-Jackson, and perhaps the updated versions of Freeman's and Fish's laws (Milgrom 1983a, Milgrom 1989a). It can also account for the stability of disks (Milgrom 1989b). It can explain the dynamics of groups and clusters of galaxies using only the observed luminous matter (Milgrom 1983b). It can also explain the correlation between x-ray luminosity and cluster velocity dispersion found by Melnick and Quintana (1982). Despite these and other successes there are some difficulties, the most important among them are the limited scope of validity of the present theory, its violation of the strong equivalence principle at a very deep level (a conflict with General Relativity), and a couple of observational claims of the impossibility to fit all the data with a single value of a_0 (Lake 1989), and to explain the elongated nature of the X-ray emission from one cluster of galaxies (Buote & Canizares 1994).

3. The MOND and Newtonian Dynamical Frictions Compared

Using the Fourier transforms formalism of Kalnajs (1972) to solve Boltzmann's equation the author recently showed that within MOND a similar effect to Chandrasekhar's frictional deceleration is found (Selman 1996).

The Newtonian and MOND expressions can best be compared if we write them side by side, in the following way:

$$\vec{a}_{Ch} = (\tfrac{GM}{R_{max}^2} \log \Lambda) \; \left[4\pi (\tfrac{\tau_c}{\tau_d})^2 \xi(\tfrac{V_0}{\sqrt{2}\sigma}) \hat{V}_0 \right]$$

$$\vec{a}_M = (\tfrac{1}{32} a_0) \; \left[4\pi (\tfrac{\tau_c}{\tau_d})^2 \xi(\tfrac{V_0}{\sqrt{2}\sigma}) \hat{V}_0 \right]$$

where I have written τ_c for the crossing time of the system by the perturber (the object been decelerated, in our case a globular cluster), given by $\tau_c = R_{max}/V_0$; and τ_d for the dynamical time of the system, given by $\tau_d = 1/\sqrt{G\rho_0}$. M and V_0 are the mass and speed of the perturber, respectively. R_{max} is the size of the system were the perturbing object moves, and σ its velocity dispersion. ξ is a known increasing function of its argument, expressible in terms of the error function and its first derivative, and varies monotously between zero and one. The terms in square brackets are near unity. We see that the main differences between the two types of friction are the mass dependency and the runaway character of the Newtonian one, versus the steady and mass independent character of the frictional deceleration in MOND. The latter leads to **dynamical friction without mass segregation**.

4. The Evidence for MOND from Dynamical Friction Effects

The following is a summary of the successes of MOND from the point of view of dynamical friction. The most important consideration that distinguishes the predictions of MOND from those of the dark matter hypothesis (and the way to decide wether the missing ligh puzzle is due to new physics or to the existence of large quantities of exotic matter) is the fact that one can not add arbitrary quantities of matter to a galaxy with impunity; its effects will be felt via dynamical friction and should be observable through mass/luminosity segregation:

- The 2-point correlation function of galaxies is featureless down to scales that correspond to their *optical* sizes and not to the sizes of their putative dark-halos (Gott 1979).
- The distribution of isolated physical pairs of galaxies show no evidence of luminosity segregation (White et al. 1983).
- Clusters of galaxies show no evidence of luminosity segregation (e.g., Oemler 1974, Thompson & Gregory 1993).
- MOND can account for the luminosities of cD envelopes (Selman 1996).
- MOND can account for the X-ray luminosities of clusters of galaxies utilizing only the available kinetic energy of the member galaxies as the source for heating the intergalactic medium (Selman 1996).

5. Globular Clusters as Pet Food

It was discussed in the introduction that for the case of M87, our best studied case, dynamical friction can not account for the core radius of its GCS. Moreover, there are not enough missing globular clusters to account for the mass of the MDO at its center. With MOND we can derive the followin expression for the core radius of the GCS:

$$r_c = 8kpc \left(\frac{R_{max}}{3kpc}\right)^{2/3} \left(\frac{\sigma_{stars}}{250km/s}\right)^{-1} \left(\frac{a_0}{2 \times 10^{-8}cm/s^2}\right)^{1/3} (H_0 t)^{1/3}$$

We can see that this value compares rather well with the observed value of $6.7kpc$. Although the dependence of this value on R_{max} introduces a large degree of uncertainty, the agreement is nevertheless remarkable.

The other argument, namely, that the number of missing globulars is not enough to account for the mass of the MDO can be answered by pointing out that the process of cannibalizing can be an almost steady one that continuously brings objects to the center, but that for a given spherical *shell* in the GCS as many objects can be leaving for lower shells as can be coming in from higher ones. Thus, the detailed balance we can make in a given instant of time is of no consequence for the problem at hand.

The observed lack of mass segregation is natural within the MOND context.

6. Conclusions

The mass independent dynamical friction deceleration found for MOND implies that globular clusters are still viable candidates to enrich the center of galaxies and thus provide food for the central engines.

MOND provides a framework in which many hitherto unexplained astrophysical facts find a natural explanation (e.g., the Tully-Fisher relation). Its prediction of dynamical friction without mass segregation might have a strong impact in our understanding of many cosmic phenomena, including the history of our Galaxy's center. It is particularly tempting to speculate that the present status of the GCS of a galaxy is the result of the interplay of two opposite phenomena: in one hand we have the formation of globular clusters through merging events, enriching the GCS as it has been recently proposed (Whitmore & Schweitzer 1995); and in the other we have the depletion of the GCS through the combined effect of dynamical friction and their tidal breakdown once sufficiently close to the center. One could explain with such an scenario why there is no correlation between the total number of globulars and the total luminosity of the parent galaxy.

Acknowledgments. My sincere thanks to ESO for allowing me to use the facilities at the Vitacura office.

References

Buote, D. A., & Canizares C. R. 1994, ApJ, 427, 86
Gott, J. R., & Turner, E. L. 1979, ApJ, 232, L79
Kalnajs, A. J. 1972, in Gravitational N-body Problem, M. Lecar, IAU Colloquium 10, Dordrecht: Reidel, 13
Kent, S. A. 1987, AJ, 93, 816
Lake, G. 1989, ApJ, 345, L17
Lauer, T. R., & Kormendy, J. 1986, ApJ, 303, L1
McLaughlin, D. E. 1986, ApJ, 303, L1
Milgrom, M. 1983a, ApJ, 270, 371
Milgrom, M. 1983b, ApJ, 270, 384
Milgrom, M. 1989a, A&A, 211, 37
Milgrom, M. 1989b, ApJ, 338, 121
Milgrom, M. 1989c, Comments Astrophys., 13, 215
Selman, F. J. 1996, Preprint
Thompson, L. A., & Gregory, S. A. 1993, AJ, 106, 2197
White, S. D. M. et al. 1983, MNRAS, 203, 701
Whitmore, B. C., & Schweitzer, F. 1995, AJ, 109, 960

Part 3. The Central Engine and High Energy Phenomena

Section A. Sagittarius A*

The Discovery of the Radio Source Sagittarius A (Sgr A)

W. M. Goss

NRAO, P. O. Box O, Socorro, NM 87801, USA

R. X. McGee

CSIRO, Radiophysics Laboratory, P.O. Box 76, Marsfield, NSW, 2121 Australia

Abstract. The discovery of Sagittarius A, a discrete radio source associated with the nucleus of the Milky Way, was made by Piddington and Minnett (1951) at 1210 MHz. McGee and Bolton (1954), however, brought to the attention of the astronomical world the association of Sgr A with the Galactic Nucleus. The paper was published in *Nature* and many well known astronomers (Baade, Oort, van de Hulst, Pawsey, Mills, Kerr, and Shain) were involved in the events leading to publication. We summarize the preparation of this paper, re-examining some details published by Morton (1985). In addition we summarize the significance of five publications from CSIRO, Sydney, Australia, in the period 1951 to 1954.

1. Introduction

In contrast to the discovery of the early strong radio sources such as Taurus A, Virgo A, Cygnus A, Centaurus A and Cassiopeia A, the story of the discovery of the radio source Sagittarius A at the center of the Milky Way is not well known. Bolton, Stanley and Slee (1949) suggested optical identifications of Taurus A (the Crab nebula), Virgo A (M87) and Centaurus A (NGC 5128) in a famous but brief (about 600 words) paper in *Nature*. Sea interferometers in Australia and New Zealand had been used at 100 MHz to determine the positions. The positional accuracy was 10′, a substantial improvement over the 5- 10° errors in the earlier sea interferometer results of Bolton (1948), who describes the first six radio sources observed at Dover Heights (Sydney) with the original sea interferometer consisting of a two element 100 MHz Yagi antenna (a photo can be seen in Bolton, 1982, Figure 1).

An example of the existing misconceptions concerning Sgr A can be found in David H. Levy's biography (*The Man Who Sold the Milky Way: A Biography of Bart Bok*, 1993):

"By 1955 radio telescopes were peering into the center of the galaxy and finding deuterium, or heavy hydrogen, at the 91.6 cm wavelength. In the same year Americans R. X. McGee and J. G. Bolton found a bright radio source known as Sagittarius A and proposed, somewhat prematurely, that it could be the actual center of the galaxy."

Of course, deuterium (the DI line) has never been detected by radio astronomers despite repeated searches, and the Australian (McGee) and Englishman (Bolton) never changed their citizenships!

In this paper we summarize the early CSIRO (Australia, "Commonwealth Scientific and Industrial Research Organization") observations in the period 1951 to 1955 made by the Division of Radiophysics in Sydney, New South Wales. We discuss the discovery paper by Piddington and Minnett (1951) of the source Sgr A, and detail the circumstances of the 1954 *Nature* paper by McGee and Bolton. We re-examine some details concerning this publication as discussed by Morton (1985).

A short summary of the discovery of Sgr A has been given by Burke (1964). Slee (1994), in the J. G. Bolton memorial volume, also has described his role in the 80-foot antenna Galactic center survey.

Prior to about 1950, optical observations of stellar kinematics gave a description of the Milky Way as a disk galaxy of radius 8-10 kpc whose center was inferred to be in the direction of Scorpius-Sagittarius. Prior to the radio observations the difficulty in determining the Galactic center position was due to the extinction arising from interstellar dust.

Most of the early papers use the terminology "the Galactic center source". The first reference to Sagittarius A that we have found is by Kraus and Ko (1954), a report on the 250 MHz all sky image made at Ohio State and discussed at the June 1954 American Astronomical Society Meeting in Ann Arbor, Michigan. Also in 1955, Priester used the term *Sgr A* in a paper discussing the possibility that Sgr A is an HII region at a distance of 3 kpc. By 1958-59, the name *Sgr A* was in common usage (eg. Westerhout, 1958 and Drake, 1959).

2. Pre-1950 Observations

In Jansky's 20.5 MHz data of 1932 (Jansky 1935 and Sullivan 1978), made with a resolution of 24° x 35°, a prominent concentration of the Galactic background in the direction of the Galactic center is obvious; Jansky's data do not quite extend to the southern declination of the Galactic center, however. Reber's initial all-sky radio images at 160 MHz (beamwidth, $12°.5$, 1944) and 480 MHz (beamwidth 4°, 1948) show prominent concentrations in Sagittarius (RA 17^h56^m, Dec -25°, 1950 coordinates).

3. Piddington and Minnett (1951)

Jack Piddington and Harry Minnett observed at the high frequencies (in 1951) of 1210 and 3000 MHz. They initially used a 10 foot and later an 18 by 16 foot prime focus antenna located at Potts Hill Reservoir in Sydney, shown in Figure 1. This aerial had been used earlier for solar work at 200, 600 and 1200 MHz (Lehany and Yabsley, 1949). At 1210 MHz the noise temperature was 3400K, about 100 times modern values, and the beamwidth was $2°.8$.

Piddington and Minnett(1951) describe a "......new, and remarkably powerful, discrete source" at 1210 MHz, the "Sagittarius-Scorpius" source (the position was close to the border of the two constellations). These authors identify this new source with the center of the galaxy:

"An interesting feature....... is that it lies close to the Galactic plane and very close.... to the plane defined by the maximum level of radio emission. It also lies very close to the centre of the Galaxy and to the maximum of Galactic radiation...... the significance of the position may be considerable. If the source were relatively close to the Sun, it could lie in any direction with equal probability. If, on the other hand, it were at a distance from the Sun which was a considerable fraction of the Galactic diameter, it would be more likely to lie in a direction close to the plane of the galaxy."

Piddington and Minnett then discuss the luminosity of the Sagittarius - Scorpius source assuming a distance of 10 kpc, as well as the presence of a flat radio spectrum between 100 and 1210 MHz. An analogy is drawn with the flat spectrum of Taurus A and the suggestion is made that the source might be an optically thin thermal gas; however, the authors are already aware of the problem of this interpretation in those "pre-synchrotron emission" days.

In a footnote to his book on the history of Australian radio astronomy, Robertson (1992) has given credit to Piddington and Minnett for the discovery of the Galactic nucleus; the main text in the Robertson book is concerned with the McGee and Bolton (1954) paper, however.

4. Mills (1952)

In a fascinating paper in 1952, Mills describes an all sky survey made at 101 MHz with a three element, phase-switched interferometer. The spacings were 270m (lobe spacing 40') and 60m (3° lobe); seventy-seven sources were in this catalog. Mills identifies two classes of sources: class I, concentrated to the Galactic plane and class II, the sources randomly distributed over the sky. In addition Mills shows that the extended radio Galactic background cannot be the integrated output of the class I sources. In addition, "....it might be said that there is a suggestion, but no definite evidence, that some of the class II sources are extragalactic." This paper also includes a log N - log S (number counts as a function of source flux density) analysis for both types of sources.

Source 17 - 2 B is undoubtedly Sgr A; the position is a few degrees displaced, probably due to the large primary beam of the single elements of the interferometer (24° x 14°). Mills makes no mention of a possible identification with the Galactic center. The vastly different flux densities between the shortest spacing (3000 Jy) and the longest (300 Jy) suggested an angular size of 35'. In addition, the flat radio spectrum between 101 MHz and 1210 MHz is discussed, suggesting a "thermally emitting thin gas." However, the much larger flux density at 18 MHz (57,000 Jy) observed by Shain and Higgins (1954) indicates that the spectrum may be more complex and thermal emission must "be discounted."

5. Bolton et al. 1954a

Bolton, Westfold, Stanley and Slee (1954a) subsequently investigated the radio sky looking for large sources, sizes $> 1°$ at 110 MHz. The primary instrument was a two element sea interferometer with variable azimuth spacings of 12 to 60m. An intense source ("L") near the Galactic center with a large angular size (12° x 2°) was detected. In addition, the 72- foot reflector, the prototype of the final 80-foot reflector, was used to study the source at 160 MHz (see Figure 7 in Bolton, 1982). The position ".... is close to the accepted position of the Galactic centre. .... There is an extended physical object at the centre of the galaxy, which is an unusually intense source of radio noise."

6. Bolton et al. 1954b

Bolton, Stanley and Slee (1954b) used a sea interferometer to image the declination range +50° to -50° with an approximate lobe separation of 1°. The structural elements of the 9-Yagi array were used to build a 12-Yagi array on an azimuth mounting at the cliff edge at Dover Heights on Defence Department property (see the cover of the Proceedings of the Astronomical Society of Australia, Vol. 4, No. 4, 1982). As Slee (1994) has pointed out, the rusty remains of the frame work of the azimuth mounting are all that are to be found in the mid 1990s. The survey was made in 1951-52 and was the final, most sensitive sea-interferometer survey. The positions had a precision of $\sim 0.5°$ in right ascension and $\sim 1°$ in declination. The final source list had 122 sources down to a limiting flux density of 40 Jy. Source "68" is Sgr A (although Bolton et al place it in Scorpius!).

For the Dover Heights scene in the mid- 1950s , see the photo (Figure 1c) published by Slee, 1994. In July 1995, Goss photographed this scene; only the rusty remains of the rusting radar turntable remains. A small brass plaque commemorating the 40th anniversary of the first paper on the optical identifications published by Bolton, Stanley and Slee (1949) (see Slee 1994) is mounted nearby on a concrete slab at ground level.

7. The 80-foot Telescope - McGee and Bolton 1954

John Bolton has described his efforts and those of his colleagues to dig a hole in the sandy soil on top of the cliff at Dover Heights, Sydney, to construct first a 72-foot diameter paraboloid to observe at 160 MHz, and later an extension to 80 feet, concreted and surfaced with chicken wire, to operate at 400 MHz (Bolton 1982). During the preparation of the 400 MHz receiving equipment, Bolton was transferred to the Rain Physics group of the Radiophysics Laboratory in mid 1953. He was to spend a year there before taking up an appointment to establish radio astronomy at the California Institute of Technology. Thus Dick McGee, a newcomer to experimental radio astronomy, was left to make a sky survey between declinations -17° and -49° ably assisted by Gordon Stanley and Bruce Slee. The survey and its analysis were completed towards the end of January, 1954. The 80-foot antenna is shown in Figure 2 (see also Bolton, 1982, Figure 11). McGee is shown in this photo from 1953. Another view of this telescope can be found as Plate 1 in McGee et al (1955). A construction photo, with has modern parallels with the construction of the Leighton mm telescopes, is shown in Figure 10 in Bolton (1982).

Don Morton (1985) has told the story of the 400 MHz Sgr A source (detected in this survey) including Baade's excitement over the map: "Frankly I jumped out of my chair the moment I saw what it meant..." Henk van de Hulst writes: "...The position agrees quite well with the best we can do on the basis of the 21 cm observations.". Jan Oort's letter replying to Baade concludes "...I have been excited by Pawsey's diagram..." However some of the details in the account could not be remembered by McGee even when Morton showed him two strongly written letters from John Bolton in 1985. They portray Dr. Joe Pawsey in a rather poor light. In an effort to set the record straight, McGee recently has obtained from the Australian Archives his own observing log, his analysis log, the reductions and maps and, most importantly, his publications file of the *Nature* paper "Probable Observation of the Galactic Nucleus at 400 Mc/s."

His version of the events and how the paper received its title follow. Late one evening in late 1953 or early 1954 the radio astronomy group leader, Joe Pawsey, called into the Dover Heights field station just as McGee was completing the 400 MHz contour map in pencil. He was immediately excited by the prominent appearance of the Sgr A source and asked McGee to redraw the map in Galactic coordinates so that he could send a copy to Walter Baade at the Palomar Observatory. Baade had discovered what he described as the nucleus of the Andromeda (M31) galaxy and in fact had suggested to John Bolton as early as 1951 that a radio search ought to be made for the nucleus of our own galaxy.

On 22 February 1954 Pawsey sent a memorandum to:

"Mr. J. G. Bolton, Mr. R. X. McGee. The attached is Baade's reply to my letter telling him of the source at the galactic centre. What do you think about a note to *Nature* or *Observatory* outlining the various wavelength surveys? The subject is of wide interest, signed J. L. Pawsey, Assistant Chief of Division."

The Baade letter of 16 February 1954 is quoted extensively by Morton (1985). In addition to his excitement (see above), Baade concludes:

"It is very improbable that the coincidence between inferred and observed position of the nucleus is accidental."

Morton also has quoted in detail the letter of Henk van de Hulst (he was in Pasadena in 1954) to Pawsey as well as Jan Oort's reply to Baade concerning the location of the Galactic nucleus.

McGee wrote a letter for *Nature* entitled "The Galactic Nucleus." Pawsey looked over the first draft, changed the title to "Radio Observation of the Galactic Nucleus" and reduced the text by nearly 40 per cent. It was suggested to John Bolton (by McGee) that he should be listed as senior author, but he pointed out that he was no longer involved and eventually compromised to accept second authorship. He was far too busy rainmaking to contribute in any way to the writeup. On 29 March the draft went to be refereed by the famous publications committee of the Radiophysics Laboratory with Frank J. Kerr as chairman. The committee did not take the optimistic view of the leading astronomers mentioned above. To give some examples of the caution they displayed:

Kerr:
"In view of the uncertainty as to whether this is the nucleus, or a source near, or in the direction of the nucleus, the wording wants to be carefully phrased..."

The paper was passed to B. Y. Mills for perusal.

Mills:
"The results should obviously be published quickly as they are important. However the general tone of the paper seems far too dogmatic-even the title! And positive identification with the Galactic nucleus is impossible at present..." An interesting comment which perhaps indicates the state of the art at that time: Mills: "In view of the importance of deciding whether Pop II systems emit radio waves it seems important that a definite identification with the nucleus should not be claimed. (there is also the possibility that the position is a pure fluke - see Haddock APJ Jan 1954.)" (This paper, by Haddock, Mayer, and Sloanaker, appears in the March 1954 Astrophysical Journal.)

On reading through a draft of the current paper in April, 1996, B. Y. Mills, commenting on his views expressed above, writes:

"After several months spent at Caltech (in late 1953 - early 1954) I no longer looked on eminent astronomers as minor deities and I would certainly have taken Baade's reaction with a grain of salt... Also I had been impressed by the observations at NRL (Haddock et al.) which showed a number of sources along the galactic plane, including Sgr A, and I had discussed them at length with Fred Haddock. Why no comment from Baade then? He certainly knew of them. Perhaps the NRL emphasis on emission nebulae [*comment by Goss and McGee: there was, of course, the possibility that Sgr A could have been an HII region along the line of sight to the Galactic center*] made the difference. ...Both Baade and Minkowski in 1954 still believed that the source of galactic radio emission was a type of Pop II star (Baade later told me [*ie Mills*] that he only changed his mind after my failure to detect emission from globular clusters with the Cross in 1955) and galactic nuclei were then believed to be concentrated Pop II systems - AGNs were unheard of. I must have seen the connection and Baade's enthusiasm for the identification could have been the result of suddenly realising that it fitted nicely with his overall view, a view that I did not share. Re-reading a few articles has convinced me that most radio astronomers regarded the determination of a reliable distance as essential for a definite identification and this did not happen until the later H-line observations. In my PASP review of 1959 (Mills, 1959) I wrote : 'For some time it has been known that a strong radio source exists very close to the direction of the galactic centre... and there has been much speculation whether this source, Sagittarius A, might represent emission from the nucleus. As a result of 21 cm line observations by Rougoor and Oort (1959) it now appears very probable that this is so; at least their observations show conclusively that the source must be located in the central regions of the Galaxy. '"

The detailed reader on the publications committee was C. A. Shain who suggested that the title should be "Observations of the Region of the Galactic Nucleus at 400 Mc/s" together with several pages of quite helpful criticisms. Quoting one general remark:

Shain: "The title and the statement towards the end of page 2 indicate that you are making a definite claim that the 'hump' on your contours is the galactic nucleus and, if only by inference, that the position of this 'hump' is the best available determination of the nucleus."

McGee incorporated most of the committee's suggestions into the text which went back to Mills and finally Kerr for further comments and, after a required change by R. N. Bracewell on the contours, final acceptance. The extraordinary interest of senior members of the Radiophysics Laboratory in papers for publication can surely only be gratefully commended. Nevertheless McGee's confidence in claiming observation of the Galactic nucleus was severely shaken.

On 14 April, McGee took the paper to Pawsey for his final approval. Pawsey immediately changed the title back to "Observation of the Galactic Nucleus at 400 Mc/s". McGee suggested that the word "Possible" before "observation..." might be appropriate in view of the committee's attitude. But Pawsey, buoyed up by the support from Baade, van de Hulst and Oort, compromised with "Probable etc." Thus it appears that John Bolton's recall of these events as related to Don Morton was not accurate. Certainly the watered-down title cannot be blamed on Pawsey. The *Nature* paper appeared in the May 22, 1954 issue; the acknowledgements mention Westfold, Stanley, Slee, and Baade. A full account of the Galactic center survey was presented by McGee, Slee and Stanley (1955).

As C. M. Wade has pointed out to us, the general acceptance of Sgr A as the Galactic nucleus was still a controversial point of lunch time conversation at Radiophysics in the period 1957 - 1959. Only later does it appear that the astronomical community accepted the association of Sgr A with the Galactic center with no reservations. (See the above comments by Mills and also a critical discussion by Mills, 1956, who suggests that, based on the lack of spherical symmetry in the non-thermal radio source, the radio emission cannot be related to the spherically symmetric Pop II system of stars in the Galaxy.) In his classic textbook of 1960 (apparently updated in September 1958), I.S. Shklovsky suggests :

".... let us say that today we can consider it definitely established that Sagittarius A is indeed the Galactic radio nucleus."

John Bolton attended the 1955 IAU General Assembly in Dublin and proposed that, in view of the findings on Sgr A, the Galactic coordinate system should be revised. However, the new system was mainly based on 21-cm hydrogen line surveys at Leiden and Sydney. This system was introduced in a series of five papers in 1960 by Blaauw et al. Nevertheless Blaauw, Gum, Pawsey and Westerhout write in 1960 ("The New I.A.U. System of Galactic Coordinates"):

"We shall.......assume that Sagittarius A is located at the galactic centre."

In the final paper in this series by Oort and Rougoor ("The Position of the Galactic Centre," 1960), they conclude:

"The position of Sagittarius A has been discussed.... This position agrees so precisely with the direction of the galactic centre..... that this by itself makes it almost certain that Sgr A is situated at the centre of our Galaxy."

8. Acknowledgements

The National Radio Astronomy Observatory is a facility of the National Science Foundation operated under cooperative agreement by Associated Universities, Inc. The initial impulse for this article was the lecture given by Goss at Sydney University in August 1995 for the Sydney Association for Astrophysics. We thank B.Y. Mills, Letty Bolton, C. M. Wade, O.B. Slee, K. I. Kellermann, D.H.Levy, D.C. Morton, W.T. Sullivan III, H.C. Minnett, David Finley, Douglas Roberts, W. N. Christiansen, R.L.Brown, Andrew M.Goss, P.J.Napier, Tony Beasley and H.F. Weaver for comments and Gwen Anne Manefield for assistance in the preparation of the Safa lecture.

References

Blaauw, A. ,Gum, C.S. ,Pawsey, J.L. ,and Westerhout, G. 1960, MNRAS,121,123
Bolton, J.G. 1948, Nature, 162, 141
Bolton, J.G. ,Stanley, G.J. ,and Slee, O.B. 1949, Nature, 164, 101
Bolton, J.G. ,Stanley, G.J. ,and Slee, O.B. 1954b, Aust.J.Phys. ,7, 110
Bolton, J.G. ,Westfold, K.C. ,Stanley, G.J. ,and Slee, O.B. 1954a,Aust.J.Phys. 7, 96
Bolton, J.G. ,1982, Proc. Atron. Soc. Aust. ,4,349
Burke, B.F. 1965, Annual Rev. AA, 3, 275
Drake, F. D. 1959, AJ, 64, 329
Haddock, F.T., Mayer, C.H. , and Sloanaker, R.M. 1954, ApJ, 119, 456
Jansky, K.G. 1935, Proc. IRE , 23, 1158
Kraus, J.D. and Ko, H.C. 1954, Sky and Telescope, 14, 22
Lehany, F.T. and Yabsley, D.E. 1949, Aust. J. Sci. Res. ,A2, 48
Levy, D.H. 1993 ,*The Man Who Sold the Milky Way: A Biography of Bart Bok*, Univer. of Arizona Press
McGee, R.X. and Bolton , J.G. 1954, Nature, 173, 985
McGee, R.X. , Slee, O.B. and Stanley ,G.J. 1955, Aust. J. Phys., 8, 347
Mills, B.Y. 1952, Aust. J. Sci. Res., A5, 266
Mills, B.Y. 1956, Obs. ,76, 65
Mills, B.Y.,1959, PASP, 71, 267
Morton, D.C. 1985, Aust.Physicist, 22, 218
Oort, J.H. and Rougoor, G.W. 1960, MNRAS, 121, 171
Piddington, J.H. and Minnett, H.C. 1951, Aust. J. Sci. Res. ,A4, 495

Priester, W. 1955, Zeit. f. Astro.,38,73
Reber, G. 1944, ApJ, 100, 279
Reber, G. 1948, Proc. IRE ,36, 1215
Robertson, P. 1992, *Beyond Southern Skies- Radio Astronomy and the Parkes Telescope*, Cambridge Univer. Press
Shain, C.A. and Higgins, C.S. 1954, Aust. J. Phys., 7, 130
Shklovsky, I.S. 1960, "Cosmic Radio Waves", Harvard Univer. Press
Slee, O.B. 1984, Aust. J. Phys., 47, 517
Sullivan, W.T. III, 1978, Sky and Telescope, 56, 101
Westerhout, G. 1958, BAN, 14, 215

Figure 1. One of the antennae used by Piddington and Minnett(1951). It was an 18 x 16 foot paraboloid. The observing site was at the Potts Hill Reservoir field station some 20 km south west of Sydney. Photograph provided by W.N. Christiansen. The first detection of Sgr A was made with this antenna.

Figure 2. The north aspect of Dover Heights in 1953. The 80-foot paraboloid - the hole in the ground antenna - occupies the foreground. This was a transit telescope that was pointed in declination by tilting the mast. The front end of the 400 MHz receiver was mounted immediately above the dipoles and reflector feed. R. X. McGee is holding one of the mast guy ropes. South and North Sydney Heads can be seen in the central background. Today this site is a sports field in Rodney Reserve. This radio telescope was used for the observations described by McGee and Bolton in their 1954 *Nature* paper.

The Spectrum of Sgr A* and the Central Parsec

Peter G. Mezger

Max-Planck-Institut für Radioastronomie, Auf dem Hügel 69, 53121 Bonn, F.R. Germany

Abstract. This paper reviews recent results related to the spectrum and morphology of the compact synchrotron Sgr A*, considered to be a candidate for an underfed Black Hole located at or close to the dynamical center of our Galaxy. Sgr A* is surrounded by a region of very high surface brightness at all wavelengths. It therefore requires high angular resolution to separate Sgr A* from its background emission. We show that the spectrum of the Galactic Center, observed e.g. from the distance of M 31 (D~700Kpc) with an angular resolution of ~1 arcsec, would mimick a weak Seyfert I AGN of ~$10^8 L_\odot$ rather than the obviously non-thermal spectrum of Sgr A*.

1. Circum Nuclear Disk (CND) and Central Cavity (CC)

An assembly of rather irregular shaped clumps of together $M_{H_2} \sim 10^4 M_\odot$ rotates around the Galactic Center. Its inner edge at a distance R~1.7pc is well defined and confines the HII region Sgr A West with its ionized Minispiral. At the transition region CND/HII region hydrogen column densities drop from values as high as $10^{23} cm^{-2}$ by factors ~10 and more and one therefore refers to the inner 3.4pc also as Central Cavity (CC).

Figure 1 shows an overlay of dust emission associated with the CND on an H-band ($\lambda 1.6 \mu m$) image obtained with the ESO/MPIA 2.2-m telescope on La Silla. 24 Hot and massive stars ($T_{eff} \sim 25\,000K$), at least 300 cool (~4000K) intermediate mass supergiants and more than one million of low-mass, low-luminosity ($M_*/L_* \sim 3$) MS stars contribute comparable fractions to the NIR flux density integrated over the central 30″. But only the central cluster of hot stars with a core radius of ~0.17pc contributes significantly to the total luminosity of ~$10^8 L_\odot$ of the central pc (Mezger, Duschl and Zylka, 1996; hereafter MDZ 96). The dust emission from the CND correlates well with regions of low H-band surface brightness indicating that the core radius of $\sim 30'' = 1.2pc$ for the cool stars could be underestimated due to extinction from dust in the CND.

2. The morphology of Sgr A*

The compact synchrotron source Sgr A*, located at or close to the dynamical center of our Galaxy, is the best candidate for a starving Black Hole (BH) of a few $10^6 M_\odot$. Its presence has been predicted on theoretical grounds by Lynden-

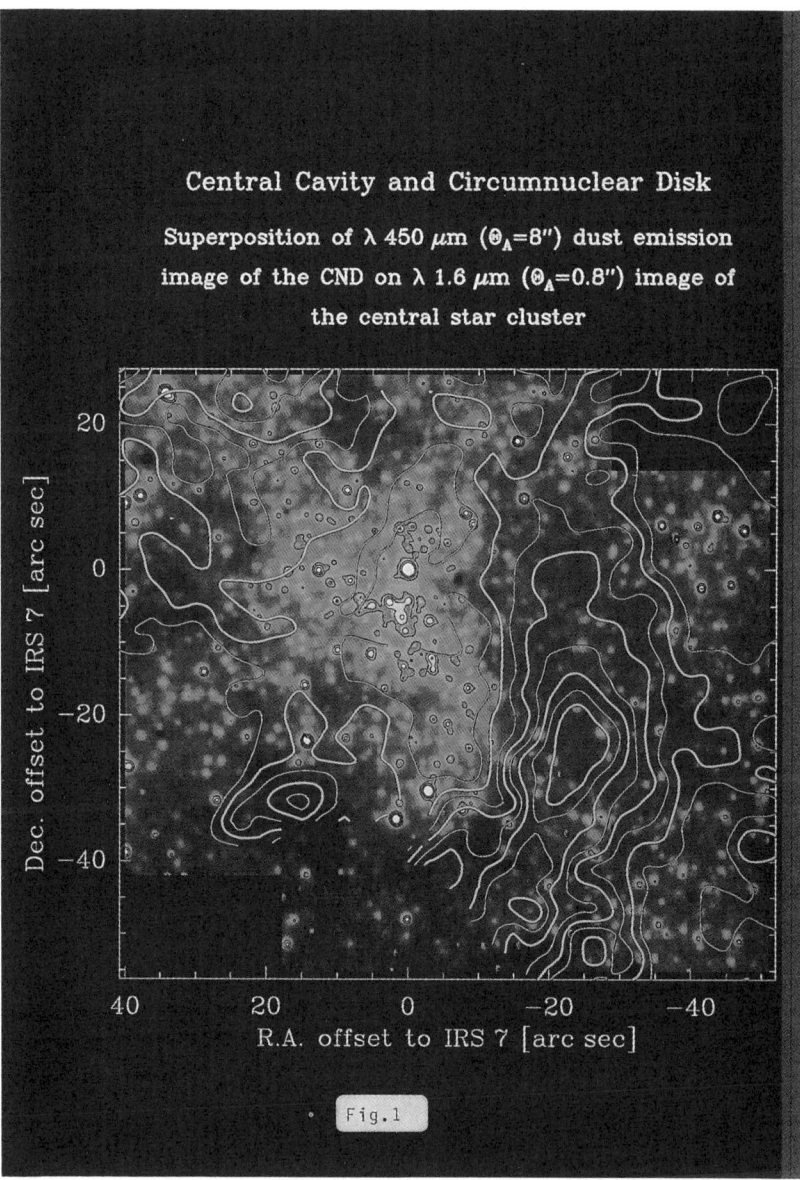

Figure 1. Overlay of a 450μm contour line image of the warm dust emission associated with the Circum-Nuclear-Disk (CND) on an H-band ($\lambda 1.6\mu$m) false color image of the stars in the central 2x2pc^2. The bright spot is IRS7. Contours are in 1Jy (8"-beam)$^{-2}$, which corresponds - for $Z/Z_\odot$=2, b=1.9 and T_d=50K - to a hydrogen column density of $N_H = 6.5 \cdot 10^{21}$ cm^{-2} and a visual extinction of A_v=7 mag (Philipp, Zylka et al., in prep.).

Bell and Rees (1971); it was definitely detected three years later by Balick and Brown (1974) using the Green Bank interferometer.

Since then Sgr A* became a favorite object of VLBI observations. At λ1cm its apparent size decreases $\propto \lambda^2$, indicating source broadening due to electron scattering. The interpretation of recent VLBI observations at mm wavelengths (Krichbaum et al., 1993 and 1994; Backer et al., 1993; Rogers et al., 1994) is still somewhat controverse regarding the question of the wavelength dependence of the source size. Especially the 3mm data seem to indicate a source size which is only marginally consistent with and perhaps slightly larger than the extrapolated scattering size, in which case the actual structure of Sgr A* could be imaged by future VLBI observations at short mm-wavelengths.

Figures 2a and b (Krichbaum, pers.comm., see also MDZ 96 for a detailed reference list) summarize the present state of high-resolution imaging of Sgr A*. Figure 2a shows the geometric mean of the major and minor axis of the elliptical image plotted versus wavelengths. The best fit slope is 2.04± 0.01. There are indications that at λ <7mm the measured sizes of ~0.2/0.3 mas fall above the predictions based on an extrapolation of the data at λ1cm. The corresponding linear sizes are 2.5 - 4 10^{13}cm. Except for λ3mm, where the ellipticity of the structure has not yet been determined, Sgr A* seems to be elongated in east-west direction along the P.A. 90-100° (Figure 2b). This orientation is independent of wavelength. The major and minor axis of an elliptical Gaussian component used here to describe the basic structure of Sgr A* also follow a λ^2 dependence. The data exhibit a remarkable consistency over the entire wavelength range.

3. The radio/IR spectrum of Sgr A*

Figure 3 shows the most recent radio/IR spectrum of Sgr A* (Beckert et al., 1996). The flux densities at $\lambda \gtrsim$ 3mm are variable (see MDZ 96 for details). For wavelengths λ <350μm there are at present only upper flux densitiy limits available. The spectrum has a high frequency cut-off at ~2-4 10^3GHz and a low-frequency turnover at ~0.8GHz. In between the spectrum increases $S_\nu \propto \nu^{1/3}$. Beckert et al. explain this spectrum as optically thin synchrotron emission from quasi-monoenergetic relativistic electrons, with the low-frequency turnover caused either by free-free absorption or by synchrotron self absorption (see also Duschl, these proceedings). The integrated luminosity of the radio/IR spectrum amounts to a few $10^2 L_\odot$.

The surface brightness of optically thin dust emission increases with $I_\nu \propto \nu^4$ for $\lambda \gtrsim 100\mu$m while the flux density of Sgr A* increases only $\propto \nu^{1/3}$. Hence, with the CSO 10-m-telescope at λ350μm and an angular resolution of ~ 10″ only an insignificant upper limit of ~10Jy for the flux density of Sgr A* could be obtained. In the IR (for $\lambda \lesssim 60\mu$m), the dust surface brightness actually begins to decrease and the angular resolution of ground-based large telescopes in the MIR is as high as $\lesssim$ 1″, so that background emission from hot dust should be efficiently suppressed. Todate, however, only upper limits of MIR flux densities of Sgr A* have been obtained which, nevertheless, significantly constrain the IR part of the spectrum, which appears to decrease exponentially.

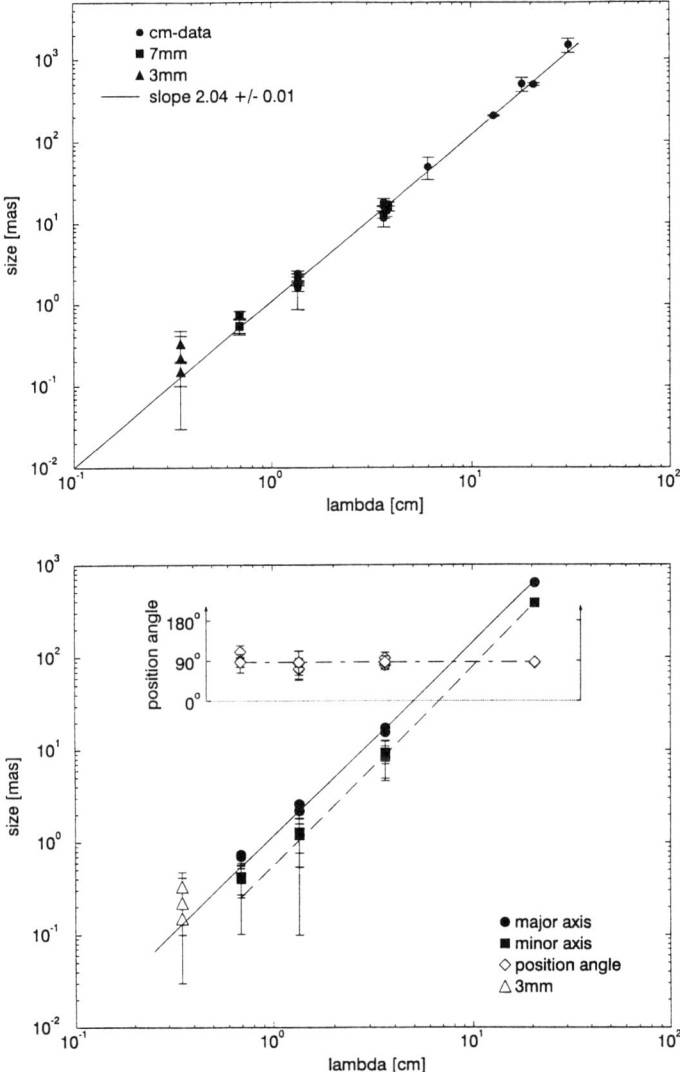

Figure 2. Observed source shape of Sgr A* as a function of wavelength. Data are from references cited in the text. a) Geometric mean of major and minor axis; the observed size at mm wavelengths could be marginally larger than the expected scattering size, as extrapolated from the data at $\lambda >1$cm. b) Details of the morphology: Major and minor axis and the position angle of the elliptical shape of the compact source Sgr A* are plotted versus wavelength λ. At λ3mm so far only a cicular (Gaussian) size of the source structure could be determined.

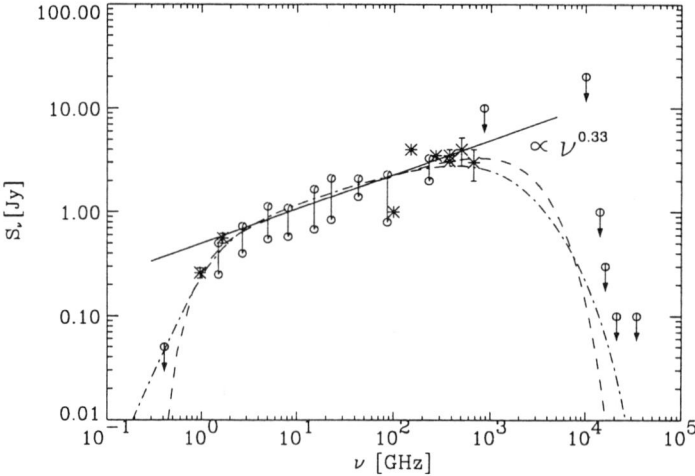

Figure 3. The observed radio spectrum of Sgr A* compared with best-fit model spectra (symbols as in Zylka et al., 1995; note that bars with symbols at both ends denote the variabiliy range and not error bars of individual observations; additional rhombs at low frequencies are from Davies, Walsh and Booth 1976). The low-frequency turnover can be either explained by free-free absorption (dashed curve) or by synchrotron self-absorption (dash-dotted curve; Beckert et al., 1996).

4. The optical/UV/X-ray spectrum of Sgr A*

Interstellar gas and dust, amounting to $A_v \sim 31$ mag, makes the Galactic Center inaccessible to direct observations between $\lambda 1 \mu m$ and the soft X-ray regime $E \lesssim 1$ KeV. At $\lambda 2.2 \mu m$ Eckart et al. (1995) find a cluster of 6 star-like objects one of which - with $S_{2.2\mu m} \cdot 0.01$ Jy - could coincide with Sgr A* (Radio). In the soft X-ray regime 1.2-2.5KeV Predehl and Trümper (1994) detected a point source with a - for $A_v = 31$ mag dereddened - flux density of $3 \; 10^{-7}$ Jy.

If Sgr A* were a massive but underfed Black Hole surrounded by an accretion disk we would expect its spectrum to be composed of a synchrotron spectrum (possibly associated with a jet-like object) and a thermal spectrum originating mainly close to the the inner edge of the accretion disk. While the synchrotron spectrum is well determined (see Figure 1) the thermal disk emission - if it exists at all - is only vaguely defined by the NIR and X-ray flux densities mentioned above.

5. The Galactic Center as seen from M31 with 1 arcsec resolution

Figure 4, from MDZ 96, compares the spectrum of Sgr A* with the spectrum integrated over the central parsec. To the radio/IR spectrum of Sgr A* shown in Figure 3 we have added the two upper flux density limits at $\lambda 2.2 \mu m$ and 1.2-2.5KeV mentioned above and fitted to these points the theoretical spectrum of an accretion disk surrounding a BH of $\sim 10^6 M_\odot$. It is found that the NIR flux

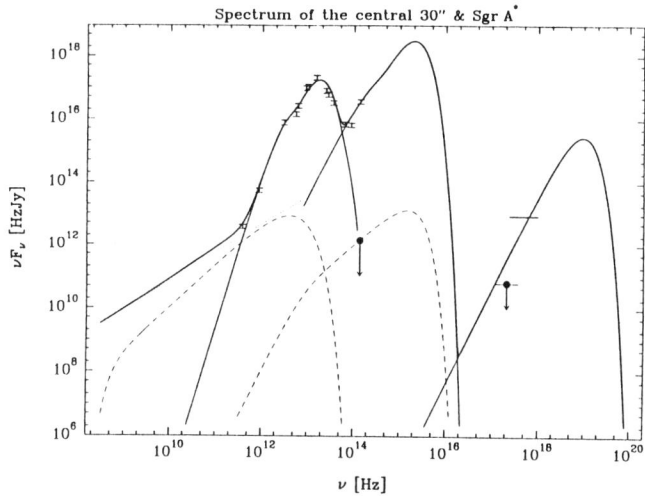

Figure 4. The radio through UV spectrum of the central 30″ (=1.2pc) as seen from the Galactic Poles (heavy curve). Free-free emission dominates the spectrum for $\nu < 2 \cdot 10^{11}$ Hz, dust emission for $2 \cdot 10^{11} < \nu/\text{Hz} < 3 \cdot 10^{13}$, stellar radiation for $3 \cdot 10^{13} < \nu/\text{Hz} < 10^{16}$ and a diffuse X-ray emission of unknown origin at still higher frequencies. Note that the stellar flux densities relate to the central parsec (i.e. $\sim 24''$) only. The observed spectrum of Sgr A* together with a disk spectrum fitted to the upper limit determined by the observed $\lambda 2.2\mu$m flux density and the mass $M_{BH} \sim 10^6 M_\odot$ are shown as dashed curves (MDZ 96).

density and the BH mass are the limiting factors. It is of interest to note that the (well determined) synchrotron spectrum and the (only vaguely defined) thermal spectrum of the accretion disk account for comparable luminosities of a few $100 L_\odot$. Such a division of thermal and non-thermal radiation has been suggested by Falcke, Mannheim and Biermann (1993) to hold for all BH/accretion disk "engines".

Details how the spectrum of the central parsec was constructed can be found in MDZ 96. Note that in Figure 4 we use a νS_ν representation of both spectra. The spectrum of the central parsec for $\nu < 10^{11}$Hz is dominated by free-free emission with S_{15GHz} -7.9Jy, for $\lambda < 10^{14}$ Hz by optically thin dust emission, for $\nu < 10^{16}$Hz by black-body emission (mainly from stars with $T_{eff} \sim 4000$K and $\sim$25 000K) and in the energy range 0.8-3KeV ($\nu > 10^{16}$Hz) by diffuse X-ray emission of 3 10^{-5}Jy, which originates in a hot (KT$\sim$10KeV) and extended (- $80 \lesssim R_{pc} \lesssim 80$) plasma. The integrated luminosity of the central parsec amounts to $\sim 10^8 L_\odot$ and is dominated by a central cluster of hot stars with a core radius of $\sim$0.17pc, rather than by the BH candidate Sgr A* which has a luminosity of only $< 10^3 L_\odot$. Less than 10% of the stellar emission in the central parsec is absorbed by dust and reradiated in the MIR/FIR (see MDZ 96).

Observed from M31 (D$\sim$700Kpc) with an angular resolution of $\sim 1'' \approx 3.4$pc the spectrum of the nucleus of our Galaxy would look similar to the integrated spectrum in Figure 4 but would have to be scaled by $(8.5/700)^2$. It would mimick a thermal (or radio-quiet) spectrum similar to that of a weak Seyfert 1 AGN, rather than the obviously non-thermal spectrum associated with the compact synchrotron source Sgr A*.

6. Comparison with AGN spectra

It appears that a radio spectrum $S_\nu \propto \nu^{1/3}$ is typical for the core emission of AGNs. Slee et al.(1994) find compact radio continuum cores in $\sim 70\%$ of radio-emitting elliptical and S0 galaxies which have a flat ot inverted spectrum with a median spectral index of 0.3. In this case a radio/IR spectrum similar to that of Sgr A* (Figure 3) would always be associated with a BH/accretion disk fueled AGN. But the comparison of the spectra of Sgr A* and the central parsec, respectively, shown in Figure 4, warns us that radio emission from nearby extended sources may mask such a synchrotron spectrum if the angular resolution of the observations is insufficient to separate compact core and extended emission regions. This, for example, would be the case with the spectrum of the center of our Galaxy if it were observed from the solar system with an angular resolution of only $24'' - 30''$.

Observations of the Seyfert 2 Galaxy NGC 1068 tend to support this view. With MERLIN and an angular resolution of $\sim$60mas ($\hat{\approx}$ 4pc at D=15Mpc) Muxlow et al. (1996) detected a compact core with an inverted ($S_\nu \propto \nu^{0.31}$) spectrum between 5 and 22GHz located close to the starting point of a flaring jet. This compact core is surrounded by four more sources with steep radio spectra which have masked the inverted spectrum in earlier observations of lower angular resolution.

An example of a positive detection of an inverted core spectrum relates to the compact core of M81, whose radio spectrum is very similar to that of Sgr A* but with a $\sim 10^4$ higher integrated luminosity (Figure 5; Reuter and Lesch, 1996). The fact that this spectrum could be separated, although the angular resolution at λ1.2mm was only $\sim 12''$, can be explained by the relatively low gas and dust content in the central region of M81.

7. Conclusions

i) Probably all compact AGN cores have inverted radio spectra $S_\nu \propto \nu^{1/3}$ (e.g. Slee et al. 1994).

ii) Falcke, Mannheim and Biermann (1993) hypothesize that AGN spectra consist of non-thermal (synchrotron) emission from the jet and thermal(Planck) radiation from the accretion disk and suggest that both mechanisms contribute proportionally to the total luminosity. The total Sgr A* spectrum would fit into this picture.

iii) In this case the intrinsic spectra of all BH/accretion disk " engines" would be similar, but the observed spectra would depend on extrinsic factors such as the interaction of the central engine with its surroundings, the distribution of ISM core of the AGN and - most important - the angular resolution of the telescope(s).

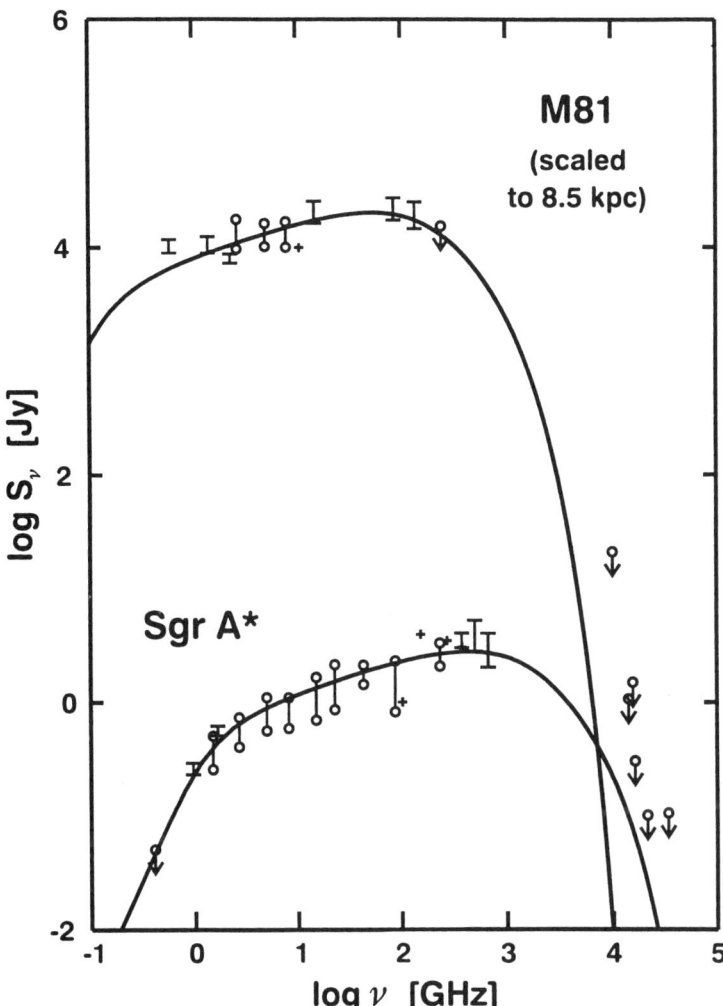

Figure 5. A comparison between the radio/IR spectrum of Sgr A* (Figure 3) and the core spectrum of M81 reduced to the distance of the GC (Reuter and Lesch, 1996).

iv) Recent observations of NGC 1068 (Muxlow et al., 1996) and M 81 (Reuter and Lesch, 1996) tend to support this hypothesis.

Acknowledgments. I want to thank my colleagues W.J. Duschl and R. Zylka for helpful discussions and remarks and for preparing the Figures, and H. Falcke and H. Lesch for valuable suggestions. This paper is based on a review by Mezger, Duschl and Zylka (1996, in press).

References

Balick B., Brown R.L., 1974, ApJ 194, 265
Backer D.C., Zensus J.A., Kellermann K.I., et al.., Science 262, 1414
Beckert T., Duschl W.J., Mezger P.G., Zylka R., 1996, A&A 307, 450
Davies R.D., Walsh D., Booth R.S., 1976, MNRAS 177, 319
Eckart A., Genzel R., Hofmann R., Sams B.J., Tacconi-Garman L.E., 1995, ApJ 445, L23
Falcke H., Mannheim K., Biermann P.L. 1993, A&A 278, L1
Krichbaum T.P., Zensus J.A., Witzel A., Mezger P.G., Standke K.J., Schalinski C.J., Alberdi A., Marcaide J.M., Zylka R., Rogers A.E.E., Booth R.A., Rönnäng B.O., Colomer F, Bartel N., Shapiro I.I., 1993, A&A 274, L37
Krichbaum T.P., Schalinski C.J., Witzel A., Standke K.J., Graham D.A., and Zensus J.A., 1994, " First Detection of Sgr A* with VLBI at 7mm and 3mm wavelength" , in: The Nuclei of Normal Galaxies: Lessons from the Galactic Center, NATO Advanced Science Institutes Series C, Vol. 445, eds. R. Genzel and H. Harris, Kluwer Academic Publishers, Dordrecht, p. 411-414
Lynden-Bell D., Rees M.J., 1971, MNRAS 152, 461
Mezger P.G., Duschl W.J., Zylka R., 1996, AAR (in press)
Muxlow T.W.B., Pedlar A:, Holloway A.J., Gallimore J.F., Antonucci R.R.J., 1996, MNRAS 278, 854
Predehl P., Trümper J., 1994, A&A 290, L29
Slee O.B., Sadler E.M., Reynolds J.E., Ekers R.D., 1994, MNRAS 269, 928
Reuter H.P., Lesch H., 1996 A&A (in press)
Rogers A., Doeleman S., Wright M., et al., 1994, ApJL 434, L59
Zylka R., Mezger P.G., Ward-Thompson D., Duschl W.J., Lesch H., 1995, A&A 297, 93

The Interpretation of the Sgr A* Spectrum

Wolfgang J. Duschl[1] and Thomas Beckert

ITA, Institut für Theoretische Astrophysik, Heidelberg, Germany

Abstract. We give a brief overview of our current understanding of the spectrum of Sgr A*.

1. Introduction

Balick and Brown (1974) using the NRAO Green Bank interferometer combined with a 14 m telescope located about 35 km southwest of the other telescopes, discovered at λ 11 cm the compact synchrotron source Sgr A* whose possible presence at the Galactic Center (GC) had been predicted three years earlier by Lynden-Bell and Rees (1971). Since then Sgr A* became a prime target of array and, especially, VLBI observations. Davies, Walsh and Booth (1976), who observed Sgr A* at 0.408, 0.96 and 1.66 GHz with the Jodrell Bank interferometer, were the first to note that the apparent size of Sgr A* depends on the wavelength $\propto \lambda^2$. This is the sign of source broadening due to electron scattering (see the paper by Mezger in these proceedings).

To our best current knowledge, Sgr A* is a very compact object, presumably a black hole of a few million solar masses (see papers by Genzel and by Eckart in these proceedings).

The radio spectrum of Sgr A* was found to have a low-frequency cut-off at $\nu \sim 1.4$ GHz, to be inverted ($S_\nu \propto \nu^{1/3}$) at higher frequencies, attaining a maximum at $\nu \sim 600$ GHz and decreasing steeply at still higher frequencies (Figure 1).

Sgr A* is seen against the compact GMC Sgr A East Core. Its spectrum is rather flat as compared with that of the dust emission from the East Core, which increases $\propto \nu^4$ at submm wavelength and somewhat less steeply at shorter wavelength. This renders flux density measurements in the submm/FIR/MIR range increasingly difficult. Actually only upper limits have been obtained for $\nu > 800$ GHz. Extinction due to the ISM between GC and Sun (which in the visual amounts to $A_V \sim 31^m$) prevents any direct observations of Sgr A* between the NIR at $\lambda \sim 1\,\mu$m and the soft X-ray regime $E < 1$ keV. Although weak point sources at $\lambda 2.2\,\mu$m and $E \sim 1 - 2$ keV are found in the immediate vicinity of the radio position of Sgr A* their identification with Sgr A* is uncertain; the observed emission could as well be contributed by the central cluster of massive HeI/HI stars detected by Krabbe, Genzel, Drapatz et al. (1991).

---

[1] MPIfR, Max-Planck-Institut für Radioastronomie, Bonn, Germany

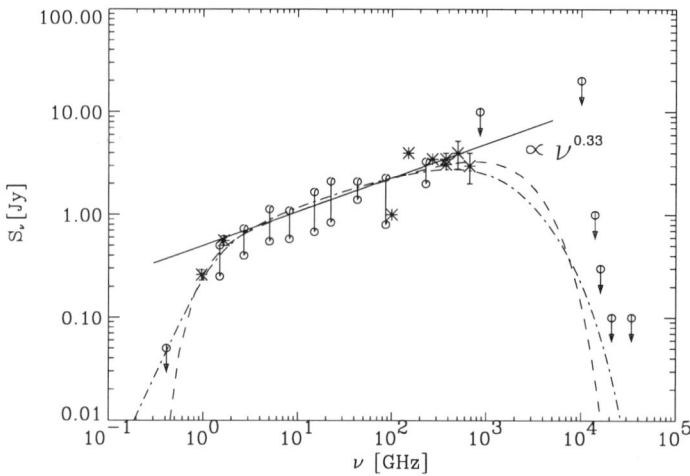

Figure 1. A comparison of the observed radio spectrum of Sgr A* with best-fit model spectra (Beckert, Duschl, Mezger et al. 1996). Note that bars with symbols at both ends denote the variability range and not error bars of individual observations; additional points at low frequencies are from Davies, Walsh and Booth (1976). The low-frequency turnover is due to either free-free absorption (dashed curve) or synchrotron self-absorption (dashed-dotted curve).

The variability of the radio emission from Sgr A* is now well established (Zhao, Goss, Lo et al., 1992). These authors suggest that refractive interstellar scintillation causes most of the variability at $\nu < 8.5$ GHz while intrinsic variations on time scales ~ 0.5 yr dominate at higher frequencies. Between 1990.2 and 1991.2 three radio outbursts have been observed with the VLA at 15 and 22.5 GHz. The variability at mm/submm wavelengths is still disputed (Zylka, Mezger, Ward-Thompson et al. 1995 = Z95).

2. Sgr A*: Spectrum, time variation and morphology

2.1. The radio/IR spectrum

The most recent spectrum of Sgr A* shown in Figure 1 is from Beckert, Duschl, Mezger et al. (1996 = B96) who have added to the Z95 spectrum three flux densities at 0.408, 0.96 and 1.66 GHz, derived nearly twenty years ago by Davies, Walsh and Booth (1976). Also added are two more upper limits obtained at $\lambda 350\,\mu$m by Serabyn and Lis (quoted in Mezger 1994) and at $\lambda 30\,\mu$m by Telesco, Davidson and Werner (1996). The gap in the spectrum between $\lambda 450$ and $30\,\mu$m allows - as an upper limit - the presence of $\sim 3\,M_\odot$ of ISM in a 10″-beam, with a dust temperature of ~ 50 K. Such a low dust temperature would be hard to reconcile with the high intensity of the radiation field in the GC where dust reaches temperatures of $\sim 200-300\,\mathrm{K}$ (Z95 and Mezger, Duschl and Zylka 1996 = MDZ96).

We follow here the spectral analysis by B96 keeping in mind, however, that other synchrotron emission models have been suggested (see, e.g., the paper by Falcke in these proceedings). The spectrum of Sgr A* has a high-frequency cut-off $\nu_c \sim 2-4\,10^3\,\mathrm{GHz}$ and a low-frequency turnover $\nu_t \sim 0.8\,\mathrm{GHz}$. In between the time-averaged spectrum increases $S_\nu \propto \nu^{1/3}$ (Duschl and Lesch 1994). The integrated radio luminosity is $L_\mathrm{radio,IR}\,(1-10^4\,\mathrm{GHz}) \sim 3\,10^2\,\mathrm{L}_\odot$. One possible explanation of this spectrum is optically thin synchrotron emission from quasi-monoenergetic electrons (Z95; B96, Beckert and Duschl 1996) combined with free-free absorption. Model parameters for the synchrotron source are an electron density $n_\mathrm{e,rel} \sim 3\,10^4\,\mathrm{cm}^{-3}$ and a magnetic field strength $B \sim 11\,\mathrm{G}$. The low-frequency turnover would be caused by an absorbing thermal plasma, with an emission measure of $E_t \sim 10^6\,\mathrm{pc\,cm}^{-6}$ for an electron temperature $T_e \sim 6\,000\,\mathrm{K}$ (Figure 1, dashed curve).

How does this model concur with observations? Sgr A West, the H{\sc ii} region surrounding Sgr A*, consists of an extended component and the minispiral, whose Central Bar overlaps in position with Sgr A*. If Sgr A* is located at the center of Sgr A West the emission measure provided by one half of the extended component, $0.5\,E_\mathrm{ext} \sim 10^6\,\mathrm{pc\,cm}^{-6}$ could account for the free-free absorption. The Bar, however, with an estimated emission measure of $9\,10^6 - 4\,10^7\,\mathrm{pc\,cm}^{-6}$ at the position of Sgr A* would have to be located behind the source.

In the second interpretation the low-frequency turnover in the spectrum is attributed to synchrotron self-absorption in Sgr A*. It should be noted that both alternatives have already been considered by Davies, Walsh and Booth (1976). However, thanks to the now much better determined spectrum of Sgr A* and source parameters of Sgr A West, B96 arrive at firm conclusions regarding the structure of Sgr A West and Sgr A*.

Synchrotron self-absorption can in fact explain the low-frequency turnover. The required magnetic field strength of $\sim 11\,\mathrm{G}$ is the same as in the above case and is in accordance with independent estimates of the equipartition magnetic field strength in the inner regions of an accretion disk accreting into a Black Hole of a few $10^6\,\mathrm{M}_\odot$ with a mass flow rate of $10^{-7\ldots-6}\,\mathrm{M}_\odot/\mathrm{yr}$. The source diameter of $\sim 2.4\,10^{13}\,\mathrm{cm}$ required by this model is $\sim 30\,R_S$ with $R_S \sim 9\,10^{11}\,\mathrm{cm}$ the Schwarzschild radius of a $3\,10^6\,\mathrm{M}_\odot$ black hole. It should be noted that - within the observational errors and the uncertainties of the model parameters - this predicted source size is compatible with source sizes of $\leq 4\,10^{13}\,\mathrm{cm}$ obtained from mm-VLBI observations of Sgr A*.

Both effects, i.e. free-free absorption by Sgr A West and synchrotron self-absorption in the compact source Sgr A* must affect the Sgr A* spectrum. With the observations available todate one cannot yet decide which is the dominating effect at the observed low-frequency turnover ν_t. Array observations at or below $\nu_t \leq 1.5\,\mathrm{GHz}$ may eventually resolve the question.

The NIR through X-ray spectrum Eckart, Genzel, Krabbe et al. (1992) using high-resolution NIR imaging, detected at $\lambda 2.2\,\mu\mathrm{m}$ an emission ridge of size $\sim 1''$ with a dereddened flux density of $0.06\,\mathrm{Jy}$. Assuming optically thick free-free emission from a source with disk-shaped brightness distribution of electron temperature $T_e \geq 2\,10^4\,\mathrm{K}$ yields a luminosity of

$$\frac{L}{L_\odot} \sim 1.2\,10^5 \left(\frac{T_e}{3\,10^4\,\mathrm{K}}\right)^3. \tag{1}$$

With effective temperatures in the range $T_{\mathrm{eff}} \sim 2-4\,10^4\,\mathrm{K}$ this yields luminosities of $L_* \sim 3.6 - 29\,10^4\,L_\odot$. More recently, however, with the remarkable K-band resolution of $\sim 0\overset{''}{.}15$ Eckart, Genzel, Hofmann et al. (1995) resolved the extended source into a cluster of six (probably stellar) sources of which only one can coincide with Sgr A*. Hence, the above estimates have to be decreased by $\sim 1/6$, yielding upper limits of the optical/UV luminosities of Sgr A* of $\leq 5\,10^4\,L_\odot$.

This central cluster of IR sources is referred to by Eckart, Genzel, Hofmann et al. (1995) as Sgr A* (IR) to discriminate against the compact radio source Sgr A* (R). These authors conclude that Sgr A* (IR) represents a small local clustering of moderately luminous stars near or at the position of Sgr A* (R). With one exception their polarization is similar to other sources in its vicinity. Eckart et al. conclude that Sgr A* (R) is probably dark; its M/L ratio at present has a lower limit of $\sim 100\,M_\odot/L_\odot$.

Several groups claim to have detected X-ray emission from Sgr A*, most recently Predehl and Trümper (1994) with ROSAT in the energy range 1.2 - 2.5 keV, and Koyama (1994) and Tanaka (priv.comm.) with ASCA in the energy range $\sim 1 - 10$ keV (for a review of previous X-ray observations of the Galactic Center see, e.g., Skinner (1993) and Predehl, Genzel, Trümper et al. (1994)). The point source detected by ROSAT with an angular resolution of $25''$ coincides with the radio position of Sgr A* to within $10''$, i.e., the positional accuracy of ROSAT observations. Previous detections at higher energies (3 - 30 keV) with coded mask telescopes on Spacelab-2 and ART-P but with considerably lower positional accuracy were reported by Skinner, Willmore, Eyles et al. (1987) and Pavlinsky, Grebenev and Sunyaev (1994), respectively.

Here we follow the arguments by B96. The ART-P source exhibits a non-thermal spectrum. If the ROSAT flux density is corrected for an X-ray absorption corresponding to the standard extinction $A_V \sim 31^m (\widehat{=} N_H \sim 5.4\,10^{22}\,\mathrm{cm}^{-2})$ between Sun and Galactic Center its intensity averaged over the energy range 0.8 - 2.5 keV is $1 - 2\,10^{34}\,\mathrm{erg\,s}^{-1}$. This lies ~ 2 to 3 orders of magnitude below the extrapolated ART-P spectrum. To bring the ROSAT flux densities in agreement with the ART-P spectrum requires an additional absorption correction corresponding to a column density of insterstellar matter $N_H \sim 1-1.5\,10^{23}\,\mathrm{cm}^{-2}$ and hence to a visual extinction of $A_V \sim 56^m - 83^m$. This could be explained if the ROSAT source were located deep inside the Sgr A East Core GMC against which Sgr A* and Sgr A West are seen in projection. In this case, however, the ROSAT source could not be identical with Sgr A*, which our observations place at the center of the HII region Sgr A West and in front of the Bar of the ionized Minispiral. Sgr A West, on the other hand, is located in front of the extended synchrotron source Sgr A East which, in turn, is located in front of the Sgr A East Core GMC (Mezger, Zylka, Salter et al. 1989).

Alternatively, Predehl and Trümper suggest that the ROSAT source in fact coincides with Sgr A* but is subject to a very local or intrinsic absorption. NIR observations of the central 0.5 pc do not indicate localized dust absorption on scales of $\sim 1''$ (Krabbe, Genzel, Eckart et al. 1995). To comply with these observations the X-ray absorption would have to be provided by an ionized shell

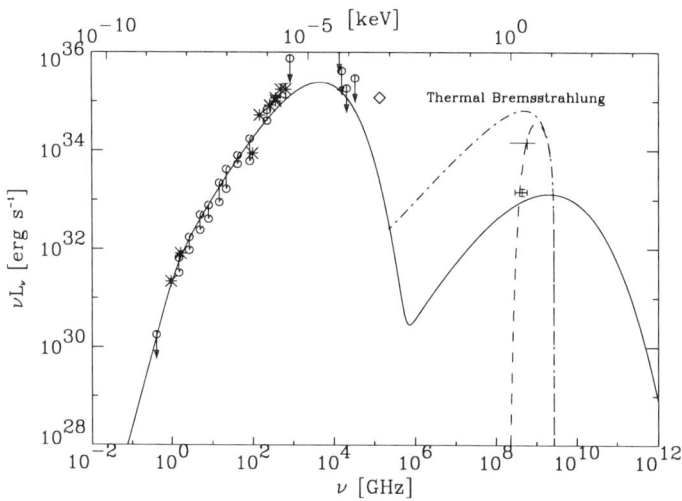

Figure 2. The spectrum of Sgr A* from the radio to the X-rays: synchrotron radiation and synchrotron-self-comptonized emission (full line). The dash-dotted line is additional thermal bremsstrahlung of $kT = 5$ keV. Corrected for standard extinction (broken line) this provides a possible explanation for the x-ray measurements by ROSAT and the Einstein observatory.

surrounding Sgr A*, where most of the central dust has evaporated. B96 show that this is not consistent with observations.

Nonetheless the synchrotron model implies the existence of a synchrotron self-comptonized flux which appears in the X-rays. The luminosity due to inverse Compton photons is lower than the synchrotron luminosity by two orders of magnitude if the assumed optically thin synchrotron radiation is produced by relativistic electrons with energy equipartition between electrons and magnetic field. The model predictions are shown in Figure 2. An additional emission by thermal bremsstrahlung from the vicinity of Sgr A* possibly provides the lacking flux (dash-dotted curve in Figure 2). We have assumed a standard extinction corresponding to $N_H = 5.4\ 10^{22}$ cm^{-2} to account for the measurements with ROSAT and the Einstein satellite (dashed curve).

Above (see Eq. 1) we derived an upper limit for the luminosity of Sgr A* assuming black-body emission with a Planck curve fitted through the $\lambda 2.2\,\mu$m measurement. The upper limit for the soft X-ray emission obtained by Predehl and Trümper allows a more sophisticated estimate. If both the NIR upper limit with $S_{2.2\,\mu m} \sim 0.01$ Jy derived from Eckart, Genzel, Krabbe et al. (1992) and the soft X-ray source with $S_{1\,keV} \sim 1\ 10^{34}$ erg s^{-1} observed by Predehl and Trümper (1994) close to the position of Sgr A*(R) are taken as uper limits to the emission by an accretion disk surrounding the central BH one can fit the spectrum of such an accretion disk[1]. It is found that the spectrum is constrained by the NIR flux density and the mass of the BH for which we adopt $M_{BH} \sim 2\,10^6\,M_\odot$. In this way an upper limit to the luminosity of an accretion disk in Sgr A* of $\sim 200\,L_\odot$ is obtained.

Acknowledgments. This paper is absed on a forthcoming review article by Mezger, Duschl and Zylka (1996).

References

Balick B., Brown R.L., 1974, ApJ 194, 265
Beckert T., Duschl W.J., 1996, A&A (submitted)
Beckert T., Duschl W.J., Mezger P.G., Zylka R., 1996, A&A 307,450 (B96)
Davies R.D., Walsh D., Booth R.S., 1976, MNRAS 177, 313
Duschl W.J., Lesch H., 1994, A&A 286, 431
Eckart A., Genzel R., Krabbe A., et al., 1992, Nature 355, 526
Eckart A., Genzel R., Hofmann R., Sams B.J., Tacconi-Garman L.E. 1995 ApJ 445, L23
Koyama K., 1994, in: New Horizon of X-ray Astronomy - First Results from ASCA, Universal Academy Press, Tokyo, Japan, p. 181
Krabbe A., Genzel R., Drapatz S., Rotaciuc V., 1991, ApJ 382, L19
Krabbe A., Genzel R., Eckart A. et al., 1995, ApJ 447, L95
Lynden-Bell D., Rees M.J., 1971, MNRAS 152, 461
Mezger P.G., 1994, in: The Nuclei of Normal Galaxies - Lessons from the Galactic Center, Kluwer Academic Publishers, Dordrecht, The Netherlands, p. 415
Mezger P.G., Duschl W.J., Zylka R., 1996, A&AR, *in press* (MDZ96)
Mezger P.G., Zylka R., Salter C.J., Wink J.E., Chini R., Kreysa E., Tuffs R., 1989, A&A 209, 337
Pavlinsky M.N., Grebenev S.A., Sunyaev R.A., 1994, ApJ 425, 110
Predehl, P., Genzel R., Trümper J., Zinnecker H., 1994, in: The Nuclei of Normal Galaxies - Lessons from the Galactic Center, Kluwer Academic Publishers, Dordrecht, The Netherlands, p. 21
Predehl P., Trümper J., 1994, A&A 290, L29
Skinner G.K., 1993, A&AS 97, 149
Skinner G.K., Willmore A.P., Eyles R.D., et al., 1987, Nature 330, 554
Telesco C.M., Davidson J.A., Werner M.W., 1996, ApJ 456, 541
Zhao J.-H., Goss W.M., Lo K.-Y., Ekers R.D., 1992, ASP Conf. Series 31, 295
Zylka R., Mezger P.G., Ward-Thompson D., Duschl W.J., Lesch H., 1995, A&A 297, 83 (Z95)

---

[1] We have taken a steady standard accretion disk which extends from 3 to 100 schwarzschild radii assuming that it radiates locally like a black body. Given all the uncertainties in the model (unknown inclination against LOS, stationarity, local black body spectrum etc.), we estimate this model to be good to an order of magnitude.

Activity and Radiative Characteristics of the Central Engine in the Galactic Center

F. Melia[1]

Physics and Astronomy Departments, University of Arizona, Tucson, AZ 85721

Abstract. If Sgr A* is a massive ($\sim 10^6\,M_\odot$) black hole embedded in a region with strong gaseous outflows, as suggested by stellar kinematic studies and the observation of He I, Brα and Brγ line emission, it should be accreting from its environment via the Bondi-Hoyle process, unless it too is a source of powerful mass ejection. We discuss the consequences of this activity, including the expected mass and angular momentum accretion rate onto the black hole, and the resulting observable characteristics. The latest infrared images of this region appear to rule out the possibility that this large scale flow settles down into a standard α-disk at small radii. We discuss some possible scenarios that might account for this, including strong advection in the disk or the presence of a massive, fossilized disk. Not all of the gas affected in this way by Sgr A*'s strong gravitational field becomes bound. Some of it is redirected into a focused flow that in turn interacts with other coherent gas structures near the black hole. We suggest that the mini-cavity (to the south-west of Sgr A*) may be formed as a result of this activity, and argue that the characteristics of the mini-cavity lend some observational support for the presence of a concentrated mass near Sgr A*. We show, however, that as far as the mini-cavity is concerned, this concentrated mass need not be in the form of a point mass, but may instead be a highly concentrated cluster of stellar remnants.

1. Introduction

The uniqueness of the radio source Sgr A*, together with its low proper motion (< 40 km s^{-1}; Backer & Sramek 1987) and its location near the dynamical center of the galaxy, suggest that it may be a massive point-like object dominating the gravitational potential in the inner 0.5 pc region. Its discovery followed a prediction that such an object might be visible at radio wavelengths (Lynden-Bell & Rees 1971), though the exact mechanism powering the radiative emission has been difficult to identify.

Over the past several years, however, multi-wavelength observations of the galactic center region have provided a much more detailed picture of the gas dynamics in the vicinity of Sgr A*, pointing to the accretion model as perhaps the

---

[1] Presidential Young Investigator

most likely physical process producing the spectral properties we now attribute to this blackhole candidate. Most, perhaps all, of Sgr A*'s radiative characteristics may be due to the energy liberated by a galactic center wind as it accretes down the deep potential well. This gaseous flow is inferred to have a velocity $v_w \approx 500-700$ km s^{-1} and to constitute a mass loss rate $\dot{M}_w \approx 3-4 \times 10^{-3}$ $M_\odot$ yr^{-1} (Hall, Kleinmann & Scoville 1982; Geballe et al. 1991; Serabyn et al. 1991; Yusef-Zadeh & Melia 1991).

On the other hand, the nature of the Bondi-Hoyle accretion onto a point like object also presents somewhat of a challenge in understanding what happens to the gas as it settles down into a planar configuration close to the event horizon. Even if the ambient flow is uniform, fluctuations beyond the bow shock (located at roughly the accretion radius $R_{acc} \approx 10^5$ R_g, where $R_g \equiv 2GM/c^2$ is the Schwarzschild radius) produce a transient accretion of net angular momentum that ought to result in the formation of a temporary (albeit small) disk. More realistically, the inflow itself carries angular momentum, so that the formation of a disk-like structure at small radii (i.e., $r \approx 10^{2-3}$ R_g) is difficult to avoid. In this paper, we will discuss the characteristics expected of the large scale flow should Sgr A* be a point-like object with the mass ($\sim 1-2 \times 10^6$ $M_\odot$; Haller et al. 1996) implied by the stellar kinematics, and we will consider several possibilities for the flow at smaller radii.

Coupled to this, is the question of whether the gaseous motions near Sgr A* are responsible for producing a feature known as the minicavity located in one of the ionized streamers (the Bar) appearing in radio continuum maps to the southwest of the dynamical center (Yusef-Zadeh, Morris & Ekers 1989). A chain of blobs of emission appearing in more recent, high-resolution radio images seem to lead away from Sgr A* toward the cavity, providing some morphological evidence that the two are physically connected. Spectroscopy and 1″ narrow-band imaging of near-infrared line emission in the central 0.3 pc of the Galaxy showed that the mini-cavity is a source of bright 2.2178μm line emission that has been unambiguously identified as [FeIII] (Lutz, Krabbe & Genzel 1994). This followed earlier detections of 2.217μm emission toward the Sgr A*/IRS 16 region by Eckart et al. (1992), who speculated that the radiation originates within an expanding bubble of ionized, hot iron driven by an outflow from the direction of the massive black hole candidate. This picture is supported by the derived fractional abundances of [Fe]/[H] $\geq 2.5 \times 10^{-5}$ and [Fe^{++}]/[Fe$^+$]≈ 1 which, together with the observed spatial distributions and line profiles of [FeIII] and HI Brγ, suggest that an expanding gas bubble is being created in the partially neutral gas streamer by a fast (~ 1000 km s^{-1}) gas flow from one or several sources within a few arcsecs of the present location of the mini-cavity. Lutz et al. (1994) conclude that the most likely physical configuration of the mini-cavity is that of a shocked gas heated and compressed by this interaction, which destroys dust grains and releases iron into the gas phase. The dense postshock gas then forms an HII region, maintained at about 7000 K electron temperature via photoionization by the central UV field. The idea that dust is being destroyed is supported by the finding of Gezari & Yusef-Zadeh (1991) that the ratio of 12μm dust continuum to radio continuum flux has a minimum at the position of the mini-cavity.

Thus, in addition to considering the large scale infall onto the mass concentration near Sgr A*, we shall also address the issue of whether the "excess",

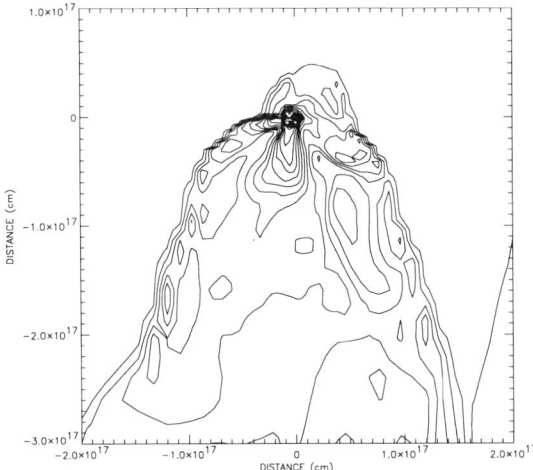

Figure 1. Cross-sectional contour plot of the density ρ in the Bondi-Hoyle bow shock at Sgr A* (positioned at $0,0$). The image has 25 contours increasing as the square root of ρ from 4.8×10^3 to 2.1×10^5 cm^{-3}. The galactic center wind entering from the top is described in the text and it flows past a black hole with mass 10^6 $M_\odot$. The inner boundary is at 0.1 R_{acc}.

highly focused, hydrodynamic stream past Sgr A* has the correct distribution to be consistent with the gaseous flow required in the detailed modeling of the structure of the mini-cavity (Lutz et al. 1994). Knowing the IRS16 wind characteristics and Sgr A*'s mass provides an unambiguous profile of the collimated post-bow shock wind. We shall examine this situation both in the case where the mass concentration is in the form of a point-like object and where it is represented by a highly condensed cluster of stellar remnants. The latter may be of potential interest to an alternative scenario for the activity in this region, in which the concentrated mass may be in the form of a collapsed core in a cluster of white dwarfs (see, e.g., Haller et al. 1996). Although we make no attempt here to justify such a distribution on dynamical and evolutionary grounds, we do nontheless explore its viability in accounting for large scale features such as the minicavity. In this picture in which no massive black hole is present in the region, Sgr A* might then simply be the hot, magnetized gas trapped in the core of the stellar distribution, not unlike the X-ray emitting plasma in the potential well of clusters of galaxies (Melia & Haller 1996).

2. Bondi-Hoyle Accretion Onto Sgr A* and Radiative Emission

The length scale (R_{acc}) for accretion onto a compact object such as Sgr A* is set by determining the distance from the accretor at which the potential energy density of the flowing gas is matched to its kinetic energy density. In Figure 1 we show the cross-sectional density contours resulting from a 3-D hydrodynamic

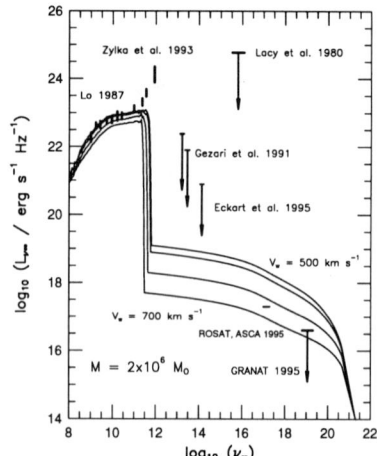

Figure 2. Bremsstrahlung and Magnetic Bremsstrahlung spectrum from the post bow-shock infall toward the black hole. The various curves range from a wind velocity of 500 km s^{-1} (upper) to 700 km s^{-1} (lower).

simulation with a domain of solution spanning a region of 30 R_{acc} in either direction, using the black hole and galactic center wind characteristics discussed in § 1 above (see also Ruffert & Melia 1994; Melia, Coker & Yusef-Zadeh 1996; Coker & Melia 1996). Compression of the gas and magnetic field dissipation in the converging flow toward the black hole heat the gas, which in turn cools primarily by bremsstrahlung and magnetic bremsstrahlung emission with an emissivity characteristic of the particle density n, temperature T and magnetic field intensity (Melia 1994). For the conditions in the galactic center, n scales roughly as $r^{-3/2}$ with a value of about 2×10^{11} cm^{-3} at $r = R_g$, and T varies from about 10^6 K behind the shock to as high as $\sim 10^{10}$ K in the very inner region. Figure 2 shows the broad-band spectrum expected from this post-shock flow, in which the emissivity shortward of $\sim 10^{12}$ Hz is due primarily to thermal cyclotron/synchrotron radiation, whereas the flux at higher energies results from bremsstrahlung processes. In this figure, the ROSAT data point is consistent with the ASCA measurement when the $N_H \approx 2.2 \times 10^{23}$ cm^{-2} of column density in the accreting flow is taken into account in estimating the line-of-sight absorption. Since the soft X-rays are produced gradually as the gas descends toward the event horizon, the overlying column density in the Bondi-Hoyle flow adds to the overall N_H associated with this emission. We shall return to the question of what happens to this gas when it approaches the black hole in § 4.

3. Formation of the Mini-Cavity

The focused "excess" flow behind the accretor carries mass and mechanical energy flux into the Bar to the southwest of Sgr A*. In Figure 3, we show a

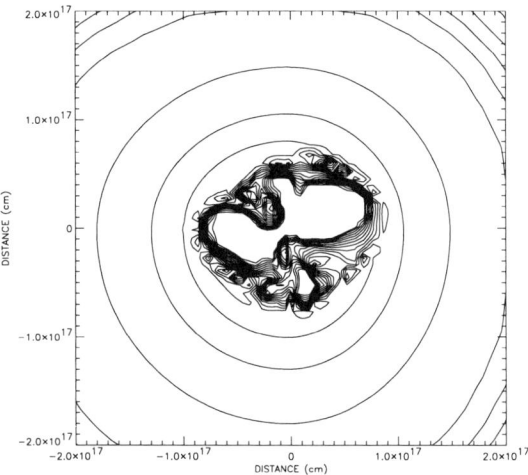

Figure 3. Cross-sectional contour map of the mechanical energy flux flowing into the minicavity. The image shows 20 contours increasing as the square root of the energy flux from $1,100$ to $16,500$ ergs cm^{-2} s^{-1}. Note that the inner portion is low in flux due to the removal of the gas by accretion onto the black hole.

contour map of this kinetic energy flux perpendicular to the flow direction at the observed location of the minicavity. The radius of the post-bow shock flow corresponds very closely with the observed radius ($\sim 1''$, see Yusef-Zadeh et al. 1990) of the mini-cavity, suggesting that the energy required to dynamically control the activity in this hole is brought in by the mechanical flux. Indeed, Lutz et al. (1993) conclude in their detailed analysis of the cavity's emission that the power required to sustain its luminosity is $L_{cav} \approx 6 \times 10^{37}$ ergs s^{-1} for reasonable values of the surrounding medium parameters. Our calculated value of the collimated flow's mechanical luminosity ($L_{shell} \approx 2.5\, L_{cav}$) is therefore about right to be the dominant energizing agent (Melia, Coker & Yusef-Zadeh 1996). We note that if the black hole mass is larger than the $\approx$ one million solar masses assumed here, then the focusing power of the accretor is greater and we would expect a larger mechanical luminosity and mass flux into the streamer. In addition, since the accretion radius scales as the mass, we would then also expect a larger size for the mini-cavity.

4. Accretion Onto the Black Hole

As one might guess by looking at the density profile in Figure 1, even when the upwind medium is uniform, the downwind region exhibits fluctuations in $\dot{M}$ that carry a net angular momentum down to the black hole. Detailed simulations (e.g., Ruffert & Melia 1994) suggest that at any given time, the accreting gas should circularize between 5 and 10 R_g. If the upwind medium has gradients in ρ or v_w, as might arise from a distributed wind source, this circularization radius

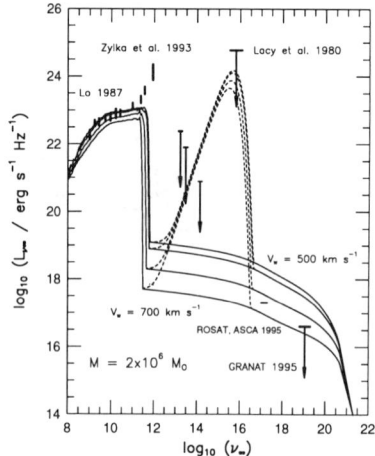

Figure 4. Same as Fig. 2, but now with the additional component (dotted curves) due to an α-disk forming at smaller radii. The disk size and accretion rate are determined by the mass and angular momentum inflow rates from the Bondi-Hoyle shock.

might be even bigger. We have calculated the spectral component expected from a sum of blackbody emissions due to this disk (under the assumption that it is a standard α model), and show it as dotted curves in Figure 4. Since the infrared data are seemingly inconsistent with this additional spectral component, we conclude that either (1) the circularized flow does not form an α-disk, but rather advects most of its dissipated energy through the event horizon (Narayan & Mahadevan 1995), (2) the Bondi-Hoyle flow merges into a massive, fossilized disk, storing most of the deposited matter at large radii (Falcke & Melia 1996), or (3) Sgr A* is not a point-like object (see below).

The first of these possibilities seems very promising. Every disk is advective to some degree. The question is whether the conditions in and around Sgr A* will permit a self-consistent disk solution with a very large advected fraction when the essential physics, including non-local effects and non-Keplerian flows, is incorporated into the analysis. In the second scenario, the wind infall must have a very large specific angular momentum, for it will otherwise settle onto the disk at small radii, where its kinetic energy will be thermalized and radiated away. A fossilized disk would thus need to function as a mass storage device, undergoing episodic accretion instabilities, not unlike those observed in X-ray Novae. The third possibility is discussed in the next section.

5. A Compact Cluster of Stellar Remnants ?

It is not unreasonable to presume that some $2-3$ million white dwarfs could have formed and settled within the central cluster (Haller et al. 1996). Whether this distribution of stellar remnants could have undergone core collapse, however, is still not known. The reason this is of potential interest to our understanding of

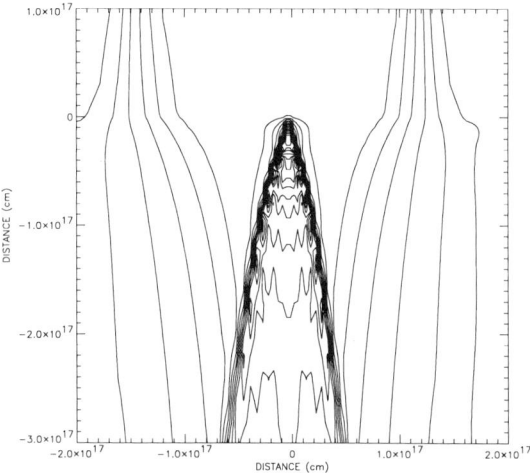

Figure 5. Gas condensation resulting from a Galactic center wind flowing past a cluster of 10^6 one solar mass stellar remnants concentrated within a radius of $2R_{acc}$ at the location of Sgr A*. The image shows 25 contours scaling as the square root of the density from 1.3×10^2 to 7×10^4 cm^{-3}.

Sgr A* is that a concentration of $1-2$ million solar masses within $1-2$ R_{acc} could in principle mimic the environmental impact of a point object. In that case, Sgr A* itself might simply be the trapped hot, magnetized plasma within the cluster (Melia & Haller 1996). We have carried out several 3D hydrodynamic simulations of a galactic center wind flowing through such a condensed (isothermal) cluster of one solar mass objects and the density profile of one of these is shown in Figure 5. Of course, unless the enclosed mass is $M_{cluster} = 2G/v_w^2\, R_{cluster}$, no shock will form since $R_{cluster}$ will then exceed R_{acc}. The case shown in Figure 5 is for $R_{cluster} = 2R_{acc}$. We see, however, that even in this case a condensation of the mass flow past the central mass has several features in common with the downwind Bondi-Hoyle flow we saw in Figure 1. When we examine the size and luminosity of the cavity produced in the Bar due to this type of flow (see Figure 6), the expected characteristics are not so different that we can immediately rule this picture out. To be sure, the minicavity should then be smaller and the luminosity is about half of the value we inferred in Figure 3. Nonetheless, this scenario is sufficiently intriguing that it merits further study. The most important test will be to see if the trapped gas can in fact account for the observed radio (and possibly X-ray) spectrum from this source.

Acknowledgments. This work was supported by NSF grant PHY 88-57218 and NASA grant NAGW-2822.

References

Backer, D.B. & Sramek, R.A. 1987, in AIP Proc. 155, ed. D.B. Backer, 155

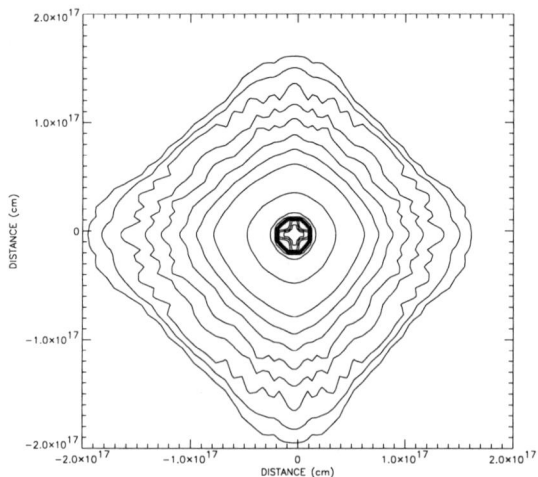

Figure 6. Cross-sectional contour map of the mechanical energy flux flowing into the minicavity for the case shown in Fig. 6. The image shows 20 contours increasing as the square root of the energy flux from 23 to $8,190$ ergs cm^{-2} s^{-1}.

Coker, R.F. & Melia, F. 1996, these proceedings.
Eckart, A., Genzel, R., et al. 1992, *Nature*, 355, 526
Eckart, A., et al. 1996, these proceedings
Falcke, H. & Melia, F. 1996, ApJ, submitted
Gezari, D. & Yusef-Zadeh, F. 1991, in Astrophysics with Infrared Arrays, ed. R. Elston, A. S. P. Conference Series Vol. 14, p. 214
Hall, D.N.B., Kleinmann, S.G. and Scoville, N.Z. 1982, ApJ, 260, L63
Haller, J.W., Rieke, M.J., Rieke, G.H., Tamblyn, P., Close, L. & Melia, F., 1996, ApJ, 456, 194
Lo, K.Y. 1987, in *AIP Proc. 155*, ed. D.C. Backer, (AIP: New York), 155, 30
Lutz, D., Krabbe, A. & Genzel, R. 1993, ApJ, 418, 244
Lynden-Bell, D. & Rees, M. 1971, MNRAS, 152, 461
Melia, F. 1994, ApJ, 426, 577
Melia, F., Coker, R.F. & Yusef-Zadeh, F. 1996, ApJ, 460, L33
Melia, F. & Haller, J.W. 1996, ApJ, in preparation
Narayan, R. & Mahadevan, R. 1995, *Nature*, 374, 623
Ruffert, M. & Melia, F. 1994, *A.A.*, 288, L29
Serabyn, E., Lacy, J.H., & Actermann, J.M. 1991, ApJ, 378, 557
Yusef-Zadeh, F. & Melia, F. 1991, ApJ, 385, L41
Yusef-Zadeh, F., Morris, M. & Ekers, R.D. 1990, *Nature*, 348, 45

Bondi-Hoyle Accretion onto Sgr A*

R. F. Coker

Physics Department, University of Arizona, Tucson, AZ 85721

F. Melia[1]

Physics and Astronomy Departments, University of Arizona, Tucson, AZ 85721

1. Introduction

Multi-wavelength observations (Geballe et al. 1991) of the inner ~ 0.3 pc region of the Galaxy provide strong evidence for the presence of an ambient Galactic center wind of 3×10^{-3} $M_\odot$ per year flowing near the unique variable, nonthermal point source Sgr A*, which, based on kinematic studies of nearby stars (Haller et al. 1996) might have a mass of $\sim 10^6$ $M_\odot$. The conditions suggest that the luminosity and spectrum of Sgr A* may be primarily due to its accreting this $v \sim 700$ km s^{-1} and $n \sim 5.5 \times 10^3$ cm^{-3} wind (Melia 1992) via the Bondi-Hoyle Process. Bondi-Hoyle accretion and its accompanying bow shock will occur whenever a gas flowing in a potential well has a kinetic energy density smaller than its gravitational potential energy density. In the case of Sgr A*, this condition is met at the accretion radius, $R_{acc} \equiv 2GM/v^2 \sim 5 \times 10^{16}$ cm $\sim 1/2''$. Using the finite-differencing 3D hydrodynamical (HD) code ZEUS (Stone & Norman 1992), we have carried out basic simulations of this activity in order to determine the temperature, density, and velocity profiles of the resulting flow as well as the mass and angular momentum accreted by Sgr A*. We show that this Bondi-Hoyle accretion, which is responsible for an infall of up to $\sim 10^{22}$ g s^{-1} onto Sgr A*, also produces a downstream, focused flow with a radius and mechanical luminosity that may be the cause of the $\sim 1''$ mini-cavity (and its associated "blobs") in the streamer 3.5" to the southwest of Sgr A* (Yusef-Zadeh et al. 1989). We show that either a highly concentrated cluster of stellar remnants or a point mass can produce such a collimated flow.

2. The Simulations

Two 3D HD simulations have been carried out, one for a 10^6 $M_\odot$ point mass and the other for an isothermal distribution of 10^6 1 $M_\odot$ stellar remnants. In both cases, the cube of solution is 32 R_{acc} on a side with 115 zones per side in the single point mass case and 75 zones per side in the isothermal distribution run (with higher resolution nearer the central region). The wind is modeled as a Mach 10 planar flow entering the volume of solution from one side through a 0.1 pc

---

[1]Presidential Young Investigator

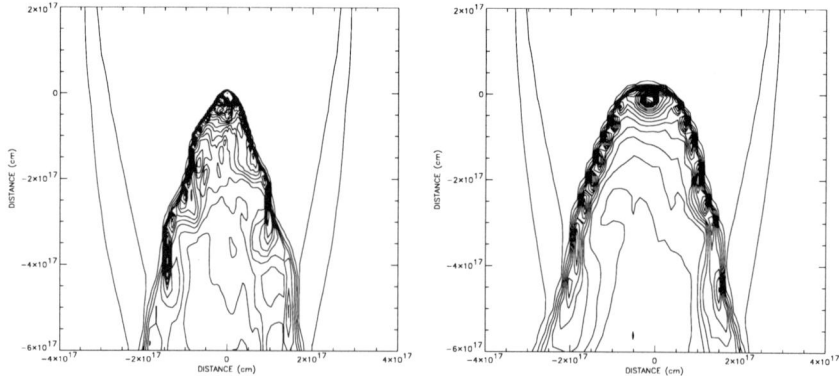

Figure 1. Cross-sectional contour plot of the density ρ in the Bondi-Hoyle bow shock for both runs (the single point mass case on the left and the isothermal distribution on the right). There are 25 contours increasing linearly with ρ from 2.3×10^3 to 5.8×10^4 cm^{-3}. Sgr A* is located at 0,0, and the wind enters from above.

radius circle while the other edges of the cube permit only gas outflow. The code uses van Leer advection and fully explicit time integration, determined by the Courant condition. Although the dissipation of the magnetic field entrained by the highly ionized flow is expected to be an important heating agent (Melia 1992, 1994), it has not as yet been included. Also, bremsstrahlung and, especially, magnetic bremsstrahlung emission become significant toward smaller radii; they too have not yet been included. Instead, the medium is assumed to be an adiabatic polytropic gas. See Stone & Norman (1992) for more details.

The 10^6 $M_\odot$ point mass accretor is modeled as a totally absorbing sphere with a radius of 0.03 R_{acc}; its interior is kept effectively empty by resetting the density in the interior zones to very small values. The mass and specific angular momentum accretion rates are determined by the gas flowing through this inner boundary. The isothermal distribution is truncated at 1 R_{acc} and given a velocity dispersion of 600 km/sec so that the total integrated mass is correctly 10^6 $M_\odot$. Note that if the velocity dispersion is less than ~ 600 km/sec (or, equivalently, the 10^6 1 $M_\odot$ objects have a core radius larger than $R_{acc}/2$), no shock front forms and no Bondi-Hoyle accretion occurs at all.

3. Resulting Profiles

In Figure 1 we show the cross-sectional density contours for the two simulations. The bow shock is clearly evident in both cases, although it is sharper, more focused, and more erratic in the black hole case. The turbulent post-shock flow is evident, as well as the post-shock focusing of the gas into a "tube". Both configurations yield similar profiles and produce a fairly stable shock front at $\sim 1 R_{acc}$ while having central densities and temperatures more than 100 times that of the inflowing wind.

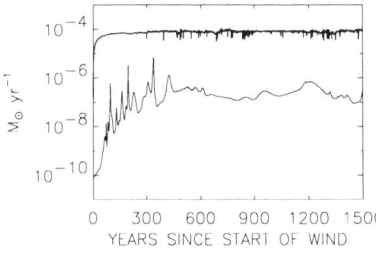 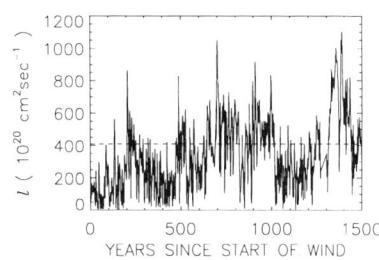

Figure 2. On the left is the mass accretion rate versus time for the two runs; the single point mass case is the thick upper line while the distributed masses case is the thin lower line. Note that equilibrium requires ~ 500 years. On the right is the specific angular momentum accreted by the black hole versus time. The dashed line is the time averaged equilibrium value.

4. Specific Angular Momentum and Mass Accretion Rates

It has long been known that spherical accretion onto a point mass will yield the Hoyle-Lyttleton (1939) accretion rate, $\dot{M}_{HL} \equiv 4\pi(GM)^2 \rho_{wind}/v^3$. In the case of Sgr A*, this is $\sim 1/30$th of the total Galactic wind or $\sim 10^{-4} M_\odot$ yr^{-1}. In Figure 2 we show the resulting total specific angular momenturm and mass accretion rates. $\dot{M}$ for the black hole run agrees with the Hoyle-Lyttleton theory and earlier results (Ruffert & Melia 1994) to within $\sim 25\%$ while the distributed masses accrete less due to the M^2 dependence of $\dot{M}$. However, the latter's time variability is much larger due to the smaller length scales involved. The $\dot{M}$ for the isothermal distribution was calculated assuming that all of the 1 $M_\odot$ objects are local perfect Hoyle-Lyttleton accretors.

The average specific angular momentum accreted by the black hole is 4×10^{22} cm^2 sec^{-1}. This yields a "circularization" radius ($\equiv 2l^2/[c^2 R_g]$ where $R_g \equiv 2GM/c^2$) of $\sim 50\ R_g \sim 1$ A.U. Thus, most of the infalling angular momentum is cancelled in the post shock flow and any resulting accretion disk, in the black hole scenario, is likely, even in the case of highly non-uniform inflow, to have a radius of less than a few A.U. In the case of the distributed masses, the radio flux that is identified with Sgr A* may be coming from hot, magnetized gas trapped in the core of the stellar remnant distribution. If the core radius of the distributed masses is more than $\sim R_{acc}/2$ then the mass accretion rate falls to $\sim 10^{-10}\ M_\odot$ yr^{-1} and no gas gets trapped within the cluster.

5. Creating the Mini-Cavity

As discussed elsewhere (Melia, Coker & Yusef-Zadeh 1996; Melia 1996), the focused post-shock flow carries mass and mechanical energy flux into the Bar 3.5" to the southwest of Sgr A*. The radius of this flow ($\sim 1''$) closely matches the observed radius of the mini-cavity (Yusef-Zadeh et al. 1990), suggesting that

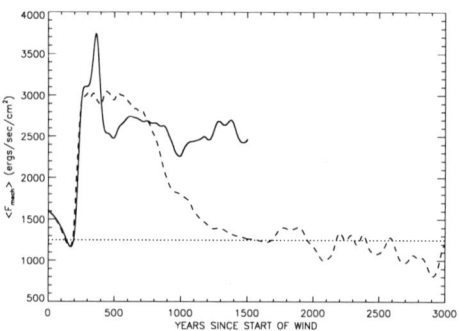

Figure 3. Assuming the mini-cavity has a radius of 1″ and is located 3.5″ behind Sgr A*, these curves show the average mechanical flux for the two runs (solid = black hole; dashed = isothermal distribution; dotted = minimum required to power the mini-cavity).

the mechanical energy brought in by this gas flow is providing the power to sustain the mini-cavity's luminosity of $\approx 6\times 10^{37}$ ergs sec^{-1} (Lutz et al. 1993). In Figure 3 we show the calculated mechanical flux ($\equiv \rho v^2 v_z/2$) at the location of the mini-cavity. The black hole case clearly has sufficient mechanical energy to power the mini-cavity while the isothermal distribution is marginal due to its less efficient focusing. In addition, the mass transfer rates and variability are consistent with forming the "blobs" connecting Sgr A* and the mini-cavity. However, the inclusion of heating and cooling may change this picture significantly. Thus, in principle, either a $10^6 M_\odot$ black hole or a concentration of 10^6 1 $M_\odot$ stellar remnants (e.g., white dwarfs) within 1 R_{acc} could account for the mini-cavity. Further study is required to determine if the gas trapped by the cluster can account for the observed radio spectrum of Sgr A*.

References

Geballe, T. R., Krisciunas, K., & Bailey, J. A. 1991, ApJ, 370, L73
Haller, J. W., Rieke, M. J., Rieke, G. H., Tamblyn, P., Close, L., & Melia, F. 1996, ApJ, 456, 194
Hoyle, F. & Lyttleton, R. A. 1939, Proc. Cam. Phil. Soc., 35, 405
Lutz, D., Krabbe, A., & Genzel, R. 1993, ApJ, 419, 244
Melia, F. 1992, ApJ, 387, L25
Melia, F. 1994, ApJ, 426, 577
Melia, F. 1996, these proceedings
Melia, F., Coker, R. F., & Yusef-Zadeh, F. 1996, ApJ, 460, L33
Ruffert, M. & Melia, F. 1994, A&A, 288, L29
Stone, J. M., & Norman, M. L. 1992, ApJS, 80, 753
Yusef-Zadeh, F., Morris, M. & Ekers, R. D. 1989, in The Center of the Galaxy, M. Morris, Dordrecht: Kluwer, 443
Yusef-Zadeh, F., Morris, M., & Ekers, R. D. 1990, Nature, 348, 45

The First Detection of a Source Coincident with Sgr A* at 8.7 μm

Susan R. Stolovy, T. L. Hayward, and Terry Herter

Department of Astronomy, Cornell University, Ithaca, NY 14853

Abstract. We present the first mid-infrared detection of a source coincident with Sgr A* in an image of the inner parsec with 0″.7 resolution and rms noise of 1.6 mJy/beam. Images were obtained at 8.7 μm with SpectroCam-10 at the 5-m telescope at Palomar in June and July 1994. The source is apparent in the raw data, but is in poor contrast to the local diffuse dust emission. Several deconvolution techniques were applied to the data, and a source coincident with Sgr A* within the ±0″.3 uncertainty in locating Sgr A* relative to IRS 7 was apparent in each case. The source appears as a peak superposed on a narrow NW-SE ridge and is estimated to have a flux density of ∼ 25 mJy. The extinction-corrected flux density of ∼ 100 mJy is greater than what is predicted from dustless accretion disk models. It is likely that emission from warm dust contributes significantly to the 8.7 μm flux.

1. Introduction

Attempts to detect a source coincident with black hole candidate Sgr A* at mid-infrared (MIR) wavelengths have previously yielded only upper limits: Gezari 1992 and Gezari et al. 1994a have set 3 σ limits of 100 mJy at 12.4 μm and 300 mJy at 20.0 μm. In contrast to the crowded field of stellar photospheric emission seen in the near-infrared (NIR) continuum, the MIR is dominated by diffuse, warm dust emission around these stars. Unfortunately, Sgr A* is situated on a steep gradient of such diffuse emission. Therefore, optimal seeing, registration, and flat fielding are more crucial to a detection than simply signal-to-noise.

2. Observations

We obtained deep continuum images of the central parsec of the Galaxy at 8.7 μm with SpectroCam-10 on the 200-inch Hale telescope at Palomar in June and July 1994. The camera mode of SpectroCam-10 illuminates a 64×64 pixel subsection of a 128×128 Rockwell International Si:As BIB array (Hayward et al. 1993). Each pixel corresponds to 0″.25 on the sky. Sgr A* is expected to be a relatively blue object in comparison to the dust emission, so we chose the shortest wavelength filter available with good atmospheric transmission. The data were taken at airmasses ranging from 2.2 to 2.5. The 8.7 μm, 1 μm bandpass filter also avoids the deepest part of the interstellar silicate absorption feature at 9.7

µm. Two independent methods of observation were employed in June and July 1994.

On June 17 and 19 short exposures (1 second on-source) were taken in order to freeze seeing motions. The telescope was "dithered" by several arcseconds every few minutes in order to optimize the flat-fielding of the array. The flux calibrators were α Lyr and α Boo. Observations of α Boo at an airmass similar to that of the Galactic Center observations provided the point spread function (PSF). The total on-source integration time in June was 11 minutes. On July 23 and 25 a tip-tilt secondary system was installed on the Hale telescope. This enabled us to take much longer (10 second on-source) exposures. The standards α Boo and α Lyr were again used as flux calibrators and VX Sgr was observed at the same airmass as the Galactic Center to determine the PSF. The total on-source integration time in July was 27 minutes.

3. Data Reduction

The short exposures were mosaicked together by aligning the bright sources in each frame with a reference frame after expanding each pixel by linear interpolation into 9 subpixels. The mosaic procedure weighted each pixel according to its total integration time, replaced bad pixels, and achieved a relative image registration accuracy of $\pm 0\rlap{.}''1$. We have retained the small pixel size for the remainder of the discussion here. Point-spread functions were constructed in a similar way from images of the calibrator stars. A total of 642 and 159 seeing-selected images were used for the June and July mosaics, respectively. Figure 1 displays the mosaicked raw data, combining the June and July data sets for the best signal-to-noise. The noise level in this mosaic is ~ 1.6 mJy beam^{-1} (1 beam = 0.385 arcsec^2), as measured in the "gap" south of IRS 7 where the image overlap is maximized and there are relatively few infrared sources.

IRS 7, which is a source common to both the radio and infrared, has traditionally served as the coordinate reference to determine the infrared position of Sgr A*. Although a centroid on the bright stellar peak of IRS 7 is easily determined to $0\rlap{.}''1$ at 8.7 µm, this is not the case in the radio. High-resolution $0\rlap{.}''4$ resolution radio images at 2 cm from Yusef-Zadeh & Melia 1992 (YM92) reveal a bow shock structure and a "tail" around IRS 7. The bowshock introduces a $\sim 0\rlap{.}''3$ shift of the lower-resolution centroid of IRS 7 reported in Yusef-Zadeh & Morris 1991 (YM91) to the north. The southern part of this bowshock agrees to within $0\rlap{.}''1$ with the astrometric 1.65 µm (Becklin et al. 1987) and far-red (Rosa et al. 1992) stellar positions for IRS 7. YM92 argue that the relatively weak stellar wind of IRS 7 should place the star close to this southern shock front. There is no morphological evidence for a bowshock in the 8.7 µm data. We therefore adopt the B1950 IRS 7 position of RA=$17^h42^m29^s.32$, Dec=$-28°59'13''.0$ ($\pm 0\rlap{.}''2$) from Becklin et al. 1987. In the figures, we place the origin of the coordinate system and position of Sgr A* $0\rlap{.}''1$ W and $5\rlap{.}''4$ S of IRS 7, derived from the radio coordinate system of YM91, with a $\pm 0\rlap{.}''3$ uncertainty based on errors in centroids and comparing reference frames. We also include in each figure a $2'' \times 2''$ box centered at (0,0) for reference. Fortunately, the position of Sgr A* will be significantly improved by the recent detection of Galactic Center maser sources that are coincident with several of the stellar NIR sources (including

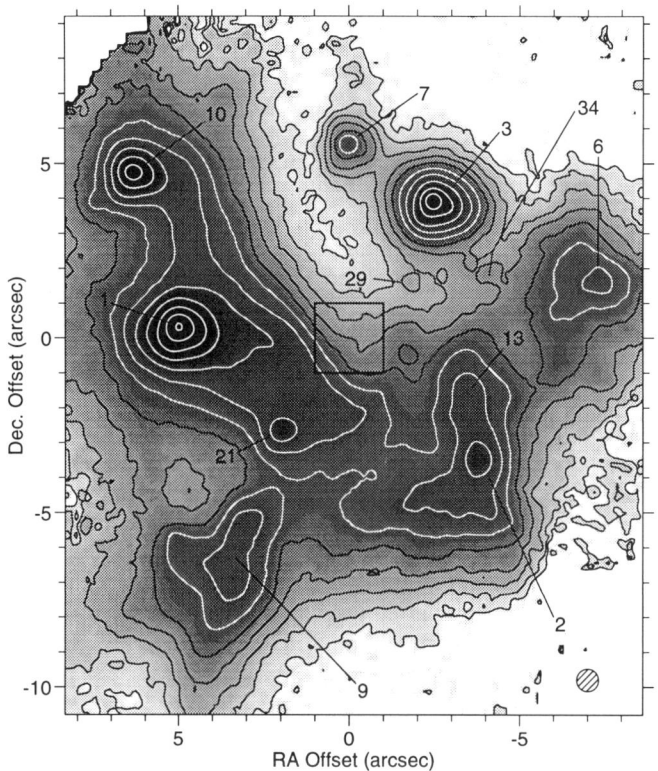

Figure 1. Combined June and July Raw Mosaic. The box centered on Sgr A* at (0,0) is $2'' \times 2''$ in size. The beam size is shown at lower right, and IRS sources are labelled. Contours are spaced by a factor of 1.58, from 0.15 to 0.90 Jy arcsec^{-2} (black) and from 1.4 to 23 Jy arcsec^{-2} (white).

IRS 7) reported at this conference (and M. Reid and K. Menten 1996, private communication).

A notable feature in the mosaic is the first MIR continuum observation of the "tail" extending northward of IRS 7. This tail was first observed in the radio continuum (YM91,YM92) and is thought to be a cometary feature created by the interaction of strong stellar winds from the IRS 16 cluster with the expanding envelope of mass-losing supergiant IRS 7. These winds may also be responsible for providing mass inflow to the purported black hole at Sgr A*. Another feature of interest is the fact that IRS 3 is extended in our 8.7 μm image. IRS 3 is a single source with a featureless spectrum in the NIR, and has no radio counterpart. This implies that it is dust-enshrouded, possibly young star.

4. Analysis

A faint source at Sgr A* is weakly evident above the steep gradient of background emission as can be seen as a "kink" in the contour near the center of the box in Figure 1. In order to estimate the flux of this feature, the PSF was scaled to various flux levels and then subtracted from the raw image at the position of Sgr A*. This method served only to provide an upper limit of 100mJy (a value above this produced a "hole") that is well calibrated in flux. The diffuse emission near Sgr A* varies by a factor of five within only $\pm 0.5''$ of Sgr A* and is on average ~ 500 mJy arcsec^{-2}. Attempts to reveal the Sgr A* source by subtracting a model background interpolated from nearby regions were unsuccessful due to fluctuations in the diffuse emission on subarcsecond scales.

4.1. Deconvolution

We used deconvolution techniques to improve the spatial resolution to the diffraction limit and to solidify the detection of the source at Sgr A*. The FWHMs for the June and July PSFs are $0''.63$ and $0''.71$, respectively, whereas the diffraction limit is $0''.44$ (1.22 λ/D). The tendency for deconvolution to redistribute flux from the wings of bright sources to the central peak works well for isolated point-like sources, but can produce artifacts at low levels in the wings around bright but extended sources such as IRS 1 and IRS 3. Fortunately, Sgr A* is located sufficiently far away from the bright sources to avoid such problems.

We deconvolved the June and July data sets independently in order to provide a validity check on the flux and location of Sgr A*. We applied four different deconvolution techniques: Richardson-Lucy (R-L), Maximum Entropy (MEM), Sigma Clean, and a Wiener filter. These methods differ in emphasis and implementation; for instance, MEM suppresses and Sigma Clean emphasizes point sources, and the Wiener filter is non-iterative. Despite these differences, all four methods revealed a peak on a ridge that agreed to within one detector pixel of the expected position of Sgr A*, as shown in Figure 2 for the July data set. The dashed box indicates the $\pm 0''.3$ error in locating Sgr A*. The deconvolution method that we preferred was the R-L method as implemented in the IRAF STSDAS "lucy" routine. This routine minimizes edge effects, includes a noise threshold parameter, and allows for an input weighting mask for each pixel. The compact source is situated on a $\sim 0''.5 \times 1''.5$ ridge of emission that is extended along a position angle of $155 \pm 10°$ East of North. A void $1''$ south of Sgr A* is also evident in all data sets and deconvolutions. The June and July deconvolutions are similar in general flux and morphology, but have some differences. In the July image, which has a factor of 1.6 times higher signal-to-noise, the Sgr A* source appears more point-like and detached from the ridge than in the June image, which has slightly superior seeing. The peaks in the June and July images agree within one detector pixel and are located $0''.0$ W and $5''.6$ S of IRS 7 and $0''.2$ W and $5''.5$ S of IRS 7, respectively. These positions are slightly to the south of the expected position of Sgr A*, but are coincident within one detector pixel.

We have found an excellent correlation between 8.7 μm sources enhanced by deconvolution and known NIR sources, such as the multiple nature of IRS 10, IRS 34, and IRS 13. The MIR ridge extends from IRS 16SW toward Sgr A*,

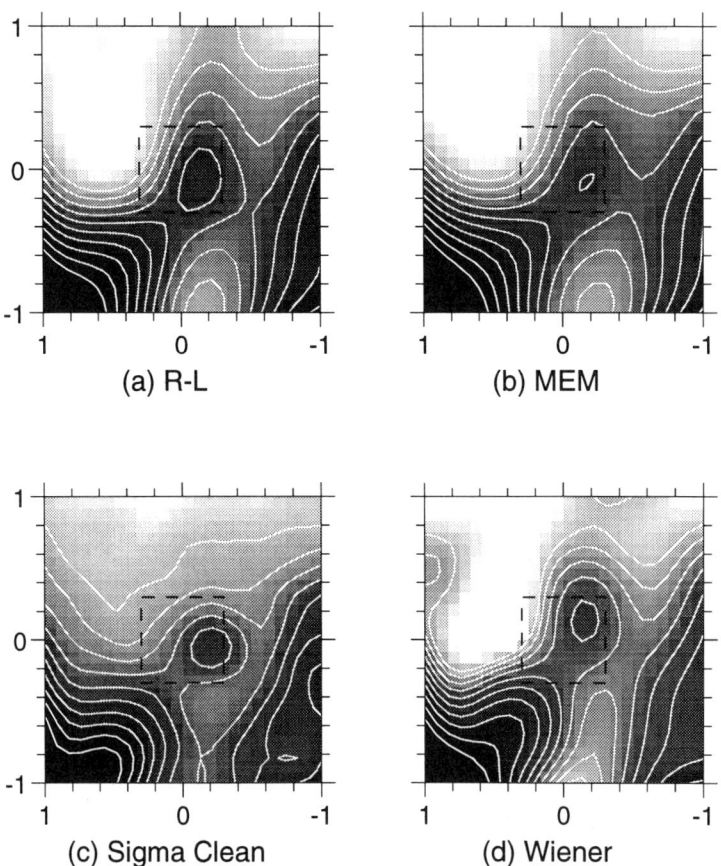

Figure 2. Comparison of inner $2'' \times 2''$ of the July data deconvolved with different techniques at approximately diffraction-limited ($\sim 0\rlap{.}''5$) resolution. Contours are from 0.21 to 1.2 Jy arcsec^{-2} for all images.

then northward towards IRS 16 NW, beyond which it drops off rapidly. This ridge has also been observed in the NIR; Herbst et al. (1993) refer to a "spur" at K, L', and M bands at the same orientation. The "spur" appears to dim toward the northwest, indicating the the ridge is relatively red near Sgr A*. The coincidence of the 8.7 μm ridge with a similar structure in raw NIR images is further evidence that the ridge is real. The blue IRS 16 cluster is conspicuously absent in the 8.7 μm image. This should be contrasted to IRS 13(E), which has similar colors and is situated in a similar region of high diffuse background to the IRS 16 cluster but which has a clear MIR counterpart. The lack of excess MIR emission around the IRS 16 cluster may indicate that the cluster is in the foreground relative to the majority of the dust. Images displaying the June and July deconvolved data and the correspondence of NIR sources with the July deconvolved data for the inner ~ 0.5 pc are presented in Stolovy et al. 1996.

4.2. Flux Determination

Direct photometry of the Sgr A* source from the raw mosaics was impractical due to the background emission; therefore, we measured its flux from the deconvolved images. We attempted to model the ridge as a generally smooth structure and Sgr A* as a point-like object superimposed on it. We found that the most trustworthy method to extract the flux of the Sgr A* source was to subtract a scaled PSF from the raw image at the position of Sgr A*, then deconvolve to determine what flux level just made the source on the ridge disappear. This method was relatively insensitive to the number of iterations, allowed the use of full $3'' \times '' 3$ images of the PSF stars, and avoided the problem of different convergence rates of isolated versus embedded sources. We found that the source at Sgr A* above the ridge has a flux density of 25 ± 5 mJy uncorrected for extinction. The error is dominated by the uncertainties in flux estimation rather than noise or flux calibration.

The interstellar extinction towards the Galactic Center at 8.7 μm is estimated to be 1.5 magnitudes, as determined by interpolating the extinction law from Rieke, Rieke & Paul (1989) between 8.5 and 9.0 μm and adopting a mean $A_V = 27$ measured toward the IRS 16 cluster (DePoy & Sharp 1991). The extinction correction uncertainty is $\sim \pm 35$ % at 8.7 μm based on variations and measurement errors in A_V. Our measured flux density of 25 mJy implies an extinction-corrected flux of 100 mJy (± 40 mJy) and a source specific luminosity of 8.7×10^{21} erg s^{-1} Hz^{-1}, assuming a Galactic Center distance of 8.5 kpc.

5. Discussion

Is the compact 8.7 μm source consistent with the emission expected from an accretion disk? At radio through mm wavelengths, the observed synchrotron emission from Sgr A* follows a $F_\nu \sim \nu^{0.3}$ power law, but is expected to decline sharply between mm and MIR wavelengths (e.g. Zylka et al. 1995). The 8.7 μm flux density can provide an upper limit to the high energy end of the synchrotron spectrum. Current models of accretion disk spectra around a $\sim 10^6 M_\odot$ black hole have been constrained by near infrared observations with typical observed K magnitudes of $\sim$ 12-14. The extinction-corrected 8.7 μm flux generally exceeds (but is marginally consistent with some of) the current model predictions. The recent resolution of the K-band source into multiple fainter ($K \sim 14$ -15) components (Eckart et al. 1995) calls for revisions of these models, and will make the MIR excess even greater. Many models predict the radiation from a hot plasma accretion disk to exhibit a $F_\nu \sim \nu^{2.0}$ spectrum in the NIR-MIR regime (e.g. Close et al. 1995). If one assumes $A_v \sim 30$ and an observed K magnitude of 15, then a $\nu^{2.0}$ power law predicts an 8.7 μm flux of *less than 1% of the observed value*. Equivalently, if all 25 mJy at 8.7 μm is emission from an accretion disk, $K_{\text{observed}} = 9.6$ is predicted, which is far brighter than any observation. If one assumes that the NIR flux is masked by local extinction (such as what might be expected from a dusty torus typically predicted in black hole models of extragalactic cores) and $K_{\text{observed}} = 15$, then $A_v \sim 120$ is required to be compatible with a $\nu^{2.0}$ law for 25 mJy at 8.7 μm. Reradiation at infrared through mm wavelengths from an object at Sgr A* with such a high local extinction (of $A_v \sim 90$) is not consistent with the observed spectrum of Sgr A*.

Thus, it is likely that an emission component from warm dust of typical effective temperature in the ~ 200 to 400K range is present which has not previously been considered in accretion disk models for Sgr A*.

However, the 8.7 μm source could simply be a clump of warm dust along the line-of-sight that is not physically associated with Sgr A*. The dust could be heated externally and even formed by nearby luminous stellar outflow sources. For instance, IRS 16NE alone has a luminosity of log $L/L_\odot = 6.13$ (Krabbe et al. 1995) and can provide enough UV/optical photons to heat dust to peak at MIR wavelengths (assuming the true distance to the dust is similar to the projected distance to Sgr A* of 0.12 pc). The illumination geometry provided The IRS 16 cluster and IRS 13 on opposing sides of the ridge can also explain the high polarization (29%) by scattering seen in one of the sources along the ridge at 2.2 μm (Eckart et al. 1995). Alternatively, the radiation from the series of faint sources lying along the ridge as seen in images from Eckart et al. (1995) combined with the NIR sources near Sgr A* could provide enough flux to heat dust in the ridge assuming the dust is close to the stars. Our spatial resolution is not sufficient to resolve correlations with these sources. A more complete presentation and analysis of these data can be found in Stolovy et al. 1996.

References

Becklin, E. E., Dinerstein, H., Gatley, I., Werner, M. W., Jones, B. 1987, in AIP Conf. Proc. 155, The Galactic Center, ed. D. C. Backer, AIP, NY, p. 162.

Close, L. M., McCarthy, D. W. J., & Melia, F. 1995, ApJ, 439, 682

DePoy, D. J., & Sharp, N.A. 1991, AJ, 101, 1324

Eckart, A., Genzel, R., Hofmann, R., Sams, B., & Tacconi-Garman, L. 1995, ApJ, 445, L23

Gezari, D. 1992, The Center, Bulge and Disk of the Galaxy, ed. L. Blitz, Kluwer Academic Publishers (Netherlands), 23

Gezari, D., Ozernoy, L., Varosi, F., McCreight, C., and Joyce, R. 1994, The Nuclei of Normal Galaxies, ed. R. Genzel and A. Harris, Kluwer Academic Publishers (Netherlands), p. 427 (1994a) and p. 343 (1994b)

Hayward, T. L., Miles, J. W., Houck, J. R., Gull, G. E., & Schoenwald, J. 1993, Proc. SPIE, 1946, 334

Herbst, T., Beckwith, S., & Shure, M. 1993, ApJ, 411, L21

Krabbe, A., et al. 1995, ApJ, 447, L95

Rieke, M. J., Rieke, G. H., & Paul, A. E. 1989, ApJ, 336, 752

Rosa, M. R., Zinnecker, H., Moneti, A., & Melnick, J. 1992, A&A, 257, 515

Stolovy, S. R., Hayward, T. L., & Herter, T. 1996, ApJ(Letters), in press.

Yusef-Zadeh, F. & Melia, F. 1992, ApJ, 385, L41 (YM92)

Yusef-Zadeh, F. & Morris, M. 1991, ApJ, 371, L59 (YM91)

Zylka, R., Mezger, P., Ward-Thompson, D., Duschl, W., & Lesch, H. 1995, A&A, 297, 83

ROSAT / X-ray Studies of the Galactic Center

P. Predehl
Max-Planck-Institut für extraterrestrische Physik, Giessenbachstrasse, D-85748 Garching, Germany

H. Zinnecker
Astrophysikalisches Institut Potsdam, An der Sternwarte 16, D-14482 Potsdam, Germany

Abstract. The Galactic Center region has been observed with ROSAT several times. In a 50ksec exposure, we could resolve the X-ray source 1E 1742.5-2859, the source which had been commonly associated with Sgr A*, into three individual objects. One of these (RXJ 1745.6 – 2900) is coincident with Sgr A* within 10". The flux derived from the ROSAT observation is much less than one would expect on the basis of previous measurements at higher energies. In order to explain this discrepancy, an X-ray absorption much in excess of $N_H = 6 \times 10^{22} cm^{-2}$, the expected value for the Galactic Center, has been assumed. This assumption is now supported by the fact that ROSAT could detect radiation only above 2 keV from RXJ 1745.6 – 2900.

1. Introduction

At X-ray energies above 1keV the interstellar medium is sufficiently transparent to allow direct observations of the Galactic Center region. In particular, at soft X-ray energies between 0.5 and 3 keV the absorption cross section of the interstellar medium changes by more than two orders of magnitude and therefore the soft X-ray range is ideally suited for probing the interstellar absorption.

The first X-ray cartography of the Galactic Center region has been made with the Imaging Proportional Counter (IPC) onboard the *Einstein Observatory* in the 0.5 - 4 keV band (Watson et al., 1981). 12 sources could be detected within a field of about one square degree. One of these sources, 1E 1742.5 - 2859, was found to be coincident with Sgr A West within the accuracy of the IPC ($\approx$ 1 arcmin). Assuming an interstellar absorption $N_H = 6 \times 10^{22} cm^{-2}$ and a distance d = 8.5 kpc, both typical for the Galactic Center, a luminosity $L_x = 1.1 \times 10^{35} ergs\ s^{-1}$ was derived making the core of our Galaxy a rather weak representative of the class of galactic nuclei.

The Galactic Center has been observed later with a variety of instruments working a higher energies. Therefore, these observations were not as much affected by the interstellar absorption. 1E1742.5 - 2859 was detected in an energy band between 3 and 30 keV with a coded mask camera onboard the Spacelab 2 mission (SL2-XRT, Skinner et al., 1987) and associated with Sgr A* itself.

The derived luminosity, however, was eight times greater than expected based on the earlier measurement. Although such a dicrepancy could be explained by a time variability of the source, it has been concluded that the spectrum assumed by Watson et al. (1981) was in error: similar discrepancies were found for other sources, too. Skinner et al. suggested that the interstellar absorption is $N_H = 3 \times 10^{23} cm^{-2}$ rather than $N_H = 6 \times 10^{22} cm^{-2}$. Such a high value, however, cannot be explained easily by the interstellar absorption if the source is at the Galactic Center.

The ART-P telescope onboard the russian *GRANAT* satellite observed the region several times since 1990 (Pavlinskii, Grebenev, and Sunyaev, 1992). The measured luminosities were even higher than the value derived by Skinner (1987). The measurements suggest a variability within a factor two.

Recently, the japanese X-ray satellite ASCA has observed the Galactic Center (Koyama et al., 1995). ASCA found an extended high temperature plasma associated with $K\alpha$ lines of highly ionized atoms. Fluorescent X-ray emissions from cold iron in molecular clouds were also found, due possibly to an irradiation by an X-ray source which was bright in the past but is dark at present. Two pointlike objects were detected in addition: a 'soft' source at the position of the Center and a 'hard' source 40" south-west of Sgr A* which probably is a transient source (A1742 – 289) found by the Ariel satellite in 1875 (Branduardi, 1976). It has been suggested by Koyama (1994) that this source is coincident with the object found by ART-P and SL2-XRT. However, this would require a positioning error of both ART-P and SL2-XRT of almost 1 arcmin which seems to be quite unlikely (Skinner, Sunyaev, priv. comm.)

2. ROSAT Observations

Apart from the all-sky survey, the Galactic Center was observed by ROSAT several times: an area of 14 square degrees around the Galactic Ceter has been observed with the Position Sensitive Proportional Counter (PSPC) in the energy range between 0.5 and 2.5 keV (Predehl et al., 1994). The observation was assembled from 41 individual pointings, set in a 9×5 raster within a 4.3 degrees long and 3.2 degrees wide strip across the galactic plane. 62 individual X-ray sources could be detected out of which only six were previously known as X-ray emitters. No X-ray source at the position of Sgr A* was detected although the accumulated observing time exceeded 25000 sec in the central field. This result was explained by a possible source variability or, in agreement with the earlier findings of Skinner et al. (1987), by assuming an interstellar absorption higher than $N_H = 2 \times 10^{23} cm^{-2}$. In addition to the 62 sources, a ridge of apparently diffuse emission with a size of 25-30 arcsec in galactic longitude and in 15-20 arcsec in galactic latitude. This is in agreement with the *Einstein* (Watson et al., 1981) and ASCA (Koyama et al., 1995) results.

Nothing particular could be detected in 27000 sec observation using the ROSAT High Resolution Imager but a 50 ksec exposure of the Galactic Center with the PSPC revealed the central source 1E1742.5 – 2859 being assembled from at least three individual X-ray sources (Fig. 1) out of which the northern one emits also below 1.2 keV and, therefore, must be considered as a foreground object (Predehl and Trümper, 1994).

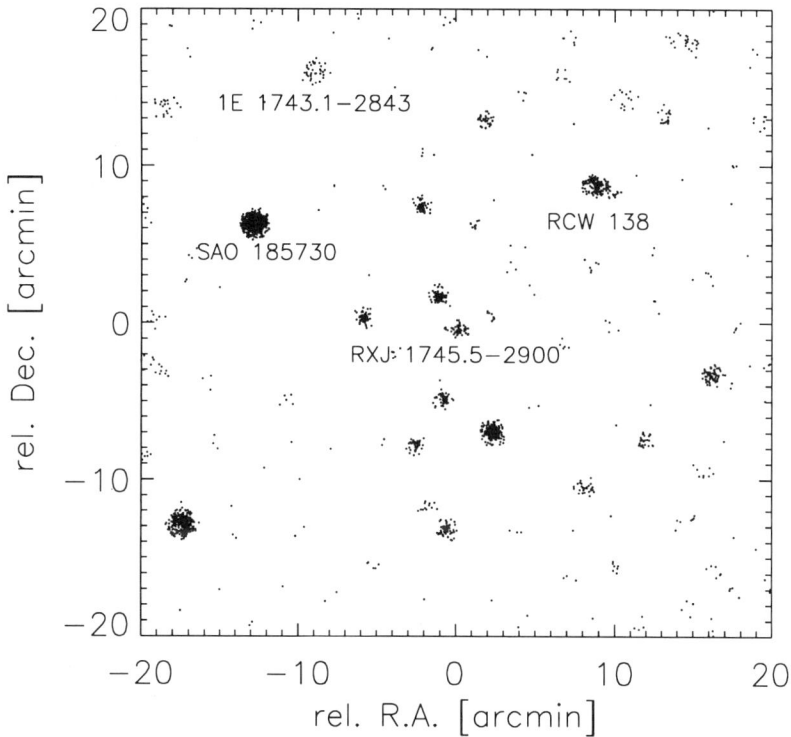

Figure 1. X-ray sources in the inner half square-degree around the Galactic Center. The background including the diffuse emission have been subtracted on a photon basis (A detailed description of this method will be given by Scheingraber and Predehl). RXJ 1745.5 − 2900 and A 1743.1 − 2843 are the two highest absorbed sources in the field.

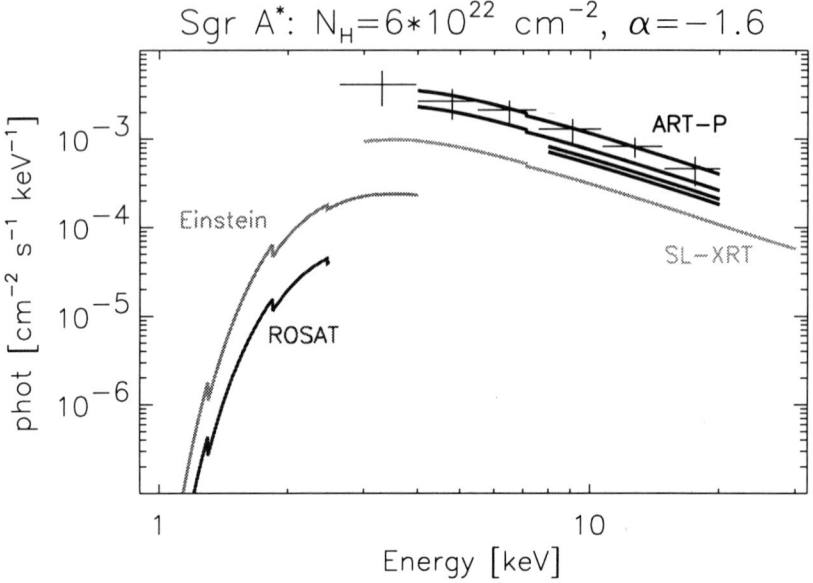

Figure 2. Comparison of the flux from RXJ 1745.5 – 2900 measured with various instruments. ART-P has observed the Galactic Center region several times. The measurements suggest a slight source variability. A power law spectrum with a photonindex $\alpha = $ -1.6 (as determined by ART-P) and an interstellar absorption according to an equivalent hydrogen column density $N_H = 6 \times 10^{22} cm^{-2}$, the canonical value for the Galactic Center, has been assumed. Both ROSAT and *Einstein* fluxes are far below the fluxes measured with ART-P and SL2-XRT. This is explained by an underestimation of the interstellar absorption. ROSAT could resolve the source found by *Einstein* into three individual objects, therefore its flux is somewhat lower.

The southern source (RXJ 1745.6 – 2900) coincides with the radio position of Sgr A* within 10 arcsec, representing a factor of ≈ 40 improvement in the error box size compared with earlier measurements. This makes the identification with Sgr A* quite plausible. When comparing the flux from this source with the flux at higher energies as measured by SL2-XRT (Skinner et al., 1987) and ART-P (Pavlinskij et al., 1994), we also found the ROSAT source to be underluminous: The derived X-ray luminosity is $L_x = 1 - -2 \times 10^{34} ergs^{-1}$ within the 0.8 – 2.5 keV energy range assuming $N_H = 2 \times 10^{23} cm^{-2}$. This is by far too low in order to coincide with the the flux measured above 3 keV (Fig 2.) A detailed spatial and spectral analysis (Scheingraber and Predehl, in preparation) reveals that two sources (1E 1743.1 – 2843 and RXJ 1745.6 – 2900) can be found in the ROSAT data only when selecting photons events above pulse height channel 200 which roughly corresponds 2 keV photon energy. It is already known that 1E 143.1 – 2843 has an absorption of more $N_H = 10^{23} cm^{-2}$ (Kawai et al. 1988). Therefore we conclude that the same holds also for RXJ 1745.6 – 2900.

3. Discussion

The important result of the ROSAT observation is the detection of a source coincident with Sgr A* within 10". The variable and therefore compact source found at higher energies (3-30 keV) coincides in position with Sgr A* within 1 arcmin (Skinner et al., 1987, Pavlinskii, Grebenev, and Sunyaev, 1992). This suggests that at hard X-rays most of the flux comes from RXJ 1745.6 – 2900 because there is no other counterpart of the hard X-ray source in the ROSAT image.

The comparison of the high energy data with the low energy data requires an absorption of $N_H = 1.5 - -2 \times 10^{23} cm^{-2}$ (Predehl and Trümper, 1994). A result based on the comparison of data from different instruments might be questioned especially if there are indications for a source variability. The new result of the spectral analysis uses ROSAT data alone. Therefore, we regard this as a confirmation that the absorption towards RXJ 1745.6 – 2900 is really much higher than the value typically assumed for the Galactic Center region. The interstellar extinction towards the Sr A* region is $A_V = 27 - 31$ mag (Eckart, priv comm.) which can be translated into an X-ray absorption $N_H = 4.8 - 5.6 \times 10^{22} cm^{-2}$ (Predehl and Schmitt, 1995).

RXJ 1745.6 – 2900 has the most absorbed spectrum among all sources in the region including also the ridge of diffuse emission. Therefore, it is really more absorbed than typically for the Galactic Center or all other sources including the ridge of diffuse emission lie in the foreground and have, therefore, nothing to do with the Galactic Center region.

The extra absorption of at least $N_H = 1 \times 10^{23} cm^{-2}$ could be possbly due to the fact that RXJ 1745.6 – 2900 is an unrelated background source. Just behind the Galactic Center lies the giant molecular cloud Sgr A East Core providing enough column density. However, our analysis puts the source within 10" distance from Sgr A*. This makes the coincidence with Sgr A* or at least its immediate environment rather likely. ASCA has detected a 'soft' source at this position but its angular resolution does not allow to resolve the various objects found by ROSAT. Therefore, it is more likely that the 'soft' source is a mixture of several sources together with a contribution from extended emission as also found by ROSAT. There is no ROSAT counterpart for the 'hard' source (A1742 − 289) which suggests that this source has indeed a transient behaviour.

But there remains one problem: a conclusion that RXJ 1745.6 − 2900 is identical with Sgr * requires an enormous amount of intrinsic absorption which obviously can be excluded on the basis of radio measurements (Duschl et al., 1996).

References

Branduardi G., Ives T. C., Sanford P. W., Brinkman A. C., and Maraschi L. 1976, Mon. Not. R. astr. Soc. 175, 47

Dusch W. J., Beckert T., Mezger P. G., and Zylka R., 1996 in " Röntgenstrahlung from the Universe", Zimmermann H.U., Trümper J., Yorke H. (eds.), MPE report 263, 651

Kawai N., Fenimore E. E., Middleditch J., Cruddace R. G., Fritz G. G., Snyder W. A., Ulmer M. P., 1988, ApJ 330, 130

Koyama K., 1994, New Horizon of X-ray Astronomy, Universal Academy Press, Tokyo

Pavlinskii M.N., Grebenev S.A., Sunyaev R.A., 1994, ApJ 425, 110

Predehl P., Trümper J., 1994, A&A 290, L29

Predehl P., Schmitt J.H.M.M., 1995, A&A 293, 889

Skinner G.K., Willmore A.P., Eyles C.J., Bertram D., Church M.J., Harper P.K.S., Herring J.R.H., Peden J.C.M., Pollock A.M.T., Ponman T.J., Watt M.P., 1987, Nature 330, 544

Watson M.G., Willingale R., Grindlay J.E., Hertz P., 1981, ApJ 250, 142

Part 3. The Central Engine and High Energy Phenomena

Section B. High Energy Phenomena

ASCA Observations of the Galactic Center
$--$ X-ray view of the interstellar matter $--$

Yoshitomo. Maeda and Katsuji. Koyama

Dept. of Physics, Kyoto University, Sakyo-ku, Kyoto 606-01, Japan

Abstract. With the wide X-ray band imaging and spectroscopic capabilities of ASCA, we have made, for the first time, a fine X-ray image of the Galactic center in the 1.5-10.0 keV band. We found an extended emission with K-shell transition lines from highly ionized silicon, sulphur, argon, calcium and iron, in which the region near Sgr A is distinctly bright. We also found an extended emission of 6.4-keV lines of low ionized irons. The 6.4-keV line fluxes are well correlated with some cool clouds. We inferred that the 6.4-keV line is due to florescence from the cool clouds irradiated by bright X-ray source(s).

1. Introduction

The third Japanese X-ray satellite Ginga discovered intense emission lines from highly ionized iron in a thin high temperature plasma of spatial extension of about 100 pc. Since the thermal energy of this plasma was estimated to be extremely large (10^{53}–10^{54} $ergs$) and its dynamical time scale was found to be very short (10^4–10^5 yr), Koyama *et al.* 1989 suggested that the thin hot plasma is originated from a past violent and variable activity of the Galactic center.

The next satellite ASCA has a capability to observe the Galactic center with higher spatial and spectral resolution than Ginga in the energy band up to 10 keV (Tanaka, Inoue and Holt 1994). Early reports on the diffuse X-ray emissions near the Galactic center are found in Koyama 1994 and Koyama et al. 1996. In this paper, we report on the ASCA results of the hot gas in the inner 150 pc, the surface brightness, temperature and metalicity, and show two indications for the presence of cool interstellar matter near the Center; one is the soft X-ray absorption by cool clouds, and the other is the fluorescent 6.4-keV line from cool iron atoms. We adopted a distance of the Galactic center to be 8.5 kpc.

2. Observations & Results

2.1. X-ray map

The ASCA satellite observed the Galactic center region from 30 September to 7 October in 1993. Figure 1 shows the mosaic image of the SIS instrument, covering about 1° × 1° in the 1.5-10.0 keV band. We see several bright spots as well as extended emissions (figure 1). Three brightest spots are found

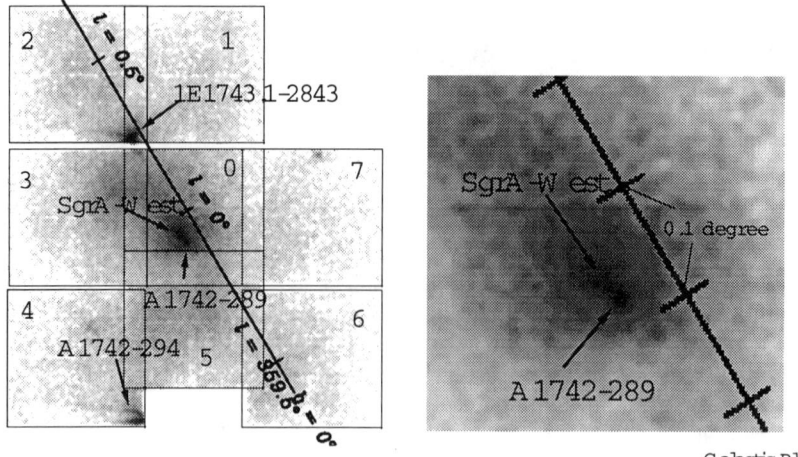

Figure 1. The SIS image with 1.5-10.0 keV band (a mosaic of X-ray images of eight adjacent fields) is shown in the left panel. An enlargement around the SgrA region is shown in the right panel. Numbers in the left panel show eight consecutively observed regions. A Chris-cross structure is the gap of each CCD chip (one CCD camera consists of 4 adjacent CCD chips). No correlation of the non-uniform efficiency over the detector field was made on the images.

to be point-like, which are already catalogued, and are named as A1742-289, A1742-294 and 1E1743.1-2843. Details of the X-ray properties for the three sources observed with ASCA are given by Maeda and Koyama 1994 (A1742-294 and 1E1743.1-2843), Maeda et al. 1996 (A1742-289) and Nishiuchi et al. 1996 (1E1743.1-2843).

Besides these X-ray sources, we found a diffuse-like structure at north-east of A1742-289. The peak position of the structure coincides with SgrA-West within systematic errors($\sim$ 0.4 arcmin). Although it is unclear whether this structure is due to unresolved point sources, a diffuse emission or both, this structure is likely to be an X-ray counterpart of SgrA-West(SgrA*). Details of the X-ray properties for SgrA-West(SgrA*) are given by Koyama et al. 1996. We also found another diffuse emission extending largely over the all eight observed fields.

2.2. Diffuse X-ray spectrum

Figure 2 shows the SIS spectrum of the diffuse emission over the entire $\sim$1 square degree field. Its spectrum contains many emission lines, which are identified to be K-lines from helium-like(He-like) and hydrogen-like(H-like) ions of silicon, sulphur, argon, calcium, and iron. These highly ionized ions are likely from the hot gas discovered with Ginga. In addition, strong 6.4-keV iron lines due to fluorescence from neutral Fe or Fe ions below XVII are found.

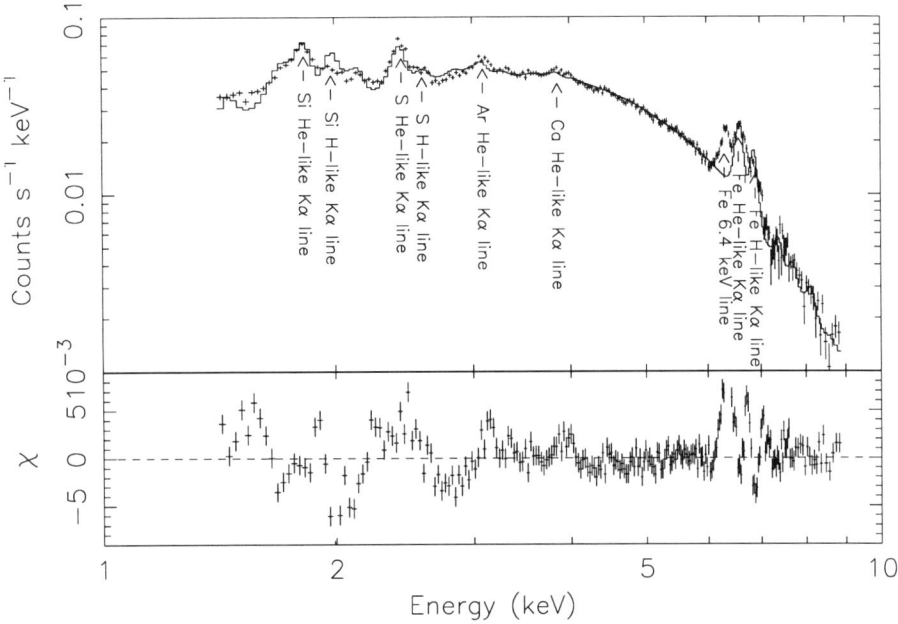

Figure 2. The SIS spectrum of diffuse emission (sum of spectra of all the eight fields after excluding spectra of the three luminous X-ray binaries). Luminous emission lines are identified with K-lines from He-like and H-like ions of silicon, sulphur, argon, calcium, and iron. A significant 6.4-keV iron line, a fluorescent line from neutral atoms or ions below XVII, is also present.

2.3. Line-Map

Figure 3 & 4 show narrow energy band maps with the center energies of three luminous emission-lines: He-like sulphur, He-like iron and iron 6.4 keV line. The former two lines have a bright spot around SgrA-West (figure 3). The iron He-like line is symmetrically extending along the Galactic plane on either side of SgrA-West, while the spatial distribution of the He-like sulphur is highly asymmetric. The latter 6.4-keV line-map appears entirely different from the former two maps (figure 4). There are two bright regions; the brightest region lies near the prominent molecular cloud, Sgr B2 and the other bright region is located between Sgr A and the Radio Arc, which is also associated with another dense molecular cloud. Remarkably, there is no bright region at SgrA-West.

3. Analysis & Discussion

3.1. Hot gas

To investigate the distributions of the temperature and metalicity of the hot plasma, we separately extracted the line fluxes and equivalent widths from

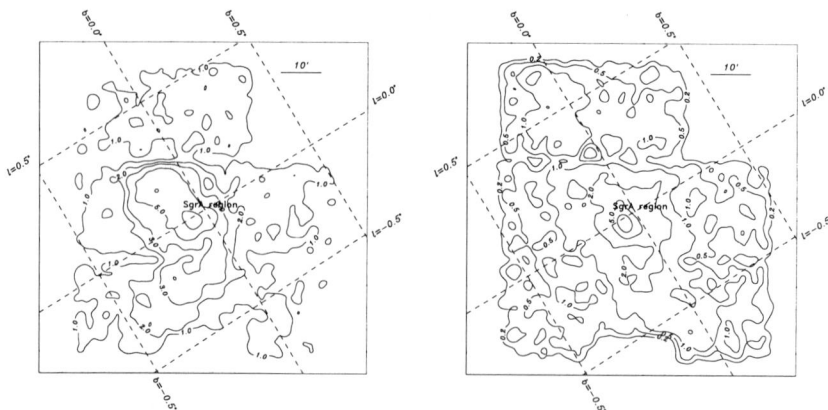

Figure 3. Brightness-distributions of two luminous emission-lines: He-like sulphur(left panel), He-like iron(right panel). In the He-like sulphur and He-like iron line-map, there is a bright spot around SgrA-West. Only in the iron He-like line-map, there is a symmetricity along the Galactic plane. No correlation of the non-uniform efficiency over the detector field was made on the images.

the eight observed fields. Figure 5 shows the fluxes of the four luminous lines: K-lines from He-like and H-like ions of both sulphur and iron. The fluxes of the four lines vary from field to filed, but the $\frac{\text{H-like}}{\text{He-like}}$ flux-ratios of sulphur and iron K-lines are roughly constant. The energy of the K-lines from He-like ions is not largely different from that of H-like ions of the same element. Thus the $\frac{\text{H-like}}{\text{He-like}}$ flux-ratios of the same atoms are not largely affected by absorption. Therefore, the constant flux-ratios of sulphur and iron imply that the temperature of the hot gas is nearly constant in every observed field. Assuming an ionization-equilibrium in the hot gas, the temperature inferred from the $\frac{\text{H-like}}{\text{He-like}}$ flux-ratio of sulphur is found to be about 1 keV, which is largely different from the 9-keV temperature inferred from the iron line-ratio. Therefore we inferred that the hot gas is in multi-temperature with uniform mixing ratio across $1° \times 1°$.

Figure 5 shows the equivalent widths of the four lines at each observed field. The equivalent widths are also constant from place to place. Because the temperature is found to be constant, abundances of sulphur and iron should be also constant. Thus, we inferred that the metallicity in the hot gas is uniform across $1° \times 1°$.

Since the K-lines from the highly ionized ions come from the hot gas, their surface brightness give an indication of the spatial distribution of the hot gas. X-ray photons with the energy larger than the iron K-line are not largely absorbed by the interstellar or circumnuclear gas toward the Galactic center. Therefore, the observed iron K-line flux should be proportional to the emission-measure of the hot gas. As we already noted, the flux of the He-like iron line is concentrated toward SgrA-west and symmetrically extends along the Galactic plane on either side of the Center. Thus, we suggest that the hot gas is also concentrated toward our Galactic nucleus and symmetrically extends along the Galactic plane.

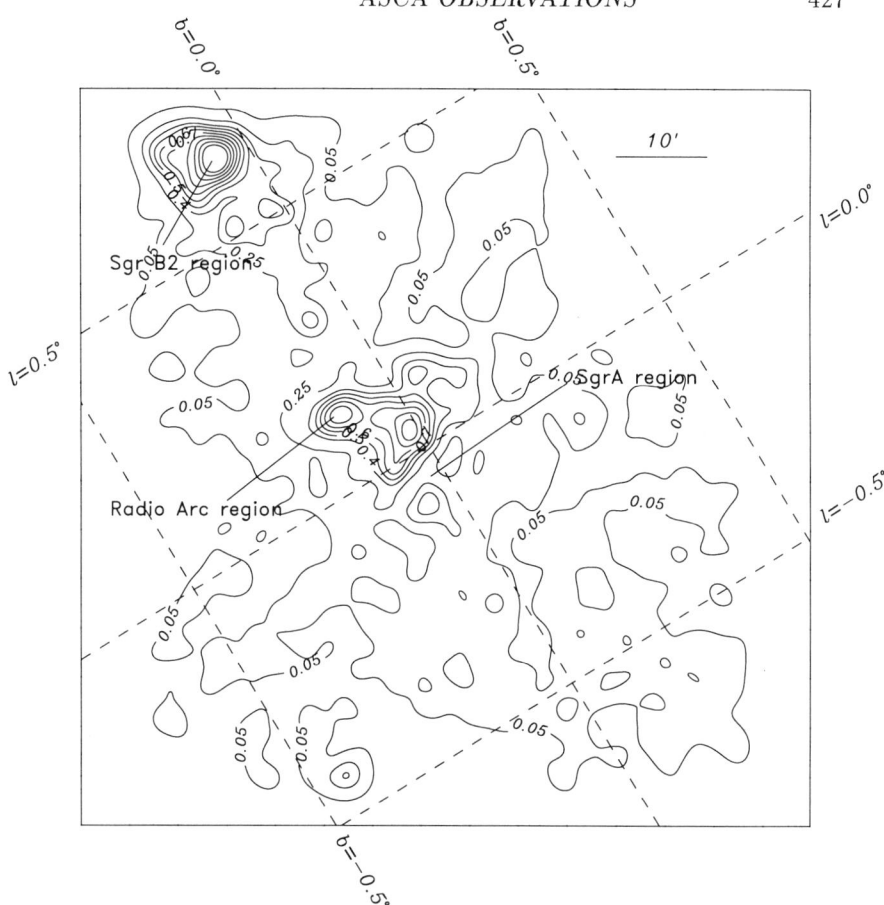

Figure 4. Brightness-distribution of iron 6.4 keV line(upper bottom panel). In the iron 6.4 keV line-map, there are two bright regions. One is northern bright spot (upper-left), which is located near the SgrB2 cloud. The other (middle) between SgrA and the Radio Arc appears to be associated with another dense molecular cloud. No correlation of the non-uniform efficiency over the detector field was made on the images.

3.2. Soft X-ray absorption

As is shown in figure 3, the surface brightness of the He-like sulphur is not symmetric along the Galactic plane, which may be due to local absorption near the Center because the energy of this line is critical value for the low energy-cut by the relevant interstellar or circumnuclear absorption toward the Galactic center of $\sim 10^{23}$ H cm^{-2}. We tried to compare ^{12}CO maps with two different velocities taken by Hasegawa et al. (private communication) at Nobeyama 45m radio telescope. Bright regions of the ^{12}CO map with velocities around -150 km s^{-1} is located on the negative side of the galactic latitude. These regions are anti-

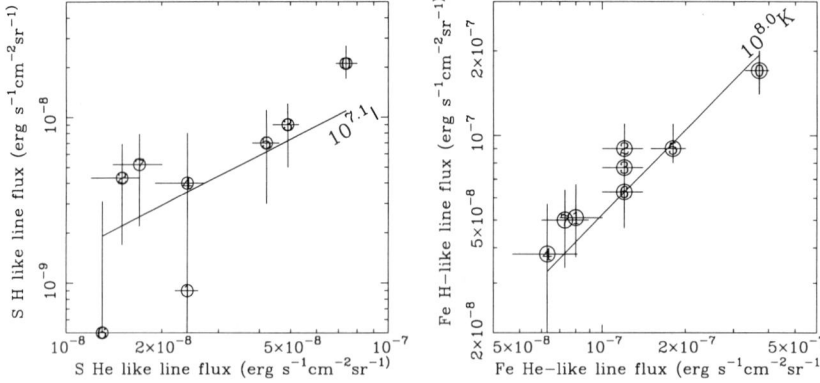

Figure 5. Correlations between fluxes of two K-lines from He-like(left panel) and H-like(right panel) ions of sulphur and iron at each observed region, respectively. Numbers of observed regions are corresponding to the numbers of figure 1. We draw solid lines, which mean line-ratios (kT = $10^{7.1}$ and $10^{8.0}$K, respectively) predicted from the plasma model in ionization equilibrium calculated by Mewe, Gronenschild and Van den Oord (1985).

correlated with the brightness-distribution of the He-like sulphur line. Suppose that a molecular cloud is located in front of the hot gas, the sulphur line flux will be absorbed by the cloud. We also found that there is no apparent correlation between the sulphur He-like line-map and the ^{12}CO map with about +100 km s^{-1}, which is bright at the positive side of the galactic latitude. Thus, this cloud may be located behind the hot gas. Since the hot gas is symmetrically extended around the Galactic center, we found an indication that some molecular clouds at negative latitude are in front and others at positive latitude are in behind the Galactic center.

3.3. 6.4-keV line

A new discovery with ASCA is a strong 6.4-keV line from the Galactic center. This line is due to neutral or low ionization states of iron. Therefore we need cool gas and high energy source to excite K-shell electrons of the cool irons.

As the brighter region of 6.4-keV line-map (figure 4) is well correlated with some dense molecular clouds, the source of the cool irons would be the molecular clouds. The observed flux of the 6.4-keV line near the Sgr B2 region (in the circle of $\sim 3.5'$ radius) is 1.7×10^{-4} photons cm^{-2} s^{-1}. The required incident flux of the Sgr B2 cloud for producing this line flux is given as a function of the column density and iron abundance of the Sgr B2 cloud. We assume that the total mass of the Sgr B2 cloud to be 2×10^6 $M_\odot$ (Lis and Carlstrom 1994) with solar abundance of iron, and the radius to be $\sim 3.5'$ (~ 7 pc). Then the column density of the Sgr B2 cloud is estimated to be 4×10^{23} H-atoms cm^{-2}.

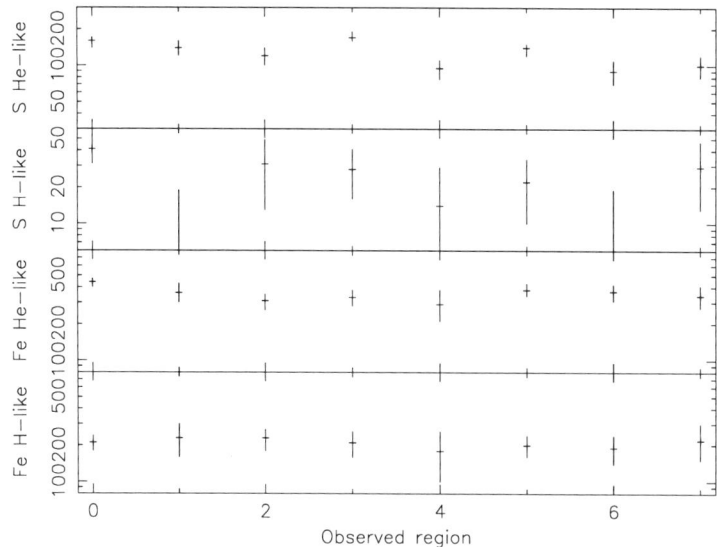

Figure 6. Line-equivalent widths of the K-lines from He-like and H-like ions of sulphur and iron at each observed region. Numbers are corresponding to the numbers of figure 1.

One possible high energy source is the existing hot plasma. However, the integrated X-ray flux is found to be far insufficient, by more than a factor of ten, to account for the observed 6.4-keV line flux. Another possibility is nearby transient sources. In fact, a transient source, 1E1743.1-2843, lies only about 0.4 degrees from Sgr B2. During the present observation, its luminosity was $\sim 2 \times 10^{36}$ ergs s^{-1} (2-10 keV) after correction for the measured absorption. Whereas, the time-averaged luminosity required to account for the observed 6.4-keV line is $\sim 10^{38}$ erg s^{-1} which is near the Eddington limit of an accreting neutron star. The other bright region in the 6.4-keV line near the Radio Arc also requires an equally luminous X-ray source(s) in its neighborhood. In view of the small duty ratios of bright transients, this possibility is rather unlikely. We found no luminous X-ray source to explain the observed 6.4-keV line flux, hence we infer that there is unknown(hidden) luminous X-ray source somewhere in the Galactic center.

4. Summary

ASCA detected the diffuse X-ray emission over $1° \times 1°$. In the diffuse X-ray spectrum, we found many K-shell transition lines. Using these emission lines, we investigated interstellar matters in the Galactic center.

- We have detected the K-lines from helium-like and hydrogen-like ions of silicon, sulphur, argon, calcium, and iron, which is due to the hot gas.

Using the brightness-distribution of the iron K-line from helium-like ions, we inferred that the hot gas is concentrated toward our Galactic nucleus and symmetrically extends along the Galactic plane. Using the flux-ratio and the equivalent widths of the sulphur and iron K-lines from helium-like and hydrogen-like ions, we found that the temperature and metalicity of the hot gas are nearly homogeneous over 1° × 1°.

- The surface brightness of the sulphur He-like lines show an asymmetric structure, which may be due to a spatially different absorption. We tried to compare the ^{12}CO maps with the two different velocities and found an indication that some molecular clouds at negative latitude are in front and other with positive latitude are in behind the Galactic center. This implies that the molecular clouds are expanding from the Center.

- A significant 6.4-keV iron line due to the fluorescence from neutral irons or ions below Fe XVII is present. There are two bright regions in the 6.4-keV line, which are well correlated with the dense molecular clouds. Thus, we found that the cool interstellar matter is irradiated by the luminous X-ray source. However, we found no luminous X-ray source to explain the flux of the 6.4-keV line observed at the Sgr B2 region.

Acknowledgments. The authors thank all the member of the ASCA team, who have made this study possible. We owed much to Mr. T. Sonobe and Prof. Y. Tanaka for the data reduction and discussion. YM is supported by the Hayakawa Satio Fundation to attend this work-shop. The authors also thank Prof's T. Hasegawa, M. Tsuboi and their Co-investigators for their providing the radio data before publication and useful discussions.

References

Koyama, K., Awaki, H., Kunieda, H., Takano, S., Tawara, Y., Yamauchi, S., Hatsukade., Nagase, F. 1989, Nature , 339, 603

Koyama, K., Maeda, Y., Sonobe, T., Takeshima, T. Tanaka, Y., Yamauchi, S. 1996, PASJ, accepted

Koyama, K. 1994, in the proceedings of "New Horizon of X-Ray Astronomy" eds, F. Makino and T. Ohashi (University Academy Press, Tokyo), p181

Lis, D. C., Carlstrom, J. E. 1994, ApJ, 424, 189.

Maeda, Y., Koyama, K. 1994, in the proceedings of "New Horizon of X-Ray Astronomy" eds, F. Makino and T. Ohashi (University Academy Press, Tokyo), p497

Meada, Y., Koyama, K., Sakano, T., Takeshima, T., Yamauchi, S. 1996, PASJ, accepted

Mewe, R., Gronenschild, E. H. B. M., Van den Oord, G. H. J. 1985. A&AS,62,197

Nishiuchi M. et al. 1996 in preparation

Tanaka, Y., Inoue, H., Holt, S. S. 1994, PASJ, 46, L37

Six Years of Hard X-Ray and Soft γ-Ray Observations of The Galactic Center with The Sigma Telescope

M. Vargas, A. Goldwurm, J. Paul, J. Ballet

CEA/DSM/DAPNIA/SAp, CEA-Saclay, 91191 Gif-sur-Yvette Cedex, France

L. Bouchet, J.P. Roques, E. Jourdain, V. Borrel

Centre d'Etude Spatiale des Rayonnements, 9 avenue du Colonel Roche, BP 4346, 31029 Toulouse Cedex, France

R. Sunyaev, M. Gilfanov, E. Churazov, S. Kuznetsov, S. Trudolubov, M. Revnivtsev, N. Khavenson, A. Dyachkov

Space Research Institute, Profsouznaya 84/32, Moscow 117296, Russia

Abstract. The coded-mask telescope Sigma onboard the *Granat* Space Observatory has been operating successfully for six years in the energy band from 35 keV to 1.3 MeV. Since its launch on December 1st, Sigma has performed a long deep survey of the Galactic Bulge. During this survey, some fifteen hard X-ray/soft γ-ray sources have been detected and located with an accuracy of about 3'. The source closest to the Galactic Center is situated $\approx 9'$ away. No evidence of hard X-ray/soft γ-ray emission from the radio source SgrA* has been recorded until now. We present the summed image of the Galactic Bulge derived from the Sigma database (complete to September, 1995) and we report preliminary upper limits of the high energy emission from SgrA*. Preliminary results (discovery of GRS 1739−278 and detection of GRO J1744-28) of the most recent Sigma observation campaign (March 1996) are also reported.

1. Introduction

The French coded-mask telescope Sigma provides high resolution images in the energy band from 30 keV to 1.3 MeV, with a 20 hour exposure sensitivity (2σ) of ≈ 26 mCrab in the 35–150 keV energy range, in a total field of view of $18° \times 17°$ (Paul et al. 1991). The telescope angular resolution is $\approx 15'$, but point sources can be located with a accuracy of 30" to 5', depending on the signal-to-noise ratio.

Mounted aboard the Russian *Granat* Space Observatory, Sigma has observed until now the $18° \times 17°$ region around the Galactic Center (GC) on spring and fall of each year from 1990 to 1994 and also in September, 1995. This unprecedent database contains 161 observations, for a total of 2578 hours of effective exposure time.

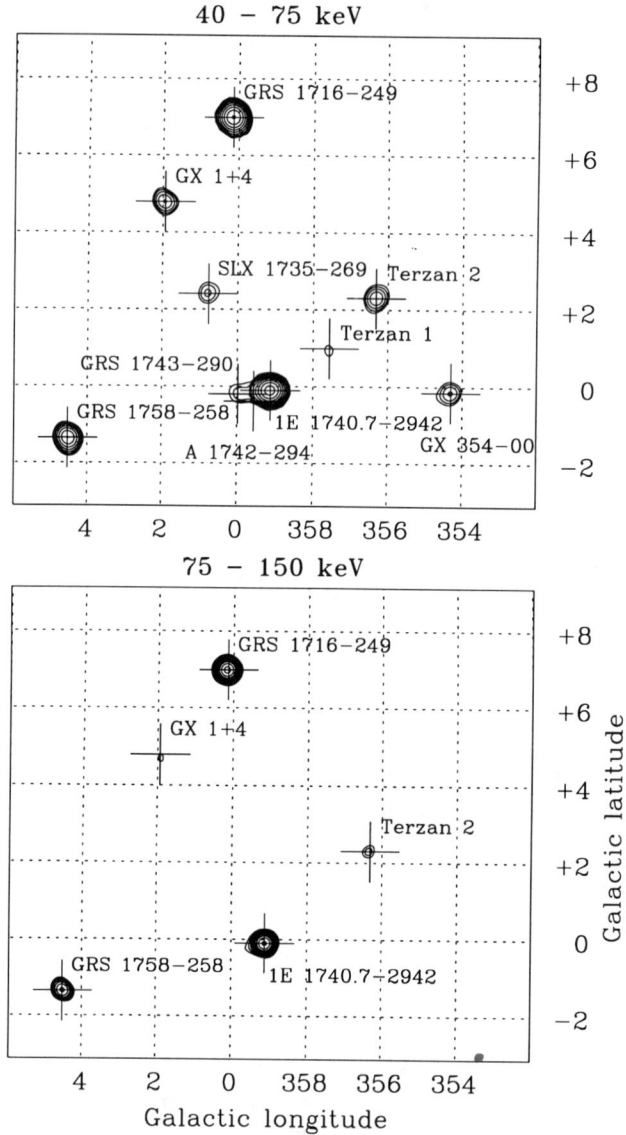

Figure 1. Maps of the GC region derived from the Sigma database (complete to September, 1995) in the 40–75 keV band (top) and in the 75–150 keV band (bottom). Contours show the significance of source detection logarithmically scaled from 5 to 44 σ. Crosses indicate the positions of the sources detected by Sigma

2. Hard X-Ray Map of the Galactic Bulge Region

The GC maps derived from the Sigma database, complete to September, 1995, are presented in Figure 1 for the 40–75 keV and for the 75–150 keV energy bands. At low energy, the map reveals a clusters of ten sources above 5 σ confidence level. Five other sources (KS 1731−26, Barret et al. 1992; GRS 1734−292, Churazov et al. 1992; GRS 1741.9−2853, Churazov et al. 1993; GRS 1747−341, Cordier et al. 1994; GRS 1730−312, Vargas et al. 1996a) have also been detected by Sigma in this region, but due to their weakness and transient variable behaviour, they do not appear in the summed image. Three of the ten sources are also very bright at high energy, which implies that we can distinguish two groups of sources: the soft and the hard ones.

Among the soft sources, seven are known at other wavelengths. Four have been identified with binary systems containing an accreting neutron star, including three X-ray bursters: KS 1731−26 (Barret et al. 1992), GX 354−00 (Claret et al. 1994), and A 1742−294 (Churazov et al. 1995), and one X-ray pulsar: GX 1+4 (Laurent et al. 1993). The two sources associated with globular clusters, X 1724−308 ($\equiv$ Terzan 2, Barret et al. 1991), XB 1732−304 ($\equiv$ Terzan 1, Borrel et al. 1996), may also harbour an accreting neutron star. The nature of the seventh (SLX 1735−269, Goldwurm et al. 1996) remains unknown. The spectra of soft sources, which can generally be described by a power-law $I_\gamma \propto E^{-\alpha}$ with $\alpha \geq 3$, extend up to 100-150 keV. Different kinds of long term temporal behaviours have been observed: flaring activities for GX 354−00 or GX 1+4, rather persistent emissions from Terzan 1, Terzan 2 or SLX 1735−269 (Goldwurm et al. 1995).

On the other hand, the Sigma spectra of the hard sources (1E 1740.7−2942, GRS 1758−258 and GRS 1716−249) extend up to 300 keV with flatter slopes ($\alpha \leq 2.3$). Their close spectral similarities with Cygnus X-1, one of the best candidate for an accreting stellar mass black hole (BH), suggest they might also be BH candidates. The clear identification of the GC hard X-ray emission with 1E 1740.7−2942 and the discovery of the similar source GRS 1758−258 were the first important results of Sigma GC observations. These sources have dominated the emission for the most part of the Sigma GC survey and showed very peculiar temporal variabilities (Cordier et al. 1994). On October 13th, 1991, 1E 1740.7−2942 underwent a spectacular outburst beyond 300 keV up to 600 keV. This spectral structure was represented by a Gaussian line centered at 480 keV with a FWHM of 240 keV and interpreted as a positron-annihilation radiation produced in a hot medium of temperature $\approx$ 40 keV. The derived line flux is $1.3\,10^{-2}$ photon cm^{-2} s^{-1} (Bouchet et al. 1991). Simultaneous Sigma and radio observations led to the identification of the radio conterpart of this source and to the discovery of a double sided jet structure (Cordier et al. 1992, Mirabel et al. 1992). GRS 1716−249 is a hard X-ray nova discovered by Sigma on September 25th, 1993, showing a strong outburst. The emission of this source, which certainly does not belong to the galactic bulge, lasted until the end of this campaign, but was off in the spring of 1994 and was again active in the fall of 1994 (Vargas et al. 1996b, Revnivtsev et al. 1996).

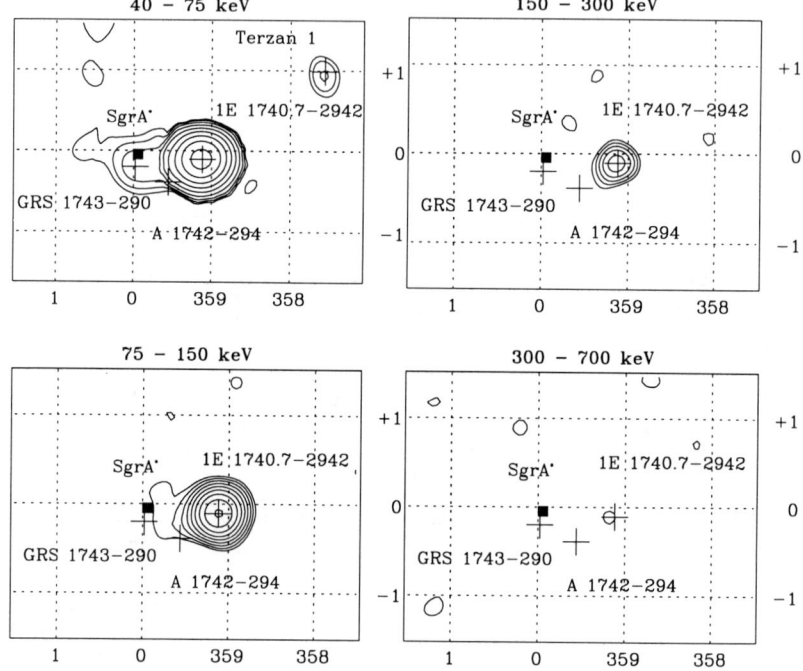

Figure 2. Maps of a 3° × 4° region around the Galactic Nucleus derived from the Sigma database (complete to September 1995), in four different energy bands between 40–700 keV. Contours show the significance of source detection logarithmically scaled from 3 to 44 σ.

3. Sigma observation of the Galactic Nucleus

Figure 2 shows the 3° × 4° region surrounding the radio source SgrA* as seen by Sigma in its six year survey. Four maps are traced corresponding to four energy ranges, from 40 to 700 keV. The source 1E 1740.7−2942 dominates the emission of this region in each map and particularly at high energies. At energies > 75 keV there is no risk of source confusion and we were able to set the upper limits to the SgrA* average emission as reported in Table 1.

In the lower energy band, we note a structure which can not be explained by the emission of 1E 1740.7−2942 alone, considering the telescope Point Spread Function. It suggests that contribution of other probably variable point sources is present. Two point sources were indeed detected during the Sigma survey. One of them is GRS 1743−290 discovered in the spring of 1991, when 1E 1740.7−2942 was in a low state. GRS 1743−290 is located $\approx 9'$ from SgrA* (Goldwurm et al. 1994). An other excess in this region was detected in the fall of 1992, and was attributed to the X-ray burster A 1742−294 (Churazov et al. 1995). The low energy maps of these periods are presented in Figure 3. These variable contributions make difficult to estimate upper limits for SgrA* at low energy. The values reported in Table 1 for the 40–75 keV range have been then estimated assuming that the whole emission, which elongated the contours of 1E 1740.7−2942 toward the left, is attributed to GRS 1743−290 and A 1742−294. A detailled work to derive more reliable upper limits at low energy is in progress. Note finally that an upper limit of $2.3\,10^{-4}$ ph cm^{-2} s^{-1} have been reported by Malet et al.(1995) for the SgrA* 511 keV emission.

The idea that a accreting massive black hole (MBH), identified with the non-thermal radio source SgrA*, is present in the very center of the Milky Way is currently prevalent. However, results reported in Table 1 show that SgrA* is very quiet at energies higher than 40 keV since the total 40–300 keV luminosity is $\leq 8\,10^{35}$ erg s^{-1}. This value is very low compared to AGN emission or even to Cygnus X-1 or other stellar BH candidates, as observed by Sigma (Jourdain et al. 1992; Bond et al. 1996; Salotti et al. 1992) We reported in the Figure 4 the Sigma values on the soft X-ray spectrum observed from SgrA* by the ART-P telescope in the fall of 1990 (Pavlinsky et al. 1994). The observed spectrum requires a break around 30 keV. Theoretical models must take into account this observational material.

Table 1. 2 σ upper limit for SgrA* emission

Energy range	
40- 75 keV	$1.4\,10^{35}$ erg s^{-1} (a)
75-150 keV	$2.0\,10^{35}$ erg s^{-1}
150-300 keV	$4.2\,10^{35}$ erg s^{-1}
300-700 keV	$2.1\,10^{36}$ erg s^{-1}

[a]see text

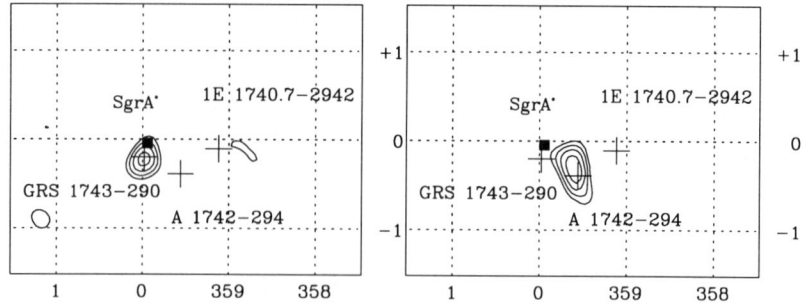

Figure 3. Left: map of the GC obtained from Sigma data collected in the spring 1991, in the 40–75 keV energy range. Right: map of the GC obtained from Sigma data collected in the fall 1992 in the 40–75 keV energy range after substration of 1E 1740.7−2942. Contours show the significance of source detection from 3 σ and spaced by 0.5 σ.

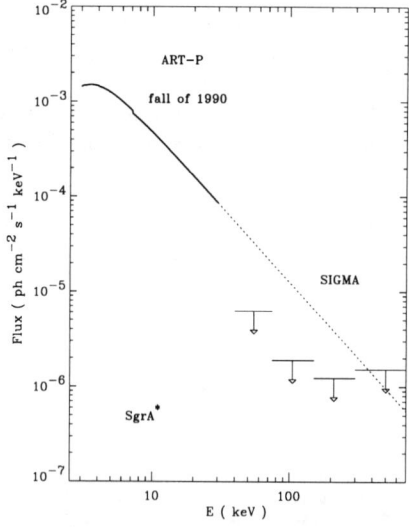

Figure 4. Photon spectrum of SgrA* derived from X-ray observation performed by the ART-P telescope (Pavlinsky et al. 1994). The dotted line represents its extrapolation in the Sigma domain. The Sigma upper limits are at 2 σ.

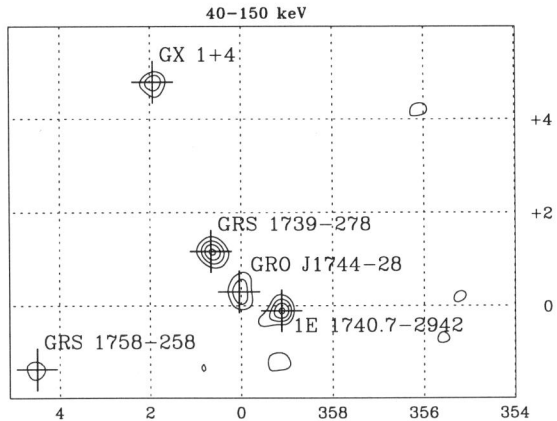

Figure 5. Map of the GC region obtained from Sigma data collected from March 15 to March 31, 1996, in the 40–150 keV energy band. Contours show the significance of source detection from 3.5 σ and spaced by 2 σ.

4. Results of the March 1996 campaign

During the Fourth CTIO/ESO workshop, the Sigma telescope started, on March 15th, a new campaign of GC observations, which lasted until the end of the month. Figure 5 shows the total summed map in the 40–150 keV energy band derived from the data collected during this campaign. Four of the sources seen in Figure 1 (1E 1740.7−2942, GRS 1758−258, Terzan 2 and GX 1+4) showed activity, while two other new sources have been detected.

The first one is the accreting binary pulsar GRO J1744−28 (Bouchet et al. 1996) discovered by Batse on December 2nd, 1995 (Fishman et al. 1995). It was detected during the whole campaign with an average flux of 150 mCrab in the 40–75 keV energy band, and it features a soft spectrum with a power-law photon index of $\alpha \approx 4$.

The second one, GRS 1739−278, is a newly discovered hard X-ray transient source located $\approx 1°$ from GRO J1744−28 and $\approx 1.4°$ from the GC (Paul et al. 1996). Its 40–75 keV flux increased in two days and decreased slowly during the 13 following days, whereas its 75–150 keV flux did not show significant evolution. GRS 1739−278 appears on the summed map (Figure 5) with a confidence level of 9.5 σ. Its averaged 40–150 keV intensity is 62 mCrab during the whole period. The source showed activity well beyond 100 keV with a hard power-law photon index $\alpha \approx 2.3$.

References

Barret, D., Mereghetti, S., Roques, J.P., et al., 1991, ApJ379, L21
Barret, D., Bouchet, L., Mandrou, P., et al., 1992, ApJ394, 615
Bond, I.A., Ballet, J., Denis, M., et al., 1996, A&A 307, 708
Borrel, V., Bouchet, L., Jourdain, E., et al., 1996, ApJ, in press
Bouchet, L., Mandrou, P., Roques, J.P., et al., 1991, ApJ383, L45-L48
Bouchet, L., Paul, J., Gilfanov, M., Sunyaev, R., 1996, IAUC 6343
Claret, A., Goldwurm, A., Cordier, B., et al., 1994, ApJ423, 436
Churazov, E., Gilfanov, M., Cordier, B., Schmitz-Fraysse, M.C., 1992, IAU Circ. 5616
Churazov, E., Gilfanov, M., Sunyaev, R., et al., 1993, A&AS 97, 173
Churazov, E., Gilfanov, M., Novickov, B., et al., 1995, ApJ, 443, 341
Cordier, B., Paul, J., Goldwurm, A., et al., 1992, A&A272, 277
Cordier, B., Paul, J., Ballet, J., et al., 1994, *The Gamma Ray Sky with Compton GRO and SIGMA* , NATO ASI Series, 461, Ed. Kluwer Academic Publishers
Fishman, G.J., Kouveliotou, C., Van Paradijs, J., et al., 1995, IAUC 6272
Goldwurm, A., Cordier, B., Paul, J., et al. 1994, Nat. 371, 589
Goldwurm, A., Denis, M., Paul, J., et al. 1995, Adv. Space Res., 15, No.5, 41
Goldwurm, A., Vargas, M., Paul, J., et al., 1996, A&A, in press
Jourdain, E., Bassani, L., Bouchet, L., et al., 1992, A&A 256, L38
Laurent, P., Salotti, L., Paul, J., et al., 1993, A&A 273, 444
Malet, I., Roques, J.P., Bouchet, L., et al. 1995, ApJ, 444, 222
Mirabel, I.F., Rodriguez, L.F., Cordier, B., et al., 1992, Nat. 358, 215
Paul, J., Mandrou, P., Ballet, J., et al., 1991, Adv. Space Res. 11, No.8, 289
Paul, J., Bouchet, L., Churazov, E., Sunyaev, R., 1996, IAUC 6348
Pavlinsky, M.N., Grebenev, S.A., Sunyaev, R.A., 1994, ApJ, 425, 110
Revnivtsev, M., Gilfanov, M., Churazov, E., et al., 1996, to be submitted
Salotti, L., Ballet, J., Cordier, B., et al., 1992, A&A 253, 145
Vargas, M., Goldwurm, A., Denis, M., et al., 1996a, A&A, accepted
Vargas, M., Goldwurm, A., Paul, J., et al., 1996b, Proc. of the *3rd Compton Symposium*, A&AS, accepted

A Search for Infrared Positronium Line Emission from the Great Annihilator near the Galactic Centre

P.J. Puxley

Gemini 8m Telescopes Project, 950 N. Cherry Ave., Tucson, AZ 85726, USA.

G.K. Skinner

School of Phyics & Space Research, University of Birmingham, Edgbaston, Birmingham B15 2TT, UK.

Abstract.

The region around the compact source of 511keV gamma rays near the Galactic Centre (the "great annihilator") has been surveyed for the positronium recombination line emission expected to be produced by positron-electron systems before annihilation. The search was performed using the CGS4 spectrometer on UKIRT and centered on the 2.18μm line which is analogous to the Paschen gamma transition of hydrogen. Upper limits to the line flux and positron production rate are given.

1. Background: The Positronium Spectrum and the Galactic Centre

Positronium (Ps) is formed by the recombination of an electron and positron. The Ps wave function and energy levels are analogous to hydrogen once the mass is replace by $(m_e/2)$, the reduced mass of the system. If e^+ and e^- recombine in an excited state then they will cascade down through the energy levels radiating photons with wavelengths twice those of the corresponding hydrogen transitions (e.g. Ps(Hα) at 1.3126μm and Ps(Brγ) at 4.332μm) since the radiative lifetime is shorter than the annihilation lifetime. Eventually the 'atom' reaches the ground state followed by annihilation into two 511keV gamma rays or three 0-511 keV photons. (By analogy to helium there are both singlet and triplet systems because of the two 'electrons'; the triplet system is forbidden from two-photon decay).

The 511keV gamma ray line was first detected from the Galactic Centre region in the 1970s (e.g. Leventhal & MacCallum 1982) and is known from subsequent studies to undergo outbursts with the line flux varying by up to an order of magnitude (e.g. Leventhal et al. 1986, Leventhal et al. 1989). Increasingly high angular resolution observations have shown the enigmatic object 1E1740.7–2942 to be identified as the dominant source of hard X-rays and soft gamma rays in this region (e.g. Skinner et al. 1991). Radio measurements with the VLA have shown a double-sided radio source to be coincident with the high energy source and for the radio and hard X-ray fluxes to be correlated (Mirabel et al. 1992). Misra & Melia (1993) have modeled the lobe emission as synchrotron radiation from e^+ and e^- shock acceleration within the outflow.

Mirabel et al. (1991) and Bally & Leventhal (1991) have observed CS and HCO+ emission from a molecular cloud coincident with 1E1740.7 and proposed this as the fuel source for the emission mechanism. Moreover the cool molecular cloud might act to slow the relativistic particles (by Bremsstrahlung or collisional ionization) to the thermal velocities required for Ps formation. No near-infrared counterpart to 1E1740.7 has been detected in H (1.65μm) or K (2.2μm) imaging surveys of the field (Skinner 1992, personal communication; Haller & Melia 1994). Previous radio recombination line searches for Ps near the galactic centre have also been unsuccessful (Anantharamaiah et al. 1989).

2. Near-IR Spectroscopic Observations

Although no near-IR counterpart to 1E1740.7 has been detected in broadband imaging surveys, the possibility remains of detecting a narrow line source with a recombination spectrum. Therefore a spectroscopic search was made on 1993 April 5 using the facility cooled grating spectrometer (CGS4) on UKIRT.

The 3 arcsec-wide slit was centred on the radio continuum core [RA(1950)= $17^h40^m43.01^s$, $\delta(1950)=-29°$ 43' 25.5"] with a position angle aligned along the radio jets (161.1° E of N). The 150 l/mm grating (R=1400) was centred on the Ps Paschenγ transition at 2.18μm. The telescope was nodded to random positions between 10 and 40 arcsec away from the nominal position with an integration time of 4×20s at each position for a total on-source integration time of 800s. Wavelength calibration was performed using a Krypton arc lamp and the star BS6494 (B9IV) was observed before and after 1E1740.7 to provide atmospheric calibration. This was assumed to have K=6.0, based on its visual magnitude V=6.0, for the purpose of flux calibration.

The field in which 1E1740.7 lies contains a great many near-IR sources (e.g. Haller & Melia 1994). Consequently each object and sky position of the slit contained several, typically 1 to 3, unwanted continuum objects. The usual technique for reducing imaging data of such fields, that of forming a background frame by median filtering a number of offset images, was found not to be particularly useful in this instance because of the residual telluric OH line emission. Instead, averages of the object minus sky positions were formed from groups of eight adjacent frames with each group having residual OH lines removed by fitting perpendicular to the dispersion direction and continuum sources removed by fitting a low order polynomial along the spectrum. The final frame was formed from the average of these groups.

3. Discussion

Figure 1 shows the flux calibrated spectrum extracted from the position corresponding to the core radio source. No line emission is apparent in the data which is not attributable to noise, nor at any other position along the slit. The noise is significantly greater than that theoretically achievable with this instrument because of the surface density of sources and telluric OH emission which varied appreciably during the course of the observation. Indeed the Ps Paγ line lies very close in wavelength to one of the brighter lines in the OH spectrum, as can be seen in the raw sky spectrum in Figure 2.

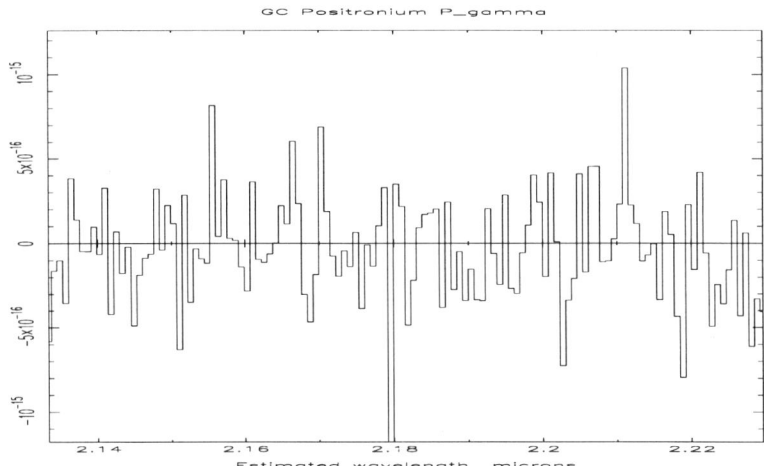

Figure 1. Spectrum around the Ps Paγ ($\lambda 2.18\mu$m) line at the position of the core radio continuum source. The flux density units are Wm$^{-2}\mu$m^{-1}. No significant spectral features other than residual telluric OH emission are apparent.

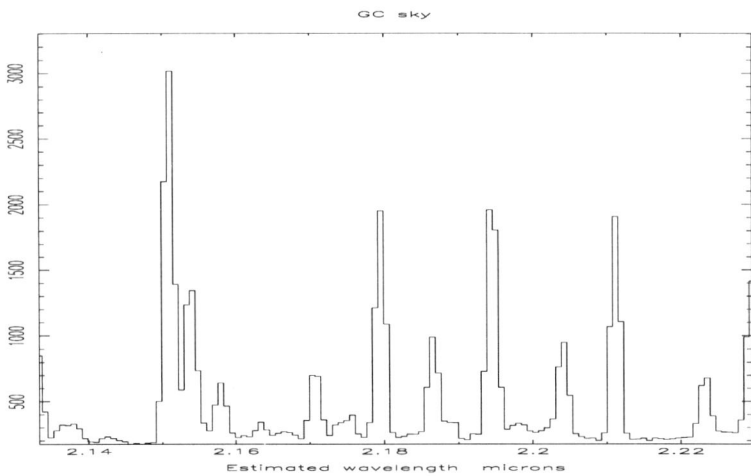

Figure 2. Spectrum of the night sky taken from the Ps dataset (arbitary units). The wavelengths of strong telluric OH lines correlates with residual features in the Ps spectrum (Figure 1).

The average noise level in the Ps spectrum corresponds to a 3σ upper limit for an unresolved line of 3×10^{-19} Wm^{-2}. At a distance to the Galactic centre of 8.5kpc, this corresponds to a Paγ production rate of less than 3×10^{42} s^{-1}, which increases to 6×10^{43} s^{-1} if we assume an extinction to the source of $A_K = 3$mag. This upper limit corresponds to the positron production rate if the (over)simplification is made that all recombinations pass through the Paγ transition and may be compared with the upper limit from Anantharamaiah et al. (1989) of $< 3.1 \times 10^{43}$ s^{-1}, and the flux estimated from the 511keV detection in outburst of 4×10^{43} s^{-1}.

Our new IR spectroscopy cannot place any additional constraints on the Ps emission mechanism however the limits are approaching the expected flux level. We believe that the prospect for observations at the higher angular resolution and collecting area of forthcoming 8m class telescopes such as Gemini and the VLT is promising.

Acknowledgments. Thanks to Ruth Doherty and Stuart Lumsden for assistance with the observations and to Toby Moore, Philip James, Suzanne Ramsay-Howat and Matt Mountain for listening to a crazy idea.

References

Anantharamaiah, K.R. et al. 1989, IAU, The Center of the Galaxy, p607
Bally, J. & leventhal, M. 1991, IAU Circ. No. 5228
Haller, J.W. & Melia, F. 1994, ApJ, 423, L109
Leventhal, M. & MacCallum, C.J. 1982, ApJ, 225, L11
Leventhal, M., MacCallum, C.J., Huters, A.F. & Stang, P.D. 1986, ApJ, 302, 459
Leventhal, M. et al. 1989, Nature, 339, 36
Mirabel, I.F. et al. 1991, A&A, 251, L43
Mirabel, I.F. et al. 1992, Nature, 358, 215
Misra, R. & Melia, F. 1993, ApJ, 419, L25
Skinner, G.K. et al. 1991, A&A, 252, 172

G357.1−00.2: A Peculiar Nonthermal Radio Source near the Galactic Centre

Andrew D. Gray

Dominion Radio Astrophysical Observatory, Herzberg Institute of Astrophysics, National Research Council Canada, P.O. Box 248, Penticton, BC, V2A 6K3, Canada

Abstract. The Galactic Centre is home to a number of unusual nonthermal radio emitting objects, such the filaments and threads near Sgr A, the Snake, and G357.7−0.1. This paper presents recent high resolution radio observations of an object first catalogued at low resolution nearly 20 years ago, which reveal it to have unsuspected properties. The possible classifications of this object are discussed.

1. Introduction

G357.1−00.2 was first catalogued (as G357.2−0.2) in the $4'$ resolution Parkes 5 GHz radio continuum survey of the southern Galactic Plane (Haynes, Caswell & Simons 1978, 1979), where it appeared as an unremarkable, amorphous feature. It was later listed as a supernova remnant (SNR) candidate based on a low infrared-to-radio flux density ratio (Broadbent, Haslam & Osborne 1989). An image obtained at $1'$ resolution in the MOST 843 MHz radio continuum survey of the Galactic Centre (Gray 1994a) revealed an unusual "S"-shaped core structure surrounded by a weak halo. The radio pulsar PSR B1736−31 (listed in Taylor, Manchester & Lyne 1993) was noted to lie about $1'$ away (Gray 1994b), but no conclusive classification of the object or its relationship, if any, with the pulsar could be made. This paper presents the preliminary results of a new high resolution investigation aimed at clarifying these issues.

2. Observations and Reduction

Radio continuum data were obtained using the VLA at 4860 MHz ($\lambda = 6.17$ cm) in the DnC configuration on 1995 January 23, and at 1490 MHz (20.1 cm) in the CnB configuration on 1996 January 28. To image the whole object, four overlapping fields were observed at 6 cm, while at 20 cm only a single field was necessary. 3C286 was used for primary and polarisation angle calibration, while J1751−253 was used for secondary calibration. The data were reduced in NRAO's $\mathcal{AIPS}$ package using standard methodologies, including maximum entropy deconvolution, with "mosaicing" of the four fields at 6 cm. The final images had spatial resolutions of approximately $13''$, with rms noise of $\sim 25\,\mu$Jy/beam.

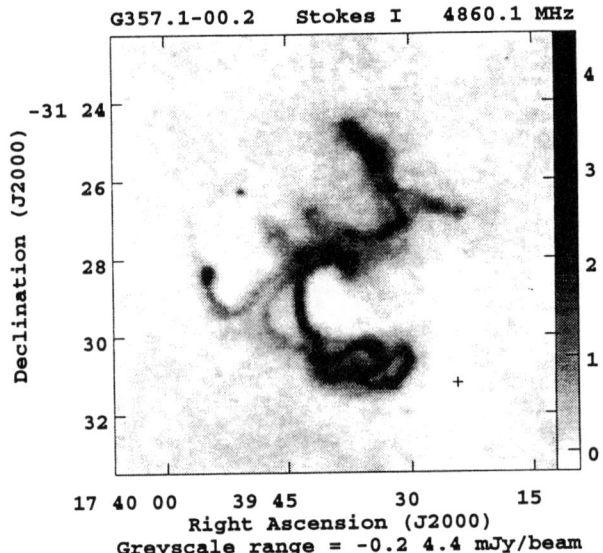

Figure 1. Total intensity at 6 cm. The cross marks PSR B1736−31.

3. Results

Figure 1 shows the 6 cm total intensity image (the 20 cm image is similar). The object is revealed to be composed of several compact components embedded amongst curved and kinked filaments, with no obvious centre of activity. PSR B1736−31, not visible in Figure 1 owing to its $\alpha = -2.47$ ($S_\nu \propto \nu^\alpha$) spectrum, is marked with a cross. The halo detected at 843 MHz is also absent owing to the lack of interferometer spacings short enough to image that component.

At 6 cm most of the object is highly polarised (see Figure 2), with the polarised fraction peaking at ∼60% on the N-S filament near the centre of the object. However, the compact components show no polarisation, and there is only weak polarisation from the bright southern filaments. At 20 cm very little polarisation is seen, suggesting the presence of strong depolarisation (a rotation measure of ∼2000 rad m^{-2} is inferred). The observed polarisation angles at 6 cm are thus probably not intrinsic to the emission. Nonetheless, it is noted that they tend to be aligned along the filaments in the north-western part of the object, and perpendicular to them in the eastern region, with the most strongly polarised emission having angles at ∼45° to the filament.

An examination of spectral index shows that most of the object has $\alpha \simeq -0.5$, although the three "spurs" at the northern edge are noticeably steeper at $\alpha \simeq -0.6$, and there is a region of flattening to $\alpha \simeq -0.3$ over the eastern region of the source, where fractional polarisation is also highest. Despite their lack of polarisation, the embedded compact components have nonthermal spectra. Other compact sources in the field also have nonthermal spectra and are probably background sources.

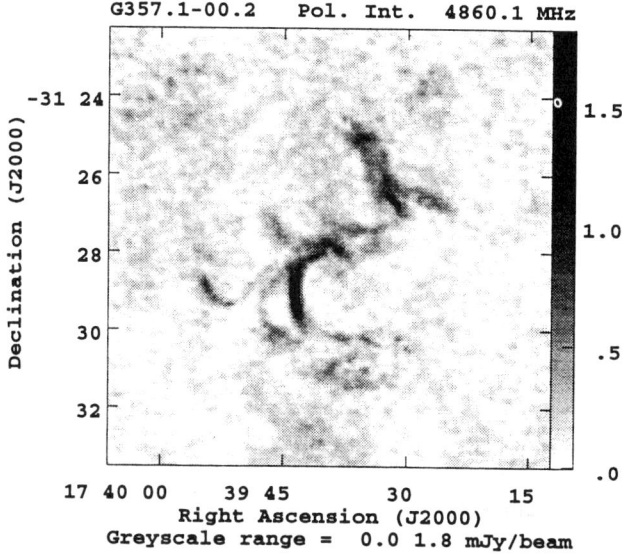

Figure 2. Polarised intensity at 6 cm (see text for details).

4. Discussion

The morphology of this object is unusual for an SNR, at least in the sense of a shock-wave expanding in the interstellar medium. However, despite its $\alpha = -0.5$ spectrum, it is possible that it is a Crab-like object, as the flat spectrum commonly associated with such objects only holds below a spectral break. Landecker et al. (in preparation) show that in Crab-like SNR G76.9+1.0 the spectral break is ~1 GHz, and further argue that the break frequency in such objects is correlated with surface brightness. For G357.1−00.2 the predicted break is also near 1 GHz. Although PSR B1736−31 lies close to G357.1−00.2, its characteristic age of ~0.5 Myr and location well outside the nebula probably rule it out as the powering object[1]. However, only two Crab-like objects have had their central pulsar identified (the Crab and Vela X), so this does not rule out this possible classification.

A second possibility is that G357.1−00.2 is a radio-galaxy, or perhaps several superimposed galaxies. However, it does not appear to be "normal" for this scenario either, although many unusual morphologies are known among that class of objects, so such a subjective argument is hardly conclusive. Nonetheless, none of the compact components in G357.1−00.2 are flat spectrum, which is a property common to core sources in radio galaxies, and there is no obvious core-jet-lobe structure readily identifiable in this object. It is also not clear how the diffuse halo component seen at 843 MHz would be interpreted in this model.

---

[1] It is interesting to note, however, that the distance to PSR B1736−31 is estimated at ~7.7 kpc, putting it close to the Galactic Centre, which is home to several other unusual radio sources.

A third possibility is that G357.1−00.2 is another member of a loosely defined class of peculiar nonthermal phenomena which inhabit the Galactic Centre region, including the "threads" and "filaments" near Sgr A (e.g. Yusef-Zadeh, Morris & Chance 1984; Morris & Yusef-Zadeh 1985), "the Snake" (G359.1−00.2; Gray et al. 1995), and "the Tornado" (G357.7−0.1; Becker & Helfand 1985; Stewart et al. 1994). Most of these objects are still poorly understood, and in the case of the Tornado, as for G357.1−00.2, it is still unclear whether the object is an SNR, or even if it is Galactic!

5. Conclusion

G357.1−00.2 is an intriguing nonthermal object in the Galactic Centre region, showing very strong polarisation from narrow, tangled filamentary structures, with a normal nonthermal spectrum. No compelling evidence has been found to associate the object with the adjacent radio-pulsar PSR B1736−31; indeed, with the data available from the short observational history of this feature it is still not possible to make a conclusive classification of the object itself. A more thorough analysis of the preliminary results shown here is planned, and data with higher spatial resolution (probably also at a higher frequency to defeat depolarisation) as well as H I absorption data may be acquired to help resolve this nature of this object.

Acknowledgments. The MOST data were collected and analysed while supported by an Australian Postgraduate Research Award. The Very Large Array is operated by Associated Universities Incorporated under a co-operative agreement with the U.S. National Science Foundation.

References

Becker, R. H., Helfand, D. J. 1985, Nature, 313, 115
Broadbent, A., Haslam, C. G. T., Osborne, J. L. 1989, MNRAS, 237, 381
Gray, A. D. 1994a, MNRAS, 270, 822
Gray, A. D. 1994b, MNRAS, 270, 835
Gray, A. D., Nicholls, J., Ekers, R. D., Cram, L. E. 1995, ApJ, 448, 164
Haynes, R. F., Caswell, J. L., Simons, L. W. J. 1978, Aust. J. Phys. Astrophys. Suppl., 45, 1
Haynes, R. F., Caswell, J. L., Simons, L. W. J. 1979, Aust. J. Phys. Astrophys. Suppl., 48, 1
Landecker, T. L., Yi-jia, Z., Xi-zhen, Z., Higgs, L. A., in preparation
Morris, M., Yusef-Zadeh, F. 1985, AJ, 90, 2511
Stewart, R. T., Haynes, R. F., Gray, A. D., Reich, W. 1994, ApJ, 432, L39
Taylor, J. H., Manchester, R. N., Lyne, A. G. 1993, ApJS, 88, 529
Yusef-Zadeh, F., Morris, M., Chance, D. 1984, Nature, 310, 557

The Interaction of the G359.54+0.18 Nonthermal Filaments with the Ambient Medium

J. Staguhn and J. Stutzki

Universität zu Köln, 1. Physikalisches Institüt Zülpicher Str. 77, 50937 Köln, Germany

F. Yusef-Zadeh

Dearborne Observatory, Northwestern University, 2131 Sheridan Road, Evanston, IL 60208, USA

K. I. Uchida

Max Planck Institut für Radioastronomie, Auf dem Hügel 69, 53121 Bonn, Germany

Abstract. We present a study of the Galactic Center nonthermal filament system G359.54+0.18 located to the north of the Sgr C region. We find evidence in support of the suggestion by Serabyn & Morris that the nonthermal filaments are the manifestations of large-scale vertical magnetic field lines illuminated by collisions with molecular clouds. Included in the study are observations of, (1) the 3 mm emission lines of CS, HCO^+ and other molecular species with the SEST, (2) several lines of ^{12}CO and ^{13}CO with the 3-m KOSMA antenna, (3) 5 GHz radio continuum emission with the VLA, and (4) $H79\alpha$ recombination line emission with the 100-m antenna at Effelsberg.

The G359.54+0.18 nonthermal filament system and its potentially associated molecular clouds are well suited for this study: It is the system furthest off the Galactic plane and thus suffers least from source confusion. Our high resolution 5 GHz continuum image of the G359.54+0.18 system (Figure 1, contours) reveals a diffuse and somewhat clumpy emission structure near the easternmost tip of the filaments. Additionally, the molecular maps of the region, made with the SEST [1], show a localized cloud (identified as "Cloud A", seen east of the radio feature in Figures 1 and 2) east of, and directly adjacent to, the clumpy continuum component. $H79\alpha$ recombination line emission, at velocities (v_{lsr} = 106 km s^{-1}) consistent with the velocity gradient of the nearby molecular gas, is detected at the position of the diffuse continuum component (α,δ = 17:41:10.0, -29:14:00), indicating that it is thermal in nature and linking it kinematically to the adjacent molecular Cloud A. A strong velocity gradient in Cloud A, directed towards the interface region, implies interaction between the two (Figure 2). The cloud emitts stronger in the high density tracing transition of HCO^+(1-0) than in

---

[1] The SEST telescope is operated by the Swedish National Facility for Radio Astronomy, Onsala Space Observatory and by ESO.

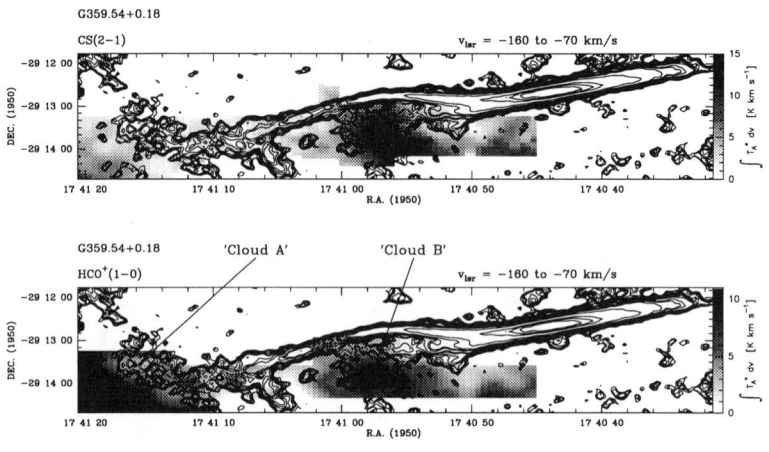

Figure 1. 5 GHz continuum flux ranging from 0.3 to 7 mJy/beam (contour lines) superimposed on CS(2-1) (top) and HCO$^+$(1-0) (bottom) integrated intensities between -160 and -80 km s^{-1} (grey scale). Cloud A with the associated clumpy radio continuum emission can be seen at the eastern tip of the radio structure, cloud B is situated at a position where the nonthermal filaments appear to bend.

CS(2-1). At the central position of Cloud A we have detected emission from the HNCO 5(0,5)-4(0,4) transition, a transition possibly pumped by the IR emission from warm dust or tracing high density clumps within the cloud with $n_{crit} \sim 10^5$ cm^{-3} (Armstrong & Barret, 1985).

On larger scales, both the position and velocity of Cloud A suggests that it is part of the northern edge of the "negative velocity feature" (Bally et al., 1988). The negative velocity feature has highly "forbidden" velocities and displays one of the steepest velocity gradients in the Galactic Center region.

A second molecular cloud (identified as "Cloud B", lower center of Figure 1 and Figure 3) is observed along the curved portion of nonthermal filament. This cloud also exhibits a velocity gradient towards the nonthermal filaments. The CS(2-1) emission from Cloud B is relatively strong. In fact, the brightness ratios of CS to HCO$^+$ are very different in the two clouds described here. Keeping the slightly lower critical density of HCO$^+$(1-0) in comparison to CS(2-1) in mind, this finding could indicate a different chemistry in both clouds. D. Jansen (1995) e.g. argues that HCO$^+$ emission in comparison to most other neutral molecules, appears to be more enhanced in diffuse molecular clouds which can be penetrated by ambient UV fields. Sternberg (1995) points out that HCO$^+$ is efficiently produced in the hot HI/H$_2$ regions of PDRs whereas the formation of CS is more efficient in the more inner cold SI layers of PDRs in which the molecules are more shielded from the ambient UV field. Cloud A, with the

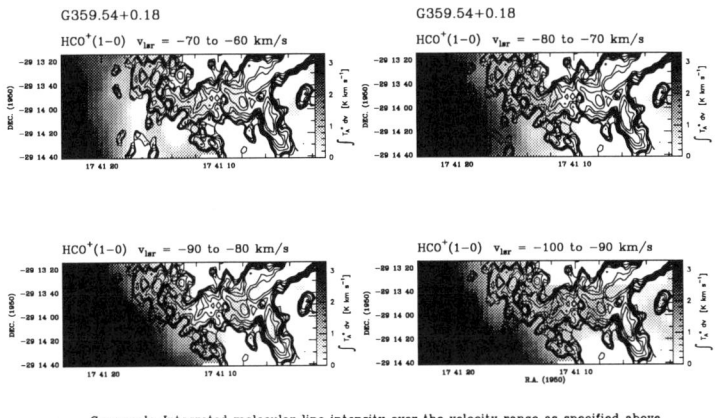

Figure 2. 5 GHz continuum flux (contour lines) superimposed on HCO$^+$(1-0) channel maps of cloud A, integrated in the intervals as indicated. Note the velocity gradient in the molecular line emission towards the interface region of the 5GHz continuum source. The H79α recombination line velocity of -106 km s^{-1} fits well to the the observed molecular velocity gradient.

smaller of the CS to HCO$^+$ intensities, is situated closely adjacent to the diffuse HII region at the filaments tip, whereas Cloud B is located near the nonthermal filaments itself.

Cloud B also appears to be physically associated with the filaments. It has been previously suggested by Bally & Yusef-Zadeh (1989) that the bend in the nonthermal filaments is in response to a collision with a molecular cloud. The location of Cloud B within the bend of the filaments makes it a prime candidate for the colliding partner. As with the case of Cloud A, the molecular line data towards this region shows possible kinematic evidence of a collision — a steep velocity gradient, directed toward the nonthermal filaments, is observed within Cloud B (Figure 4). The role that Cloud B possibly plays in the illumination of the nonthermal filaments, however, is yet to be explored.

Our observations of the G359.54+0.18 filament system support the scenario, proposed by Serabyn & Morris (1994), whereby the nonthermal filaments are illuminated by synchrotron emitting electrons accelerated by the process of magnetic field line reconnection instigated by colliding molecular clouds. HII regions are also involved in this process — as is observed toward at least three other potential cloud/filaments pairs. Some degree of preionization is possibly needed in order that the cloud is able to effectively interact with the magnetic field component and/or, perhaps, a nearby HII region is required to provide enough free electrons for subsequent acceleration along the filaments. We find evidence of all three components requisite for the cloud-collision/magnetic-reconnection scenario: (1) a set of nonthermal filaments supposedly delineating large scale

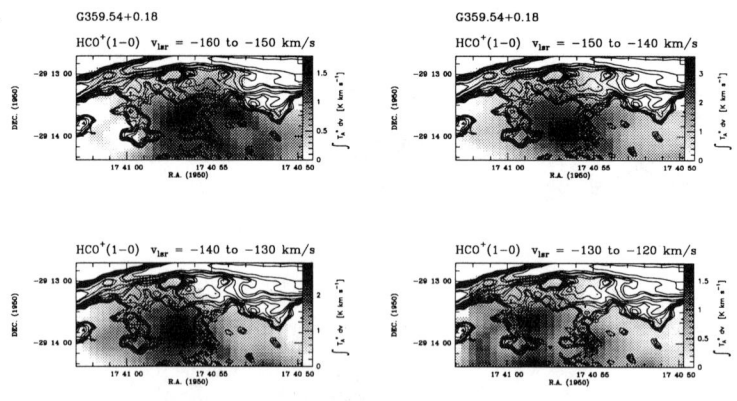

Figure 3. 5 GHz continuum flux (contour lines) superimposed on $HCO^+(1\text{-}0)$ channel maps of cloud B, integrated in the intervals as indicated. The molecular cloud appears to be situated at a position where the nonthermal filaments appear to bend.

GC magnetic field lines, (2) two molecular clouds adjacent to the filaments, and (3) an HII region at the tip of the filaments, evidenced by structure in the radio continuum images and by the detection of H79α recombination line emission. Moreover, a direct association between molecular Cloud A and the HII region is suggested by their similar H79α recombination line and molecular emission line velocities.

References

J.T. Armstrong & A.H. Barret, 1985, APJS, 57, 535

J. Bally, A.A. Stark, R.W. Wilson, & C. Henkel, 1987, APJS, 65, 13

J. Bally & F. Yusef-Zadeh, 1989, APJ, 336, 173

D. Jansen, Thesis, Leiden 1995

A. Sternberg, 1995, in "The Physics and Chemistry of Interstellar Molecular Clouds', Proceedings of the 2nd Zermatt Conference, ed. by G. Winnewisser and G.C. Pelz, Springer Verlag, Heidelberg 1995

E. Serabyn & M. Morris, 1994, APJ, 424, L91

Part 3. The Central Engine and High Energy Phenomena

Section C. Comparison with other Nuclei of Galaxies

Sgr A* and its Siblings in nearby Galaxies

Heino Falcke
Department of Astronomy, University of Maryland, College Park, MD 20742-2421, USA (hfalcke@astro.umd.edu)

Abstract. We have proposed previously that Sgr A* is simply a scaled down AGN with a black hole, an accretion disk and a radio jet operating at a very low power. It appears as if M81* – the nuclear source in the nearby galaxy M81 – is an ideal laboratory to study a Sgr A*-like source at a higher power level. The jet/disk model can explain M81* in great detail with no basic changes in the model parameters other than the accretion rate. Radio cores in other LINERs may be explained by the same model and they appear to be low-power counterparts to radio-loud quasar cores. For Sgr A*, models without a supermassive black hole are facing difficulties – some of which are discussed here, but a persistent puzzle in any scenario are the non-detections and low flux limits for Sgr A* from IR to x-rays. Especially the IR limits are a threat to accretion models. I discuss whether a thin molecular disk (as seen in NGC 4258) around Sgr A* could intercept infalling material before it reaches the black hole.

1. Introduction

The Galactic Center (GC) is a unique place in our galaxy, however, it is not necessarily a unique place in the universe. For this reason the GC has often been used as an analogy for other galaxies, and the GC can help us to understand what we do not understand in more distant places. But for some aspects of the GC itself, the GC is not necessarily the best place to look for answers.

The latter is especially true for the central compact radio nucleus Sgr A*. While its basic radio properties are known for quite a while, the search for counterparts in other wavelength regimes has been largely unsuccessful. This is mainly due to the intrinsic weakness of the Galactic Center and the obscuration in the Galactic plane. Those difficulties have in part driven the developments of many new instruments and techniques – the GC is often among the first objects to be observed with new cameras. And once in a while this has led to a detection of Sgr A* at frequencies inaccessible to radio astronomers (e.g. IR, NIR, X-rays, 511 keV line, etc.), but whenever the next generation of instruments provided higher sensitivity and resolution, it was shown that this emission was due to stellar objects and not due to the suspected supermassive black hole in the very center. This means that any successful model for Sgr A* has not only to be self-consistent but must also be stable against the annual variations in detections of Sgr A* which are a function of wavelength and spatial resolution. Fortunately, at least the evidence from dynamical estimates for the presence of a dark mass of $M_\bullet = 2 \cdot 10^6 M_\odot$ in the center of the Galaxy has become more and more convincing (Genzel 1996, Rieke & Rieke 1996, Haller et al. 1996) in recent years and now seems well established.

The presence of such a large concentration of dark mass in the very nucleus of the Galaxy places the GC in line with many other galaxies where similar or even much higher dark mass concentrations have been found (Kormendy & Richstone 1995). Another similarity to other Galactic Nuclei is the presence of a compact flat spectrum radio core which is found in all radio loud active galaxies, as well as in many other active galaxies like radio quiet quasars, Seyferts, LINERs, in many elliptical galaxies, and also in some spiral galaxies. In this respect is the Galactic Center fairly typical and therefore should not be considered as an isolated case. In this paper I will therefore not only discuss possible explanations for Sgr A* and their difficulties, but also apply our Sgr A* jet/disk model to other weakly active galaxies, specifically to the nucleus of M81.

2. Modelling Sgr A*

2.1. What we see...

Sgr A* is constrained by what we see and also by what we do not see. The size and spectrum of the radio core are the primary input data to all models, but even though this is one of the few things we see, both are controversial in some details. The size of Sgr A* is dominated by scatter broadening at frequencies at least up to 22 GHz and the smallest sizes reported so far are 1.7-2.8 AU (Doeleman et al. 1996, Rogers et al. 1994, Krichbaum et al. 1994) at $\lambda 3$mm. The overall spectrum of Sgr A* is inverted. While Duschl & Lesch (1994) claim a spectral shape of $\nu^{1/3}$ from cm to submm wavelengths, Mark Morris presented during the conference a spectrum of Sgr A* which was taken within 2 weeks by the CSO/JCMT/OVRO collaboration and the VLA, indicating that the cm part may be fitted by a single powerlaw, while the submm part shows a submm-excess. The possibility of a submm-excess has been around for quite a while (Zylka et al. 1992) but due to the variability of Sgr A*, which is well established in the radio regime, it was never unambigiously proven. A clear experiment to demonstrate this, would be truly simultaneous cm/mm/submm observations of the GC, and preparations for such a campaign are on the way.

The importance of the submm-excess is that, if it exists, it implies synchrotron self-absorption at frequencies around 100 GHz, and as shown e.g. in Falcke (1996a) this requires an ultra compact region of ~ 0.1 AU. Given the mass of $2 \cdot 10^6 M_\odot$ this size translates into a region which is only 2-3 Schwarzschildradii in diameter. A proper determination of the mm/submm spectrum of Sgr A* could fix this number to a relatively high degree. Consequently, if there really is a black hole, a future global submm VLBI experiment would be able to probe a region which is strongly affected by General Relativistic effects. Light bending and asymmetries due to the Kerr Metric could in principle be directly imaged. Even though the technical realization of such an experiment may be decades ahead, the principal feasibility warrants a lot of excitement and motivation for future work in the GC – we do not know many other places, if any at all, where such an experiment might ever become possible.

Observations of the better known cm part of the spectrum reveal kinks (Wright & Backer 1993) and strongly varying spectral indices (Zhao et al. 1996, in prep.) which requires synchrotron self-absorption also at lower frequencies and argues for a stratified medium in Sgr A* rather than a single component model. In fact, this is what most models for Sgr A* actually imply; and because the observed quantities of Sgr A* are so few, but very constraining, the published Sgr A* models do not really differ in their underlying physical processes. The basic ingredients of these models are a black hole and an accretion flow, a certain conversion factor between the accretion power and the non-thermal emission, and various equipartition arguments. In the Bondi-Hoyle accretion model of Melia (1992 & 1994) and in the advection dominated disk model by Narayan et al. (1995) the radiative efficiencies are fairly low and the emission mechanism for Sgr A* is cyclotron/synchrotron emission, while in the jet/disk-symbiosis model by Falcke et al. (1993a) the radiative efficiency is fairly high and the emission mechanism is pure synchrotron radiation. In the former cases one has pure inflow, in the latter case one has inflow and outflow. Duschl & Lesch (1994) proposed a stationary, homogenous blob of synchrotron radiating monoenergetic electrons, which they qualitatively link to an accretion disk.

2.2.   ... and what we do not see

As mentioned above, the basic theme for all models so far is accretion onto a supermassive black hole and it is very difficult to avoid strong thermal radiation from the accretion flow. Not long ago it was thought that a substantial fraction of the central luminosity of $10^7 L_\odot$ in the GC is produced by this process. Alas, it is now apparent that the stars we see are enough to produce the bulk of this luminosity and the heating of the ambient medium and the CND (e.g. see Latvakoski et al. 1996). The hope that some thermal emission from Sgr A* had been discovered at least in the NIR (Eckart et al. 1992) also faded recently with improved astrometry and resolution (Genzel 1996). Hence, there is currently *no direct evidence for any thermal emission from Sgr A**. That may not mean much, because there are also many luminous O/B stars in the GC which we cannot see, just because of obscuration. Nevertheless, the constraints on accretion disks are severe. The current NIR limits and mass estimates of the black hole require accretion rates to be $< 10^{-8} M_\odot$/yr for a standard accretion disk (see Falcke et al. 1993a). On the other hand Melia (e.g. 1992) has argued that a supermassive black hole in the GC should accrete of the order $10^{-4} M_\odot$/yr from stellar winds. He suggested spherical accretion and the formation of only a small transient disk, Narayan et al. (1995) have proposed that the accretion disk is advection dominated and therefore more than 99.9% of the energy is advected and never radiated. With the ever decreasing limits on the dereddened Sgr A* flux of less than 20mJy (inferred from Eckart et al. 1995), however, the "inefficiency levels" for the latter two models become uncomfortably high, while the accretion rate for standard accretion disk models also become uncomfortably low (even though I have to admit that "comfort" is not really a well defined physical quantity).

2.3.   Spherical accretion and fossil disk

Recently we have proposed another alternative (Falcke & Melia 1996) which builds up on the suggestion by Falcke & Heinrich (1994) that Sgr A* might well be surrounded by a fossil accretion disk — a remain of past activity. Such a fossil disk (or ring) may be very optically thick, very stable over a long time and could in principle capture infalling matter at a large radius, possibly without producing the amount of luminosity usually expected if the matter were to fall into the very center. The disk would not be disrupted by the infalling wind because it could still be very massive. The dynamical timescale of the fossil disk is much shorter than the time scale for the infall, which is given by the ratio $\dot{\Sigma}_{\rm wind}/\Sigma_{\rm disk}$ between the mass deposition rate per unit area and the surface density of the disk.

To study this process in more detail, we have modified the standard accretion disk equations to allow for matter and angular momentum deposition, and have calculated the time evolution and spectrum of such an accretion disk. In a second step we have coupled this accretion disk model with 3D hydrodynamical calculations of a Bondi-Hoyle flow (Coker et al. 1996). This allowed us to test which scenario might be compatible with current observations. The boundary condition was that the high mass inflow that had been intercepted at a large radius should not propagate into the center within a couple million years — the presumed age of the high mass-loss star — and the luminosity should not exceed the current IR/NIR and total luminosity limits.

The first result is that it is *not* possible to hide any strong inflow with zero angular momentum. While indeed a lot of the wind is captured by the fossil disk at larger radii, the remaining part of the wind is still too large and would produce an enormous amount of luminosity. Another problem with a zero angular momentum wind is that even if it is absorbed at a large radius, the kinetic energy dissipated in the impact will produce strong emission.

Therefore, one has to invoke a non-zero angular momentum wind, which circularizes at a large radius. We find that a minimum radius for the circularization need to be of the order $R_{\rm circ} \gtrsim 10^{16}$ cm (i.e. 0.1"), but could of course be larger. The viscous timescale for changes in such a structure could be as long as several 10-100 Million years. The resulting spectrum differs substantially from a normal accretion disk spectrum and shows a strong peak in the IR. A complication for the modelling is that the accretion radius of the Bondi-Hoyle flow is 10^{17}cm and therefore any fossil structure of the size discussed here might already start to influence the whole Bondi-Hoyle structure and the bow shock. Another problem is that the Bondi-Hoyle spherical accretion solution changes if one adds angular momentum to the flow, the wind which is finally captured does not retain the same specific angular momentum it had before it encountered the black hole. Therefore one can not easily translate an uneven, or rotating source distribution into an angular momentum of the infalling wind.

¿From Maser observations of NGC 4258 (Miyoshi et al. 1995) we know that molecular disks exists on such small scales. But the structure we have discussed for Sgr A* so far is pure fantasy and was born out of the need to explain what we do not see, and is not based on any positive detection. Nevertheless, if present a fossil disk should of course have observational consequences, especially in the IR. Stolovy et al. (1996) have announced the detection of a source at 8.7μm with dereddened flux of $\sim$ 100 mJy at the position of Sgr A*, but at the current resolution a direct association with Sgr A* in this crowded and tricky field is by no means certain. Such a flux would correspond to a black-body disk with radius 10^{15} cm, inclined by 80° at a temperature of 350°K (not quite room temperature but close). A disk with $R \sim 10^{16}$cm at that temperature would already produce too much flux. This shows that the current limits for such a configuration are already very tight.

2.4. Are there alternatives?

With the problems currently troubling accretion models one is tempted to ask whether there are alternative models for Sgr A*. Is there a life without a black hole? Can we replace the black hole with an ultradense cluster of stellar remnants? Well, it wouldn't be much fun in the first place. Secondly, while one may find alternative solutions by just considering the emission properties it becomes very difficult if one takes the whole context into account. There is for example the observation by Backer (1996) that Sgr A* does not move w.r.t. to the Galactic Center, while any low mass object should (as we now see directly in the stars). The total mass of Sgr A* therefore needs to be at least several 100 $M_\odot$ and this mass can obviously not be in the synchrotron radiating gas, so that we need an anchor of at least several hundred, possibly thousand, stellar remnants. On the other hand we need at least $10^{3.5} L_\odot$ for the synchrotron radiation alone. If we assume that this energy comes from accretion onto this hypothetical central cluster we have to have at least an accretion rate of $10^{-4} M_\odot$/yr. The minimum size scale for Sgr A* is 10^{12}cm (see Falcke 1996a) and the radiative efficiency for a 1000 $M_\odot$ object at this scale is only $6 \cdot 10^{-5}$, because we are not very deep in the potential well. However, if the accretion continues onto the stellar remnants, they would inevitably turn into strong x-ray emitters. In fact, a fraction of only 10^{-6} of this accretion rate would be enough to violate all current x-ray limits (see Koyama et al. 1996, Maeda et al. 1996).

Can we then power Sgr A* without accretion, e.g. if the plasma is heated in some mysterious way by the kinetic or potential energy of the supposed stellar remnants? The problem here, as well as for the accretion scenario above, is that the pressure one derives for the synchrotron radiating gas in Sgr A* (especially in the submm where we have $n \sim 10^4 \mathrm{cm}^{-3}$, $B \sim 10\mathrm{G}$, $r \sim 10^{13}$ cm, see e.g. Falcke 1996a) would require a central mass of $10^{8-9} M_\odot$ to keep it in the center — otherwise it would literally be blown away within seconds, just like in the jet-model (see also an earlier discussion in Reynolds & McKee 1980). A cluster of stellar remnants would never have enough potential energy to keep the gas in the center. Consequently, even if there are a bunch of stellar remnants throughout the central star cluster as suggested by Haller et al. (1996), Rieke & Rieke (1996), and Saha et al. (1996) it appears very unlikely that they have anything to do with Sgr A* itself.

3. The siblings

3.1. M81*

We are strongly limited in our modelling of Sgr A* by two important effects: scatter broadening and obscuration. Thus we know neither the intrinsic shape and size of Sgr A*, nor its optical/UV properties. However, as mentioned in the beginning, any model for Sgr A* should be invariant to translation by at least a few Mpc. Therefore, it seems as if the best place to learn more about Sgr A* is the nucleus of M81 (see Falcke 1996b). This is a spiral galaxy, classified as a LINER, where we are not strongly affected by obscuration. In the nucleus we find a compact flat-spectrum radio core (which we call M81* in analogy to Sgr A* and M31*) with a size of 550 AU at 22 GHz and an inverted spectrum ($F_\nu \propto \nu^{0.2\pm0.2}$, see Reuter & Lesch (1996) and references therein). Unlike Sgr A*, this core is resolved with VLBI at various frequencies and shows a size proportional to $\nu^{-0.8\pm0.05}$ (Bietenholz et al. 1996, and references therein), hence it is not scatter broadened. Moreover, the core is elongated and one finds structure with the VLA at a much larger scale in a similar direction. The most likely explanation for this observation is the presence of a jet.

The bolometric luminosity of the M81 nucleus has been estimated to be of the order $10^{41.5}$ erg/sec (Ho et al. 1996) and Bower et al. (1996) recently discovered broad double-peaked Hα emission from M81, which is either due to an accretion disk or a bi-polar outflow.

With this information it was of course tempting to apply the jet/disk symbiosis model we developed initially for Sgr A* (Falcke et al. 1993b) to M81*, at least here it is much easier to argue that a mini-AGN with jet and accretion disk really is present. Especially the detailed VLBI informations allow a more detailed test of the model.

The first important point is the frequency dependence of the size, with a size index $m = -0.8 \pm 0.05$ ($r \propto \nu^m$) and an inverted spectrum. The frequency dependence of the size was one of the basic predictions of the jet model, while in homogenous, optically thin models (e.g. Duschl & Lesch 1994) a constant size is expected – this reflects the main differences between homogenous and inhomogenous (i.e. with gradient in magnetic field) models.

However, the extremely simplified jet emission model also does not fit perfectly, as it predicts a flat ($\alpha = 0$) rather than an inverted spectrum, and the predicted size index is $m = -1$, thus slightly steeper than observed in M81*. It is of course fairly easy to modify the jet model to fit those values, e.g. by imposing a certain non-conical jet shape (as it is frequently done for quasar cores). Such a non-conical shape would imply external confinement or acceleration of the jet. On the other hand those models usually lack a physical justification for the acceleration or collimation (especially with the high internal pressures involved) and it makes one feel uncomfortable to just add another arbitrarily chosen input parameter for each new observed quantity. Fortunately, it turned out that there is a slight inconsistency in the canonical Blandford & Königl (1979) jet model used previously (e.g. Falcke & Biermann 1995), where one usually neglects the dynamical effects of the pressure gradient on the velocity field of the jet flow. If calculated self-consistently this pressure gradient will indeed lead to a slight acceleration of the jet. In terms of the velocity structure this is a weak effect, however, if one starts with a fully relativistic gas (i.e. sound speeds of the order $0.6c$ – something necessary to escape from the inner parts of a black hole) it is just enough to make the jet mildly relativistic ($\gamma_j \simeq 2 - 3$). Due to the boosting effect, the emission at lower frequencies, coming from more distant regions, will be Doppler-*dimmed* w.r.t. the higher frequencies for most aspect angles and thus yield an inverted spectrum and a flatter size index.

For the given luminosity of M81* and the jet-power/disk-luminosity ratio we found for quasars (see Falcke et al. 1995), the whole Sgr A* jet/disk symbiosis

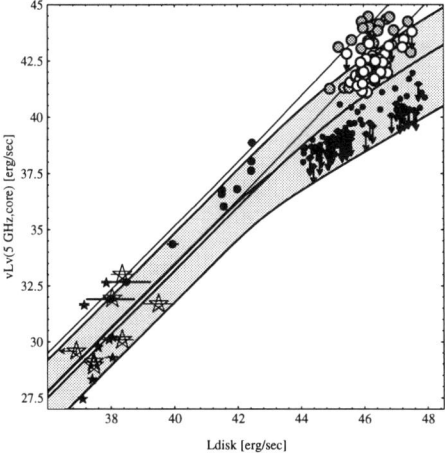

Figure 1. Radio core vs. bolometric nuclear luminosity for a variety of known and putative jet/disk systems (from Falcke & Biermann 1996a) and the predicted distributions from the jet/disk symbiosis model. The stars in the lower left are galactic jet sources and x-ray binaries, circles and dots in the upper right are radio loud and radio quiet quasars. Black dots below $L_{\rm disk} = 10^{44}$ erg/sec are LINERs and Sgr A* and M31* – it appears as if LINERs could be the missing link between highly active radio loud quasars and almost inactive, yet radio-luminous, nuclei like Sgr A*.

model can then be boiled down to a two parameter model, where we need only the electron Lorentz factor $\gamma_{\rm e}$ and the inclination angle i as an input parameter, which on top of that, are both fairly well constrained.

And in fact, for $\gamma_{\rm e} = 220$ and $i \simeq 30 - 40°$ the model predicts the observed size (550 AU at 22 GHz), flux (110 mJy at 22 GHz), spectral index ($\alpha = 0.17$), and size index ($m = -0.9$) for M81* reasonably well. Sgr A* can be explained by the same model for an assumed $L_{\rm disk} \sim 10^{39}$ erg/sec with $i \sim 60° - 70°$ and $\gamma_{\rm e} = 140$, the predicted average spectral index is $\alpha = 0.23$ and $m = -0.9$ — the size of the major axis should be around 6 AU at 7mm.

3.2. The rest of the family

It is interesting to note that we had to use a *radio-loud* model (defined by the radio/$L_{\rm disk}$ ratio) to explain M81* (the same is true for Sgr A* and possibly M31*). Could it be that the cores of radio loud quasars have their low-luminosity counterparts in LINERs and other weakly active galaxies? For this reason we have begun to revisit the radio properties of some nearby galaxies with signs of nuclear activity; quite a few have compact flat-spectrum radio cores similar to Sgr A* and M81*, e.g. like the Sombrero galaxy (M104). Unfortunately, due to the low level of activity, the determination of a bolometric or "disk"-luminosity for the nuclei can be very difficult. Bearing that in mind, we have plotted the radio core fluxes of a small sample of prominent galaxies versus what we estimate to be their disk luminosity in Figure 1. Those results are of course very preliminary and need further refinement, nevertheless, it is quite interesting that the cores of those LINER nuclei all seem to fall on the radio-loud branch of the jet/disk symbiosis model, and some of them, like NGC 1097, do indeed have well known jets. This could mean that basically all those radio cores in LINERs are the bases of radio jets and they could be the missing link between Sgr A* and radio loud quasars. Further study of those radio cores in the Galactic Center and elsewhere might therefore not only reveal something about the true nature of Sgr A*, but also help us to understand the radio-loud/radio-quiet dichotomy in quasars.

4. Conclusion

All the models proposed for Sgr A* have a certain appeal. The jet/disk model – with and without monoenergetic electrons – offers a scope that goes far beyond the GC and has survived a series of critical tests in a variety of very different source classes with compact flat spectrum cores, including Sgr A*. Advection-dominated and fossil disks may help to explain why the optical luminosity of Sgr A* is so low, and Bondi-Hoyle accretion is a process that seems to be unavoidable at a certain level. Applying all these concepts to the nuclei of nearby galaxies may help us to sort out which process dominates in which regime. Until then we perhaps could agree on a "theorists-for-galactic-peace-model" for Sgr A*: a jet of monoenergetic electrons, produced by an advection dominated disk coming from a fossil ring which is fed by Bondi-Hoyle accretion.

Acknowledgments. This work was supported by NASA under grants NAGW-3268 and NAG8-1027. I am grateful for many discussions during and after the conference.

References

Backer D.C. 1996, in "Unsolved Problems of the Milky Way", IAU Symp. 169, L. Blitz & Teuben P.J. (eds.), Kluwer, Dordrecht
Bietenholz, M.F., Bartel, N., Rupen, M.P., Conway, J.E., Beasley, A.J. et al. 1996, ApJ 604, 28
Blandford R.D., Königl A. 1979, ApJ 232, 34
Bower, G.A., Wilson, A.S., Heckman, T.M., Richstone, D.O. 1996, AJ 111, 1901
Coker R., Melia F., Falcke H. 1996, to be submitted to ApJ
Doeleman S., Rogers A.E.E., Bower G.C., Wright M.C.H., Backer D.C. 1995, BAAS 187, #12.14
Duschl W.J., Lesch H. 1994, A&A 286, 431
Eckart, A., Genzel, R., Krabbe, A., Hofmann, R., van der Werf, P. P., Drapatz, S. 1992, Nat 355, 526
Eckart, A., Genzel, R., Hofmann, R., Sams, B.J., Tacconi-Garman, L.E. 1993, ApJ 407, L77
Eckart, A., Genzel, R., Hofmann, R., Sams, B.J., & Tacconi-Garman, L.E. 1995, ApJ 445, L23

Falcke H., 1996a, in "Unsolved Problems of the Milky Way", IAU Symp. 169, L. Blitz & Teuben P.J. (eds.), Kluwer, Dordrecht, p. 163
Falcke H., 1996b, ApJ Letters, Vol. 464 (June 10)
Falcke H., Biermann P.L. 1995, A&A 293, 665
Falcke H., Biermann P.L. 1996, A&A 308, 321
Falcke H., Heinrich O. 1994, A&A 292, 430
Falcke H., Melia F. 1996, to appear in ApJ
Falcke H., Biermann P. L., Duschl W. J., Mezger P. G. 1993a, A&A 270, 102
Falcke H., Mannheim K., Biermann P. L. 1993b, A&A 278, L1
Falcke H., Malkan M., Biermann P.L. 1995, A&A 298, 375
Genzel R. 1996, this volume T. 1991, ApJ 381, L43
Haller J., Rieke, M., Rieke, G., Tamblyn, P. Close, L. & Melia, F. 1995, Ap.J., in press
Ho L.C., Filipenko, A.V., & Sargent, W.L.W. 1996, ApJ 462, 183
Krichbaum T.P., Schalinski C.J., Witzel A., Standke K., Graham D.A., and Zensus J.A. 1994, in "The Nuclei of Normal Galaxies: Lessons from the Galactic Center", eds. Genzel R. & Harris A.I., Kluwer, Dordrecht
Kormendy J., Richstone D. 1995, ARA&A 33, 581
Koyama K., Maeda Y., Sonobe T., Takeshima T., Tanaka Y., Yamauchi S. 1996, PASJ 48, in press Astronomy - first results from ASCA, Universal Academy Press, p. 181
Latvakoski H., Stacey G., Hayward T., Gull G. 1996, this volume
Maeda Y., Koyama K., Sakano M., Takeshima T., Yamauchi S. 1996, PASJ 48, in press
Melia F. 1992, ApJ 387, L25
Melia F. 1994, ApJ 426, 677
Miyoshi M., Moran J., Herrnstein J. et al. 1995, Nat 373, 127
Narayan R., Yi I., Mahadevan R. 1995, Nat 374, 623
Reuter, H.-P., & Lesch, H. 1996, A&A subm
Reynolds S.P., McKee C.F. 1980, ApJ 239, 893
Rogers A.E.E., Doeleman S., Wright M.C.H. et al. 1994, ApJ 434, L59
Stolovy S.R., Hawyard T., Herter T. 1996, this volume
Wright M.C.H., Backer D.C. 1993, ApJ 417, 560
Zylka R., Mezger P.G., Lesch H. 1992, A&A 261, 119

Stars, Stellar Remnants, and the Black Hole in the Galactic Center

G. H. Rieke and M. J. Rieke

Steward Observatory, University of Arizona, Tucson, AZ 85721 USA

Abstract.
We catalog and describe the constituents of the Galactic Center for comparison with models of the nuclei of other galaxies. A detailed mass distribution has been determined using radial velocity measurements for more than 200 stars. These measurements agree closely with the mass distribution from measurements of gas dynamics, both indicating the presence of a very compact mass of $\sim 2 \times 10^6 M_\odot$ within the central 0.2 pc. The close agreement among a number of independent techniques strongly supports the presence of a central black hole of this mass. Luminosity functions and spectra of individual stars in the 10 pc region centered on the Galactic Center demonstrate the presence of a substantial population from a starburst that occurred no more than $\sim$ 100 million years ago. The occurrence of this starburst is reinforced by direct determination of a ratio of $M/L \leq 0.3$ in this region. The trend of M/L with radius shows evidence for a component of extended dark matter; the case for this component is strengthened substantially if allowance is made for the low M/L of the young stars. A plausible identification for the dark matter is stellar remnant white dwarfs.

1. Introduction

We generally describe galactic nuclei in simple terms. They contain a population of old stars, possibly complemented by a stars of "intermediate" age. Exceptional cases are dominated by a massive black hole and associated active nucleus and/or by a starburst. How complete are these abstractions? We could test them by probing the gravitational potentials on a subparsec scale or by classifying individual stars. However, such observations are far beyond our capabilities except in special cases.

The most favorable of these special cases is the nucleus of the Milky Way, which is sufficiently close that a full range of emerging techniques can be used to probe with the necessary resolution and detail. A decade ago, we used abstractions similar to those applied to external nuclei to describe the Galactic Center. We now know that it is far more intricate. It appears to harbor a massive, dark object, but one that is mysteriously dim. It has a complex history of star formation over a variety of physical scales. It apparently derives most of its current luminosity from a very young and compact cluster of extremely massive stars, normal individually but each of a rare type and hence peculiar collectively.

Although we have begun to lose our naivete about the Galactic Center, we retain it virtually undiminished for other galactic nuclei. Indeed, since we cannot probe the complexities of other nuclei observationally, we need to characterize the Galactic Center with such thoroughness that we can transfer a physical understanding to similar but not identical environments. With this goal in mind, in this paper we will catalog and describe the constituents that we now identify in this region.

2. An Extended Starburst

Near-infrared images of the central 10 parsecs (2.5 arcmin radius) of the Milky Way are dominated by late type stars and can be converted to a bolometric luminosity function by making use of the properties of such stars (Rieke 1987; Rieke 1993). The distribution of $H-K$ colors in these images peaks at $H-K = 2.3$. We can assume an average intrinsic $H-K = 0.3$, comparable to that for the red stars in Baade's Window (Frogel and Whitford 1987). Then the apparent color of $H-K = 2.3$ implies an extinction of $A_V = 31.8$ magnitudes, similar to the extinction deduced from individual, well-studied stars.

We can use these nominal values along with the bolometric corrections from K magnitudes to derive stellar luminosities. If carried out as described below, this derivation is very robust; deviations from the nominal values have very little effect on the derived result. For example, if the intrinsic $H-K$ of a star were 0.5, then the derived extinction assuming $H-K = 0.3$ would be too large, but the bolometric correction would be too small. The two effects largely cancel; the net error in bolometric magnitude for any intrinsic $H-K$ between 0.1 and 0.5 is less than 0.2.

Figure 1 shows the luminosity function derived from the large scale image described in Haller et al. (1996). All stars with $H-K < 1.5$ have been eliminated from the sample under the assumption they must be foreground objects, as is verified by modeling the distribution of stars along the line-of-sight to the Galactic Center. The remaining stars were dereddened by $A_V = 31.8$ and with a bolometric correction of 3 magnitudes appropriate to $H-K = 0.3$. The distance to the GC was assumed to be 8.5 Kpc. A luminosity function for cool stars in Baade's Window is also shown in Figure 1 (Frogel & Whitford 1987). The normalization of the two luminosity functions is set according to a model that extends the Baade's Window luminosity function to very faint stars (Tiede, Frogel, & Terndrup 1995), accounts for the surface brightnesses in the two regions, and requires that the luminosity function for the young population in the Galactic Center field be reasonably smooth across the transition near $M_{Bol} = -2$. For stars brighter than this value, the observations should be virtually complete; for fainter stars, they are incomplete and the remainder of the luminosity function must be filled in to satisfy the surface brightness constraint. For example, if too little of the Baade's Window population were included, a large increase in the young stellar luminosity function would occur just fainter than $M_{Bol} = -2$. Figure 1 demonstrates clearly the population of very luminous and young red stars in the Galactic Center, stars which do not have counterparts in Baade's Window (Rieke 1987; Rieke 1993; Blum, Sellgren & DePoy 1996, this conference). The derived normalization indicates that approximately 2/3 of the

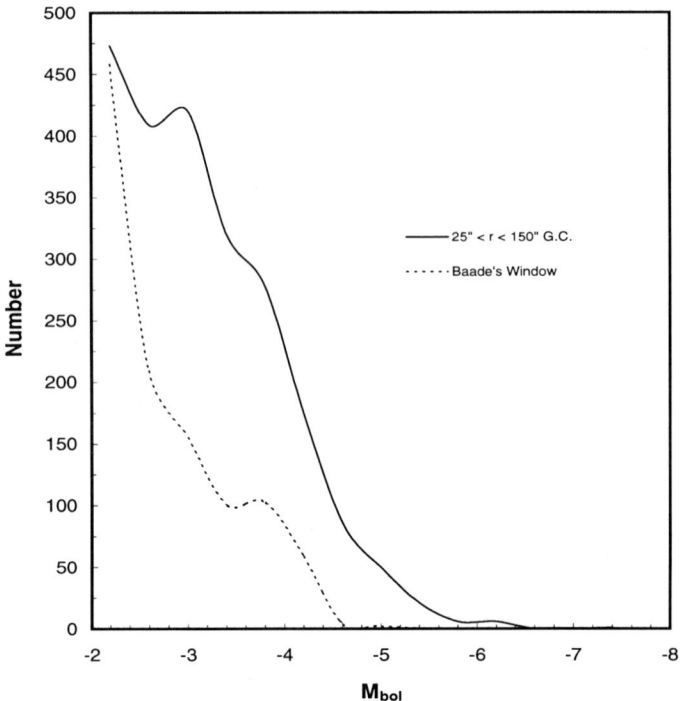

Figure 1. Luminosity Functions for Baade's Window and for an Annulus from 1 to 6 pc from the Galactic Center. The luminosity functions have been normalized using a model that accounts for the K surface brightnesses observed in the two regions and requires the Galactic Center luminosity function to behave smoothly from $M_{Bol} < -2$ toward fainter values (where the GC observations become incomplete).

surface brightness in the Galactic Center field is produced by the young stars and the remainder by a bulge population similar to that in Baade's Window.

We have obtained spectra of more than 50 of the luminous stars that form the high luminosity extension of the Galactic Center luminosity function in Figure 1. Sample spectra are shown in Figure 2, where we compare one of the most luminous Baade's Window stars ($M_{Bol} \sim -6$) with Galactic Center stars of similar luminosity. It is noteworthy that the GC stars do not have the strong steam absorption near $2\mu m$ that identifies the Baade's Window star to be on the extreme asymptotic giant branch (AGB). As a result, the GC stars are likely to be relatively massive, $> 5\ M_\odot$, and hence no more than 100 million years old (Schaller et al. 1992; Meynet et al. 1994). These results show that a substantial episode of star formation, a starburst, has occurred around the GC within roughly the last 100 million years.

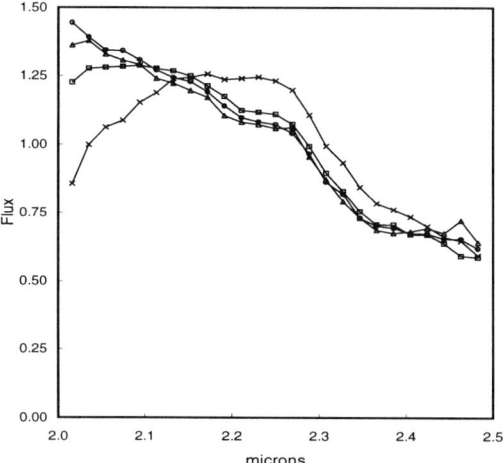

Figure 2. Comparison of Spectra of Galactic Center Stars with a Baade's Window Star of Similar Luminosity. The Baade's Window star, BMB 239, is indicated with x's; it is of $M_{Bol} = -6.0$. Although its CO absorption at $\lambda \geq 2.3\mu$m is similar to that of the Galactic Center stars, it has very strong steam absorption near 2.0μm. The GC stars shown for comparison are S22, open squares, $M_{Bol} = -7.1$; a star at 17:42:25.2, -28:59:25, open triangles, $M_{Bol} = -6.3$; and a star at 17:42:28.8, -28:57:30, filled circles, $M_{Bol} = -6.0$.

3. The Central Cluster of Young Stars

Our luminosity function is incomplete within 25" (1 pc) of SgrA* because of crowding. Even without correcting for the incompleteness, the central region contains a large number of stars with $M_{Bol} < -6$, suggesting that very recent star formation has occurred in this region. Tamblyn, Melia, & Rieke (1996, this conference) discuss the extremely luminous blue stars clustered in the central half parsec of the Galaxy. Even if minimal temperature constraints are imposed, these objects must be of very high mass, $M > 40 M_\odot$. They have been modeled by Navarro (1995), who deduces temperatures and luminosities, and hence masses, even higher than the minimum indicated by Tamblyn et al. Such massive stars must have formed in the last few million years.

Tamblyn et al. (1996) have attempted to account for these stars by conventional starburst models and with a conventional initial mass function (IMF). It appears to be very difficult to explain the concentration of extremely massive, evolved stars within such a small volume with these initial assumptions. A plausible explanation would be that the formation of extremely massive stars is strongly favored within the central parsec of the Milky Way, compared with conventional IMFs. For more details, see Tamblyn et al. (1996).

4. Dark Matter

Both gaseous and stellar motions have been used to probe the mass distribution in the GC. It has been difficult to demonstrate that the results consistently and convincingly demand the presence of a central massive dark object (MDO) (e.g., Phinney 1989). Recently, this situation has improved dramatically. The radial motions from [NeII] emission have been measured and modeled with increasing sophistication (Lacy, Achtermann, & Serabyn 1991); consistent measurements but with an alternate model have recently been reported with HI (Goss, Roberts, & Yusef-Zadeh 1996, this conference). The conclusion from both approaches is that a central dark mass of $2 - 3 \times 10^6 M_\odot$ is concentrated within the central 0.2 pc. Haller et al. (1996) have reported a detailed study of stellar radial velocities. Because many of the measured points include more than one star, the total number of velocities included in their study exceeds 200. They find excellent agreement with the conclusions from gas motions, obtaining a central mass of $1.5 - 2 \times 10^6 M_\odot$. This agreement of stellar and gaseous dynamics is confirmed by Genzel (1996, this conference) with a suite of individual stellar velocity measures. Because gaseous and stellar velocities have differing strengths and shortcomings for determining the mass distribution, the dual confirmation of results using both techniques represents a substantial increase in our confidence in the existence of a central massive dark object.

Kormendy & Richstone (1995) discuss the evidence for MDOs in the nuclei of eight galaxies, including the Milky Way. The mass deduced for the MDO in the Milky Way falls close to the general correlation they find between MDO and bulge masses. It is therefore likely that the object we detect in the Galactic Center is similar to those found in the other galaxies.

We observe the MDOs of the other galaxies with a resolution of only 1 to 10pc. Thus, as Kormendy and Richstone point out, only indirect arguments bear on whether the MDOs in these galaxies are truly massive black holes or are some other form of dark matter such as a compact cluster of stellar remnants. This question can be addressed much more effectively in the Galactic Center.

In Figure 3, we show the differential trend of M/L_K from the data of Haller et al. (1996). There appears to be an extended region of increased M/L within a radius of one parsec of the GC. This result is confirmed by Saha et al. (1996).

The increase in M/L is unlikely to result from a radially dependent change in stellar population. As discussed in Sections 2 and 3, with decreasing radius, progressively younger stellar populations are encountered, and hence M/L should decrease rather than increase. We can make this argument more quantitative as follows. We integrate the exponential fit to the bulge surface brightness of Kent, Dame, & Fazio (1991) to obtain the total K flux from the central 300 pc (radius) of the Galaxy. Comparing with the enclosed mass within this radius (Burton & Liszt 1978; Liszt & Burton 1978), we find $M/L_K = 1.0$. For comparison, McWilliam & Rich (1994) estimate for the stars in Baade's Window (at a radius of just over 300 pc) an average metallicity of $[Fe/H] = -.25$ and an age of ~ 8 GYr. Worthey's (1994) models for these parameters would predict $M/L_K = 1.1$, in excellent agreement with the measured value.

The data in Figure 1 allow a determination of the relative amount of K emission from the younger stars compared with that from the old population at a radius of about 4 pc from the GC. Since the young population extends in mass

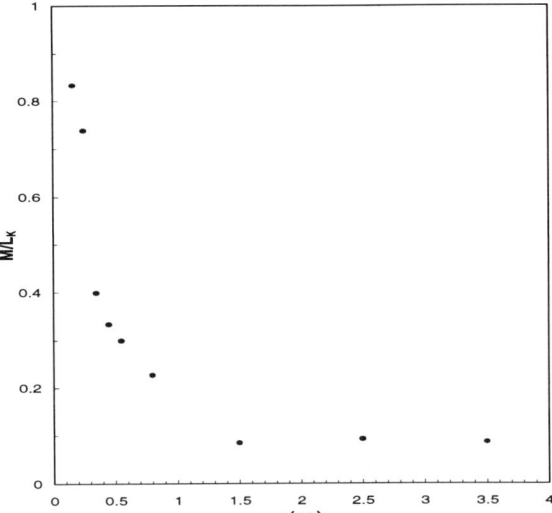

Figure 3. Trend of M/L_K with radius. The annular masses have been estimated by smoothing the measurements of Haller et al. (1996). K luminosities are estimated by assuming azimuthal symmetry and measuring the relevant surface brightness at all regions within the azimuthal zone that are not affected by localized extinction above the general level seen toward the Galactic Center.

up to $\sim 10 M_\odot$, its M/L_K will be much less than 1. Therefore, making use of the normalization derived in Figure 1, we estimate a net $M/L_K \sim 0.3$ at 4 pc radius. Stars which appear to be members of the young population are found in large numbers at smaller radii, indicating that the stellar M/L_K should not increase above 0.3. We omit the extremely young and luminous stars from this discussion, and they are not included in Figures 1 and 3.

We can also measure M/L_K directly. The mass in any annular region can be estimated by fitting a smooth curve through the data points in the enclosed mass diagram of Haller et al. (1996). The K surface brightness must be estimated in a manner that avoids the effects of the non-uniform extinction in the area. We have assumed that the K surface brightness is azimuthally symmetric and have estimated its value for each annular zone from measurements in regions without strong localized extinction (we have also avoided the large region of heavy extinction to the southeast). The measured value is then dereddened by an amount corresponding to $A_V = 31.8$ as derived for the region in Section 2. We find a nominal value of $M/L_K \sim 0.1$ for the annulus between 3 and 4 pc from Sgr A* and similar values for the adjacent annuli. A value of 0.2 would also be consistent within the errors. Given the uncertainties in the two methods, this result is in satisfactory agreement with that derived in the preceding paragraph.

Haller et al. (1996) make a rough calculation of the production of stellar remnants in the Galactic Center over the age of the Milky Way. They find that a significant portion of the mass in this region may be in the form of white dwarfs, too faint for detection. Their calculation includes only the remnants from the old, Baade's-Window-like population; the younger populations we see superimposed suggest that the rate of generation of remnants may be higher than they calculated. This remnant population could be expected to yield an extended dark mass distribution and appears to be the most likely explanation for the observed trend in M/L, particularly if dynamical evolution has helped concentrate remnants into the central parsec.

Alternately, the trend in Figure 3 would be removed if we have significantly underestimated the central point mass. The data could be explained by a point mass of $> 4 \times 10^6 M_\odot$ and little extended dark matter; however, in this case the innermost mass determinations from both stellar and gaseous dynamics would for some reason be systematically low by the same amount. Moreover, such an explanation would leave the puzzle of the disposition of the stellar remnants that should have formed over the life of the system.

Viewed on the physical scale we can achieve in observing other galactic nuclei, the extended dark component in the Galactic Center would appear as part of the massive central dark object. The total extended dark mass in the Galactic Center is comparable to that in the central 0.2 pc, which is plausibly associated with a central massive black hole. Hence, the example of the GC suggests that the MDOs detected in other galaxies may contain a significant component of stellar remnant mass. Since most of these cases have substantially larger MDOs than the GC, the fraction of mass in remnants may, however, be smaller than in the GC.

5. Conclusions

Many components have been identified in the Galactic Center, as follows:

1. The dominant mass component on a large scale is the bulge stellar population, as shown by the agreement in M/L within 300 pc and the value of this parameter appropriate for the sample of the bulge seen through Baade's Window.

2. The light ~ 4 pc from the nucleus is dominated by a younger stellar population, of age $\leq 10^8$ years.

3. Within the central parsec, there are extremely massive stars that can be no more than a few million years old.

4. There is a mass concentration of $2 \times 10^6 M_\odot$ within the central 0.2 pc that is plausibly associated with a massive black hole.

5. There is extended dark matter within the central 1 pc radius that may be due to stellar remnants from the massive stars that have formed in this region over the lifetime of the Galaxy; this explanation is particularly likely if dynamical evolution has tended to concentrate the remnants toward the center.

Acknowledgments. Data reported in this paper were obtained at Cerro Tololo Inter-American Observatory. We thank the conference organizers for their hospitality. Our travel was supported by the NSF, the University of Arizona, and the conference organizers.

References

Blum, R. D., Sellgren, K., & DePoy, D. L. 1996, this conference
Burton, W. B., & Liszt, H. S. 1978, ApJ 225, 815
Frogel, J. A., & Whitford, A. E. 1987, ApJ 320, 199
Genzel, R. 1996, this conference
Goss, W. M., Roberts, D. A., & Yusef-Zadeh, F. 1996, this conference
Haller, J. W., Rieke, M. J., Rieke, G. H., Tamblyn, P., Close, L., & Melia, F. 1996, ApJ 456, 194
Kent, S. M., Dame, T. M., & Fazio, G. G. 1991, ApJ 378, 131
Kormendy, J., & Richstone, D. 1995, ARA&A 33, 581
Lacy, J. H., Achtermann, J. M., & Serabyn, E. 1991, ApJ 380, 71
Liszt, H. S., & Burton, W. B. 1978, ApJ 226, 790
McWilliam, A., & Rich, M. 1994, ApJS 91, 749
Meynet, G., Maeder, A., Schaller, G. Schaerer, D., & Charbonnel, C. 1994 A&AS 103, 97
Najarro, P. 1995, Ph.D. thesis, Ludwig-Maximilian University, Munich
Phinney, E. S. 1989, in The Center of the Galaxy, ed. M. Morris (Dordrecht: Kluwer), p543
Rieke, M. J. 1987, in Nearly Normal Galaxies from the Planck Time to the Present, ed. S. M. Faber (Springer-Verlag: New York), p90
Rieke, M. J. 1993, in Back to the Galaxy, eds. S. Holt & F. Verter (AIP Press: New York), p37

Saha, P., Bicknell, G. V., & McGregor, P. J. 1996, preprint
Schaller, G., Schaerer, D., Meynet, G., & Maeder, A. 1992, A&AS 96, 269
Tamblyn, P., Melia, F., & Rieke, G. H. 1996, this conference
Tiede, G. P., Frogel, J. A., & Terndrup, D. M. 1995, AJ 110, 2788
Worthey, G. 1994, ApJS 95, 107

Comparison between the Galactic Center and Activity in NGC 4258 (M106)

E.M. Burbidge

Center for Astrophysics and Space Sciences and Department of Physics, University of California, San Diego, La Jolla, California 92093-0111, U.S.A.

Abstract. The active galactic nucleus of the H_2O megamaser galaxy NGC 4258 may display phenomena similar to the activity studied in the Galactic Center, on a greatly magnified scale because NGC 4258 has a much more active nucleus than that of our Galaxy. These phenomena in NGC 4258 are described, including evidence for directed outflow of ionized gas and charged particles from the nucleus, and the likely ejection of X-ray-emitting low-redshift QSOs. Preliminary data on the similar but more distant H_2O megamaser galaxy NGC 2639 is described.

1. Introduction

The observational data presented at this Workshop on the center of our Galaxy are extensive and detailed at all wavelengths that can be observed from our position close to the Galactic plane. I therefore thought it would be interesting to compare and contrast the relatively modest activity of our Galactic center with that in a nearby very active spiral galaxy, NGC 4258, for which interesting new data have become available. Powerful H_2O maser emission at 1.35cm wavelength (cf Claussen et al. 1984, Miyoshi et al. 1995, Greenhill et al. 1995a,b) has been observed from the nuclear region of that galaxy. Also Pietsch et al. (1994) have discovered two compact X-ray sources located equidistant along the minor axis of this galaxy, suggesting that these sources were ejected from the active nucleus, and my own observation (Burbidge 1995) has shown that these X-ray sources are low-redshift QSOs. I believe that the complex phenomena observed at radio, optical, and X-ray wavelengths in NGC 4258 are due to periodic explosive ejection of high-energy particles and ionized gas from the active nucleus. This suggestion is in line with earlier interpretations of the complex velocity field observed in the ionized gas of NGC 4258, and of its continuum radio structure.

The discovery that the two compact X-ray sources on the minor axis of NGC 4258 mapped by ROSAT are QSOs has been suggested to be "yet another isolated case" of an apparent connection between galaxies and QSOs at different redshifts, rather than indicating the existence of an intrinsic component of QSO redshifts. I will describe briefly another galaxy, NGC 2639. It also has an H_2O maser in its nucleus. It has also been mapped with the ROSAT PSPC instrumentation revealing again two compact X-ray sources on either side of the galaxy, aligned fairly closely along its minor axis. And again these X-ray blobs have turned out to be centered on low-redshift QSOs. Thus this is "yet another isolated case" of a connection between such galaxies with low redshifts and QSOs (Arp, Burbidge & Radecke 1996).

In §2 I discuss the early optical spectroscopic and radio work on NGC 4258 that led to the idea that mass ejection was taking place in bursts from the nucleus. In §3 I describe the H_2O maser observations which have led to a model of a rotating ring of masering blobs circulating about a massive black hole, and suggest an alternative model. In §4 I describe the ROSAT identification of the compact X-ray blobs as blue stellar objects, and the observations which revealed these to be QSOs, followed by a description of the work underway on the X-ray sources near NGC 2639.

In the concluding section, §5, I relate these observations to the work by Hoyle, Burbidge & Narlikar (1993,1994a,b) in which they present their theory of the physics of matter in a very strong gravitational field, leading to massive eruptive phenomena from galactic nuclei.

2. Optical Spectroscopy and Radio Continuum in NGC 4258

The strong emission lines in the early Mount Wilson spectra of the nucleus led to this galaxy being included in Seyfert's original list, but it has more recently been classified as a LINER galaxy (low-energy nuclear emission-line regions). The long prominent knotty spiral arms, well defined by luminous HII regions, and its large angular size, made it a good candidate for long-slit rotation curve measurement and mass determination (Burbidge, Burbidge & Prendergast 1963; BBP hereafter). $H\alpha$ and [NII] emission-line velocities were measured by BBP on the major axis (P.A. 157°) and in P.A. 126° and these velocities clearly demonstrated the existence of non-circular motions (see Figures 2 and 5 of BBP). Earlier, Courtès & Cruvellier (1961) noted peculiar ("anomalous") arms visible in $H\alpha$ emission without the normal optical stellar continuum expected from stars in spiral arms, and at a considerable angle to the prominent spiral arms of the galaxy (i.e. in P.A. 125°).

Measurement of the radio continuum at 1415 MHz at Westerbork (van der Kruit, Oort & Mathewson 1972) showed smooth curved ridges of continuum emission which differ in position, shape, and continuity from the optical arms, but which inside 5 kpc from the center coincide precisely with the filamentary $H\alpha$ arms discovered by Courtès & Cruvellier. Their study of the optical and 1415 MHz data led them to the interpretation that there had been ejection from the nucleus of clouds in two opposite directions about 18 million years ago, at velocities ranging from about 800 to 1600 km s^{-1}.

Rubin & Graham (1990) using echelle spectrograph observations with the Kitt Peak 4-m telescope studied the velocity field in P.A. 150°, 125°, and 60° within ±100 arcsec of the center. Their Fig 1, from an image published by Chincarini & Walker (1967), shows well the HII and dust arms of the galaxy and their slit orientations, and their Fig 2 shows the complexity of the velocity field revealed at the resolution obtainable with the KPNO echelle spectrograph. They concluded that their data "support a model in which the anomalous arms are produced by two double-sided jets emanating from a low-level active nucleus". In P.A. 125°, they suggested that "we may be viewing a galactic chimney...where hundreds of supernovae produce a superbubble." The systemic velocity V_o is 465 km s^{-1}, and Rubin & Graham measured an extremely complex multi-valued velocity distribution in P.A. 125° (the direction of the "anomalous arms"). Ford et al. (1986) using narrow-band Hα images had confirmed the presence of continuum-free emission-line arms which coincide with the radio arms, and whose morphology suggested the presence of two double-sided jets that on the SE side fork into two arms which appear to braid or wrap around one another. ROSAT observations of the soft X-ray emission led Cecil et al. (1995) to conclude that this is produced by hot shocked gas along the jets (their Fig 2c shows the feathery appearance of the radio arms at 6 cm).

3. The H$_2$O Masers in NGC 4258

H$_2$O maser emission at λ=1.35 cm was first reported by Claussen et al. (1984) at velocities within $\pm$150km s^{-1} of the systemic velocity. Nakai et al. (1993) discovered maser emission at velocities offset by $\pm$750 to 1000 km s^{-1}, bracketing the emission at the systemic velocity. In our Galaxy H$_2$O maser activity is associated with regions of rapid star formation. The much stronger emission from NGC 4258 and a few other AGN galaxies has led to these being called megamasers.

Greenhill et al. (1995a,b) measured a linear gradient of velocity of the emitting blobs with position, and VLBI imaging of these by Miyoshi et al. (1995) showed a run of velocities with position that has led to the interpretation that the masering blobs lie in a rotating disk of about 0.5 pc outer diameter, with rotation speeds of up to $\sim$1100 km s^{-1}. These various groups of authors have modeled the observations as being produced by rotation of a thin disk seen edge-on about a central black hole of mass $\sim 3.6 \times 10^7 M_\odot$.

How the masers are excited is not understood. The picture is complicated by the measurement of a linear drift of 9.5 km s^{-1} yr^{-1} in those features with velocities $\pm$150 km s^{-1} of the systemic velocity, yet the overall velocity range of the complex remains stationary over time (Greenhill et al. 1995a). Haschick et al. (1994) concluded that all the masering clouds are being accelerated systematically at a rate slightly increasing with velocity.

Another puzzling feature of the H$_2$O maser observations is that the emission occurs in clumps or blobs, which seem to lie at regular distance intervals. An interesting suggestions has been put forward by Maoz (1995) – that the blobs lie in a spiral structure in the circum-nuclear disk, produced by the instability of such a configuration to radial perturbations.

Despite the eagerness with which the astrophysical world has accepted the H$_2$O maser observations as "proof" of the existence of a central black hole, it must be remembered that the Schwarzschild radius of a $3 \times 10^7 M_\odot$ black hole is only 3×10^{12} cm, or 10^{-6} pc. There is no doubt that NGC 4258 possesses a large central mass; it is indeed a massive spiral galaxy as BBP showed. Gravitation produces a central concentration; massive star formation and rapid evolution produces elements from C,N,O upward, depending on the stellar masses and types of supernovae produced. The Maxwell-Boltzmann distribution of stellar velocities will lead to some stars having velocities exceeding the escape velocity, and as they leave such a dense innermost star cluster, like that described during this Workshop at the center of our Galaxy, the remainder will contract further, leading eventually to an approach to the canonical black hole.

It is here that I part company with the traditional view. There is no physical theory of the behavior of matter under very strong gravitational forces. Ideas such as those involving multidimensional space may be relevant. At the moment, I am attracted by the ideas of Hoyle, Burbidge & Narlikar (1993, 1994a,b), to which I return in §5, and I have been led by the observations described in §4 to return to the model of explosive ejection described in §2.

4. Observations of ROSAT X-ray Sources Near AGN Spiral Galaxies

4.1. NGC 4258

Figure 1 shows the ROSAT PSPC images of two compact X-ray sources mapped by Pietsch et al. (1994) (their numbers 8 and 26). The figure is a reproduction of their Fig 2, with the identification numbers indicated. Following their identification of a ~ 20 mag blue stellar object at each of these sources, and their note that these are outstanding in terms of their X-ray brightness and symmetrical and equidistant position with respect to the nucleus of NGC 4258, spectra were obtained with the Kast double spectrograph on the 3-m Shane telescope at Lick Observatory. The results (Burbidge 1995) were that both objects are QSOs with redshifts 0.398 for the W object, no. 8, and 0.654 for the E object, no. 26, as indicated on Fig 1.

As Pietsch et al. (1994) pointed out, the line connecting the two X-ray sources touches the ends of the anomalous arms described in §2 for which Courtès et al. (1993) presented further imaging data. The similarity of the optical spectra of the two QSOs, the details of their discovery by the X-ray astronomers, and in particular their location with respect to the nucleus and peculiar structure of this AGN galaxy led me to conclude that the QSO redshifts must be made up of shifts due to the velocities of ejection and the intrinsic component, i.e.

$$(1 + z_{obs}) = (1 + z_{doppler})(1 + z_{intrinsic}).$$

If they were indeed ejected from NGC 4258, they must lie at its distance of $D = 7$ Mpc, and have absolute magnitudes ~ -10. I think the data and my interpretation were unwelcome to the majority of the astrophysics community, but they caused me to reach a watershed in thinking about the relationship between galaxies and QSOs and the possible physics of non-cosmological redshifts.

4.2. NGC 2639

Despite the difficulty in obtaining observing time for a controversial project, it was obvious that the problem should be pursued. The ROSAT group had mapped another AGN spiral galaxy with a Seyfert-like nuclear spectrum, also described as a LINER, NGC 2639. The distance of this galaxy is ~ 7 times that of NGC 4258, so there is less detail available about its structure. However, Arp (1980) had already drawn attention to the fact that several QSOs lie fairly close to NGC 2639 in the plane of the sky, and he presented spectra of these. Braatz et al. 1994) discovered that NGC 2639 is another H_2O megamaser galaxy whose maser emission comes from a very compact region.

Wilson et al. (1995), monitoring the H_2O emission with the Max-Planck-Institute 100m radio telescope at Effelsberg, made the exciting discovery that the nuclear region resembles that of NGC 4258 in another respect: the brightest 1.34 cm H_2O feature is drifting redward, with an acceleration similar to that found for the near-systemic features in NGC 4258.

The ROSAT group provided data from their mapping of the field around NGC 2639, together with their identification of two strong compact X-ray sources with blue stellar objects, at positions close to the minor axis of the galaxy, but further away from the nucleus than was the case with NGC 4258. It was exciting that Arp (1980) had already obtained spectra of one of these, among his group of QSOs in the field, and found its redshift to be $z \sim 0.30$.

Spectra were obtained in January, 1996, for both the ROSAT compact sources, nos. 37 and 5, with the Kast double spectrograph on the 3-m Shane telescope at Lick Observatory. The results are that the objects are QSOs with redshifts $z = 0.306$ for no. 37 and $z = 0.325$ for no. 5. It was exciting to confirm Arp's data on ROSAT #37, and to discover that the other compact X-ray source is also a QSO at almost the same redshift. Partially reduced data of the red end of the spectra are shown in Figure 2. The emission lines of $H\beta$, [OIII], and $H\alpha$ clearly yield almost identical redshifts, and MgIIλ2800 appears in the blue spectra of each. Although the spectra are only partially reduced (e.g. the night-sky A and B bands at $\lambda 7600, 6900$ Å have not been removed), the similarity of the two spectra and their resemblance to NGC 4258 ROSAT #8 is remarkable. Despite the greater distance of these objects from the nucleus of NGC 2639, and the lack of detailed information on the nucleus of NGC 2639 itself, it appears that

- both spiral galaxies have AGN LINER nuclei,

- both have nuclear megamaser H_2O emission,

- acceleration outward of the strongest H_2O features occurs in both galaxies.

A paper on this object (Arp, Burbidge & Radecke 1996) will be submitted for publication as soon as the QSO spectra are fully reduced.

5. Discussion, Conclusions

I have referred to the center of our Galaxy as the site of fairly modest activity as compared with that in galaxies such as NGC 4258 and NGC 2639, but the detail afforded because the Galactic Center is 800 times closer than NGC 4258 can give clues to what is taking place in the "megamaser" galaxies. Central star clusters must exist, analogous to what the IR observations of the Galactic Center have revealed. Dynamical studies of the Galactic Center stellar distribution and stellar motions should be pointers to what happens in megamaser galaxies.

However, the step from this comparison to the interpretation suggested for the existence of X-ray emitting QSOs ejected from galactic nuclei requires a new interpretation of the effect on matter of a very strong gravitational field. The work by Hoyle, Burbidge & Narlikar on a negative energy field (the C field analogous to the field postulated by inflationary big-bang cosmologies) suggests that matter is created in the galactic nucleus in the form of Planck particles, whose breakup into baryons and leptons also leads to explosive reactions that produce the outpourings of relativistic particles and the outward flow of matter that are so characteristic of active galactic nuclei, including the standard double-lobed radio galaxies and their narrowly beamed jets.

The QSOs near to NGC 4258, if at the distance of NGC 4258, are no brighter than visual magnitude -10. Only $\sim 10^2 M_\odot$ of hot gas is required to give rise to the optical spectra though much more mass is likely to be present. Probably the gas is bound in a gravitational potential well due to much denser matter. A suggestion has been made that the pair by NGC 4258 might be two clusters of gravitational droplets, i.e. near-black-holes of comparatively small mass. In this picture, each cluster forms a well with virial motions of ~ 1000 km s^{-1}, a situation analogous to an early model of QSOs by Fowler and Hoyle when these objects were newly discovered. With a well size $\sim 10^{17}$cm, virial motions ~ 1000 km s^{-1}, and a mass of gas producing the broad emission lines of $\sim 10^2 M_\odot$, we have

$$GM = \frac{1}{2}V^2 R = \frac{1}{2}(10^8)^2 \times 10^{17}$$

i.e. $M \approx 3 \times 10^6 M_\odot$.

The breakup of a larger near-black-hole into smaller pieces is contrary to Hawking's thermodynamics of black holes, according to which black holes are accumulative, because Hawking's fields always have positive energy. With the C-field postulate, i.e. a negative energy field, similar results can be expected to hold, but in a *reversed* time sense.

There is much work to be done before the ideas put forward in this section will be seriously considered by those who accept the standard black hole paradigm. An obvious need is for study of the X-ray fields around other AGN galaxies, and optical imaging and spectroscopy of compact X-ray sources.

In conclusion, I thank Fred Hoyle and Geoffrey Burbidge for much help with the final section of this paper. Partial support for the reduction of the observational data described in §4 has been provided by NASA grant NAG5-1630, which I gratefully acknowledge.

References

Arp, H. 1980, ApJ, 236, 63
Arp, H.C., Burbidge, E.M. & Radecke, J. 1996, (in preparation)
Braatz, J.A., Wilson, A.S. & Henkel, C. 1994, ApJ, 437, L99
Burbidge, E.M. 1995, A&A, 298, L1
Burbidge, E.M., Burbidge, G.R. & Prendergast, K.H. 1963, ApJ, 138, 375
Cecil, G., Wilson, A.S., & de Pree, C. 1995, ApJ, 440, 181
Chincarini, G. & Walker, M.F. 1967, ApJ, 149, 487
Claussen, M.J., Heiligman, G.M., & Lo, K.Y. 1984, Nature, 310, 298
Courtès, G. & Cruvellier, P. 1961, Compt. Rend. Acad. Sci. Paris, 253, 218
Courtès, G., Petit, H., & Hua, C.T. 1993, A&A, 268, 419

Ford, H.C., Dahari, O., Jacoby, G.H., Crane, P.C. & Ciardullo, R. 1986, ApJ, 311, L7
Greenhill, L.J., Henkel, C., Becker, R., Wilson, T.L. & Wouterloot, J.G.A. 1995a, A&A, 304, 21
Greenhill, L.J., Jiang, D.R., Moran, J.M., Reid, M.J., Lo, K.Y., & Claussen, M.J. 1995b, ApJ, 440, 619
Haschick, A.D., Baan, W.A. & Peng, E.W. 1994, ApJ, 437, L35
Hoyle, F., Burbidge, G. & Narlikar, J.V. 1993, ApJ, 410, 437
Hoyle, F., Burbidge, G. & Narlikar, J.V. 1994a, MNRAS, 267, 1607
Hoyle, F., Burbidge, G. & Narlikar, J.V. 1994b, A&A, 289, 720
Hoyle, F., Burbidge, G. & Narlikar, J.V. 1995, proc. Roy. Soc. A, 448, 191
Maoz, E. 1995, ApJ, 455, L131
Miyoshi, M., Moran, J., Herrnstein, J., Greenhill, L., Nakai, N., Diamond, P. & Inoue, M. 1995, Nature, 373, 127
Nakai, N., Inoue, M. & Miyoshi, M. 1993, Nature, 361, 45
Pietsch, W., Vogler, A., Kahabka, P., Jain, A. & Klein, V. 1994, A&A, 284, 386
Rubin, V.C. & Graham, J.A. 1990, ApJ, 362, L5
van der kruit, P.C., Oort, J.H. & Mathewson, D.S. 1972, A&A, 21, 169
Wilson, A.S., Braatz, J.A. & Henkel, C. 1995, ApJ, 455, L127

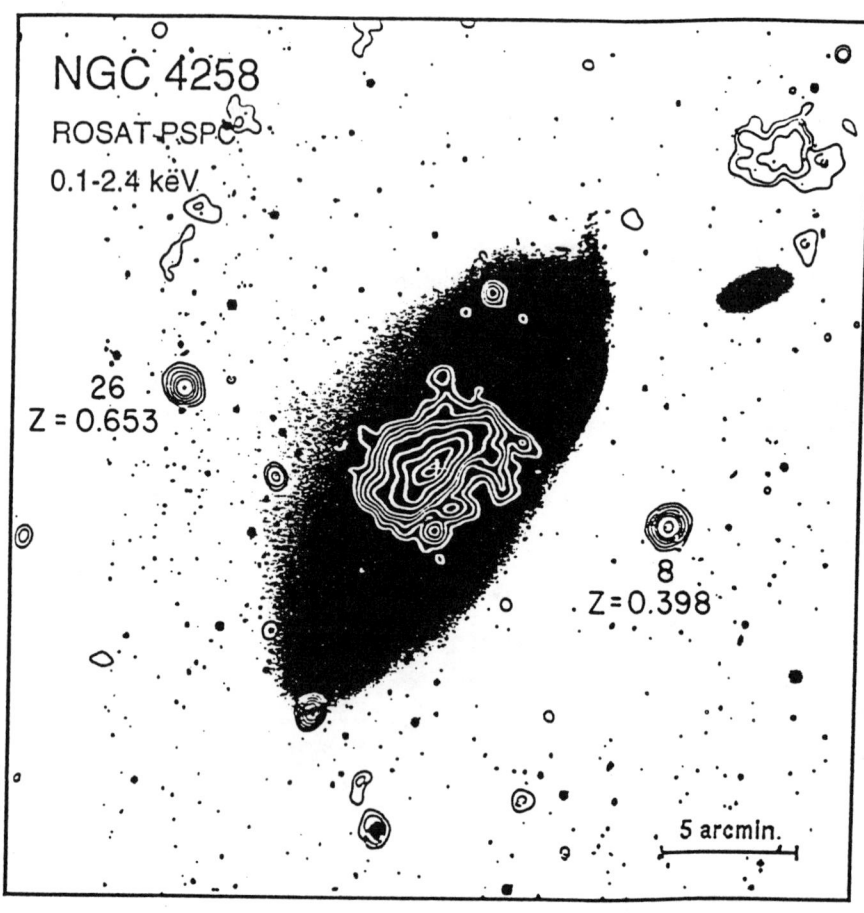

Figure 1. Reproduction of Fig 2 of Pietsch et al. (1994) showing locations of the compact X-ray sources nos. 8 and 26 relative to NGC 4258, as measured by ROSAT. Redshifts of the QSOs identified with the X-ray sources are shown.

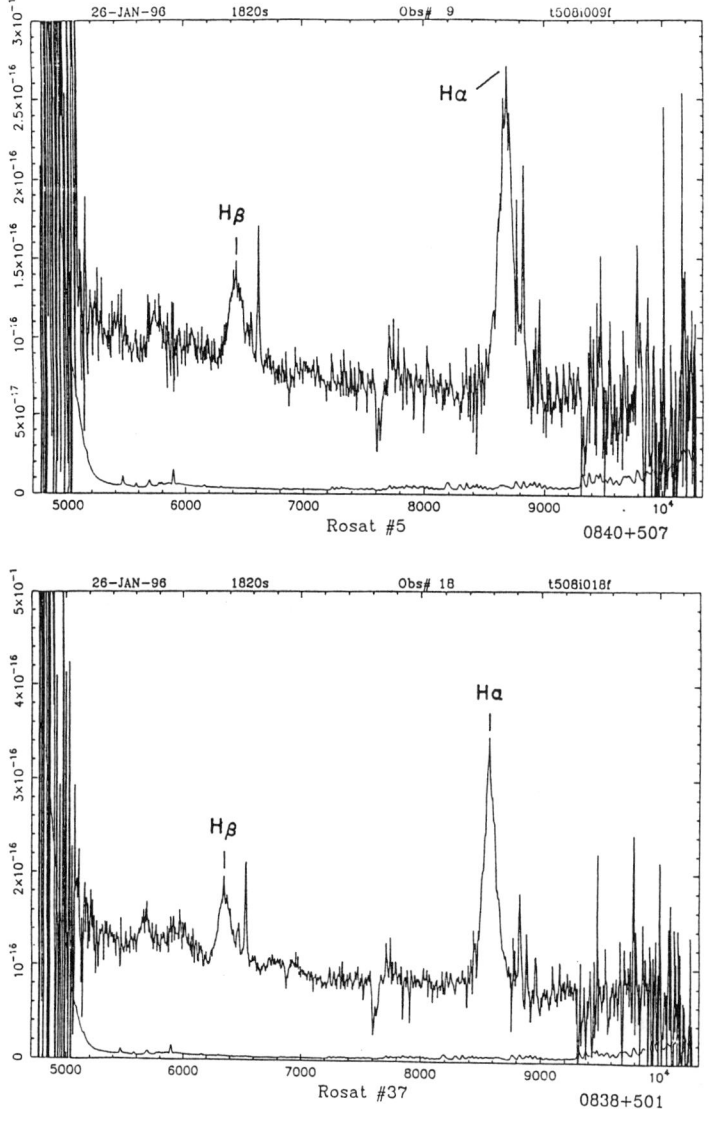

Figure 2. Preliminary (partially reduced) spectroscopic data obtained at Lick Observatory on the two QSOs identified as compact X-ray sources nos. 37 and 5 in the field of NGC 2639. Night-sky features (particularly the A and B absorption at $\lambda 6900$ and $\lambda 7600$) have not been removed. The redshifts are $z = 0.306$ for no. 37 and $z = 0.325$ for no. 5. The ordinate on each is the (preliminary) value of the flux (erg cm^{-1} sec^{-2} Å^{-1}).

The Role of the Galactic Center on our Understanding of Violent Activity in the Nuclei of Galaxies

G. Burbidge

Center for Astrophysics and Space Sciences and Department of Physics, University of California, San Diego, La Jolla, California 92093-0111, U.S.A.

F. Hoyle

102 Admirals Walk, Bournemouth BH2 5HF Dorset, England

Abstract. Possible origins of violent outbursts in galaxies are discussed. A critique of the black hole accretion disk concept is given. It is concluded that the energetic events taking place at the center of our Galaxy can largely be ascribed to stellar evolutionary processes as has been proposed by Terlevich and his colleagues. However, the weak radio source Sgr A* is probably due to accretion on to a near-black hole. This strongly suggest that the realistic near-black hole accretion disk mechanism is much less efficient than has generally been asserted, so that it cannot be responsible for violent energy release in galactic centers more powerful than our own Galactic center.

1. Introduction

There is a large body of evidence from radio, infrared, optical, x-ray and γ-ray observations that large amounts of matter and energy are being ejected from the nuclear regions of many galaxies. These phenomena are frequently called active galactic nuclei (AGN) even when an underlying galaxy of stars is not detected. This evidence, which first began to appear from optical and radio observations in the 1960s (cf Burbidge, Burbidge & Sandage 1963) shows that very powerful mechanisms are at work which can generate large fluxes of non-thermal photons over a very wide frequency range from radio waves to γ-rays, relativistic particles, high velocity clouds of plasma and, most probably the ejection of coherent objects.

There is also very strong circumstantial evidence and some observational evidence suggesting that the energy is generated in regions with sizes of order 10^{15}cm or less.

The ultimate energy sources for all of this activity remain uncertain. Three possible sources have been described (cf Burbidge 1970). They are first that the energy is gravitational in origin and due to collapse. If this is correct it is required that large amounts of energy are released with high efficiency, and this can only take place if the release is close to the Schwarzschild radius (Hoyle & Fowler 1963a,b, 1964; Hoyle, Fowler, Burbidge & Burbidge 1964). Despite the fundamental difficulty associated with this process, this class of model has been developed by Rees & Blandford (cf. Rees 1984) into the classical black hole accretion disk idea, and in this it is argued that gravitational energy *can* be extracted with an efficiency ~8% into the various energy modes in which it is seen.

A second class of model is associated with the idea that very rapid stellar evolution leading to multiple supernova outbursts in a nuclear region is able to produce the required amounts of energy. This idea was first discussed by Shklovsky (1960), Burbidge (1961) and Colgate (1967) in various forms. Another version of this general idea has been extensively worked on more recently by Terlevich and his colleagues (cf Terlevich and Melnick 1985,1988; Terlevich et al. 1992). A general difficulty with the type of model is that it does not appear to be possible to explain ejection to great distances which is required in the case of the powerful radio sources, or to explain the very large energies which are indicated in many of those sources.

In the third class of model it is proposed that the centers of galaxies are places where matter is literally being created (accompanied by a negative energy sea). In the early discussions of active nuclei such centers were described as white holes (cf Ne'eman 1964). This view that the centers of galaxies are places from which matter and energy were created was proposed long ago by Jeans (1929) in his discussion of the origin of spiral structure. Ambartsumian (1958, 1965) proposed that groups of galaxies were systems of positive energy and that all of the galaxies have come out of such centers. Hoyle & Narlikar (1964) have developed a field theory of such creation processes (the C-Field theory) using a formulation involving the adding of more terms to the standard Einstein-Hilbert action to represent the phenomenon of the creation of matter. This latter work has been developed most recently to provide an alternative to the hot big bang cosmological model which has been called the quasi-steady state cosmology (Hoyle, Burbidge & Narlikar 1993; 1994a,b; 1995). In the cosmogony associated with this cosmology all of the creation takes place in galactic nuclei. The particles which give rise to the mass and energy of the created mass are so-called Planck particles which have energies of 10^{19} Gev, and which decay into the strange particles and hence into protons, neutrons, electrons, positrons, photons and neutrinos. This is how energy appears.

We have given this lengthy introduction to stress the fact that the first of these mechanisms, the black hole-accretion disk paradigm, is not the only possible explanation of what we see in violent events in galactic nuclei. However, the observational community tends to try and interpret all of the data in terms of that model, rather than using the observations to *test* the model. Thus the popularity of the model is continuously growing although it may not be correct. It is closely related in many people's minds with the existence of massive black holes in the nuclei of galaxies. However, there is an important point to be made here:

In each of the "theories" it is required that there be a large mass concentration at the center. In other words, observational evidence that such mass concentrations are present does not support the black hole-accretion disk model alone.

Also, since in at least the black hole and creation models, the primary energy is generated at distances very close to the Schwarzschild radius $\approx 3 \times 10^5 \left(\frac{M}{M_\odot}\right)$ cm, or $10^{-13} \left(\frac{M}{M_\odot}\right)$ pc, even for a very massive central object $\sim 10^{11} M_\odot$ the real scale of interest is 10^{-2} pc so that even at a distance of 10 Mpc, where $1'' \sim 50$ pc, the angular scale of importance is 2×10^{-4} arc second. No direct measurements are therefore possible.

Thus while there are indications that large mass concentrations are present in a number of nearby galaxies (cf Kormendy & Richstone 1995) and smaller mass concentrations in nearby galaxies such as M81, and in our own Galaxy, there is no real evidence for black holes – nor in principle can there be.

Before discussing the Galactic Center and the way we view it, we look in more detail at the "black hole" problem.

2. A Critique of the Black Hole Concept

It is our contention that there are no black holes, nor can there be. Black holes are states of mind, not observable entities. Perhaps the mass aggregates that are commonly referred to as black holes will become so in the infinite future. But who has ever verified this prediction? The common use of the term black hole is indicative of an unhealthy state of affairs in which the state of mind has become more real to most people than the world of observation, which is the only real world. In the real world, as we observe it, there can be near-black holes, but not black holes. A near-black hole has a proper time barrier between its inside and outside, such that escaping signals separated by a proper time interval s on the interior are separated by γs on the outside, with γ a gravitational barrier factor ($\gamma \gg 1$). For γ large enough, events separated only by seconds, or a few tenths of seconds in the interior, can be expanded to separations of millions of years or more in the exterior.

The black hole-accretion disk model has events leading to explosive outbursts occurring all in the exterior, and so not subject to the factor γ. It is therefore a problem for these models to understand how explosive outbursts can be randomly spaced from each other by large intervals of time; the evidence suggests they are spaced by intervals of the order of millions of years. Thus the characteristic time for the separation of events across the interior of a near-black hole of mass $M_\odot$ is only $\sim 10^{-5} M_\odot$ seconds. But for events that occur on the inside of a near black hole the time separation judged by an external observer is $\sim 10^{-5} M_\odot \gamma$ seconds. With $M \approx 10^7 M_\odot$ and a time separation of a million years in the exterior, the requirement is for $\gamma \sim 3 \times 10^{11}$. Certainly a near-black hole but not a strict black hole.

Thus the over-riding mistake in the popular view of the centers of galaxies is to place their energy outbursts on the outsides of the sources of strong gravitation, instead of in the interiors. And this obvious mistake comes from loose thinking on mass concentrations themselves, whereby a theoretical concept is given precedence over the real universe. For, of course, if one errs by thinking about strict black holes, there can be no escape of energy from out of their interiors.

The factor γ can be viewed also as a redshift effect, a gravitational redshift. So as well as producing a time dilatation by γ it produces an energy reduction by γ. Whence an outburst, as is the case of some radiogalaxies, involving energies upwards of 10^{60} erg in the exterior seem at first sight as if they would translate back in the interiors into energies in excess of 10^{70} erg.

This is an impossible requirement, since an energy source of 10^{70} ergs is equivalent to a mass $\sim 10^{16} M_\odot$, far higher than any mass aggregation at the center of any galaxy. The resolution of this apparent paradox lies in the effect of energy escape on the barrier factor γ. An energy escape by the gravitational mass M would remove γ entirely ($c = 1$), while an escape by some fraction of M would remove a corresponding fraction of γ.

Hoyle and Narlikar (1964) pointed out that the existence of a negative energy field avoids the unpalatable need for the Universe to be created with a large excess of positive energy. It similarly permits creation events to take place inside objects of the kind we are discussing here. The essential point is that with the negative and positive energy forms compensating each other there is no new creation of energy. The escape outwards of the positive component is impeded by physical interactions whereas the negative component is impeded by nothing except the geometry of spacetime. And the extent to which the negative escape exceeds the positive, decides the mass concentration which becomes established. Thus the large masses at the centers of galaxies arise, not from matter falling together, as is usually supposed – without much understanding of the process ever emerging – but from a creation event in which there is a greater escape of negative energy than of positive.

It is noteworthy that the total magnitudes of negative and positive taken separately will in general be much larger than their difference. The latter may be no more than $\sim 10^7 M_\odot$, but the magnitudes of negative or positive taken separately can be very much greater than $10^7 M_\odot$. Large enough perhaps to provide for the mass of a whole galaxy, a situation which yields the beginning of an understanding of the mysterious phenomenon in which the position angle of the jet of M87 points directly to the galaxy M84 identified with NGC 4374 (Wade 1960) as if the entire galaxy had been expelled from the nucleus of M87. This understanding scarcely seems possible without the existence of the negative energy field.

Hoyle and Narlikar suggested that creation events are initiated in an interaction of negative energy quanta with the spacetime structure, the cross-section in appropriate units for the process being the so-called gravitational constant G. This will occur if the magnitude of the quantum energy is above a threshold of the order of the Planck mass (10^{19} Gev). The process, once started, takes the form of an induced discharge, which eventually becomes inhibited as negative escape exceeds positive, with a high value of γ becoming established. Then the quanta are mostly bent back into the object, being reflected at the barrier, to re-cross the object back and forth, until some among them move close to the center where they once again surmount the energy threshold, and so generate a further creation outburst. Thus within the object a sequence of creation events, separated in proper time by $\sim 10^{-5} M_\odot$ seconds, is expected. To the extent that such events lead to escape out of the object into the exterior, the time separation in the exterior is $10^{-5} \gamma M_\odot$ seconds, $\sim 10^6$ years for γ large as stated above.

Material at $\sim$ 10kpc from the center of a typical spiral galaxy possesses $\sim 10^6$ times more angular momentum per unit mass than material at the outer boundary of a near-black hole of mass $\sim 10^7 M_\odot$, even though the latter is rotating at the speed of light. This sets a problem if one seeks to argue that a central near-black hole is formed by the infall of galactic material, the problem being how the material gets rid of its angular momentum so comprehensively. This problem has never been solved satisfactorily. And if there were a process that rid the material of 999,999 parts in a million of its initial angular momentum, why not also the loss of the final one part in a million? This casts doubt on the rotating disk assumed in the black hole-accretion disk model.

The problem becomes inverted if galactic material is taken to emerge from a near-black hole. How in that case does it acquire so much angular momentum, sufficient to explain the orbital rotation of a spiral galaxy? Only if the large orbital angular momentum does not come from the central outburst, but through a division of such material into a number of extensive clouds, only a fraction of which is retained. And then through couples exerted gravitationally by expelled masses on retained masses, or by subsequent dynamical motions and collisions.

In support of the latter picture, it can be argued that in the case of giant elliptical galaxies, which display the most persuasive observational evidence of emission of material from a central object – strong radio sources – the tendency is for such galaxies to be globular, without any large external angular momentum being apparent.

3. Observational Evidence for Violent Activity

Now how is all of this related to the conditions of the center of our Galaxy? While it is much closer than any other galactic centers, it is very different from even the center of the nearest active galaxies such as M 81, NGC 4258, and NGC 5128 (Centaurus A).

The real difference lies in the fact that the center of our Galaxy is much less active than nearby galaxies, though it is hard to make direct comparisons because it is only possible to discuss the basic physics if we compare the activity in the nuclei of different galaxies using the same *linear* scales.

The evidence for violent activity in galaxies in various forms is shown in Table 1. There we have tried to indicate which processes may be due to stellar evolutionary processes, and which processes require that the energy producing mechanism involves a massive central object.

The general idea which has been adopted by most of those working in the field is to agree that there is continuity in activity all the way from the weakest sources (our own Galactic center) up to the most energetic, and then to assume that the same kind of machine is responsible in all cases. Since the view that is most popular is that a massive black hole is a necessary ingredient, searches for evidence of the existence of such objects and claims that they have been successful, have all led to the view that the paradigm is correct.

We believe that a different approach is in order. There are really two separate issues here. The first is concerned with determining whether or not large mass concentrations are present in the nuclei of galaxies.

Part 4. Late Papers

Mid-Infrared Emission and Luminous Sources in the Central Parsec

Dan Gezari

NASA/Goddard Space Flight Center, Code 685 Greenbelt, MD 20771, U.S.A.

1. Introduction

The Galactic Center contains a complex of mid-infrared (5 - 30 micron) sources so extraordinary that it would be safe to say it has no equal in the Galaxy. Sgr A West is the largest, brightest, and most luminous extended system of infrared sources observed (Figure 1). It is even more remarkable when one realizes that strong mid-infrared sources are rare in the Galaxy. While at least a few (or as many as hundreds of) 2 micron sources can be imaged in any randomly chosen 1 arcminute patch of sky, one could observe a thousand such random fields with a 10 micron camera and not detect a source in any one of the images.

The 2 micron image of the the central 30 arcmins of the Galaxy shows tens of thousands of stellar sources, many of which comprise the star central cluster. However, we can not detect these stars at 10 micron with ground-based cameras. Their photospheric emission is simply too weak, by orders of magnitude, compared to the thermal background noise level from the warm telescope and sky. Only ISO (and eventually SIRTF) have the sensitivity for a significant improvement in this situation, albeit with limited spatial resolution. A dramatic illustration of this is our 12.4 micron image of the Ney-Allen Nebula in Orion, a bright mid-infrared source heated by the Trapezium stars. These four O stars are 20 times closer than the Galactic Center, yet we do not detect their photospheric emission directly, only the warm dust surrounding them and heated by them. Another remarkable system of bright mid-infrared objects, the "Quint" sources, are also quite unique. Located roughly at the distance of the Galactic Center, the five bright point sources (actually, there are at least seven) lie within a single 30 arcsec field-of-view. Finding that many bright point sources situated so closely is a truly rare occurrence in the mid-infrared sky. The mid-infrared Quint sources bear no resemblance at all to the corresponding "Pistol" radio continuum source, which shows only the faintest hint of any structure which could be correlated with features in the mid-IR image.

In contrast, the radio and mid-infrared emission distributions in the Galactic Center Sgr A West complex are very similar, at first glance appearing nearly identical (Figure 2). The basic features of the 12.4 micron image (the northern arm, the east-west bar, the compact IRS sources) are all nearly the same in the 2-cm VLA image by Yusef-Zadeh (1989). The strong correlation suggests that the dust and gas are well- mixed within the whole central parsec, and the radiation which heats the dust and ionizes the gas may have a common origin (Gezari and Yusef - Zadeh 1990). Yet there are puzzling and potentially very important differences in the two distributions, particularly for the compact sources (Gezari 1992), which are discussed further below.

It should be noted that the very strong similarity in the radio and infrared emission from Sgr A West is another quite a rare characteristic of typical galactic mid-IR sources with significant associated radio emission. Generally the radio and mid-infrared emission are not obviously alike. And many mid-infrared sources do not have a strong radio counterpart. On a larger scale, the diffuse mid-infrared emission corresponding to the the arcminute sized spiral arm-like radio structures is proportionally much weaker and less clearly defined than the radio features. The arcs can only barely be inferred from noisy mid-infrared structure in the large mosiac image (Figure 1a).

The total infrared luminosity of the inner 2 parsecs is about 10^7 Lsun (Low et al.1969, Becklin, Gatley and Werner 1982). A cluster of recently formed O stars could account for this luminosity, and would be consistent with observations which show that the central parsec of the Galaxy is conspicuously devoid of molecular gas (Gusten et al. 1987) surrounding the infrared emission region. Excellent reviews summarizing the extensive infrared, molecular line, and radio continuum observations of this region have been presented by Brown and Liszt (1984), and Genzel and Townes (1987).

2. Mid-Infrared Observations

Our array camera system uses a 58 x 62 pixel gallium doped silicon (Si:Ga) direct readout (DRO) photoconductor detector array sensitive from 5 - 17 microns. A complete description of the array camera system is presented by Gezari et al. (1992). A set of array camera images at eight wavelengths between 4.8 and 20.0 microns has been obtained for the central 15 arcsec array field of view (24 arcsec = 1 parsec at 8.5 kpc), each with typically 5 min of integration time. A large-scale 12.4 micron mosaic image of the Galactic Center (Figure 1), covering an area of 75 x 105 arcseconds, was assembled from 50 individual array camera exposures using our MOSAIC image processing software (Varosi and Gezari 1993).

Deep array camera integrations show no indication of infrared emission from the position of the non-thermal point source Sgr A*. We have set preliminary flux density upper limits (3 sigma) for possible point sources at the position of 0.1 Janskys at 12.4 micron, 0.3 Jy at 18.1 micron, and 1.0 Jy at 20 micron, with an uncertainty estimated at about +/-0.5 Jy. These values are conservative; the Sgr A* flux densities at these wavelengths could well be an order of magnitude or more lower (see Zylka et al. 1995 for a discussion of the spectrum of Sgr A* from 1 mm to radio wavelengths). A complete discussion of the mid-infrared upper-limits for Sgr A* is presented by Gezari et al. (1994c, 1996).

3. Modeling the mid-infrared dust emission

Gezari, Dwek and Varosi (1996) have modeled the dust emission from the central using a data set consisting of seven images between 4.8 and 12.4 mm, and new 20 mm image data. The new model results show that the mid-infrared emission and derived temperature distributions are closely correlated, especially near the compact IRS sources. We find that essentially all the compact IRS sources in the central field correspond to temperature peaks or distinct temperature features in the model results. Note that no temperature enhancement is seen which can be attributed to the presence of Sgr A* itself. The significant peaks in dust emission occur between the primary IRS sources, not coincident with them, suggesting dust depletion near hot embedded stellar objects. The opacity ridge is also displaced outward compared to the infrared ridge (away from Sgr A*), which could indicate the influence of luminous sources in the central cavity. The model luminosity distribution result is very similar to the temperature distribution with the exception that the hotter sources IRS 7 and IRS 3 are not as dramatic features in the luminosity image. The total luminosity integrated over the 15 arcsec (0.6 parsec) region modeled is about 5×10^6 Lsun, roughly half the observed total luminosity of the Sgr A West complex. No luminosity enhancement is seen at the position of Sgr A*.

Gezari, Dwek and Varosi (1996 attempted to discriminate between the central engine and embedded stellar cluster heating mechanisms using the model temperature, opacity and luminosity distributions. If a central luminosity near Sgr A* was responsible for the observed IR emission, it would be expected to produce a radial temperature gradient across the surrounding infrared ridge, since the dust temperature distribution would be determined by the central radiation field which decreases at least as r^{-2}. On the other hand, embedded luminous stars would be expected to produce numerous local temperature "hot spots" throughout the extended ridge. The model results show no significant radial temperature gradient in the ridge, but the IRS sources all coincide with temperature distribution peaks. In contrast, source opacity peaks occur between the 12.4 micron IRS sources on the extended ridge, suggesting that luminous stars are embedded at the IRS source positions. The study shows that the principal IRS sources on the ridge are located in regions of locally lower dust density and higher temperature, implying imbedded luminous stars are present at the IRS source positions, creating warm, partially depleted cavities surrounding them in the extended dust ridge.

4. Discussion

Surprising discrepancies have been found (Gezari 1992) between the positions of several of the IRS sources and the ionized gas peaks shown in 2-cm VLA images (Lo and Claussen 1983, Yusef-Zadeh, Morris and Eckers 1989) as well as in Br alpha/Br gamma infrared images (c.f. Forrest et al. 1987, Reike and Rieke 1988). Pronounced positional shifts of the infrared peaks with respect to the radio continuum peaks are seen in Figure 2. In all cases, either the radio or the infrared source is double-peaked; the weaker peak of one coinciding with the brighter peak of the other, with the infrared peak located either radially outward from the corresponding radio peak (along a line through the sources and Sgr A*), or tangentially along the inside of the ridge (away from a point on the ridge closest to Sgr A*). The shifts could be evidence for the influence of a luminous source near the position of Sgr A*.

Three distinct groups of luminous stars are known to populate the central parsec of the Galaxy. These are the compact infrared (IRS) sources observed in the mid-infrared, the He I emission line stars, and the Fe[II] emission line stars. Krabbe et al. (1991, 1995) have detected a group of compact sources in the

2.06 micron emission line of HeI. These HeI emission line stars are concentrated in a 1 pc diameter region in the core of Sgr A West and are thought to be late-type blue supergiants stars (Najarro et al. 1993) with high mass-loss rates (10^{-5} to 10^{-4} Mo yr-1) and high outflow velocities ($\approx$ 1000 km s^{-1}). They have predicted individual luminosities in the range 0.1 - 2 x 10^6 Lsun, resulting in a total luminosity of 1.2 x 10^7 Lsun for the entire 1 pc cluster (Krabbe et al. 1991, 1995). Gezari (1992) showed that there was a surprisingly good correlation between the positions of the HeI line stars and mid-infrared brightness and dust color temperature features (Figure 3). A significant correlation is also found with modeled dust temperature features; in almost every instance HeI stars are associated with local temperature features (peaks or other enhancements) in the temperature contours. There is no significant correlation between the HeI star positions and the modeled source opacity (dust mass) distribution. Considering the weakness of the correlation between the stars and dust density features, we conclude that the HeI stars are located between (rather than embedded within) the dense dust clouds in the Sgr A West complex. These stars could heat the dust clouds externally, from regions of reduced dust density surrounding the HeI stars, between the IRS sources. But the HeI stars apparently are not embedded within the compact IRS sources; the IRS sources must contain other internal luminous objects. Our new dust model results indicate that the luminosity of all sources in the central 15 arcsec (0.6 pc) is 5 x 10^6 Lsun, about half of the total infrared and far-infrared luminosity observed in the central 2 parsecs of the Sgr A West complex. Thus radiation from the HeI emission line stars could account for the luminosity from the central 15 arcsec field- of-view derived from our array camera observations, as well as the total luminosity observed in the far infrared from the central few parsecs of the Galaxy.

Allen and Burton (1994) obtained a series of near infrared images of Sgr A West in [FeII] 1.644 mm and HeI 2.058 mm line emission, and were able to identify a cluster of several hundred population I stars, including about ten B[ ] stars hot enough to excite the nebular [FeII] emission. These are proported to be O-B stars, and a different cluster than the HeI line stars found by Krabbe et al. (1991, 1995). Comparison with our mid-infrared images (Figure 4) shows that the strongest [FeII] emission occurs directly south of (but not coincident with) Sgr A*, in a region between the IRS sources 1, 9, 2 and 13. This position shows a fairly strong peak in 5 GHz continuum emission, directly east of the radio "mini-cavity". [FeII] emission extends along the infrared ridge to IRS1 and IRS9. But, as with the HeI line stars, most of the mid-infrared IRS sources do not show enhanced [FeII] emission. This cluster of luminous stars, in addition to (and uncorrelated with) most of the HeI line stars and IRS sources, represents the third cluster of luminous sources in the central parsec.

5. Conclusions

From the mid-infrared array data and the model results we conclude that: 1) the compact IRS sources are all local hot spots in the Galactic Center dust clouds, and 2) also local luminosity peaks; 3) the local dust density peaks occur between the IRS sources, suggesting that 4) the IRS sources contain imbedded luminous stars; 5) the luminous HeI emission line and B[ ] stars are co-located

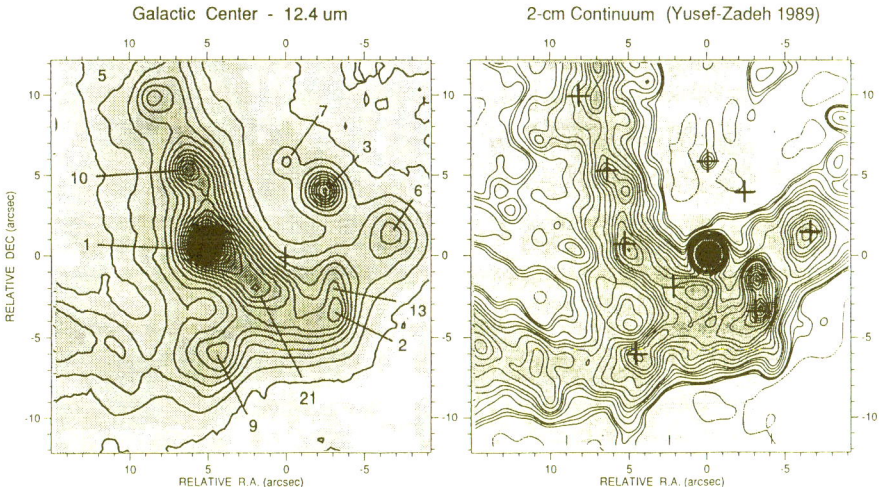

Figure 2. Intriguing discrepancies between the positions of the radio continuum peaks and the mid-infrared IRS sources seen in the comparison of our 12.4 micron continuum image of Sgr A West with the 2-cm continuum map by Yusef-Zadeh et al. (1989) using the VLA in its B configuration. The positional calibration was based on spatial registration of the two maps on a common feature, IRS 7. The infrared and radio peaks are shifted in most cases, with the infrared sources lying outward of the corresponding radio peaks, away from Sgr A*, suggesting the influence of a central object. The bright radio point source is Sgr A*is marked by a cross on the 12.4 micron image, and the IRS sources are indicated by crosses on the 2-cm map.

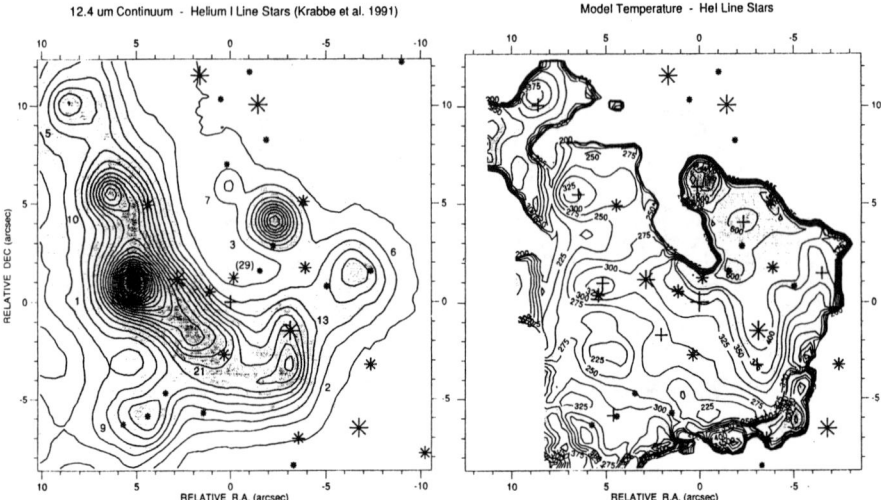

Figure 3. (Left): Positions of the more prominent Helium I emission line stars (asterisks, size proportional to brightness) in the central parsec found by Krabbe et al. (1991, 1995), shown with our 12.4 mm continuum brightness contours. Positions are measured relative to Sgr A* (cross). (Right): Helium I star positions plotted on the temperature distrubution of warm dust modeled by Gezari, Dwek and Varosi (1996). Darker regions are hotter. Note that the HeI stars are correlated with temperature enhancements, and occasionally with temperature minima in regions depleted of dust.

The second is concerned with the question of how much of the activity listed in Table 1 can be attributed to the normal processes of stellar evolution in a dense star cluster at the center of a galaxy, and how much requires completely different processes. These latter processes we attribute not to the black hole-accretion disk mechanism but to the creation of matter.

4. Central Mass Concentration

The observational evidence for the existence of central dark matter in galaxies was recently reviewed by Kormendy and Richstone (1995). The evidence comes from dynamics (stellar velocity dispersions) and potentially from studies of the light distributions (cf Wolfe & Burbidge 1970, Peebles 1972). However, Kormendy & Richstone (1995) have shown that the studies of the light distribution looking for cusps in the density distribution are all inconclusive.

The estimates of the dark masses found from stellar or gas dynamical arguments are for the Galactic Center ($\sim 2 \times 10^6 M_\odot$), M 31 ($3 \times 10^7 M_\odot$), NGC 4594 ($\sim 5 \times 10^8 M_\odot$), M 32 ($\sim 2 \times 10^6 M_\odot$), NGC 3115 ($\sim 10^9 M_\odot$), NGC 3377 ($\sim 8 \times 10^7 M_\odot$), M 87 ($\sim 3 \times 10^9 M_\odot$). We have not included NGC 4258, since the mass derived in that case stems from the assumption that the central gas clouds are moving in Keplerian orbits. However the geometry suggests that it is more likely that that gas is being ejected *along* the axis associated with the gas far out, the radio *arms* due to extended regions of synchrotron radiation, and the x-ray emitting QSO sources (Pietsch et al. 1994; Burbidge 1995); i.e. the motions determined from the gas giving rise to the maser emission is ejected gas and not gas in a rotating ring.

Leaving aside the situation in our own Galaxy, the results suggest that a dark mass concentration is present in many galaxies which show no signs of activity. The only galaxy which has been studied in which no dark mass can be discerned is M 33 where the *limit* is $5 \times 10^4 M_\odot$ (Kormendy & McClure 1993). Also where positive results have been found there is a weak correlation with the size of the galactic bulge (in spirals) and the total luminosity (in ellipticals).

5. The Nearest Dog that Did Not Bark

How does the physical state of the center of our own Galaxy fit into this picture?

In preparing this discussion we have drawn heavily on the review by Genzel, Hollenbach and Townes (1994) and Townes (1995) and the papers presented at this meeting. As we have just pointed out, there is good evidence for a dark mass $\sim 2 \times 10^6 M_\odot$ at the center.

It is possible that this dark mass is made up of a dense cluster of evolved stellar remnants, but this appears to us to be very unlikely. However, until scales much closer to the Schwarzschild radius for such a mass (6×10^{11} cm) can be probed, such a possibility cannot be completely ruled out.

Surrounding this dark mass which we assume is a near black hole, there is a dense star cluster with a radius < 2pc containing many high-mass high-luminosity stars which must have formed and be evolving in less than 10^6 years. These stars dominate the luminosity from this region. Slightly further out there is a dense torus of molecular gas. The star cluster extends out to a radius ~ 100

pc with a density falling off as R^{-2}. Practically all of the activity in this region involving gas streams, coronal winds, hot gas, etc. can be attributed to the processes of star formation and evolution involving many supernova explosions. Thus we believe that the model of Terlevich and his colleagues may very well apply for the vast majority of the activity. There are also x-ray sources which can be attributed to binary systems of the kind that are found elsewhere in the galaxy.

The only source which is unique in the Galaxy is the radio source at the center known as SgrA*. This has a size in the range $10^{12} - 5 \times 10^{13}$ cm ($\sim 1.5 - 80$ Schwarzschild radii for $M = 2 \times 10^6 M_\odot$), a radio luminosity of 10^{34} erg sec^{-1} with a spectrum with $\alpha = 0.25$ for frequencies ≤ 100 Ghz and $\alpha \geq 1$ for frequencies > 100 Ghz. The extrapolated luminosity of a nearinfrared source is of order $10^5 L_\odot$.

Since the Eddington luminosity for a $2 \times 10^6 M_\odot$ near black hole is $\sim 6 \times 10^{10} L_\odot$, this leads us to the dilemma faced by the black hole-accretion disk advocates. If energy release by that process is so efficient, and is responsible for nuclear activity in general as is widely believed, how is it that this black hole, and those in nearly all of the other nearby galaxies in which dark nuclear masses have been discovered, are not active?

We believe that the answer lies in the fact that these dark masses are creation centers and that the major activity involving large energy releases listed in Table 1 (category B) has nothing whatever to do with accretion processes.

What we do not see in our own Galaxy leads us to the conclusion that the accretion process is much less efficient than it has been claimed to be. Strenuous efforts have been made at this meeting and elsewhere (cf Rees 1988) to explain why, in this particular case, the accretion disk-black hole model is so ineffectual. Some have argued that this means that the condensed central object has a mass of only $\sim 10^2 - 10^3 M_\odot$ (Ozernoy 1987). Since we accept the view that a near black hole of $\sim 10^6 M_\odot$ is present, even in our picture there will be some accretion of surrounding gas and stars on to it. However there has not been a highly energetic outburst involving large mass and energy ejection for several million years. So what is giving rise to SgrA*? Our belief is that at this low level the radio emission is due to accretion on to a near black hole and that this shows that the whole accretion process is much less efficient than it has been assumed to be in the black hole accretion disk model as it has been developed over the years. This is not the first time that theorists have convinced each other that a much favored process will work effectively, when in reality it will not.

6. Conclusion

We believe that the Galactic center is an important key to understanding the origin of violent activity in the centers of galaxies. We suppose that most galaxies with well developed bulges contain massive near-black holes. Surrounding such an object there is a dense star cluster and much molecular gas so that a great deal of activity due to star formation and evolution is continuously taking place. This phenomenon is widespread in galaxies. However in a small fraction of the galaxies violent outbursts giving rise to large scale ejection of matter in the form of gas, coherent objects and relativistic particles are seen. We believe that this violent activity is due to creation processes in the galactic centers. However, for most of the time near black holes accrete matter at very low levels and this activity manifests itself in phenomena like Sgr A*. Such phenomena are not detectable in other galaxies because they are so weak.

References

Ambartsumian, V.A. 1958, Solvay Conf. Proc. (Stoops, R Ed. Brussels) p. 241

Ambartsumian, V.A. 1965, Structure & Evolution of Galaxies, Proc. 13th Conf. Phys. Univ. Brussels (Wiley Interscience) p. 1

Burbidge, E.M. 1995, A&A, 298, L1

Burbidge, G. 1961, Nature, 190, 1053

Burbidge, G. 1970, ARA&A, 8, 369

Burbidge, G., Burbidge, E.M. & Sandage, A.R. 1963, Rev. Mod. Phys., 35, 947

Colgate, S. 1967, ApJ, 150, 163

Field, G. 1964, ApJ, 140, 1434

Genzel, R., Hollenbach, D. & Townes, C. 1994, Rep. Progr. Physics, 57, 417

Hoyle, F., Burbidge, G. & Narlikar, J.V. 1993, ApJ, 410, 437

Hoyle, F., Burbidge, G. & Narlikar, J.V. 1994a, MNRAS, 267, 1607

Hoyle, F., Burbidge, G. & Narlikar, J.V. 1994b, A&A, 289, 720

Hoyle, F., Burbidge, G. & Narlikar, J.V. 1995, proc. Roy. Soc. A, 448, 191

Hoyle, F. & Fowler, W.A. 1963a, MNRAS, 125, 169

Hoyle, F. & Fowler, W.A. 1963b, Nature, 197, 533

Hoyle, F., Fowler, W., Burbidge, G. & Burbidge, E.M. 1964, ApJ, 139, 909

Hoyle, F. & Narlikar, J.V. 1964, Proc. Roy. Soc. A, 282, 184, 191

Jeans, J.H. 1929, Astronomy & Cosmogony, (Cambridge Univ. Press) p. 352

Kormendy, J. & McClure, R. 1993, AJ, 105, 1793

Kormendy, J. & Richstone, D. 1995, ARA&A, 33, 581

Ne'eman, Y. 1964, ApJ, 141, 1303

Ozernoy, L.M. 1987, "The Galactic Center", AIP Conf. Proc. No. 155, (ed. D. Backer) p. 181

Peebles, P.E.J. 1972, ApJ, 178, 371

Pietsch, W., Vogler, A., Kahabka, P., Jain, A. & Klein, V. 1994, A&A, 284, 386

Rees, M. 1984, ARA&A, 22, 471

Rees, M. 1988, Nature, 333, 523

Skhlovsky, I.S. 1960, Sov. Astron., 4, 885

Terlevich, R. & Melnick, J. 1985, MNRAS, 213, 841

Terlevich, R. & Melnick, J. 1988, Nature, 333, 239

Terlevich, R., Tenorio-Tagle, G., Franco, J. & Melnick, J. 1992, MNRAS, 255, 713

Townes, C. 1995, IAU Trans. IAU, The Hague, August 1994, in press.

Wade, C.M. 1960, Observatory, 80, 235

Wolfe, A. & Burbidge, G. 1970, ApJ, 161, 419

Table 1.

OBSERVED PHENOMENON	ORIGINS A[1]	B[2]
Nuclear Properties		
Strong narrow optical emission line spectrum	X	
Strong broad optical emission line spectrum	X or	X
Blue continuum	X or	X
Thermal X-ray emission	X or	X
γ-ray emission		X
Rapidly variable:		
(a) optical emission		X
(b) X & γ-ray emission		X
Highly polarized emission:		
(a) optical	X or	X
(b) radio		X
IR emission	X or	X
Non-thermal radio emission		X
Thermal radio emission	X or	X
Large non-circular motions		X
Megamaser emission	X or	X
Extended Source Properties		
Strong narrow & broad optical emission-line spectrum (radio galaxies)		X
Large scale radio emission (Radio Galaxies)		X
Luminous IR emission (IR Galaxies)	X or	X
Large non-circular motions		X
Extended synchroton emission in jets (optical/radio/x-rays)		X
Ejection of coherent X-ray sources		X
Ejection of QSOs		X

1. A – energy arising from stellar evolutionary processes
2. B – energy released due to the presence of massive objects (near black holes)

MID-IR EMISSION 499

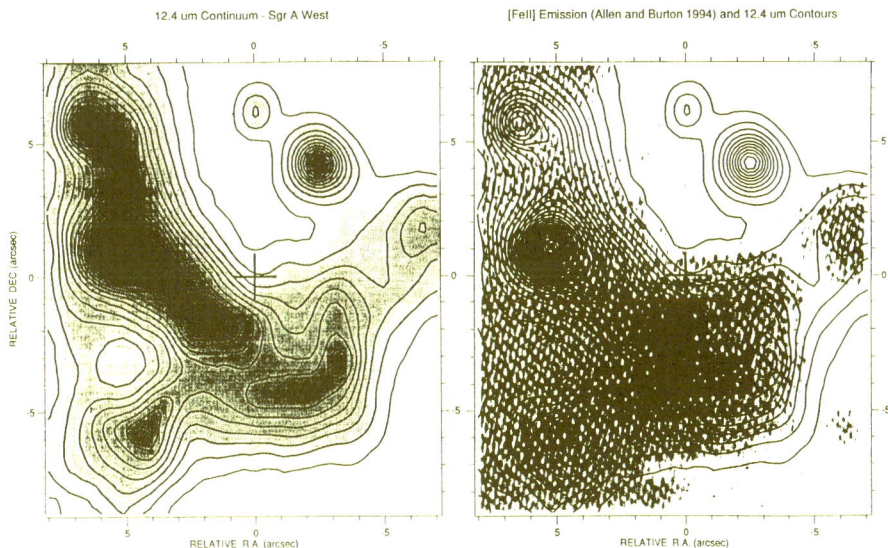

Figure 4. Comparison of the our 12.4 micron continuum image with the 1.644 micron [FeII] emission line image by Allen and Burton (1994). Note that the [FeII] emission does not peak at Sgr A* as might be assumed at first glance, but is strongest to the south of Sgr A* (cross) in the area of the radio "mini-cavity". In contrast, the 12.4 micron brightness distribution peaks at IRS 1 and is relatively weak where the [FeII]emission is strongest.

OH/IR Stars - Dynamical Studies

Anders Winnberg

Onsala Space Observatory, S - 439 92 Onsala, Sweden

Abstract. Since 20 years surveys have been undertaken to find OH/IR stars in the central parts of our Galaxy using various radio telescopes around the world. The present paper gives a review of such surveys and of dynamical studies using their data. Presently, two major survey projects are going on: a deep search in the very Centre of the Galaxy using both VLA monitoring data and new AT data and a large–scale survey of the area $|l| \leq 45°$, $|b| \leq 3°$ also using both VLA and AT data. The fact that only three parameters per OH/IR star are known – the two sky coordinates and the line–of–sight velocity – implies that severely limiting and simplifying assumptions have to be made regarding the distribution function of the stars and the gravitational potential. Usually spherical symmetry and isotropic velocity dispersion have been assumed. Recently, a project has been initiated to determine the proper motions of the OH/IR stars close to the Galactic Centre using the VLA and the VLBA on the associated H_2O and SiO masers. It is pointed out that the analysis of $l - v$ diagrams so far has overlooked the fact that they are superpositions of many $l - v$ diagrams from concentric ring areas within the total area. The slopes of the regression lines of these 'inner' $l - v$ diagrams get increasingly larger for smaller areas. The '$l - v$ slope' as a function of galactocentric distance is that expected for differential rotation in an r^{-2} mass density distribution, a fact which also is suggested by the surface density distribution of the stars.

1. Introduction

OH/IR stars are variable (pulsating) Asymptotic Giant Branch (AGB) stars. They lose mass at a high rate ($10^{-7} - 10^{-4}$ $M_\odot$/yr). Dust particles condense in the stellar wind to such an extent that the central stars become invisible and the spectral energy distribution of the objects is shifted into the far infrared. Water vapour is photodissociated by UV radiation penetrating the expanding envelope from the outside, thus forming a shell of hydroxyl (OH) molecules at a distance from the central star of $10^{16} - 10^{17}$ cm. The OH molecules are pumped by IR radiation from the surrounding dust which results in a maser–amplified OH line at 1612 MHz. As the gas expands through the OH shell, the line profile consists of two Doppler shifted components originating at the back and front sides of the shell. Therefore, the line–of–sight velocity of the star is the average of the line–of–sight velocities of the two components and the envelope expansion velocity is half the velocity difference between them. (In this paper we reserve the word 'radial' for the radial direction from the Galactic Centre.) Habing (1996) has written a review on AGB stars.

OH/IR stars are excellent objects to probe the gravitational potential in the central parts of the Galaxy because: they are insensitive to all other forces except gravitation, they probably form a dynamically relaxed system, their radio radiation is not absorbed in the interstellar dust, their OH line emission is maser amplified making them detectable throughout the Galaxy, their radial velocities can be measured accurately, their distances can be determined through a geometrical method (except for the very central parts of the Galaxy, where the interstellar scattering broadens the angular sizes of the OH shells), and their proper motions can be measured in the near future (see below).

Despite these advantages, progress in dynamical studies using OH/IR stars has been slow. One reason for this is the fact that, whereas $l-b$ and $l-v$ diagrams of various *inter*stellar gas components (as e.g. CO) delineate structures, the corresponding diagrams for OH/IR stars are scatter diagrams in which it is very difficult to see any structures, at least with today's modest number of stars. The data analysis so far has been limited to determinations of regression lines and dispersions from them. This very simplistic way of looking at the data might lead to wrong or, at least, incomplete conclusions as I will try to explain.

2. Surveys

Already in the 1970's, surveys were undertaken to find OH/IR stars in the direction of the Galactic Centre (Baud et al. 1979; Olnon et al. 1981; Habing et al. 1983). However, these surveys were either carried out with too low sensitivity or they had serious selection effects. It was shown by Habing et al. (1983) that the sky area around Sgr A had to be searched using a radio interferometer with sufficiently long baselines in order to avoid detecting strong interstellar absorption lines of OH (see below). Subsequent surveys in this part of the sky therefore were carried out using interferometers such as the Very Large Array (VLA) and the Australia Telescope (AT).

One of the major surveys in the 1980's, however, was carried out using single–dish instruments – the survey by te Lintel Hekkert et al. (1991; LCHHN). This survey was not aimed at any particular region of the sky – it is an all–sky survey – but it covers the Galactic Bulge fairly well. More than 3000 IRAS point sources were searched for 1612–MHz OH emission and about 700 were detected. The IR sources were selected according to their IR colours which were required to be similar to those of OH/IR stars with known IR counterparts. Radio telescopes on both hemispheres were used for the survey: the 25–m Dwingeloo telescope in the Netherlands, the 100–m Effelsberg telescope in Germany, and the 64–m Parkes telescope in Australia. Most of the observing time was spent on the Parkes radio telescope and therefore there is a slight bias toward southern sources. The detected sources are concentrated towards the Galactic plane and Centre but there is a distinct avoidance from the lowest Galactic latitudes due to saturation of the IRAS detectors. Therefore this survey is a good large–scale survey but it should not be used for, say, $|b| \lesssim 2°$. Another word of warning: Some of the OH spectra ($\sim 10\%$) are not typical of OH/IR stars.

Lindqvist et al. (1992a; LHWM) used the VLA to survey a small area of typical size $1.0° \times 1.5°$ covering the Galactic Centre and they detected 134 OH/IR stars. This area covers a large part of the area surveyed by Habing et al. (1983) who detected 34 OH/IR stars only. One reason for this large discrepancy, apart from the difference in sensitivity between the two surveys, is that the single dish (the 100-m telescope) used by Habing et al. detected strong interstellar absorption lines of OH which prevented the detection of weak emission lines from OH/IR stars. LHWM, on the other hand, rejected VLA visibility data from baselines shorter than 3000 wavelengths (540 m) in order to resolve out structures with angular sizes larger than about 1'. This eliminated most of the continuum flux in the Sgr A region and the spectral baselines became straight, enabling the detection of weak, narrow emission lines.

Van Langevelde et al. (1993) monitored, over a time period of 3 years, the 37 strongest OH/IR stars of LHWM. They derived periods for 13 of the stars and OH shell diameters for 3 of them. Recently Sjouwerman (Leiden/Onsala) has concatenated most of the data and searched for OH/IR stars stronger than 40 mJy. He has also made observations using the AT. These data are centred on Sgr A West and they cover a velocity range from -550 to $+600$ km s^{-1} (relative to the LSR). Their rms noise fluctuations are about 5 mJy – at least 4 times better than LHWM. Taken together (the concatenated VLA data and the AT data), about 50 previously unknown OH/IR stars have been detected (Sjouwerman et al., these proceedings).

Another survey "in the making" is the one by Sevenster (Leiden). This is a large-scale survey which is specially "designed" to complement the survey by LCHHN, i.e. to fill in the low-latitude strip in which the IRAS point source catalogue is incomplete. Another aim of this survey is to find dynamical evidence, in the stellar component, for a bar in the Centre of the Galaxy. The survey covers the area $|l| \leq 45°, |b| \leq 3°$ and uses data from both the VLA and the AT. A paper has been submitted which presents AT data (539 fields) covering the Galactic Bulge ($|l| \leq 10°, |b| \leq 3°$). The rms noise is less than 40 mJy for about 90% of the survey area. From these data 245 OH/IR stars have been detected of which 145 are previously unknown.

In this context another type of survey should be mentioned. Izumiura et al. (1994, 1995a, b) have searched for SiO $J = 1 - 0, v = 1$ and 2 maser emission from IRAS point sources toward the Galactic Bulge ($|l| < 15°, 3° < |b| < 15°$). The sources were selected according to IR colour, using criteria similar to those used by LCHHN. As a matter of fact, many sources are in common with LCHHN (about 30%) and these are certainly OH/IR stars. Probably most of the other sources are OH/IR stars as well. Izumiura et al. have detected 194 sources in total. They have tried to separate them into 'bulge' and 'disk stars' and claim that 134 of the OH/IR stars have a high probability of belonging to the Bulge and 60 to the Disk.

3. Boltzmann's equation

The relevant equation describing the relation between the OH/IR stars and the gravitational potential is the socalled *collisionless Boltzmann equation*:

$$\frac{\partial f}{\partial t} + \mathbf{v} \cdot \nabla f - \nabla \Phi \cdot \frac{\partial f}{\partial \mathbf{v}} = 0 \tag{1}$$

where $f = f(\mathbf{x}, \mathbf{v}, t)$ is the *distribution function* or *phase-space density* and Φ the *gravitational potential*. f gives the distribution of the OH/IR stars in spatial position $\mathbf{x}$ and velocity $\mathbf{v}$ at time t. Thus, f is a function of seven variables. We usually assume that the system is time independent and in that case f is a function of six variables. However, each OH/IR star is characterized by three variables only; the sky coordinates and the line–of–sight velocity. Because of this we have to make several assumptions of symmetry.

If we assume that the system is spherical, that there is no net streaming motion in the radial direction and in the direction perpendicular to the Galactic Disk, and integrate (1) over velocity, $\mathbf{v}$, we get, in spherical coordinates (r, θ, ϕ):

$$\frac{\mathrm{d}\left(n\sigma_r^2\right)}{\mathrm{d}r} + \frac{n}{r}\left[2\sigma_r^2 - \left(\sigma_\theta^2 + \langle v_\theta \rangle^2 + \sigma_\phi^2\right)\right] = -n\frac{\mathrm{d}\Phi}{\mathrm{d}r} \tag{2}$$

where $n = n(r)$ is the number density of stars and

$$\sigma_r^2 = \langle v_r^2 \rangle \tag{3}$$

$$\sigma_\theta^2 = \langle v_\theta^2 \rangle - \langle v_\theta \rangle^2 \tag{4}$$

$$\sigma_\phi^2 = \langle v_\phi^2 \rangle \tag{5}$$

are the velocity dispersions in the galactocentric radial, azimuthal, and latitudinal directions, respectively. Eq. (2) is the *Jeans equation*.

In a spherical system the gravitational potential, Φ, is related to the enclosed mass, $M(r)$, according to:

$$\frac{\mathrm{d}\Phi}{\mathrm{d}r} = \frac{GM(r)}{r^2} \tag{6}$$

where G is the *gravitational constant*. We can then solve for the enclosed mass:

$$M(r) = \frac{r\sigma_r^2}{G}\left[-\frac{\mathrm{d}\ln n(r)}{\mathrm{d}\ln r} - \frac{\mathrm{d}\ln \sigma_r^2}{\mathrm{d}\ln r} + (\lambda - 2) + \frac{\langle v_\theta \rangle^2}{\sigma_r^2}\right] \tag{7}$$

where:

$$\lambda = \frac{\sigma_\theta^2 + \sigma_\phi^2}{\sigma_r^2} \tag{8}$$

i.e. $\lambda = 2$ for an isotropic velocity dispersion distribution. Since we know nothing about the anisotropy of the velocity dispersion we usually assume that it is isotropic. This latter assumption is most probably wrong, judging from observations of other galaxies.

However, despite the fact that several adventurous assumptions have been made regarding both the gravitational potential (sphericity) and the distribution function (isotropic velocity dispersion), there are further problems in the application of eq. (7): How can we estimate the volume density, $n(r)$, the velocity dispersion, $\sigma_r(r)$, and the mean streaming motion in galactocentric azimuth, $\langle v_\theta \rangle$? We know the projected counterparts of these quantities as functions of the projected galactocentric distance, R – the surface density, $N(R)$, the line-of-sight velocity dispersion, $\sigma_{los}(R)$, and the mean line–of–sight velocity, $\langle v_{los} \rangle$. Thanks to the symmetry assumptions we can 'deproject' these quantities by solving a set of Abel integral equations (Binney & Tremaine 1987). If $N(R) \propto R^{-\alpha}$ one can show that $n(r) \propto r^{-\alpha-1}$. In particular, when $\alpha = 1$ the system is an *isothermal sphere*, a dynamically relaxed, equilibrium configuration. The line-of-sight velocity dispersion for such a configuration is dominated by the velocity dispersion in the tangential point because of the high central density concentration, $\sigma_r \approx \sigma_{los}$. A similar effect is at work with regard to the mean streaming motion in galactocentric azimuth, $\langle v_\theta \rangle$; a good approximation is the mean line–of–sight velocity, $\langle v_{los} \rangle$, at a projected galactocentric distance corresponding to the tangential distance. However, it is not particularly difficult to solve the corresponding 'de-projection integrals' numerically.

If we could determine the proper motions of the OH/IR stars close to the Galactic Centre we would know five parameters per star instead of three. This would give considerably more information on the distribution function especially the velocity dispersions. Sjouwerman and van Langevelde have started on such a project using the VLA and the VLBA on H_2O and SiO masers associated with OH/IR stars at the Galactic Centre. It is not possible to carry out such a project on the 18–cm OH masers because they are broadened in angle due to interstellar scattering (van Langevelde et al. 1992). The H_2O and SiO masers occur at 13 and 7 mm wavelength, respectively, where the broadening due to scattering is a factor of about 200 and 700 smaller, respectively.

4. Data analysis

Lindqvist et al. (1992b) analyzed their data following the procedure outlined above. In order to estimate the mean streaming motion of the stars they calculated the linear regression line through the $l - v$ diagram. In doing that, they applied a similar data analysis as did McGinn et al. (1989) on their IR data of the inner 5 parsecs around the Galactic Centre. In this way the two data sets are directly comparable and complementary.

Te Lintel Hekkert (1990) took a completely different approach in his dynamical analysis of the data of LCHHN. Since these data cover the whole Galaxy and the overall gravitational potential of the Galactic Disk is known from optical work, he adopted a gravitational potential and tried to estimate the distribution function of the OH/IR stars. He did this in collaboration with Dejonghe (Ghent) by applying a new mathematical method called *quadratic programming* invented by Dejonghe (1989). In this method the orbits of the stars are characterized by their binding energy, E, and their angular momentum, L_z. They were able to construct a probability density diagram in $E - L_z$ space. In this diagram one can see that most of the stars have nearly circular, prograde orbits except for stars in the Bulge, where there is a larger fraction of stars having elongated orbits and even stars with retrograde rotation.

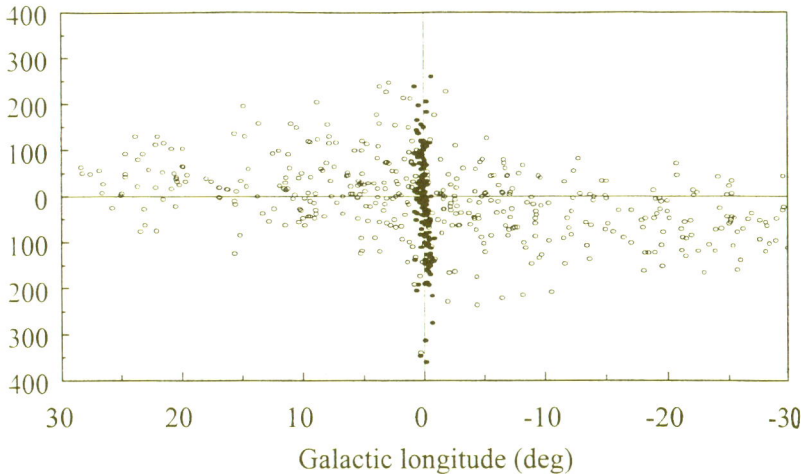

Figure 1. $l - v$ diagram of OH/IR stars in the central parts of the Galaxy. Open circles are stars from LCHHN and filled circles are stars from LHWM.

Sevenster et al. (1995) compared the data sets of LCHHN and LHWM. They used the quadratic programming technique described above and came to the conclusion that the two samples of OH/IR stars consist of two distinct populations, one that extends over the whole Galaxy and shows signs of evolution and another one which is only seen near the Galactic Centre, has high rotation and does not show signs of long evolution. However, one can ask oneself whether this approach is the most appropriate in this region of the Galaxy. Here the gravitational potential is not well known and this is precisely the reason why OH/IR stars are so valuable – they can give us information on the gravitational potential which we do not have. It seems to me that, in this region, we have more knowledge about the distribution function of the OH/IR stars than we have about the potential and that, by assuming a potential, we lose information which is contained in the data.

5. Is there a kinematic connection between the OH/IR stars in the Galactic Centre and those in the Bulge?

When comparing the $l - v$ diagrams of the LCHHN stars within $|l| < 30°$ and of the LHWM stars it is obvious that their kinematics are very different (Figure 1). This was pointed out by Dejonghe (1993) already. However, it was noted by Lindqvist et al. (1992b) that the central 15 stars of the LHWM sample obey an even steeper slope in the $l - v$ diagram, 2000 km s^{-1} deg^{-1} instead of 180 km s^{-1} deg^{-1} for the whole sample. Moreover these 15 stars are in good agreement with the data of McGinn et al. (1989). The slope of the regression line for the LCHHN stars in Figure 1 is about 11 km s^{-1} deg^{-1} only.

It is rather strange that, so far, we have concluded that the OH/IR stars in the central regions of the Galaxy rotate like a solid body whereas their surface

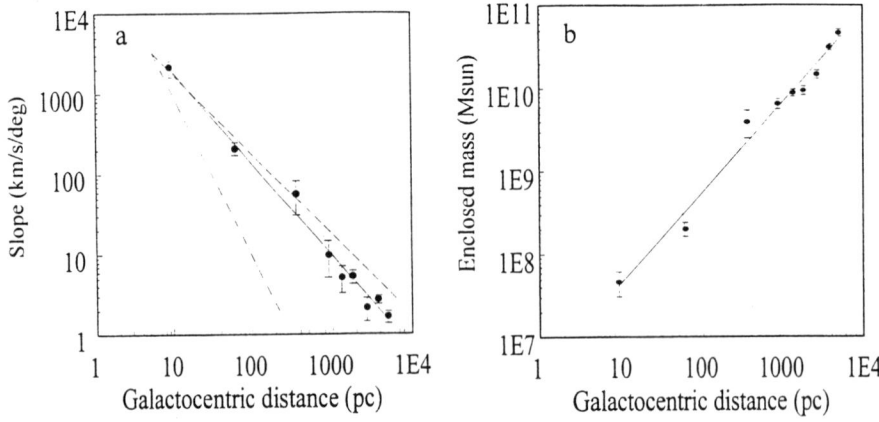

Figure 2. a. Regression–line slopes of $l-v$ diagrams for OH/IR stars plotted against galactocentric distance. b. Enclosed mass as a function of galactocentric distance.

density indicate that they have an r^{-2} volume density law. Why should the OH/IR stars have a solid-body rotation and yet show an isothermal distribution? The different $l-v$ slopes for samples of different sky extent mentioned above may give a hint of how to analyze the data.

In a still ongoing work I have divided the data of LHWM and that of LCHHN for $|l| < 30°$ into subsamples contained in concentric ellipses (Winnberg 1994). I have plotted $l-v$ diagrams of the stars within the central ellipse and within the surrounding elliptical annuli. Each $l-v$ diagram contains about 60 stars. I have then fitted linear regression lines to these $l-v$ diagrams and plotted the slopes of the regression lines as a function of the semi major axis of the outer ellipse in each annulus (Figure 2a). As you can see all the data points and their statistical errors can be described well by a straight line in this log–log plot, i.e. data points from the LHWM stars and from the LCHHN stars can be fitted well with the *same* line. For comparison I have drawn three other lines, one for a system with constant density (dotted line), one for a system with an r^{-2} density law (dashed line), and one for a system with a central point mass (dash–dotted line). These lines are valid only if the velocity dispersion is constant. The line–of–sight velocity dispersion for these stars is increasing slowly with r due to an increasing fraction of older stars.

Of course the enclosed mass can be calculated according to eq. (7) and it turns out to be somewhat higher at short galactocentric distances than previously believed and to follow a line with a slope of 1 in a log–log plot as expected for an r^{-2} density law (Figure 2b). This is what Becklin & Neugebauer (1968) claimed already a long time ago.

Admittedly it is a bit bold to carry the comparison between the two data sets this far because the corresponding surveys were very different both in the methods used and in the sensitivities reached. Also, so far, I have made no attempt at separating 'disk stars' from 'bulge stars'. However, I believe that, as long as I only deal with the kinematics of the stars, the result is not too far from reality. In any case a large-scale, homogeneous survey is now carried out (Sevenster et al. 1996) and others will no doubt follow in the near future. Therefore my hypothesis will soon be tested.

References

Baud, B., Habing, H. J., Matthews, H. E., Winnberg, A. 1979, A&AS, 36, 193
Becklin, E. E., Neugebauer, G. 1968, ApJ, 151, 145
Binney, J., Tremaine, S. 1987, Galactic Dynamics, Princeton University Press, Princeton
Dejonghe, H. 1989, ApJ, 343, 113
Dejonghe, H. 1993, in "Galactic Bulges", eds. H. Dejonghe & H. J. Habing, Kluwer Academic Publishers, Dordrecht
Habing, H. J., Olnon, F. M., Winnberg, A., Matthews, H. E., Baud, B. 1983, A&A, 128, 230
Habing, H. J. 1996, A&A, submitted (preprint Sterrewacht Leiden)
Izumiura, H., Deguchi, S., Hashimoto, O., Nakada, Y., Onaka, T., Ono, T., Ukita, N., Yamamura, I. 1994, ApJ, 437, 419
Izumiura, H., Deguchi, S., Hashimoto, O., Nakada, Y., Onaka, T., Ono, T., Ukita, N., Yamamura, I. 1995a, ApJS, 98, 271
Izumiura, H., Deguchi, S., Hashimoto, O., Nakada, Y., Onaka, T., Ono, T., Ukita, N., Yamamura, I. 1995b, ApJ, 453, 837
Lindqvist, M., Habing, H. J., Winnberg, A., Matthews, H. E. 1992a, A&AS, 92, 43 (LHWM)
Lindqvist, M., Habing, H. J., Winnberg, A. 1992b, A&A, 259, 118
McGinn, M. T., Sellgren, K., Becklin, E. E., Hall, D. N. B. 1989, ApJ, 338, 824
Olnon, F. M., Walterbos, R. A. M., Habing, H. J., Matthews, H. E., Winnberg, A., Brzezińska, H., Baud, B. 1981, ApJ, 245, L103
Sevenster, M. N., Dejonghe, H., Habing, H. J. 1995, A&A, 299, 689
Sevenster, M. N., Chapman, J. M., Habing, H. J., Killeen, N. E. B., Lindqvist, M. 1996, A&A, submitted
te Lintel Hekkert, P. 1990, PhD Thesis Ch. 7, Leiden
te Lintel Hekkert, P., Caswell, J. L., Habing, H. J., Haynes, R. F., Norris, R. P. 1991, A&AS, 90, 327 (LCHHN)
van Langevelde, H. J., Frail, D. A., Cordes, J. M., Diamond, P, J. 1992, ApJ, 396, 686
van Langevelde, H. J., Janssens, A. M., Goss, W. M., Habing, H. J., Winnberg, A. 1993, A&AS, 101, 109
Winnberg, A. 1994, in "The Nuclei of Normal Galaxies", eds. R. Genzel & A. I. Harris, Kluwer Academic Publishers, Dordrecht, p. 91

Author Index

Aitken, D. 179
van Altena, W.F. 345
Alvarez, H. 3
Aparici, J. 3, 54
Balachandran, S. C. 212
Ballet, J. 431
Bally, J. 8, 16
Beckert, T. 389
Becklin, E.E. 228
van den Bergh, S. 345
Bertelli, G. 320
Bicknell, G. V. 236
Blitz, L. 335
Blommaert, J. 289
Blum, R.D. 277, 285
Borrel, V. 431
Bouchet, L. 431
Bronfman, K. 54
Burbidge, E.M. 471
Burbidge, G. 480
Burton, M.G. 122
Carr, J.S. 212
Casement, S. 60
Caswell, J.L. 247
Chiosi, C. 320
Churazov, E. 431
Cogan, S.W.J. 122
Coker, R.F. 403
Cordes, J.M. 127
Cotera, A.S. 122
Cox, D.P. 40
Dahmen, G. 54
Dame, T.M. 54
DePoy, D.L. 277, 285
Doi, Y. 20, 101
Dufton, P.L. 271
Duschl, W.J. 389
Dyachkov, A. 431
Eckart, A. 196
Erickson, E.F. 122
Evans II, N.J. 16
Falcke, H. 453
Figer, D.F. 263
Frail, D.A. 93, 151
Franco, J. 40
Fridman, A.M. 335

Fuente, A. 47
Geballe, T.R. 163
Geis, N. 114
Genzel, R. 114, 196, 203
Gezari, D. 491
Ghez, A.M. 228
Gilfanov, M. 431
Girard, T.M. 345
Glass, I.S. 312
Goldwurm, A. 431
Goss, W.M. 93, 137, 146, 151, 369
Gray, A.D. 443
Green, A. 93, 151
Gull, G.E. 106
Hüttemeister, S.H. 54
Habing, H. 289, 294
Hayward, T.L. 106, 407
Herter, Terry 407
Hillier, D.J. 203
Hiromoto, N. 101
Ho, P.T.P. 146
Hoffmann, B. 36
Hoyle, F. 480
Jütte, M. 36
Jacoby, G.H. 298
Jaffe, D.T. 16, 28
Jourdain, E. 431
Keller, L.D. 28
Khavenson, N. 431
Khoruzhii, O.V. 335
Kimeswenger, S. 3
Klein, B.L. 228
Koczet, P. 36
Koyama, K. 423
Krabbe A. 203
Kudritzki, R.P. 203
Kuznetsov, S. 431
van Langevelde, H.-J. 289, 294
Latvakoski, H.M. 106
Lazio, T.J.W. 127
Lennon, D.J. 271
Lindqvist, M. 60, 294
Linhart, A. 54
Low, F.J. 20
Lutz, D. 203
Lyakhovich, V.V. 335

Méndez, R.A. 345
Mac-Auliffe, F. 54
Madden, S.C. 114
Maeda , Y. 423
Maillard, J.-P. 232
Majewski, S.R. 345
Martín-Pintado, J. 47, 64
Martos, M.A. 40
Matsumoto, S. 312
Mauersberger, R. 54
May, J. 3, 54
McGee, R.X. 369
McGregor, P.J. 236
McLean, I.S. 263
Mehringer, D.M. 60
Melia, F. 220, 395, 403
Meyer, K. 54
Mezger, P.G. 380
Miller, R.H. 327
Mochizuki, K. 20
Moore, T. 179
Morris, M. 228, 263
Mountain, C.M. 163
Nagata, T. 255
Najarro, F. 203
Nakagawa, T. 20, 101
Ng, Y.K. 320
Nikola, T. 114
Nishimura, T. 20
Nyman, L.-A. 60
Okuda, H. 20, 101
Omont, A. 305
Ono, T. 312
Ozernoy, L. 335
Pak, S. 28
Palmer, E.S. 54
Paul, J. 431
Planesas, P. 54
Plume, R. 16
Poglitsch, A. 114
Predehl, P. 415
Puxley, P.J. 439
Ramsay Howat, S.K. 163
Revnivtsev, M. 431
Rich, R.M. 345
Rieke, G.H. 220, 462
Rieke, M.J. 462
Roberts, D.A. 60, 93, 137, 151
Robinson, B.T. 151

Roche, P. 179
Roques, J.P. 431
Rosa, M.R. 189
Saha, P. 236
Sandqvist, Aa. 85
Santillán, A. 40
Schlosser, W. 36
Schmidt-Kaler, Th. 36
Schmidtobreick, L. 36
Sekiguchi, K. 312
Sellgren, K. 212, 277, 285
Selman, F.J. 362
Shibai, H. 20, 101
Sil'chenko, O.K. 335, 353
Simons, D.A. 232
Simpson, J.P. 122
Sjouwerman, L. 289, 294
Skinner, G.K. 439
Smartt, S.J. 271
Smith, C. 179
Sobolev, A.M. 68
Stacey, G.J. 106
Staguhn, J. 447
Stolovy, S.R. 407
Stutzki, J. 447
Sunyaev, R. 431
Tamblyn, P. 220
Tappert, C. 36
Tieftrunk, A.R. 54
Timmermann, R. 114
Townes, C.H. 114
Trudolubov, S. 431
Uchida, K.I. 60, 447
Vallenari, A. 320
Van de Steene, G.C. 298
Vargas, M. 431
van der Veen, W. 289
Verozub, L.V. 357
de Vicente, P. 47, 64
White, G.J. 171
Wiedenhöver, W. 54
Wiemann, S. 36
Wilson, T.L. 54, 64
Winnberg, A. 294, 500
Yui, Y.Y. 20, 101
Yusef-Zadeh, F. 60, 93, 137, 151, 447
Zhao, J.-H. 146
Zinnecker, H. 415

Zylka, R. 77

Subject Index

α-disk, 395
Abel integrals, 236
accretion disk, 189, 380, 389, 403, 407, 453, 480
accretion rate, 395, 403, 453
active galactic nuclei, 127, 369, 380, 431, 453, 471, 480
ambipolar diffusion, 40, 179
angular broadening, 127
angular momentum, 8, 106, 395, 403, 453, 480
annihilation radiation, 511 keV, 431, 439, 453
astrometric grid, 189
asymptotic giant branch, 196, 212, 277, 289, 294, 305, 312, 320, 462, 500
atomic lines,
    Brγ, 106, 122, 137, 163, 232, 255, 395, 439
    [FeIII], 137, 395
    H79α, 447
    H92α, 106, 137, 146
    He I, 106, 179, 196, 203, 220, 263, 312, 395
    He II, 203, 232, 255, 263
    Positronium, 431, 439
atomic fine structure lines,
    [CI], 16, 171
    [CII], 20, 101
    [NIII], 114
    [OI], 106, 114, 151
    [OIII], 114
Baade's window, 36, 277, 285, 289, 305, 320, 345, 369, 462
bar, galactic, 8, 106, 305, 345, 480
black hole, 85, 93, 137, 189, 196, 203, 236, 294, 327, 335, 353, 357, 362, 389, 395, 403, 407, 431, 453, 462, 471, 480
Boltzmann's equation, 362
Bondi-Hoyle accretion, 395, 403, 453
bow shock, 395, 403, 407, 453
Bremsstrahlung, 395, 439
bulge, galactic, 3, 8, 54, 171, 212, 277, 285, 289, 299, 305, 327, 345, 353, 431, 462, 480, 500

C-field theory, 471, 480
central stellar cluster, 137, 196, 212, 220, 228, 236, 255, 263, 305, 380, 389, 453
circumnuclear disk, 77, 85, 93, 106, 137, 151, 163, 171, 179, 335
color magnitude diagrams, 277, 320
compact objects, 8, 236, 357, 389, 395
coordinate transformation, 189
cosmic rays, 8, 40, 85
dark matter, 357, 362, 462, 480
density waves, 179, 327, 335
Eddington limit, 357, 423, 480
emission measure, 114, 127, 389
Fermi acceleration, 8
free-free radiation, 127, 380, 389
γ rays, 54, 101, 431
grains,
    Davis-Greenstein alignment, 137, 179
    destruction, 47, 395
    extinction, 20, 28, 36, 77, 106, 122, 137, 163, 179, 189, 196, 203, 212, 271, 277, 285, 289, 299, 305, 312, 345, 380, 389, 407, 415, 439, 462
    mantles, 20, 68, 106, 255, 305
    polarization, 93, 127, 137, 151, 179, 247, 255, 312, 389, 407
    silicates, 179, 255, 407
gravitational potential, 8, 40, 54, 68, 85, 137, 146, 171, 179, 220, 236, 271, 294, 327, 335, 357, 362, 403, 462, 471, 480, 500
HII regions, 3, 20, 28, 40, 47, 60, 64, 68, 106, 137, 203, 247, 305, 369, 380, 395, 447, 471
Helium-like ions, 423
high negative velocity gas HNVG, 77, 137, 146
Hydrogen-like ions, 423
inflow, 335, 353, 395, 403, 407, 453
initial mass function, 114, 212, 220, 263, 462

interstellar material, 8, 20, 28, 40, 54, 64, 77, 85, 101, 151, 171, 179, 247, 255, 271, 299, 305, 312, 369, 380, 389, 415, 443, 500
interstellar scintillation, 389
Jeans equation, 196, 236, 500
Jeans Swindle, 327
Kolmogorov spectrum, 127
Lagrangian radius, 327
LINER, 453, 471
magnetic fields, 8, 40, 68, 85, 93, 106, 114, 127, 137, 151, 179, 255, 389, 395, 403, 447, 453
   field reconnection, 8, 114, 447
   magnetic pressure, 40
   magnetic tension, 40
   poloidal field, 8, 114, 179
   Zeeman splitting, 93, 137, 151, 179
masers,
   CH_3OH, 68, 247
   H_2O, 247, 500
   H_2O mega-maser, 471
   OH, 93, 137, 151, 163, 189, 247, 289, 294, 305, 500
   SiO, 500
mass to light ratio, 236, 389, 462
metallicity, 28, 36, 114, 212, 263, 271, 277, 289, 294, 299, 320, 345, 462
Mills model, 3
mini-cavity, 137, 146, 395, 403
mini-spiral, 106, 335
molecules,
   C-H stretch, 255
   CH_3CN, 64
   CH_3OH, 68
   CN, 171
   ^{12}CO, 8, 16, 28, 54, 171
   ^{13}CO, 16, 77
   $C^{18}O$, 54, 85
   CO vibrational absorption, 236, 255, 277, 462
   CS, 8, 16, 60, 171
   H_2, 28, 163
   H_2O ice absorption, 255
   HCN, 8, 106, 114, 137, 151, 163, 171, 236
   HCO^+, 47, 85, 114, 151, 171, 439, 447
   HNCO, 54, 77, 85, 171
   NH_3, 47
   OH, 146
   PAH, 305
   SiO, 47, 171, 196
non-thermal emission, 8, 93, 101, 106, 137, 151, 189, 369, 380, 389, 431, 443, 453, 480
normal mode oscillations, 327
orbits, $x1$ and $x2$, 8
orbits, Keplerian, 327, 345, 480
Orion star forming region, 28, 88
outflow, 60, 68, 146, 151, 171, 203, 289, 327, 335, 395, 403, 407, 439, 453, 471
Parker instability, 40
photoionization, 101, 106, 299, 395
photometry, 36, 232, 255, 285, 305, 320, 327, 407
photon dominated regions (PDRs), 20, 28, 93, 101, 106, 114, 447
Planck particles, 471, 480
plasma, 77, 114, 127, 380, 389, 395, 407, 423, 453, 480
Plaut's window, 345
proper-motions, 8, 151, 189, 196, 236, 500
QSO, 471, 480
Quintuplet cluster, 114, 255, 263, 312, 491
radio-loud galaxies, 453
red giant branch, 320
relativistic particles, 8, 40, 47, 77, 357, 380, 389, 439, 453, 471, 480
Reynolds Hubble law, 236
sources: galactic,
   HD 147889, 101
   IC 342, 16
   NGC 2023, 101
   NGC 3603, 220
   NGC 6603, 320
   ρ Oph, 101
sources: galactic center,
   1E1740.7–2942, 439
   Annihilator, Great, 439
   arched filaments, 47, 114, 122, 255

E1 and E2 filaments, 114, 122
FIR 21, 60
G0.10+0.02 ('banana'), 114, 122
G0.12+0.02, 114, 137
G0.15−0.04 ('pistol'), 8, 106, 114, 122, 263, 491
G0.18−0.04 ('sickle'), 68, 106, 114, 122, 263
G1.6−0.025, 68
G357.7−0.1, 443
G359.1−0.5, 151
G359.54+0.18, 447
GRS 1716−249, 431
GRS 1730−312, 431
GRS 1734−292, 431
GRS 1741.9−2853, 431
GRS 1747−341, 431
GRS 1758−258, 431
IRS 1, 179, 407
IRS 2, 137, 179, 212
IRS 3, 179, 255, 407, 491
IRS 5, 179
IRS 7, 203, 212, 232, 236, 255, 263, 277, 285, 380, 407, 491
IRS 8, 179, 263
IRS 9, 277, 285
IRS 10, 179, 232, 407
IRS 11, 255
IRS 12, 277
IRS 13, 137, 203, 232, 407
IRS 15, 203
IRS 16, 137, 151, 163, 179, 196, 203, 228, 232, 263, 395, 407
IRS 19, 212, 255, 277
IRS 22, 212, 277
IRS 23, 255, 277
IRS 24, 212, 277
IRS 28, 277
M−0.13−0.08, 47
M0.20−0.03, 47
non-thermal filaments, 8, 47, 114, 447
OH359.855−0.078, 294
Radio Arc, 47, 85, 106, 114, 423
RXJ 1745.6−2900, 415

Sgr A*, 28, 40, 77, 106, 137, 146, 151, 179, 189, 228, 232, 236, 277, 335, 369, 380, 389, 395, 403, 407, 415, 423, 431, 453, 462, 480, 491
Sgr A East, 85, 93, 122, 145, 151, 389, 415
Sgr A West, 85, 106, 137, 146, 151, 212, 389, 491
Sgr B1, 8, 47, 60
Sgr B2, 8, 47, 60, 64, 85, 247, 423
Sgr C, 8, 447
Sgr D, 8
Terzan 1, 320, 431
Terzan 2, 431
vertical CO filaments, 8
sources: individual galaxies,
    M 31, 137, 230, 327, 353, 380, 453, 480
    M 81, 380, 480
    M 81*, 453
    M 82, 16, 28
    M 84, 480
    M 87, 327, 362, 369, 480
    NGC 253, 16, 28
    NGC 1068, 380
    NGC 2639, 471
    NGC 2685, 353
    NGC 4258, 453, 471
    NGC 4374, 480
    NGC 5128 (Cen A), 480
    NGC 7331, 353
Schwarzschild radius, 357, 389, 453, 471, 480
Seyferts, 453, 471, 480
shocks, 8, 40, 47, 64, 68, 93, 114, 163
starburst, 28, 203, 212, 220, 236, 335, 462
stars,
    CNO abundances, 212
    He I stars, 106, 179, 196, 203, 220, 263, 312, 395, 491
    LBVs, 203, 220, 263
    LPVs, 212, 277, 312
    M-giants, 20, 122, 212, 277, 285, 289, 305, 345
    OH/IR stars, 289, 294, 500
    s-process elements, 212

Wolf Rayet stars, 47, 203, 220, 255
star formation, 8, 28, 60, 64, 68, 114, 127, 203, 212, 220, 236, 247, 255, 277, 285, 305, 320, 335, 353, 462, 471, 480
stellar remnants,
   planetary nebulae, 299
   pulsar, 127, 431, 443
   supernova remnants, 3, 8, 20, 28, 93, 127, 146, 151, 203, 220, 305, 443
   white dwarfs, 196, 395, 403
stellar winds, 8, 47, 127, 137, 203, 220, 232, 407, 453
supernovae, 212, 247, 471
supershells, 8
synchrotron emission, 8, 77, 85, 137, 380, 389, 395, 407, 439, 447, 453, 480
thermal pressure, 40
tidal effects,
   tidal capture, 220, 353
   tidal stress, 8, 20, 85, 93, 106, 146, 171, 179, 236, 263, 327
turbulence, 47, 64, 85, 179, 212, 271, 277
ultracompact HII regions, 64, 247
ultraviolet radiation, 8, 16, 20, 28, 36, 64, 85, 93, 101, 106, 114, 146, 151, 163, 220, 335, 380, 389, 395, 407, 431, 447, 453, 500
velocity dispersion, 28, 68, 163, 196, 203, 236, 289, 299, 345, 362, 403, 480, 500
Virial equation, 8, 54, 171, 196, 203, 327, 357, 407, 471
X-rays, 389, 395, 415, 439, 453, 480